Computer-Based Exercises for Signal Processing Using MATLAB® 5

Jean O'Young

James H. McClellan
C. Sidney Burrus
Alan V. Oppenheim
Thomas W. Parks
Ronald W. Schafer
Hans W. Schuessler

1.1 . 1.2 . 1.3, 1.4 1.5 P126-7

MATLAB® Curriculum Series

PRENTICE HALL, Upper Saddle River, New Jersey 07458

Library of Congress Cataloging–in–Publication Data

Computer-based exercises for signal processing using MATLAB® 5 / James
H. McClellan ... [et al.].
p. cm.
Includes index.
ISBN 0-13-789009-5
1. Signal processing—Digital techniques—Mathematics. 2. MATLAB.
I. McClellan, James H.,
TK5102.9.C567 1998
621.382'2'078553042—dc21 97-41449
CIP

Publisher: Tom Robbins
Editor-in-Chief: Marcia Horton
Production Coordinator: Wanda España/WEE DESIGN GROUP
Vice President Director of Production and Manufacturing: David W. Riccardi
Managing Editor: Bayani Mendoza de Leon
Cover Designer: Design Source
Manufacturing Buyer: Donna Sullivan
Editorial Assistant: Nancy Garcia
Compositor: Techsetters, Inc.

Printed in the United States of America

10 9 8 7 6 5 4 3 2 1

ISBN 0-13-789009-5

Prentice-Hall International (UK) Limited, *London*
Prentice-Hall of Australia Pty. Limited, *Sydney*
Prentice-Hall Canada Inc., *Toronto*
Prentice-Hall Hispanoamericana, S.A., *México*
Prentice-Hall of India Private Limited, *New Delhi*
Prentice-Hall of Japan, Inc., *Tokyo*
Simon & Schuster Asia Pte. Ltd., *Singapore*
Editora Prentice-Hall do Brasil, Ltda., *Rio de Janeiro*

The MathWorks, Inc.
24 Prime Park Way
Natick, Massachusetts 01760–1500
Phone: (508) 647-7000
Fax: (508) 647-7001
E–mail: info@mathworks.com
http://www.mathworks.com

OUTLINE

CONTENTS

PREFACE

The area of digital signal processing has consistently derived its vitality from the interplay between theory and applications. Correspondingly, university courses in digital signal processing have been increasingly incorporating computer exercises and laboratories to help students better understand the principles of signal processing and experience the excitement of applying abstract mathematical concepts to the processing of real signals.

This book is a collection of computer exercises about digital signal processing. It is an outgrowth of our collective experience in incorporating computer-based projects into our signal processing courses. Each of us has been involved in teaching signal processing at our respective institutions for many years. Individually, we recognized the importance of computer demonstrations and experiments as a supplement to the theory and had been independently developing computer-based projects. Several years ago we began sharing our experiences and projects with each other and eventually decided to make our combined experiences and collection of projects more widely available.

The exercises in this book are designed to be used together with a digital signal processing textbook and on the workstations and personal computers commonly used at most universities. Students working alone or in small groups can approach the exercises in a variety of ways consistent with their individual styles of learning and at a speed and for a length of time consistent with their skills and abilities. The format and exercises hopefully encourage (even require) experimentation and learning by discovery, much as is done in engineering practice. The use of high-speed integrated graphics allows visualization, which is very valuable in learning abstract theories and methods.

In our early experiences with computer-based projects for signal processing, we used a variety of computers and software. Over a period of time, and again somewhat independently, we each began gravitating toward the use of MATLAB in our courses because we found that its interactive mathematical calculations, easy-to-use integrated graphics, simple programmability, consistent functional environment, and availability on a wide variety of hardware make it very efficient for students to focus quickly on the essential signal processing issues without getting bogged down with details of a particular machine or programming language. As a natural consequence, in combining our projects into this collection, we adopted MATLAB as the signal processing environment. The recent availability of a student version of MATLAB makes it even more attractive for educational use. There are, of course, many other excellent and well-supported computer-based signal processing environments, and we tried to make the projects in this book adaptable to these as well.

In order to provide maximum flexibility in the use of these projects, each section in a chapter contains one or more projects associated with a common theme. The intent is for each of the chapter sections to be independent of the others so that the order and selection can be as flexible as possible. When a section contains more than one project, we have again tried to keep separate projects as independent as possible, although some of the more advanced projects in a section might assume familiarity with earlier ones. Within each project are exercises which, by necessity, are somewhat interdependent, although in many cases, some of the exercises are optional.

Our individual experiences have varied in the use of these projects. In some cases, these exercises have been used to supplement lecture and homework material, while in others, the material in this book has been used in a signal processing laboratory, supplemented by real-time signal processing hardware and development systems for various DSP chips. The level of the material is appropriate for senior and first-year graduate courses. The time required to complete one exercise varies from as little as five minutes to as much as several hours. The diversity of exercises reflects the authors' different teaching styles. Some projects include considerable text material reviewing the signal processing principles involved in the associated exercises. Other projects focus on the signal processing problems to be solved and rely on the suggested reading to provide the necessary theory. Occasionally samples of special MATLAB functions are included in the text. Sometimes special files of data (e.g., speech) are referenced. The MathWorks, Inc., has arranged to supply this material from their web site.

From the outset, our goal has been to make this collection of projects available in a timely way to as wide an audience as possible. Without question, many of the projects could have been polished and expanded further. There are also many other projects that were in draft form, and with further development they could have been included. However, we felt that it was more important to proceed with publication of those contained here in their current form with the expectation that future versions of this collection can follow. Consequently, this collection should be thought of as the first version of a work in progress. We anticipate continued evolution of these projects and incorporation of others in future editions. It is also quite likely that as this book evolves through future editions, the authorship will evolve as well.

While the six of us take primary responsibility for the contents of this book, it could not have been compiled without considerable help from a long list of colleagues and students. We express our appreciation to Richard Rau of the Georgia Institute of Technology for updating all the MATLAB files used in this text to be compatible with version 5 of MATLAB. We are also indebted to Manfred Herbert, Herbert Krauss, Utz Martin, Rudolf Rabenstein, Richard Reng, Mathias Schulist, and Karl Schwarz at the University Erlangen; to Dan Burnside, Knox Carey, Bernie Hutchins, and Tom Krauss at Cornell University; to John Buck, Haralabous Papadopoulos, Stephen Scherock, Sally Santiago, Andrew Singer, Lon Sunshine, and Kathleen Wage at the Massachusetts Institute of Technology; to Ramesh Gopinath, Haitao Guo, Jan Odegard, and Ivan Selesnick at Rice University; and to Dan Drake and Diana Lin of the Georgia Institute of Technology.

We acknowledge and appreciate the support of our respective institutions and of the National Science Foundation for providing partial funding for this project. We also express our appreciation to the staff at Prentice Hall for recognizing and supporting our objectives of making this collection of projects available to a wide audience at a reasonable cost, to Wanda España for her efficient and professional managing of the final editing and production, and to MathWorks for providing technical support and for their willingness to distribute the files to users of the book.

James H. McClellan
C. Sidney Burrus
Alan V. Oppenheim
Thomas W. Parks
Ronald W. Schafer
Hans W. Schuessler

chapter 1

BASIC SIGNALS AND SYSTEMS

OVERVIEW

MATLAB is an ideal software tool for studying digital signal processing (DSP). Its language has many functions that are commonly needed to create and process signals. The plotting capability of MATLAB makes it possible to view the results of processing and gain understanding into complicated operations. In this chapter we present some of the basics of DSP in the context of MATLAB. At this point, some of the exercises are extremely simple so that familiarity with the MATLAB environment can be acquired. Generating and plotting signals is treated first, followed by the operation of difference equations as the basic class of linear time-invariant systems. An important part of this chapter is understanding the role of the numerical computation of the Fourier transform (DTFT). Since MATLAB is a numerical environment, we must manipulate samples of the Fourier transform rather than formulas. We also examine the signal property called group delay. The sampling process is studied to show the effects of aliasing and the implementation of various reconstruction schemes. Finally, a filtering method to produce zero-phase response with an infinite impulse response (IIR) filter is investigated.

BACKGROUND READING

There are many excellent textbooks that provide background reading for the projects in this chapter. We mention as examples the books by Jackson [1], McGillem and Cooper [2], Oppenheim and Schafer [3], Oppenheim and Willsky [4], Strum and Kirk [5], Roberts and Mullis [6], and Proakis and Manolakis [7]. In each section we have indicated specific background reading from Oppenheim and Schafer [3], but similar background reading can be found in other books on digital signal processing.

[1] L. B. Jackson. *Signals, Systems and Transforms*. Addison-Wesley, Reading, MA, 1991.

[2] C. D. McGillem and G. R. Cooper. *Continuous and Discrete Signal and System Analysis*. Holt, Rinehart and Winston, New York, second edition, 1984.

[3] A. V. Oppenheim and R. W. Schafer. *Discrete-Time Signal Processing*. Prentice Hall, Englewood Cliffs, NJ, 1989.

[4] A. V. Oppenheim and A. S. Willsky with I. T. Young. *Signals and Systems*. Prentice Hall, Englewood Cliffs, NJ, 1983.

[5] R. D. Strum and D. E. Kirk. *First Principles of Discrete Systems and Digital Signal Processing*. Addison-Wesley, Reading, MA, 1988.

[6] R. A. Roberts and C. T. Mullis. *Digital Signal Processing*. Addison-Wesley, Reading, MA, 1987.

[7] J. G. Proakis and D. G. Manolakis. *Digital Signal Processing: Principles, Algorithms and Applications*. Macmillan, New York, second edition, 1992.

SIGNALS

OVERVIEW

The basic signals used often in digital signal processing are the unit impulse signal $\delta[n]$, exponentials of the form $a^n u[n]$, sine waves, and their generalization to complex exponentials. The following projects are directed at the generation and representation of these signals in MATLAB. Since the only numerical data type in MATLAB is the $M \times N$ matrix, signals must be represented as vectors: either $M \times 1$ matrices if column vectors, or $1 \times N$ matrices if row vectors. In MATLAB all signals must be finite in length. This contrasts sharply with analytical problem solving, where a mathematical formula can be used to represent an infinite-length signal (e.g., a decaying exponential, $a^n u[n]$).

A second issue is the indexing domain associated with a signal vector. MATLAB assumes by default that a vector is indexed from 1 to N, the vector length. In contrast, a signal vector is often the result of sampling a signal over some domain where the indexing runs from 0 to $N-1$; or, perhaps, the sampling starts at some arbitrary index that is negative, e.g., at $-N$. The information about the sampling domain cannot be attached to the signal vector containing the signal values. Instead, the user is forced to keep track of this information separately. Usually, this is not a problem until it comes time to plot the signal, in which case the horizontal axis must be labeled properly.

A final point is the use of MATLAB's vector notation to generate signals. A significant power of the MATLAB environment is its high-level notation for vector manipulation; `for` loops are almost always unnecessary. When creating signals such as a sine wave, it is best to apply the `sin` function to a vector argument, consisting of all the time samples. In the following projects, we treat the common signals encountered in digital signal processing: impulses, impulse trains, exponentials, and sinusoids.

BACKGROUND READING

Oppenheim and Schafer (1989), Chapter 2, Sections 2.0 and 2.1.

PROJECT 1: BASIC SIGNALS

This project concentrates on the issues involved with generation of some basic discrete-time signals in MATLAB. Much of the work centers on using internal MATLAB vector routines for signal generation. In addition, a sample MATLAB function will be implemented.

Hints

Plotting discrete-time signals is done with the `stem` function in MATLAB. The following MATLAB code will create 31 points of a discrete-time sinusoid.

```
nn    = 0:30;        %-- vector of time indices
sinus = sin(nn/2+1);
```

Notice that the $n = 0$ index must be referred to as `nn(1)`, due to MATLAB's indexing scheme; likewise, `sinus(1)` is the first value in the sinusoid. When plotting the sine wave we would use the `stem` function, which produces the discrete-time signal plot commonly seen in DSP books (see Fig. 1.1):

```
stem( nn, sinus );
```

The first vector argument must be given in order to get the correct n-axis. For comparison. try `stem(sinus)` to see the default labeling.

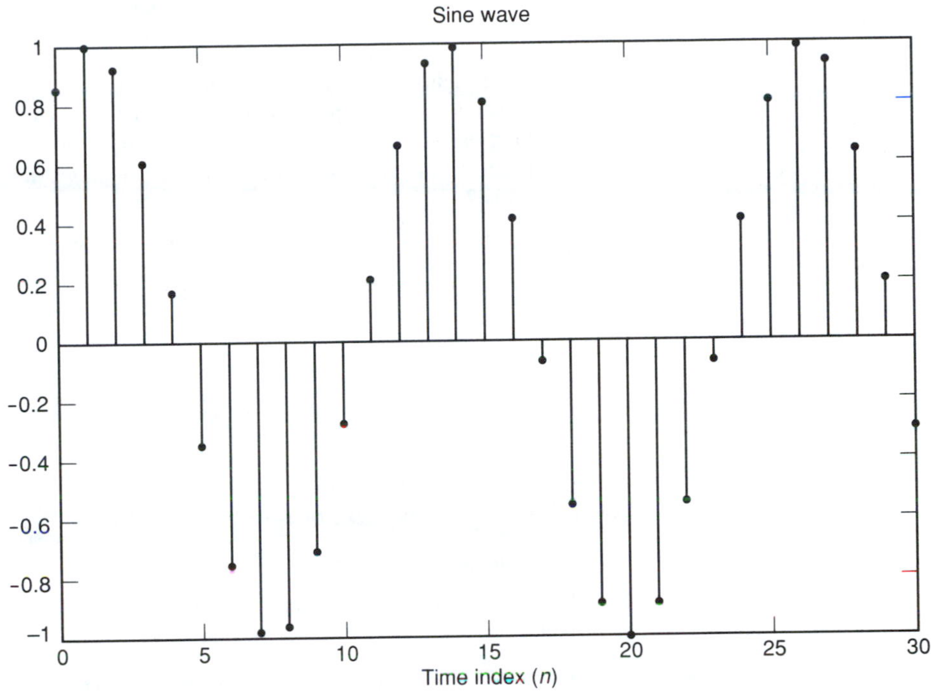

Figure 1.1 Plotting a discrete-time signal with `stem`.

EXERCISE 1.1

Basic Signals—Impulses

The simplest signal is the (shifted) unit impulse signal:

$$\delta[n - n_0] = \begin{cases} 1 & n = n_0 \\ 0 & n \neq n_0. \end{cases} \tag{1-1}$$

To create an impulse in MATLAB, we must decide how much of the signal is of interest. If the impulse $\delta[n]$ is going to be used to drive a causal LTI system, we might want to see the L points from $n = 0$ to $n = L - 1$. If we choose $L = 31$, the following MATLAB code will create an "impulse":

```
L = 31;
nn     = 0:(L-1);
imp = zeros(L,1);
imp(1)  = 1;
```

Notice that the $n = 0$ index must be referred to as `imp(1)`, due to MATLAB's indexing scheme.

a. Generate and plot the following sequences. In each case the horizontal (n) axis should extend only over the range indicated and should be labeled accordingly. Each sequence should be displayed as a discrete-time signal using `stem`.

$$\begin{aligned} x_1[n] &= 0.9\delta[n-5] && 1 \le n \le 20 \\ x_2[n] &= 0.8\delta[n] && -15 \le n \le 15 \\ x_3[n] &= 1.5\delta[n-333] && 300 \le n \le 350 \\ x_4[n] &= 4.5\delta[n+7] && -10 \le n \le 0 \end{aligned}$$

b. The shifted impulses, $\delta[n-n_0]$, can be used to build a weighted impulse train, with period P and total length MP:

$$s[n] = \sum_{\ell=0}^{M-1} A_\ell \, \delta[n - \ell P] \tag{1-2}$$

The weights are A_ℓ; if they are all the same, the impulse train is periodic with period P. Generate and plot a periodic impulse train whose period is $P = 5$ and whose length is 50. Start the signal at $n = 0$. Try to use one or two vector operations rather than a `for` loop to set the impulse locations. How many impulses are contained within the finite-length signal?

c. The following MATLAB code will produce a repetitive signal in the vector `x`:

```
x = [0;1;1;0;0;0] * ones(1,7);
x = x(:);
size(x)     %<--- return the signal length
```

Plot `x` to visualize its form; then give a mathematical formula similar to (1-2) to describe this signal.

EXERCISE 1.2

Basic Signals—Sinusoids

Another very basic signal is the cosine wave. In general, it takes three parameters to describe a real sinusoidal signal completely: amplitude (A), frequency (ω_0), and phase (ϕ).

$$x[n] = A\cos(\omega_0 n + \phi) \tag{1-3}$$

a. Generate and plot each of the following sequences. Use MATLAB's vector capability to do this with one function call by taking the cosine (or sine) of a vector argument. In each case, the horizontal (n) axis should extend only over the range indicated and should be labeled accordingly. Each sequence should be displayed as a sequence using `stem`.

$$\begin{aligned} x_1[n] &= \sin \tfrac{\pi}{17} n && 0 \le n \le 25 \\ x_2[n] &= \sin \tfrac{\pi}{17} n && -15 \le n \le 25 \\ x_3[n] &= \sin(3\pi n + \tfrac{\pi}{2}) && -10 \le n \le 10 \\ x_4[n] &= \cos\left(\tfrac{\pi}{\sqrt{23}} n\right) && 0 \le n \le 50 \end{aligned}$$

Give a simpler formula for $x_3[n]$ that does not use trigonometric functions. Explain why $x_4[n]$ is not a periodic sequence.

b. Write a MATLAB function that will generate a finite-length sinusoid. The function will need a total of five input arguments: three for the parameters and two more to specify the first and last n index of the finite-length signal. The function should return a column vector that contains the values of the sinusoid. Test this function by plotting the results for various choices of the input parameters. In particular, show how to generate the signal $2\sin(\pi n/11)$ for $-20 \le n \le 20$.

c. *Modification*: Rewrite the function in part (b) to return two arguments: a vector of indices over the range of n, and the values of the signal.

EXERCISE 1.3

Sampled Sinusoids

Often a discrete-time signal is produced by sampling a continuous-time signal such as a constant-frequency sine wave. The relationship between the continuous-time frequency and the sampling frequency is the main point of the Nyquist–Shannon sampling theorem, which requires that the sampling frequency be at least twice the highest frequency in the signal for perfect reconstruction.

In general, a continuous-time sinusoid is given by the following mathematical formula:

$$s(t) = A\cos(2\pi f_0 t + \phi) \tag{1-4}$$

where A is the amplitude, f_0 is the frequency in Hertz, and ϕ is the initial phase. If a discrete-time signal is produced by regular sampling of $s(t)$ at a rate of $f_s = 1/T$, we get

$$s[n] = s(t)|_{t=nT} = A\cos(2\pi f_0 T n + \phi) = A\cos\left(2\pi \frac{f_0}{f_s} n + \phi\right) \tag{1-5}$$

Comparison with formula (1-3) for a discrete-time sinusoid, $x[n] = A\cos(\omega_0 n + \phi)$, shows that the normalized radian frequency is now a scaled version of f_0, $\omega_0 = 2\pi(f_0 T)$.

a. From formula (1-4) for the continuous-time sinusoid, write a function that will generate samples of $s(t)$ to create a finite-length discrete-time signal. This function will require six inputs: three for the signal parameters, two for the start and stop times, and one for the sampling rate (in hertz). It can call the previously written MATLAB function for the discrete-time sinusoid. To make the MATLAB function correspond to the continuous-time signal definition, make the units of the start and stop times seconds, not index number. Use this function to generate a sampled sinusoid with the following definition:

```
Signal freq = 1200 Hz        Sampling freq = 8 kiloHz
Initial Phase = 45 deg       Starting Time = 0 sec
Amplitude = 50               Ending Time = 7 millisec
```

Make two plots of the resulting signal: one as a function of time t (in milliseconds), and the other as a function of the sample index n used in $t_n = nT$. Determine the length of the resulting discrete-time signal and the number of periods of the sinusoid included in the vector.

b. Show by mathematical manipulation that sampling a cosine at the times $t_n = n + \frac{3}{4}T$ will result in a discrete-time signal that appears to be a sine wave when $f_0 = 1/T$. Use the function from part (a) to generate a discrete-time sine wave by changing the start and stop times for the sampling.

EXERCISE 1.4

Basic Signals—Exponentials

The decaying exponential is a basic signal in DSP because it occurs as the solution to linear constant-coefficient difference equations.

a. Study the following MATLAB function to see how it generates a discrete-time exponential signal. Then use the function to plot the exponential $x[n] = (0.9)^n$ over the range $n = 0, 1, 2, \ldots, 20$.

```
function  y = genexp( b, n0, L )
%GENEXP   generate an exponential signal: b^n
%    usage: Y = genexp( B, N0, L )
%       B    input scalar giving ratio between terms
%       N0   starting index (integer)
%       L    length of generated signal
%       Y    output signal Y(1:L)
if( L <= 0 )
  error('GENEXP: length not positive')
end
nn = n0 + [1:L]' - 1;   %---vector of indices
y = b .^ nn;
end
```

b. In many derivations, the exponential sequence $a^n u[n]$ must be summed over a finite range. This sum is known in closed form:

$$\sum_{n=0}^{L-1} a^n = \frac{1-a^L}{1-a} \qquad \text{for } a \neq 1 \tag{1-6}$$

Use the function from part (a) to generate an exponential and then sum it up; compare the result to formula (1-6).

c. One reason the exponential sequence occurs so often in DSP is that time shifting does not change the character of the signal. Show that the finite-length exponential signal satisfies the shifting relation:

$$y[n] = ay[n-1] \qquad \text{over the range } 1 \leq n \leq L-1 \tag{1-7}$$

by comparing the vectors `y(2:L)` and `a*y(1:L-1)`. When shifting finite-length signals in MATLAB, we must be careful at the endpoints because there is no automatic zero padding.

d. Another way to generate an exponential signal is to use a recursive formula given by a difference equation. The signal $y[n] = a^n u[n]$ is the solution to the following difference equation when the input, $x[n]$, is an impulse:

$$y[n] - ay[n-1] = x[n] \qquad \text{initial condition: } y[-1] = 0 \tag{1-8}$$

Since the difference equation is assumed to recurse in a causal manner (i.e., for increasing n), the initial condition at $n = -1$ is necessary. In MATLAB the function `filter` will implement a difference equation. Use `filter` to generate the same signal as in part (a) (i.e., $a = 0.9$).

PROJECT 2: COMPLEX-VALUED SIGNALS

This project centers on the issues involved with representing and generating complex-valued signals. Although in the real world, signals must have real values, it is often extremely useful to generate, process, and interpret real-valued signal pairs as complex-valued signals. This is done by combining the signals into a pair, as the real and imaginary parts of a complex number, and processing this pair with other complex-valued signals using the rules of complex arithmetic. Use of signal pairs in this way is an important part of many signal processing systems, especially those involving modulation.

Complex exponentials are a class of complex signals that is extremely important because the complex amplitude (phasor notation) provides a concise way to describe sinusoidal signals. Most electrical engineering students are familiar with phasors in connection with ac circuits or power systems, but their use in radar, wave propagation, and Fourier analysis is just as significant (although the term *phasor* is not always used).

Hints

In MATLAB, the functions `real` and `imag` will extract the real and imaginary parts of a complex number. When plotting a complex vector, the defaults for `plot` and `stem` can be confusing. If `z` is complex, then `plot(z)` will plot the imaginary part versus the real part; and `plot(n, z)` will plot the real part of `z` versus `n`. However, `stem(z)` will just plot the real part. If you want to view simultaneous plots of the real and imaginary parts, the `subplot(211)` and `subplot(212)` commands prior to each `stem` command will force the two plots to be placed on the same screen, one above the other. See Fig. 1.2, which was created using the following code:

```
nn = 0:25;
xx = exp(j*nn/3);       %--- complex exponential
subplot(211)
stem(nn, real(xx))
title('REAL PART'),     xlabel('INDEX (n)')
subplot(212)
stem(nn, imag(xx))
title('IMAG PART'),     xlabel('INDEX (n)')
```

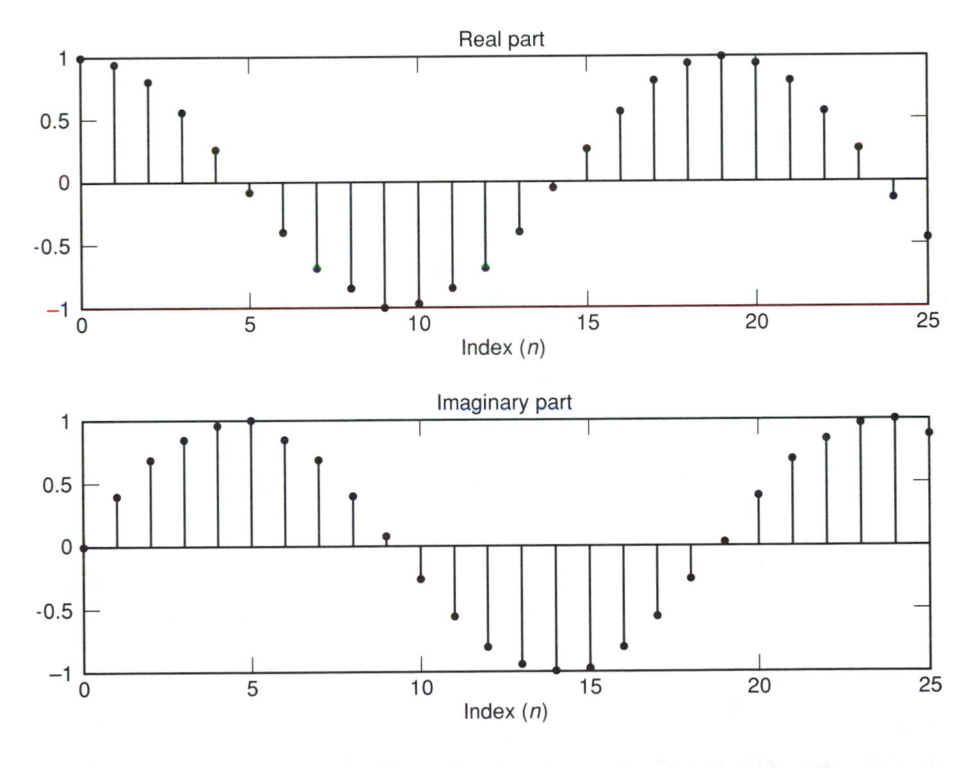

Figure 1.2 Plotting real and imaginary parts of a discrete-time signal with `subplot`.

EXERCISE 2.1

Complex Exponentials

The real exponential notation can be extended to complex-valued exponential signals that embody the sine and cosine signals. These signals form the basis of the Fourier transform.

a. In MATLAB a complex signal is a natural extension of the notation in Exercise 1.4. Thus the parameter a can be taken as a complex number to generate these signals. Recall Euler's formula for the complex exponential (in a form that gives a signal):

$$x[n] = (z_0)^n = e^{(\ln z_0 + j\angle z_0)n} = r^n e^{j\theta n} = r^n(\cos\theta n + j\sin\theta n) \quad (2\text{-}1)$$

where $z_0 = re^{j\theta} = r\angle\theta$. Use this relationship to generate a complex exponential with $z_0 = 0.9\angle 45^\circ$. Plot the real and imaginary parts of $x[n]$ over the range $0 \le n \le 20$. Notice that the angle of z_0 controls the frequency of the sinusoids.

b. For the signal in part (a) make a plot of the imaginary part versus the real part. The result should be a spiral. Experiment with different angles for θ—a smaller value should produce a better picture of a spiral.

c. Equation (2-1) is not general enough to produce all complex exponentials. What is missing is a *complex* constant to scale the amplitude and phase of the sinusoids. This is the so-called *phasor notation*:

$$G \cdot z_0^n = A\,e^{j\phi} r^n e^{j\theta n} = Ar^n e^{j(\theta n + \phi)} = Ar^n\left[\cos(\theta n + \phi) + j\sin(\theta n + \phi)\right], \quad (2\text{-}2)$$

where $G = A\,e^{j\phi} = A\angle\phi$ is the complex amplitude of the complex exponential. Generate and plot each of the following sequences. Convert the sinusoids to complex notation; then create the signal vector using `exp`. If the signal is purely real, it should be generated by taking the real part of a complex signal. In each plot, the horizontal (n) axis should extend only over the range indicated and should be labeled accordingly.

$$x_1[n] = 3\sin(\tfrac{\pi}{7}n) + j4\cos(\tfrac{\pi}{7}n) \qquad 0 \le n \le 20$$
$$x_2[n] = \sin\tfrac{\pi}{17}n \qquad -15 \le n \le 25$$
$$x_3[n] = 1.1^n \cos\left(\tfrac{\pi}{11}n + \tfrac{\pi}{4}\right) \qquad 0 \le n \le 50$$
$$x_4[n] = 0.9^n \cos\left(\tfrac{\pi}{11}n\right) \qquad -10 \le n \le 20$$

For each signal, determine the values of amplitude and phase constants that have to be used in G; also the angle and magnitude of z_0.

d. These same complex exponentials can be generated by first-order difference equations (using `filter`):

$$y[n] = z_0 y[n-1] + x[n]. \quad (2\text{-}3)$$

The filter coefficient, $z_0 = re^{j\theta}$, is a complex number. The ratio between successive terms in the sequence is easily seen to be z_0; but the correct amplitude and phase must be set by choosing a complex amplitude for the impulse which drives the difference equation (i.e., $x[n] = G\delta[n]$). Use `filter` to create the same signals as in part (c). Verify by plotting both the real and imaginary parts of $y[n]$ and comparing to the signals generated via `exp`.

e. In the first-order difference equation (2-3), let $y_R[n]$ and $y_I[n]$ denote the real and imaginary parts of $y[n]$. Write a pair of real-valued difference equations expressing $y_R[n]$ and $y_I[n]$ in terms of $y_R[n-1]$, $y_R[n-1]$, $x[n]$ and r, $\cos\theta$, and $\sin\theta$.

f. Write a MATLAB program to implement this pair of real equations, and use this program to generate the impulse response of equation (2-3) for $r = \frac{1}{2}$ and $\theta = 0$, and $\theta = \pi/4$. For these two cases, plot the real part of the impulse responses obtained. Compare to the real part of the output from the complex recursion (2-3).

DIFFERENCE EQUATIONS

OVERVIEW

Of particular importance in digital signal processing is the class of systems that can be represented by linear constant-coefficient difference equations. In this set of projects we explore the characteristics of these systems in both the time and frequency domains. Specifically, in Project 1 we consider the impulse response of infinite impulse response difference equations. Project 2 explores the steady-state response for step and complex exponential inputs. In Project 3 the frequency response is investigated.

BACKGROUND READING

In Oppenheim and Schafer (1989), Chapter 2, Sections 2.2 through 2.5, discrete-time systems, linear time-invariant systems, and linear constant-coefficient difference equations are discussed. Sections 2.6 through 2.9 cover frequency-domain representations of discrete-time signals and systems.

PROJECT 1: TIME-DOMAIN RESPONSE OF DIFFERENCE EQUATIONS

In this project you will generate the response of an IIR (infinite impulse response) filter, which is an LTI system expressed as a linear constant-coefficient difference equation:

$$\sum_{k=0}^{N_a} a_k\, y[n-k] = \sum_{\ell=0}^{N_b} b_\ell\, x[n-\ell] \tag{1-1}$$

In MATLAB, difference equations are represented by two vectors: one vector containing the feedforward coefficients, b_ℓ, for the `x` terms, and the other vector containing the feedback coefficients, a_k, for the `y` terms. The coefficient a_0 is usually taken to be 1, so that when $y[n]$ is written in terms of past values it drops out:

$$y[n] = -\frac{1}{a_0}\sum_{k=1}^{N_a} a_k\, y[n-k] + \sum_{\ell=0}^{N_b} b_\ell\, x[n-\ell]$$

In MATLAB the `filter` function will divide out a_0, so it must not be zero.

Hints

The function `y = filter(b,a,x)` implements a digital filter defined by the `a` and `b` coefficient vectors as in (1-1) to filter the data stored in `x`. If `x` is the unit impulse signal, then `y` will be the impulse response $h[n]$. Note that the function `filter` returns only as many samples into `y` as there are in `x` (i.e., the impulse response is truncated to the length of the unit impulse vector, `x`).

EXERCISE 1.1

Simple Difference Equation

a. Create vectors `b` and `a` that contain the coefficients of $x[n]$ and $y[n]$, respectively, in the following difference equation:

$$y[n] + 0.9y[n-2] = 0.3x[n] + 0.6x[n-1] + 0.3x[n-2]. \tag{1-2}$$

b. Calculate $y[n]$ analytically for $x[n] = \delta[n]$.

c. Now create a unit impulse vector, `imp`, of length 128. Generate the first 128 points of the impulse response of the filter in (1-2). Use `stem` to plot these values as a discrete-time signal versus time (see `help stem`). It may help to plot just the first 10 or 20 points.

EXERCISE 1.2

Impulse Response with `filter`

a. Use the `filter` function to generate and plot the impulse response $h[n]$ of the following difference equation. Plot $h[n]$ in the range of $-10 \le n \le 100$.

$$y[n] - 1.8\cos\left(\tfrac{\pi}{16}\right) y[n-1] + 0.81y[n-2] = x[n] + \tfrac{1}{2}x[n-1] \tag{1-3}$$

b. Also determine the impulse response analytically and confirm your results.

EXERCISE 1.3

Natural Frequencies

The impulse response of a difference equation such as (1-2) or (1-3) is known to be composed of several natural frequencies. These natural frequencies are determined by the feedback coefficients $\{a_k\}$. Each root (p_k) of the characteristic polynomial gives rise to a term in the output of the form $p_k^n u[n]$.

$$A(z) = 1 + \sum_{k=1}^{N_a} a_k z^{-k} \tag{1-4}$$

a. For the difference equation (1-3), determine the natural frequencies; see `help roots` for the MATLAB function to extract polynomial roots. If the roots are complex, the natural frequency response will be a complex exponential. Plot the real and imaginary parts of the signals $p_k^n u[n]$.

b. For a second-order difference equation, such as (1-3), there are two natural frequencies, and if these are distinct, the causal impulse response must be of the form

$$h[n] = \left(\alpha p_1^n + \beta p_2^n\right) u[n] \tag{1-5}$$

where p_1 and p_2 are the natural frequencies. In part (a), these natural frequencies were determined. For example, suppose that for a second-order difference equation, p_1 and p_2 have been obtained using `roots`, and $h[n]$ for two values of N is calculated by direct recursion of the difference equation. Write a pair of simultaneous equations for α and β. Solve these equations using MATLAB's backslash operator, \, for the difference equation (1-3). Using this result, generate $h[n]$ from (1-5) and verify that it matches the result obtained in Exercise 1.2.

PROJECT 2: STEADY-STATE RESPONSE

For certain inputs the output will take a simple form. The most notable of these is the class of complex sinusoidal inputs, in which case we can find the "steady-state" response of the difference equation.

EXERCISE 2.1

Step Response

In stable systems the natural response of the difference equation decays away to zero as n increases, because the roots $\{p_k\}$ are inside the unit circle (i.e., $|p_k| < 1$). Therefore, when the input signal is a constant for all $n \geq 0$, the output signal for large n is due entirely to the input. In fact, the output becomes a constant in this case.

a. For the system in (1-3), find the response to a step function input of amplitude 3 (i.e., $x[n] = 3u[n]$). Use a long enough section of the input signal so that the output from `filter` is nearly constant. This length can be estimated by considering the size of $|p_k|^n$ versus n. Plot the step response and determine the constant level (G_0) of the output as $n \to \infty$.

b. The constant level determined in part (a) is the steady-state response. Its precise value can be calculated by observing that both $y[n]$ and $x[n]$ become constants in the limit $n \to \infty$. Thus, $\lim_{n\to\infty} y[n] = G_0$ and $x[n] = 3$. Use these facts in (1-3) to determine G_0.

c. The variable part of the total response is called the transient response. Determine the transient response $y_t[n] = y[n] - G_0$ and plot it for $0 \leq n \leq 50$.

d. Since the filter is linear, the response to a step of different amplitude is just a scaled version of the previous result. Verify that steady-state response to $x[n] = 15u[n]$ is five times that found in part (a). Explain why similar scaling applies to the transient response.

e. Since the unit impulse signal $\delta[n]$ is just the first difference of the step signal $u[n]$, the linearity property of the filter implies that the impulse response $h[n]$ should be the first difference of the step response $s[n]$. Verify that this property holds. In MATLAB see `help diff` for a first difference operator; be careful, because `diff` reduces the vector size by one.

EXERCISE 2.2

Steady-State Response

The same decomposition into transient and steady-state response will also apply to a wider class of signals, e.g., signals of the form $e^{j\omega_\circ n}u[n]$. In this case the transient dies away, and the form of the output approaches $Ge^{j\omega_\circ n}u[n]$, where G is a complex constant with respect to n. Note that G does vary with $\omega_\circ$, so we could write $G(\omega_\circ)$. When $\omega_\circ = 0$ we have the case of the step response. If we take the real or imaginary part of the complex exponential, we have the response to a sinusoid.

a. For the system in (1-3), plot the real and imaginary parts of the response to the complex exponential, $x[n] = e^{jn\pi/3}u[n]$. Use a long enough section of the input signal so that the transient has died out. Determine the limiting value of the complex amplitude $G(\pi/3)$ of the output as $n \to \infty$. Make sure that you account for the complex exponential behavior of the output.

b. A simple derivation will yield a formula for $G(\omega_\circ)$ that is good for any $\omega_\circ$. By definition, the steady-state response is obtained as $n \to \infty$.

$$\text{INPUT} = e^{j\omega_\circ n}u[n] \qquad \Longrightarrow \qquad \lim_{n\to\infty}\left(y[n] - G(\omega_\circ)\,e^{j\omega_\circ n}\right) = 0$$

For convenience, we now drop the subscript on the frequency and write ω in place of $\omega_\circ$. Also, in steady state we can consider $x[n] = e^{j\omega n}$ and we can replace $y[n]$ with $G(\omega)e^{j\omega n}$ in the difference equation:

$$G(\omega)e^{j\omega n} - 1.8\cos\left(\tfrac{\pi}{16}\right)G(\omega)e^{j\omega(n-1)} + 0.81G(\omega)e^{j\omega(n-2)} = e^{j\omega n} + \tfrac{1}{2}e^{j\omega(n-1)}$$

Complete the derivation to obtain the following formula for $G(\omega)$, the complex amplitude of the steady-state response in (1-3) at ω.

$$G(\omega) = \frac{1 + \frac{1}{2}e^{-j\omega}}{1 - 1.8\cos\left(\frac{\pi}{16}\right)e^{-j\omega} + 0.81e^{-j2\omega}} \tag{2-1}$$

c. If $G(\omega)$ is then plotted versus ω, the resulting function is called the *frequency response* of the system. The notation $H(e^{j\omega})$ is used in place of $G(\omega)$ because of its connection with the DTFT and the z-transform. Evaluate the frequency response and plot the magnitude and phase of $H(e^{j\omega})$ versus ω (see `abs` and `angle` for magnitude and phase).

d. Check the value of $H(e^{j\omega})$ at $\omega = 0$.

e. Pick off the value of $H(e^{j\omega})$ at another frequency, say $\omega = \pi/4$, and compare to a steady-state response obtained via `filter`.

f. The total response can again be decomposed into the sum of the steady-state response and the transient response:

$$y[n] = y_{ss}[n] + y_t[n]$$

Determine the transient response and plot it for $0 \le n \le 30$. Note that it differs from the transient response determined for the step input.

g. Since the filter is linear, the response to a real sinusoid can be determined easily. Take the case of the cosine input. Since the cosine is

$$\cos\omega_\circ n\, u[n] = \tfrac{1}{2}\left(e^{j\omega_\circ n}u[n] + e^{-j\omega_\circ n}u[n]\right)$$

the steady-state response should be one half times the sum of the steady-state responses due to the complex exponentials at $+\omega_\circ$ and $-\omega_\circ$. Verify that this is true by running the filter and generating the steady-state responses for the three inputs: $\cos\omega_\circ n\, u[n]$, $e^{j\omega_\circ n}u[n]$, and $e^{-j\omega_\circ n}u[n]$. Take $\omega_\circ = \pi/3$. Does the same sort of additive combination apply to the transient responses?

h. Since the coefficients of the IIR filter (1-3) are all real-valued, an additional property holds. The steady-state response to the exponential at $-\omega_\circ$ is the conjugate of the response due to $+\omega_\circ$. Therefore, a simpler way to express the property in part (g) is that the response due to the cosine is the real part of the response due to the exponential at $+\omega_\circ$; similarly, the response due to a sine input is the imaginary part. Verify that these statements are true by applying the `real` and `imag` operators in MATLAB.

Comment. While the steady-state response could be used to generate the frequency response, in most homework problems a formula is evaluated instead. However, when dealing with an unknown system, one experimental approach to measuring the frequency response is to perform steady-state measurements at different frequencies and then plot the result.

PROJECT 3: FREQUENCY RESPONSE FOR DIFFERENCE EQUATIONS

In this project we investigate a method for directly computing the frequency response of any LTI system that is described by a difference equation. Assume that the filter coefficients, $\{a_k\}$ and $\{b_\ell\}$, are known. Then the frequency response will be computed directly by feeding the coefficients of the filter into the `freqz` function.

Consider the same difference equation as in (1-3). Since the transfer function of this system is rational:

$$H(z) = \frac{1+\frac{1}{2}z^{-1}}{1-1.8\cos(\frac{\pi}{16})z^{-1}+0.81z^{-2}} \tag{3-1}$$

the `freqz` function can be used to find its frequency response, because evaluating the z-transform on the unit circle is equivalent to finding the discrete-time Fourier transform.

Hints

The command `[H,W] = freqz(b,a,N,'whole')` will evaluate the frequency response of a filter at `N` points, equally spaced in radian frequency around the unit circle. If you do not use the `'whole'` option, `freqz` will use only the upper half of the unit circle (from 0 to π in frequency), which is sufficient for filters with real coefficients. The output vectors `H` and `W`, will return `N` frequency response samples (in `H`) and `N` equally spaced values of ω from 0 to 2π or 0 to π (in `W`).

EXERCISE 3.1

Frequency Response with `freqz`

For the difference equation (1-3), do the following frequency-domain computations:

a. Make plots of the magnitude and phase responses, with 512 frequency samples around the entire unit circle. For instance, use `plot(W,abs(H))` or `plot(W,angle(H))`.

b. Now redo the frequency response using only the upper half of the unit circle (ω ranges from 0 to π). This is sufficient because of the symmetries in the magnitude and phase response, which you should have observed in part (a).

c. Specify the type of filter defined by this difference equation: high-pass, low-pass, all-pass, bandpass, or bandstop.

EXERCISE 3.2

Experimentation

You are now encouraged to experiment with other difference equations to see what types of filters you can create. As a start, the following difference equations are provided. For each, determine the frequency response and state what type of filter it defines.

a. $y[n]+0.13y[n-1]+0.52y[n-2]+0.3y[n-3]=0.16x[n]-0.48x[n-1]$
$+0.48x[n-2]-0.16x[n-3]$

b. $y[n]-0.268y[n-2]=0.634x[n]-0.634x[n-2]$

c. $y[n]+0.268y[n-2]=0.634x[n]+0.634x[n-2]$

d. $10y[n]-5y[n-1]+y[n-2]=x[n]-5x[n-1]+10x[n-2]$

FOURIER TRANSFORM: DTFT

OVERVIEW

This set of projects will introduce *basic* properties of the discrete-time Fourier transform (DTFT). Two completely different cases are treated. The first (Projects 1 and 4) deals with finite-length signals, for which the DTFT can be evaluated exactly. The second case (Projects 3 and 5) involves infinite-length signals, which must have a special form to be evaluated: namely, exponential signals which have rational z-transforms. In Project 2, symmetries of the transform are explored.

The Fourier representation of a signal via the forward and inverse DTFT is a key part of signal analysis. Equations (0-1) and (0-2) are the analysis and synthesis equations, respectively.

$$X(e^{j\omega}) = \sum_{n=-\infty}^{\infty} x[n]e^{-j\omega n} \tag{0-1}$$

$$x[n] = \frac{1}{2\pi}\int_{-\pi}^{\pi} X(e^{j\omega})e^{j\omega n}\,d\omega \tag{0-2}$$

Similarly, the frequency response, which is the DTFT of the impulse response, provides a concise description of a LTI system when used for filtering. The DTFT $X(e^{j\omega})$ is a periodic complex-valued function of ω. The period is always 2π, and the fundamental period is usually chosen to be the domain $[-\pi, \pi)$. In the context of MATLAB, where computability is an issue, the DTFT presents two problems:

1. Its definition is valid for *infinitely long* signals.
2. It is a function of a *continuous* variable, ω.

The first point is a problem only to the extent that any signal/vector in MATLAB must be finite in length. Thus we have the problem that it is not really possible to use MATLAB to compute the DTFT of an infinitely long signal. One notable exception is when we can derive an analytic form for the transform and just evaluate it, as in the case of $x[n] = a^n u[n]$, which has a rational DTFT.

The second issue presents a frequency sampling problem. The best that we can do with MATLAB is evaluate the DTFT on a finite grid of points. We can usually choose enough frequencies so that our plots will give a smooth approximation to the true DTFT. The choice that is best for computation is a set of evenly spaced frequencies over the interval $(-\pi, \pi]$, or

$[0, \pi]$ for conjugate-symmetric transforms. With such sampling, the forward DTFT formula (0-1) becomes

$$X(e^{j\omega_k}) = X(e^{j2\pi k/N}) = \sum_{n=0}^{L-1} x[n]e^{-j(2\pi k/N)n} \qquad \text{for } k = 0, 1, \ldots, N-1 \qquad (0\text{-}3)$$

The periodicity of the DTFT means that the values for $-\pi \leq \omega < 0$ are those for $k > N/2$. This formula (0-3) is *computable* because it is a sum over *finite* limits, evaluated at a *finite* number of frequencies, $\omega_k = 2\pi k/N$. Since the signal length must be finite ($0 \leq n < L$), a case such as $x[n] = a^n u[n]$ is not covered by this summation form.

When $N = L$ this computable formula (0-3) is just an N-point discrete Fourier transform (DFT), but for the moment we ignore the DFT nature of things and concentrate on computing samples of the DTFT. Details of the DFT are treated in Chapter 2.

When sampling the DTFT, there is no requirement that N be equal to L, although it is convenient because the computation is usually carried out via the DFT (with the FFT algorithm). Indeed, if $N > L$, we only have to imagine that $x[n]$ is zero-padded to use an N-point DFT. The case where $N < L$ is much more troublesome. Correct application of the FFT in this case requires *time aliasing* of $x[n]$ prior to the N-point DFT computation. For now, you should always make sure that you evaluate the DTFT at many more frequencies than there are points in the original time sequence (i.e., $N \geq L$).

Background Reading

This basic material about the DTFT can be found in Oppenheim and Schafer (1989), Chapter 2, Sections 2.5 through 2.9, as well as in the introductory chapters of other DSP books.

Project 1: Computing the DTFT: Finite-Length Signals

In this project we consider the case of finite-length signals. This will take advantage of the `dtft` function defined below. In particular, this project deals with pulses and their DTFTs, because these sorts of signals are the easiest examples of using the DTFT. Most books use the rectangular pulse as the first example of deriving a DTFT.

Hints

We need two functions for computing the DTFT. The MATLAB function `freqz` will suffice for the infinite-length signal case, but a new function will be needed to compute the DTFT of a finite-length signal. It should be called `dtft(h,N)`, and is essentially a layer that calls `fft(h,N)`.

```
function [H,W] = dtft( h, N )
%DTFT   calculate DTFT at N equally spaced frequencies
%   usage:    H = dtft( h, N )
%      h: finite-length input vector, whose length is L
%      N: number of frequencies for evaluation over [-pi,pi)
%             ==> constraint: N >= L
%
%      H: DTFT values (complex)
%      W: (2nd output) vector of freqs where DTFT is computed
%
N = fix(N);
L = length(h);  h = h(:);  %<-- for vectors ONLY !!!
if( N < L )
   error('DTFT: # data samples cannot exceed # freq samples')
```

```
end
W = (2*pi/N) * [ 0:(N-1) ]';
mid = ceil(N/2) + 1;
W(mid:N) = W(mid:N) - 2*pi;     % <--- move [pi,2pi) to [-pi,0)
W = fftshift(W);
H = fftshift( fft( h, N ) );    %<--- move negative freq components
```

Note that you don't have to input the signal length L, because it can be obtained by finding the length of the vector h. Furthermore, since the DTFT is periodic, the region from $\omega = \pi$ to 2π is actually the negative-frequency region, so the transform values just need to be reordered. This is accomplished with the MATLAB function `fftshift`, which exchanges the upper and lower halves of a vector. Using the DTFT vector H, the $[-\pi, \pi]$ plot can be produced by noting that `H(1)` is the frequency sample for $\omega = -\pi$.

When plotting in the transform domain it would be best to make a two-panel `subplot` as shown in Fig. 1.3. The MATLAB program that produces Fig. 1.3 is given below.

```
%--- example of calculating and plotting a DTFT
%---
format compact, subplot(111)
a = 0.88 * exp(sqrt(-1)*2*pi/5);
nn = 0:40;   xn = a.^nn;
[X,W] = dtft( xn, 128 );
subplot(211), plot( W/2/pi, abs(X) ); grid, title('MAGNITUDE RESPONSE')
   xlabel('NORMALIZED FREQUENCY'), ylabel('| H(w) |')
subplot(212), plot( W/2/pi, 180/pi*angle(X) ); grid
   xlabel('NORMALIZED FREQUENCY'), ylabel('DEGREES')
   title('PHASE RESPONSE')
```

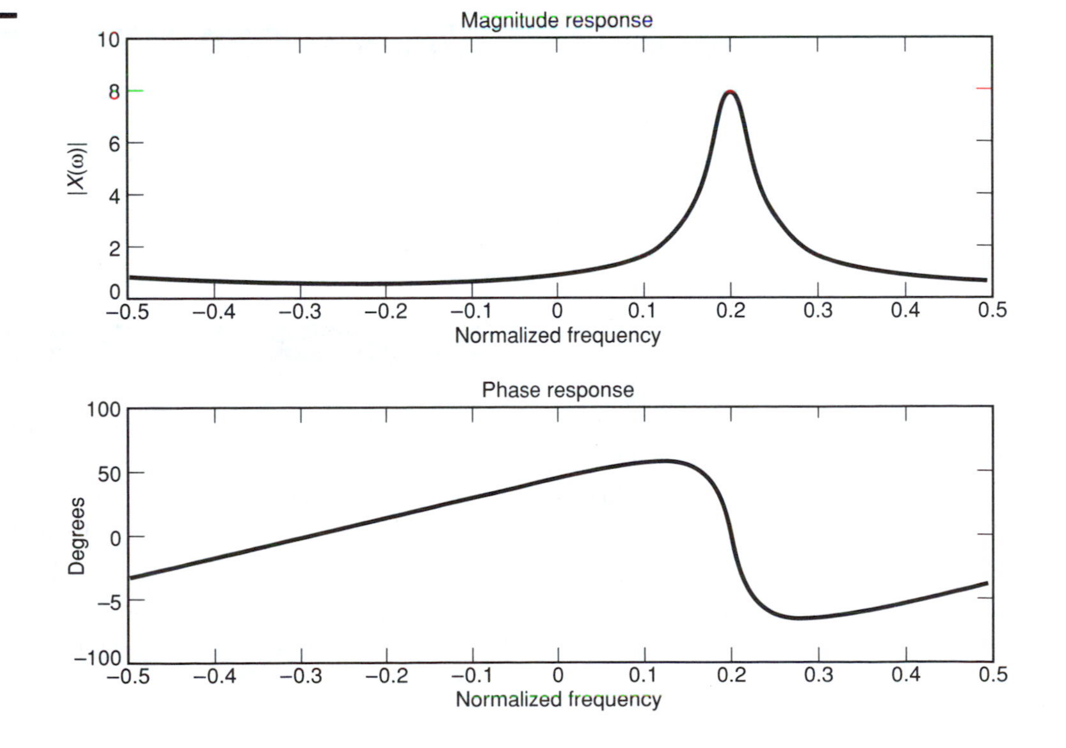

Figure 1.3 Two-panel frequency-domain plot made via `subplot`.

EXERCISE 1.1

DTFT of a Pulse

The finite-length pulse is always used as a prime example of evaluating the DTFT. Suppose that the rectangular pulse $r[n]$ is defined by

$$r[n] = \begin{cases} 1 & 0 \leq n < L \\ 0 & \text{elsewhere} \end{cases} \tag{1-1}$$

a. Show that the DTFT of $r[n]$ is given by the mathematical formula

$$R(e^{j\omega}) = \frac{\sin \frac{1}{2}\omega L}{\sin \frac{1}{2}\omega} \cdot e^{-j\omega(L-1)/2} \tag{1-2}$$

The first term in this transform has a special form that occurs quite often in conjunction with the DTFT; it will be called the *aliased sinc* function:[1]

$$\operatorname{asinc}(\omega, L) = \frac{\sin \frac{1}{2}\omega L}{\sin \frac{1}{2}\omega} \tag{1-3}$$

b. Use the function `dtft` to evaluate the DTFT of a 12-point pulse. Make a plot of the DTFT versus ω over the range $-\pi \leq \omega < \pi$. Plot the real and imaginary parts separately, but notice that these plots are not terribly useful. Instead, plot the magnitude of the DTFT (see `abs` in MATLAB). To make the plot appear smooth, choose a number of frequency samples that is 5 to 10 times the pulse length. Experiment with different numbers of frequency samples. When plotting, be careful to label the frequency axis correctly for the variable ω.

c. Notice that the zeros of the asinc function are at regularly spaced locations. Repeat the DTFT calculation and magnitude plot for an odd-length pulse, say $L = 15$. Again, check the zero locations and note the peak height.

d. Determine a general rule for the regular spacing of the zeros of the asinc function and its dc value.

EXERCISE 1.2

M-File for asinc

Write a MATLAB function `asinc(w,L)` that will evaluate $\operatorname{asinc}(\omega, L)$ on a frequency grid directly from formula (1-3). The function should have two inputs: a length `L` and a vector of frequencies `w`. It must check for division by zero as happens for $\omega = 0$.

Directly evaluate the "aliased sinc" formula (1-2) for the DTFT of a pulse. Plot the magnitude and save this plot for comparison to the result obtained with `dtft`.

EXERCISE 1.3

Phase Unwrapping in the DTFT

Since the Fourier transform (0-1) is a complex-valued quantity, not only the magnitude, but also the phase is of interest (see `abs` and `angle` in MATLAB).

a. Make a plot of the phase versus ω. In formula (1-2) the phase appears to be linear with a slope of $-\frac{1}{2}(L-1)$ versus ω. Check your plot. If the phase plot appears incorrect, consider the following: the angle is evaluated modulo–2π, so 2π jumps in the phase occur when the phase is "wrapped" into the $[-\pi, \pi]$ interval by the arctangent computed inside `angle`. In addition, the asinc term is not always positive, so there are additional jumps of π at the zeros of $R(e^{j\omega})$ which are the zeros of the numerator: $\omega = 2\pi k/L$.

---

[1] This is also called the *Dirichlet kernel*.

b. An "unwrapped" version of the phase can be produced. See the MATLAB function `unwrap`. Use this function to compute and plot the unwrapped phase of $R(e^{j\omega})$. Now it should appear linear, except for jumps of π. If not, the DTFT sampling may be too low; unwrapping requires a rather dense sampling of the frequency axis.

PROJECT 2: DTFT SYMMETRIES

Finite-length signals are often "centered" so as to have symmetries that make the DTFT simpler. For example, if the rectangular pulse is shifted to be even about the $n = 0$ point, the DTFT will be a purely real and even function versus ω. In this project, the six major types of symmetry will be reviewed and illustrated by various examples.

Hints

It will be necessary to modify the `dtft` function so that it can accept an argument that defines the starting point on the time axis for the signal $x[n]$. Ordinarily, the default starting time is assumed to be $n = 0$ by virtue of the definition of the DTFT (0-3).

Another skill that will be needed for this project is plotting the DTFT to exhibit symmetries. Ordinarily, the ω_k samples run from $\omega = 0$ to 2π. However, when studying symmetry the DTFT plot should be made from $-\pi$ to $+\pi$, which is the range returned by the `dtft` function, but not by `freqz`.

When checking a transform for the property of purely real or purely imaginary, the part of the signal that is expected to be zero might not be exactly zero. There might be some extremely small values due to roundoff noise in the computation of the DFT. If these values are on the order of 10^{-13} or 10^{-14}, it is safe to assume that the cause is rounding in the double-precision arithmetic of MATLAB (see `help eps`). However, numbers on the order of 10^{-9} are probably not due to rounding, unless a very large number of computations have been performed on the signals.

When checking whether a transform is even or odd, it would be convenient to have an operator that will flip the transform. In other words, we need a MATLAB function that corresponds to the mathematical operation of flipping the frequency axis: $Y(e^{j\omega}) = X(e^{-j\omega})$. The following function will implement this flip on a pair of vectors, `H` for the transform values and `W` for the frequency axis.

```
function  [ G, Wflipped ] = flipDTFT( H, W )
%FLIPDTFT   flip the DTFT:  G(w) = H(-w)
%      usage:
%            [ G, Wflipped ] = flipDTFT( H, W )
%
%             H = DTFT values (complex)
%             W = frequency samples
%             G = DTFT values
%      Wflipped = flipped frequency domain
%                      lies within [-pi,pi)
%
N = length(H);    %<--- works only for vectors !!!
Wflipped = -W(N:-1:1);
G = H(N:-1:1);
%---
%--- now get everything back into the [-pi,pi) interval
%---     assume that W was monotonically increasing
%---     so Wflipped is also increasing !
%---
```

```
jkl = find( Wflipped(:)' < -pi );
if( ~isempty(jkl) )
   kk = [ (length(jkl)+1):N  jkl ];
   Wflipped(jkl) = Wflipped(jkl) + 2*pi;
   Wflipped = Wflipped(kk);
   G = G(kk);
end
jkl = find( Wflipped(:)' >= (pi-100*eps) );
if( ~isempty(jkl) )
   kk = [ jkl  1:(jkl(1)-1)  ];
   Wflipped(jkl) = Wflipped(jkl) - 2*pi;
   Wflipped = Wflipped(kk);
   G = G(kk);
end
```

EXERCISE 2.1

Zero-Phase Signals

Working with zero-phase signals in MATLAB presents a difficulty—the `dtft` function from Project 1 assumes that the signal starts at $n = 0$, but any zero-phase signal must be symmetric around $n = 0$. One way to address this problem is to create a modified form of `dtft` that has an additional input argument to specify the starting index of the signal. This starting index can then be used to modify the DTFT output according to the "shifting property" of the DTFT:

$$x[n - n_0] \quad \overset{\text{DTFT}}{\longleftrightarrow} \quad e^{-j\omega n_0} X(e^{j\omega}) \tag{2-1}$$

In other words, the DTFT must be multiplied (pointwise) by a complex exponential $\exp(-j\omega n_0)$ to undo the effect of the time shift by n_0.

a. Create a new function `dtft_n0( x, n0, N )` to implement the time-shift feature by adding the argument `n0`. This should amount to a minor modification of the `dtft` function.

b. Test `dtft_n0` by taking the DTFT of a 21-point pulse that starts at $n = -10$. The resulting transform should be a purely real and even function.

c. Plot the real part of the DTFT and compare to the DTFT magnitude over the domain from $-\pi$ to $+\pi$.

d. Verify that the imaginary part is zero and that the phase is equal to either 0 or π.

Note that this symmetry works only for odd-length pulses; if the length is even, there will always be a residual phase term corresponding to a half-sample delay.

EXERCISE 2.2

Triangular Pulse

Another simple signal is the symmetric triangular pulse:

$$\Delta[n] = \begin{cases} L - n & 0 \le n < L \\ L + n & -L < n < 0 \\ 0 & \text{elsewhere} \end{cases}$$

The length of this signal is $2L - 1$, and it can be formed by convolving two L-point rectangular pulses. As a result the DTFT is an asinc-squared function. No phase term is involved, since this is a symmetric signal.

a. Make a plot of a 21-point triangular pulse over the domain $-20 \leq n \leq 20$. Then compute its DTFT with the function `dtft_n0` and plot the result over $-\pi \leq \omega < \pi$.

b. For comparison plot the squared magnitude of the DTFT of an 11-point rectangular pulse. It might be easier to make the comparison by plotting both on a log-magnitude (dB) scale.

EXERCISE 2.3

Symmetries in the DTFT

There are many symmetry properties in the time and frequency domains. One set deals with purely real, or purely imaginary signals, another with even and odd signals. For example, the DTFT of an even signal is even. Each of the following parts concentrates on one type of symmetry. Symmetry in the frequency domain can be verified by plotting the real and imaginary parts of the DTFT (or magnitude and phase) and by using the function `flipDTFT` to examine $X(e^{-j\omega})$.

a. The DTFT of a real signal is conjugate symmetric, $X^*(e^{j\omega}) = X(e^{-j\omega})$; in other words, conjugating the DTFT is the same as flipping it. For the signal $x[n] = (0.9)^n \cos(2\pi n/\sqrt{31})$, for $0 \leq n < 21$, plot the DTFT (magnitude and phase) and verify that it is indeed conjugate symmetric.

b. If the signal is purely imaginary, the DTFT will be conjugate antisymmetric. Using $x[n]$ from the previous part, define $y[n] = jx[n]$ and display its DTFT $Y(e^{j\omega})$. Use the `flipDTFT` function to compare the flipped version of $Y(e^{j\omega})$ to the conjugated version, and then check the conjugate antisymmetric property.

c. An even time signal transforms to an even function of frequency. Prove that the chirp signal $v_e[n] = \exp(j2\pi n^2/25)$ over $-30 < n < 30$ is even. Then compute and display its DTFT to verify that the transform is also even with respect to ω.

d. For the odd signal $v_o[n] = n$ over $-20 < n < 20$, verify that the DTFT is also odd.

e. The properties of real/imaginary and even/odd can be combined. Define a signal that is both purely imaginary and odd. What symmetry properties will its DTFT have? Verify by plotting the transform and by using `flipDTFT`.

PROJECT 3: DTFT OF INFINITE-LENGTH SIGNALS

It is not usually possible to *compute* the DTFT of an infinitely long signal. However, there is one important class for which the computation is easy. This is the class of exponential signals, where the DTFT is a rational function of $e^{-j\omega}$.

$$H(e^{j\omega}) = \frac{B(e^{j\omega})}{A(e^{j\omega})} = \frac{\sum_{\ell=0}^{Q} b_\ell e^{-j\omega\ell}}{\sum_{k=0}^{P} a_k e^{-j\omega k}}$$

The exponential signal $h[n] = a^n u[n]$ is one member of this class, but cannot be dealt with using the `dtft` function presented in Project 1. On the other hand, its DTFT is readily derived as a formula:

$$h[n] = a^n u[n] \quad \overset{\text{DTFT}}{\longleftrightarrow} \quad H(e^{j\omega}) = \sum_{n=0}^{\infty} a^n u[n] e^{-j\omega n} = \frac{1}{1 - ae^{-j\omega}} \quad \text{if } |a| < 1 \tag{3-1}$$

Using the rational form of $H(e^{j\omega})$, it is easy to evaluate this DTFT over a set of frequency samples. The denominator function, $1 - ae^{-j\omega}$, is evaluated at the set of discrete frequencies, and is then divided into the numerator, which is the constant 1. This strategy extends to any

DTFT that is a rational function of $e^{-j\omega}$. Furthermore, the evaluation of both the numerator and denominator can be done with the FFT, because both are, in effect, finite-length signals. Thus, evaluating the rational function amounts to doing two `dtft` calculations. This frequency-domain computation is embodied in the MATLAB `freqz` function.

Hints

The MATLAB function `freqz` is so named because it is applicable to rational z-transforms.

```
[ HH, WW ] = freqz( b, a, N, 'whole' )
```

Like `dtft`, `freqz` has two outputs: the transform values (`HH`) and the frequency grid (`WW`). The fourth input argument is optional, but when set to `'whole'` the output vector `WW`, which specifies the frequency grid, will range from $\omega = 0$ to $\omega = 2\pi$. If the fourth argument is omitted, the frequency grid consists of `N` equally spaced points over the range $0 \leq \omega < \pi$.

EXERCISE 3.1

Exponential Signal

For the signal $x[n] = (0.9)^n u[n]$, compute the DTFT $X(e^{j\omega})$ using `freqz`.

a. Make a plot of both the magnitude and the phase versus ω over the range $-\pi \leq \omega < \pi$. This will require a shift of the `[X, W]` vectors returned from `freqz`. Explain why the magnitude is even and the phase is an odd function of ω.

b. Derive formulas for the magnitude and phase from equation (3-1) for the first-order system.

c. Compute the magnitude and phase by a direct evaluation of the formulas, and compare to the results from `freqz`.

EXERCISE 3.2

Complex Exponential

If we take $a = z_0 = re^{j\theta}$ to be a complex number in (3-1), the same transform is applicable. This case is important because we can develop some insight into how the magnitude (r) and phase (θ) of the complex number affect the DTFT.

a. Take $z_0 = 0.95e^{j3\pi/11}$ and plot $x[n] = z_0^n u[n]$ for $0 \leq n \leq 30$. Plot both the real and imaginary parts versus n together in a two-panel `subplot`.

b. Again with $z_0 = 0.95e^{j3\pi/11}$, compute the DTFT and plot the magnitude versus ω. Note where the peak of the magnitude response lies as a function of ω. Relate the peak location to the polar representation of z_0.

c. If the angle of z_0 were changed to $\theta = 3\pi/5$, sketch the DTFT magnitude that you would expect. Verify by making a plot from a `freqz` calculation.

d. Change the magnitude $r = |z_0|$ and redo the DTFT plot. Use four values: $r = 0.975$, 0.95, 0.9, and 0.8. Notice that both the height and the bandwidth of the peak will change, as $|z_0|$ is moved closer to 1. Measure the numerical value of the bandwidth at the -3 dB point. Try to develop a simple formula that relates the bandwidth to r.

EXERCISE 3.3

Decaying Sinusoid

The complex-valued signal $x[n] = z_0^n u[n]$ is quite useful in representing real-valued signals that are decaying sinusoids of the form

$$y[n] = \mathrm{Re}\{Gz_0^n u[n]\} = A\,r^n \cos(\theta n + \phi)u[n] \tag{3-2}$$

where $G = Ae^{j\phi}$ and $z_0 = re^{j\theta}$. The resulting DTFT, $Y(e^{j\omega})$, is a rational function with a second-order denominator. Its transform can be derived several ways, but it is informative to do so using some properties of the DTFT. In particular, the conjugate property states that the DTFT of a complex-conjugated signal is the flipped and conjugated version of the original transform.

$$x[n] \overset{\text{DTFT}}{\longleftrightarrow} X(e^{j\omega}) \qquad \Longrightarrow \qquad x^*[n] \overset{\text{DTFT}}{\longleftrightarrow} X^*(e^{-j\omega}) \tag{3-3}$$

Thus, if we let $x[n] = Gz_0^n u[n]$, its DTFT is

$$X(e^{j\omega}) = \frac{G}{1 - z_0 e^{-j\omega}} = \frac{Ae^{j\phi}}{1 - re^{-j(\omega-\theta)}} \tag{3-4}$$

so we can apply the conjugate property to $y[n] = \mathrm{Re}\{x[n]\} = \frac{1}{2}\{x[n] + x^*[n]\}$ to get the DTFT:

$$Y(e^{j\omega}) = \tfrac{1}{2}[X(e^{j\omega}) + X^*(e^{-j\omega})] \tag{3-5}$$

$$= \frac{\frac{1}{2}Ae^{j\phi}}{1 - re^{-j(\omega-\theta)}} + \frac{\frac{1}{2}Ae^{-j\phi}}{1 - re^{+j(-\omega-\theta)}}$$

$$= \frac{1}{2}A\left(\frac{e^{j\phi}(1 - re^{-j(\omega+\theta)}) + e^{-j\phi}(1 - re^{-j(\omega-\theta)})}{(1 - re^{-j(\omega-\theta)})(1 - re^{-j(\omega+\theta)})}\right)$$

$$= A\left(\frac{\cos\phi - r\cos(\theta+\phi)e^{-j\omega}}{1 - 2r\cos\theta\, e^{-j\omega} + r^2 e^{-j2\omega}}\right) \tag{3-6}$$

Obviously, these equations start to get very messy, but the simple expedient of using complex arithmetic is easier than plugging (3-2) into the DTFT summation.

In this exercise we will use MATLAB to calculate the DTFT via this conjugate trick and compare to direct evaluation with `freqz`. The test signal is

$$y[n] = 3(0.95)^n \cos(2\pi n/7 + \pi/3)\,u[n]$$

For both, plot the resulting magnitude and phase of $Y(e^{j\omega})$ versus ω. Verify that the two approaches give the same result.

a. First, substitute the parameters of the decaying sinusoid $y[n]$ directly into formula (3-6) for $Y(e^{j\omega})$; then use `freqz(b, a, n)` with the appropriate coefficient vectors `a` and `b` to get the frequency samples.

b. Express $y[n]$ as the sum of two complex exponentials. Take the DTFT of one complex exponential (via `freqz`) and then apply the conjugate property (3-3) in the transform domain (3-5) to generate the second term in the answer.

PROJECT 4: WINDOWING FOR THE DTFT

In this project, two properties of the DTFT are illustrated: the modulation property and the windowing property. The modulation property is, in effect, a special case of windowing where the frequency-domain convolution reduces to a simple frequency shift.

EXERCISE 4.1

Modulation Property

Many DTFT properties have useful interpretations and applications. One of these is the (complex) modulation property, which finds application in communications and radar. If a signal

$x[n]$ is multiplied by a complex sinusoid, $e^{j\omega_0 n}$, the result in the transform domain is a frequency shift of ω_0; $X(e^{j\omega})$ becomes $X(e^{j[\omega-\omega_0]})$.

a. Demonstrate this property with the rectangular pulse—take the pulse length to be $L = 21$ and pick $\omega_0 = 2\pi/\sqrt{31}$. Plot the result from `dtft`. Verify that the peak of the DTFT magnitude (an asinc function) has moved to $\omega = \omega_0$. Try a value of ω_0 larger than 2π to exhibit the periodicity of the DTFT.

b. Repeat the experiment, but multiply the pulse by a cosine signal at the same frequency. This is just double-sideband AM, and it involves only real operations.

EXERCISE 4.2

Windowing Gives Frequency-Domain Convolution

The windowing property of the DTFT states that the DTFT of the (pointwise) product of two time signals (`.*` in MATLAB) is the periodic frequency-domain convolution of their Fourier transforms:

$$y[n] = x[n] \cdot w[n] \quad \overset{\text{DTFT}}{\longleftrightarrow} \quad Y(e^{j\omega}) = \frac{1}{2\pi}\int_{-\pi}^{\pi} X(e^{j\theta})W(e^{j[\omega-\theta]})\,d\theta \qquad (4\text{-}1)$$

The frequency-domain convolution will "smear" the true DTFT, depending on the exact nature of $W(e^{j\omega})$. Even when there appears to be no windowing, the very fact that the signal is finite means that a rectangular window has been applied. In this case, the window transform is an aliased sinc function.

a. It should not be possible to compute the frequency-domain convolution in (4-1) because it involves an integral. However, there is one case where the frequency-domain result can be obtained as a formula—when the unwindowed signal is a complex sinusoid [e.g., $x[n] = e^{j\theta_0 n}$, for all n]. Then $X(e^{j\omega})$ is an impulse in frequency, so the convolution evaluates to $Y(e^{j\omega}) = W(e^{j[\omega-\theta_0]})$. Of course, this is just the modulation property. The observed transform takes on the shape of the window's DTFT, shifted in frequency. Generate a windowed sinusoid with $\theta_0 = 2\pi/\sqrt{31}$:

$$x[n] = r[n] \cdot e^{j\theta_0 n}$$

where $r[n]$ is the rectangular window of length $L = 32$. The rectangular window $r[n]$ can be created via the function `ones` or `boxcar`. Plot the DTFT, $X(e^{j\omega})$, and note that the peak has been shifted to $\omega = \theta_0$ and that it has the shape of an aliased sinc function.

b. The following window function is called the von Hann (or `hanning`) window:

$$w[n] = \begin{cases} \frac{1}{2} - \frac{1}{2}\cos\frac{2\pi}{L}n & 0 \le n < L \\ 0 & \text{elsewhere} \end{cases}$$

Apply a 32-point Hann window to a sinusoid of frequency $\omega_\circ = 2\pi/\sqrt{31}$. Plot the time signal and then calculate its DTFT and plot the magnitude response.

c. The DTFT of the Hann window can be written in terms of three shifted asinc functions. This is done by viewing $w[n]$ as a rectangular window applied to the signal $\frac{1}{2} - \frac{1}{2}\cos(2\pi/L)n$, and then using the modulation property. Make a plot of the magnitude response of the Hann window $W(e^{j\omega})$ for $L = 32$. Explain how $W(e^{j\omega})$ is formed from $R(e^{j\omega})$, the DTFT of the rectangular window. Take into account the fact that both $W(e^{j\omega})$ and $R(e^{j\omega})$ are complex-valued.

d. Sometimes the following formula is mistakenly used for the von Hann window:

$$v[n] = \tfrac{1}{2} + \tfrac{1}{2}\cos(2\pi n/L) \qquad \text{for } n = 0, 1, 2, \ldots, L-1$$

The change from a minus sign to a plus is significant. Plot $v[n]$ versus n over the domain $-10 \le n \le L+10$, with $L = 32$. Notice the discontinuities at the endpoints, 0 and $L-1$.

Next, plot its DTFT magnitude and compare to the result for the Hann window. Explain how the same three asinc terms can be combined to give such a different answer for the DTFT $V(e^{j\omega})$. Take into consideration that the terms being summed are complex and all have a phase component.

EXERCISE 4.3

Convergence to True DTFT

There is one situation that has not yet been dealt with: the case of taking the DTFT of a finite portion of an infinite-length signal. This case arises often in practice because we can usually record (and store) only a small portion of a signal for analysis. Nonetheless, we would like to deduce the signal's true frequency content from a DTFT of the finite segment. In this exercise we compare a windowed DTFT of an exponential to the true DTFT.

The `dtft` function given previously is sufficient to perform the DTFT on any finite-length signal. An easy example of windowing is obtained by taking the first L points of the infinitely long exponential, $x[n] = a^n u[n]$. As the segment length increases, the resulting DTFT should converge to the analytic form given previously for $X(e^{j\omega})$.

a. Using the choice $a = 0.977$, plot the log magnitude of the DTFT for several different signal lengths: $L = 32$, 64, 128 and 256. Limits on available memory might stop this experiment. Plot all four results together with a four panel subplot. Overlay each with a plot of the "true" DTFT for the infinite-length signal as a dashed line.

b. Comment on the changes observed as the section length L is increased. Explain the differences for increasing L in terms of a window transform $W(e^{j\omega})$ operating on the true spectrum.

PROJECT 5: FREQUENCY RESPONSE OF A NOTCH FILTER

The DTFT of an impulse response is, in fact, the frequency response of the system. Therefore, the DTFT can be used to describe the filtering nature of a system, and also its steady-state behavior.

EXERCISE 5.1

Notch Filter Example

A notch filter attempts to remove one particular frequency. Suppose that a bandlimited continuous-time signal is known to contain a 60-Hz interference component, which we want to remove by processing with the standard system (Fig. 1.4) for filtering a continuous-time signal with a discrete-time filter.

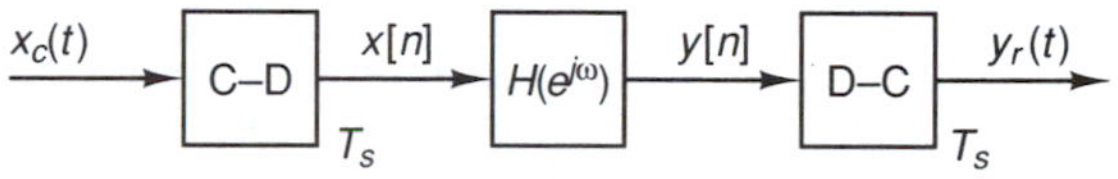

$x_c(t)$ → [C–D, T_s] → $x[n]$ → [$H(e^{j\omega})$] → $y[n]$ → [D–C, T_s] → $y_r(t)$

Figure 1.4
Standard system for implementing an analog filter by means of a cascade of an A/D converter, digital filter, and D/A converter.

a. Assume that the value of the sampling period is $T_s = 1$ ms. What is the highest frequency that the analog signal can contain if aliasing is to be avoided?

b. The discrete-time system to be used has frequency response

$$H(e^{j\omega}) = \frac{[1 - e^{-j(\omega-\omega_0)}][1 - e^{-j(\omega+\omega_0)}]}{[1 - 0.9e^{-j(\omega-\omega_0)}][1 - 0.9e^{-j(\omega+\omega_0)}]} \qquad (5\text{-}1)$$

Sketch the magnitude and phase of $H(e^{j\omega})$. Pick a trial value of $\omega_0 = 2\pi/5$ and use MATLAB to do the "sketch."

c. What value should be chosen for ω_0 to eliminate the 60-Hz component? Will the gain at other frequencies be equal to 1?

d. Make a MATLAB plot of the frequency response (magnitude only) using the value of ω_0 found in part (c).

e. Generate 150 samples of a 60-Hz sine wave sampled at $f_s = 1/T_s = 1000$ Hz. Use the function `filter` to process this input signal with the system from part (b) and the value of ω_0 from part (c). Display the output signal to illustrate that the filter actually removes the 60-Hz sinusoid.

f. Since the DTFT is a frequency response, it describes the steady-state behavior of the filter. Thus you should observe a "transient" response before the zero of the filter at 60 Hz rejects the input completely. Measure the duration of this transient (in milliseconds) from the beginning of the signal until a point where the output is less than 1% of the input signal amplitude.

EXERCISE 5.2

Steady-State Response of a Filter

In this project a sinusoidal input is filtered and the resulting output is split into two parts: the transient response and the steady-state response. The relationship of the steady-state response to the frequency response function $H(e^{j\omega})$ is explored.

a. Use the notch filter (5-1) with $\omega_0 = \pi/5$. Find the `b` and `a` coefficients for the notch filter. Make plots of the magnitude and phase responses of the filter.

b. Pick a particular frequency, say $\omega = \omega_i = \pi/4$. Compute the numerical values of the magnitude and phase of $H(e^{j\omega_i}) = |H(e^{j\omega_i})| \cdot e^{-j\phi(\omega_i)}$.

c. Use these numerical values to generate a plot of the "true" steady-state output

$$y_{ss}[n] = |H(e^{j\omega_i})| \cdot \cos(\omega_i n - \phi(\omega_i))$$

by evaluating the cosine formula directly. Let the range of n be $0 \leq n \leq 50$.

d. Generate the filter output $y[n]$ by using the MATLAB `filter` function when the input signal is the cosine: $v[n] = \cos(\omega_i n)$, $n = 0, 1, 2, \ldots, N$. This output is the sum of both the transient and steady-state responses.

e. Plot the two signals $y[n]$ and $y_{ss}[n]$, and compare them for the region where n is large. Since the system is stable, these two should be nearly the same, after all transients die out. The steady-state response is, therefore, a reasonable way to measure the frequency response at $\omega = \omega_i$, because the magnitude and phase of the output $y[n]$ will approach the magnitude and phase of $H(e^{j\omega_i})$, as $n \to \infty$.

f. Find the transient signal $y_t[n] = y[n] - y_{ss}[n]$ by subtracting the steady-state signal from the total output, and then plot the transient.

g. The steps above should be repeated for several different frequencies, including one very close to the notch, where the transient response will be much larger than the steady state.

GROUP DELAY

OVERVIEW

A convenient measure of the linearity of the phase is the *group delay*. The basic concept of group delay relates to the effect of the phase on a narrowband signal. Specifically, consider the output of a system with frequency response $H(e^{j\omega})$ for a narrowband input $x[n] = s[n]\cos(\omega_0 n)$. The signal $s[n]$ is called the *envelope*, and it must be slowly varying, which means that it has a narrow low-pass spectrum. Since it is assumed that $X(e^{j\omega})$ is nonzero only around $\omega = \omega_0$, the phase of the system can be approximated around $\omega = \omega_0$ by a Taylor series expansion up to the linear term

$$\angle H(e^{j\omega}) \approx -\phi_0 - \omega n_d \tag{0-1}$$

With this approximation, it can be shown that the response $y[n]$ to the input $x[n] = s[n]\cos(\omega_0 n)$ is $y[n] \approx s[n-n_d]\cos(\omega_0 n - \phi_0 - \omega_0 n_d)$. Consequently, the time delay of the envelope $s[n]$ of the narrowband signal $x[n]$ with Fourier transform centered around ω_0 is given by the negative of the slope of the phase at ω_0. In considering the linear approximation (0-1) to $\angle H(e^{j\omega})$ around $\omega = \omega_0$, as given above, we must treat the phase response as a continuous function of ω rather than as a function modulo 2π. The phase response specified in this way will be denoted as $\arg[H(e^{j\omega})]$ and is referred to as the *unwrapped phase* of $H(e^{j\omega})$. Then the *group delay* of a system is defined as

$$\tau(\omega) = \text{grd}[H(e^{j\omega})] = -\frac{d}{d\omega}\left\{\arg[H(e^{j\omega})]\right\} \tag{0-2}$$

An integer constant value for the group delay represents a perfect delay, such as the system $H(z) = z^{-3}$. A noninteger constant value represents a noninteger delay, which is typically interpreted in terms of bandlimited interpolation and delay. Any deviation of the group delay from a constant indicates some nonlinearity in the phase and corresponding dispersion.

The following three projects address first how to compute the group delay for a signal, and then the effect of the group delay when processing a signal through a filter. The third project demonstrates that the group delay can be negative—an apparent contradiction for a causal filter.

BACKGROUND READING

In Oppenheim and Schafer (1989), Section 5.1.2, the theory of group delay is presented and an example to demonstrate that delay can be associated with the phase response of a filter is developed in Problem 5.3.

PROJECT 1: ALGORITHM FOR COMPUTING THE GROUP DELAY

This project centers on the implementation of a MATLAB function to calculate the group delay of a rational system function or of a discrete-time sequence. For a system, the group delay is defined as the negative derivative of the phase of the frequency response (0-2); for a sequence, it is the negative derivative of the phase of the discrete-time Fourier transform. This derivative cannot be taken directly unless the phase is unwrapped to remove 2π jumps. However, the phase unwrapping can be avoided by using an alternative algorithm based on the discrete-time Fourier transform property that

$$\text{if} \quad h[n] \overset{\mathcal{F}}{\longleftrightarrow} H(e^{j\omega})$$

$$\text{then} \quad n\,h[n] \overset{\mathcal{F}}{\longleftrightarrow} j\frac{dH(e^{j\omega})}{d\omega}$$

Hints

Any algorithm implemented on a computer can evaluate the Fourier transform only at a finite number of frequencies. The MATLAB function `fft(h, N)` evaluates the Fourier transform of the sequence `h` at a set of `N` frequency points evenly spaced around the unit circle between 0 and 2π.

Looking for zeros of $H(e^{j\omega})$ is equivalent to finding roots of $H(z)$ on the unit circle. In MATLAB this is accomplished by treating $H(e^{j\omega})$ as a polynomial in $e^{j\omega}$ and using the function `roots`, followed by `abs` to extract the magnitude of any complex-valued roots.

EXERCISE 1.1

Creating a Group Delay Function

The MATLAB signal processing toolbox contains a function `grpdelay()` to compute the group delay of a rational system function. However, it is instructive to rewrite this function to have slightly different characteristics.

a. Express $H(e^{j\omega})$ in polar form as $H(e^{j\omega}) = A(\omega)e^{j\theta(\omega)}$, where $A(\omega)$ is real and prove the following property:

$$-\frac{d\theta(\omega)}{d\omega} = \text{Re}\left\{\frac{j\frac{dH(e^{j\omega})}{d\omega}}{H(e^{j\omega})}\right\} \tag{1-1}$$

b. Using the function `fft` along with any other MATLAB operations (except, of course, the MATLAB function `grpdelay`), write a function that computes the group delay of an impulse response `h` at `N` points in the interval $0 \leq \omega < 2\pi$.

c. Modify the function so that the signal represented by `h` can have a starting index other than zero. Let `nstart` be used as an additional argument to specify this starting index. Internally, the group delay function has to generate the signal $n\,h[n]$, so the starting index is needed in that calculation.

EXERCISE 1.2

Calculating the Group Delay

Use the following simple signal to test your group delay function.

a. For the impulse response $h[n] = \delta[n-4]$, analytically determine the group delay.

b. Representing $h[n] = \delta[n-4]$ as the row vector `h = [0 0 0 0 1]`, evaluate and plot the group delay of $h[n]$ using the function you wrote in Exercise 1.1.

c. Repeat the computation using the `nstart` argument to the group delay function.

EXERCISE 1.3

Dealing with Zeros

Clearly, the proposed FFT method for computing the group delay might fail if $H(e^{j\omega}) = 0$ for any of the values of ω at which we evaluate the Fourier transform.

a. For each of the following impulse responses, show that the corresponding frequency response is zero for at least one value of ω between 0 and 2π.

   i. $h_1[n] = \delta[n] + 2\delta[n-2] + 4\delta[n-4] + 4\delta[n-6] + 2\delta[n-8] + \delta[n-10]$

   ii. $h_2[n] = 3\delta[n] + \delta[n-1] - \delta[n-2] + \delta[n-3] - \delta[n-4] - 3\delta[n-5]$

b. Use your group delay function, or attempt to use it, to evaluate and plot the group delay of the impulse responses defined above as well as the following impulse response:

$$h_3[n] = 3\delta[n] + 2\delta[n-1] + 1\delta[n-2] + 2\delta[n-3] + 3\delta[n-4]$$

   How could you have predicted this result by looking at the symmetry of $h_3[n]$?

c. Determine what your computer does when it divides by zero in MATLAB, and what it does when it tries to plot the value `inf`, which represents ∞. If necessary, modify your function to deal with the possibility that some samples of $H(e^{j\omega})$ will be zero. The following fragment of code may be helpful. It divides the vector num by the vector den safely.

```
result = zeros(den);
result(den ~= 0) = num(den ~= 0)./den(den ~= 0);
result(den == 0) = inf * ones(1,sum(den == 0));
```

If there were any sequences whose group delay you couldn't find in parts (a) and (b), find them and plot them after making your changes.

PROJECT 2: EFFECT OF GROUP DELAY ON SIGNALS

In this project we will be looking at the effect of nonlinear phase, or nonconstant group delay, in the context of a filtering problem. One specific context in which constant group delay is important is in filtering to extract a narrow-time pulse from noise and then estimating, from the output, the time origin of the pulse. This often arises, for example, in radar systems, for which the range to a reflector is obtained by determining the time difference between the transmitted and received pulses. If, in the process of filtering out additive noise, the received pulse is dispersed in time, due to nonuniform group delay, the estimation of the arrival time becomes more difficult. To demonstrate the dispersion of a pulse, we use test signals that are windowed tone pulses created by multiplying a sinusoid by a Hamming window.

Hints

This project uses a data file gdeldata.mat to define the filters and signals needed for processing. If the data file is loaded via the MATLAB load command, the following filters and 256-point signals will have been predefined:

[b,a]	are the filter coefficients of an eighth-order IIR (elliptic) filter.
h	is the impulse response (filter coefficients) of a 33-point FIR filter.
x1, x2	are narrowband test signals, created as Hamming-windowed sine waves.
pulse	is a pulse starting at $n = 0$, which is relatively well localized in time.
noise	is a sample of the out-of-band noise that will be added.
pnd_1	represents the received signal and consists of a different sample of the noise added to the pulse, delayed by an amount to be estimated.
pnd_2	represents the received signal from a different reflector (i.e., it is the sum of a different noise sample and the pulse with a different delay).

For group delay computations, the file written in Exercise 1.1 should be used. Alternatively, the MATLAB function grpdelay(b, a, N) can be used to evaluate the group delay of a rational filter described by the b and a coefficients. Both rely on the FFT to evaluate the group delay at N equally spaced points around the unit circle between 0 and π, or 0 and 2π.

EXERCISE 2.1

Group Delay of the IIR Filter

a. Generate and plot the first 150 points of the impulse response of the IIR filter.

b. Compute and plot its frequency response magnitude and its group delay.

c. Plot the signals x1 and x2 and their Fourier transforms. Use these plots and the plots of the magnitude and group delay from part (b) to estimate the output you will get from running each sequence through the IIR system.

d. Verify your estimate in part (c) by explicitly computing the outputs due to `x1` and `x2` using `filter`.

EXERCISE 2.2

Group Delay of the FIR Filter

a. Plot the impulse response of the FIR filter. Then generate and plot its frequency response magnitude and group delay. How could you have anticipated from the impulse response that the group delay would be constant?

b. For the signals `x1` and `x2`, what output would you expect to get from processing each sequence with the FIR system. Verify your prediction by explicitly computing the outputs due to `x1` and `x2` using `conv` or `filter`.

EXERCISE 2.3

Pulse Distortion

Filter the signal `pulse` with the IIR filter and compare to processing with the FIR filter. Note that when you processed the narrowband signals `x1` and `x2` with the IIR filter or FIR filter, they were scaled and delayed with little distortion of the pulse shape; but when you process the signal `pulse` through the IIR filter, its pulse shape is severely distorted and dispersed in time. Explain why this happens.

EXERCISE 2.4

Filtering a Pulse from Noise

Filter the signals `pnd_1` and `pnd_2` with both the IIR and FIR filters. Since the noise occupies a different frequency band from the signal, either filter should remove the noise and make it easy to find the pulse.

a. Plot the output signals for the IIR filter, and from these plots estimate, as best you can, the time delay of the pulse in each of the two received signals. Explain your time-delay measurements in terms of the group delay curves plotted in Exercises 2.1 and 2.2.

b. Repeat part (a) with the FIR filter.

c. Describe any differences in pulse shape that you observe. Explain how the constant group delay of the FIR filter and nonuniform group delay of the IIR filter determine the quality of the output signal. Which filter, FIR or IIR, performs better for this application?

PROJECT 3: NEGATIVE GROUP DELAY

By definition a causal system cannot produce output that anticipates its input. This property might be interpreted as "a causal system always produces delay," and if the group delay had meaning as a true (physical) delay, it should never be negative. However, for many filters the group delay function versus ω will be less than zero over part of the frequency range. In this situation, the negative group delay will act on a suitable narrowband pulse so as to advance the *envelope* of the signal.

EXERCISE 3.1

Group Delay of a Minimum-Phase System

For the following filter:

$$H(z) = \frac{9.88 - 15.6z^{-1} + 6.26z^{-2}}{1 - 0.571z^{-1} + 0.121z^{-2}}$$

compute the group delay and plot it versus frequency for $0 \leq \omega \leq \pi$. Note the segment of the frequency axis where the group delay is *negative*. In fact, for a minimum-phase system such as $H(z)$, the integral of the group delay is zero, so there must always be a portion of the frequency axis where the group delay goes negative.

EXERCISE 3.2

Effect of Negative Group Delay

With an input confined to the frequency range in which the group delay is negative, we can illustrate an "advance" of the signal envelope.

a. Create the following bandlimited input signal:

$$x[n] = \tfrac{1}{2} + \sum_{k=1}^{10} \cos\left(\frac{2\pi}{256}(n - 128)k\right) \qquad \text{for } n = 0, 1, 2, \ldots, 255$$

Plot the magnitude of $X(e^{j\omega})$ and note the frequencies occupied by the input signal.

b. Calculate the output signal, $y[n]$, for this input using `filter` in MATLAB. Plot $x[n]$ and $y[n]$ on the same diagram. Note that $x[n]$ and $y[n]$ both start at the same time ($n = 0$), so the system is causal.

c. Measure the advance of the envelope (in samples) and compare to the average value of the group delay in the frequency band occupied by the input signal.

BASIC SAMPLING THEORY

OVERVIEW

This section contains three projects designed to illustrate the two basic principles of the sampling process: aliasing and reconstruction. In the first project, aliasing is investigated for sine waves and for chirp signals. If possible, these signals should be listened to before and after sampling. In the second project, the aliasing process is developed in the frequency domain by using the DTFT. The last project explores several different means by which a signal can be recovered from its samples.

BACKGROUND READING

Chapter 3 of the text by Oppenheim and Schafer (1989) is devoted to the issue of sampling.

PROJECT 1: ALIASING CAUSED BY SAMPLING

It is not easy to illustrate aliasing within a program like MATLAB, because the only types of signals in MATLAB are discrete signals represented as vectors. This project uses visual (and audio) reproductions of a signal to illustrate the nature of aliasing.

Hints

Since it is not possible to have an analog signal in MATLAB, simulation of the real-time axis t is needed. Therefore, it is important to keep straight the difference between the Δt of the simulation and the sampling period T_s under study.

In version 4 and later of MATLAB, there are M-files for playing and recording sounds on certain computers—Macintoshes and SUN workstations. For other systems, and under MATLAB version 3.5, it might be possible to play sound if special hardware is installed and/or special MATLAB MEX files are available.

EXERCISE 1.1

Aliasing a Sinusoid

Consider the formula for a continuous-time sinusoidal signal:

$$x(t) = \sin(2\pi f_\circ t + \phi) \tag{1-1}$$

We can sample $x(t)$ at a rate $f_s = 1/T_s$ to obtain a discrete-time signal

$$x[n] = x(t)|_{t=nT_s} = x(t)|_{t=n/f_s} = \sin\left(2\pi \frac{f_\circ}{f_s} n + \phi\right) \tag{1-2}$$

If we make plots of $x[n]$ for different combinations of $f_\circ$ and f_s, the aliasing problem can be illustrated. For the following, take the sampling frequency to be $f_s = 8$ kHz.

a. First of all, make a single plot of a sampled sine wave. Let the frequency of the sine wave be 300 Hz, and take samples over an interval of 10 ms. The phase ϕ can be arbitrary. Plot the resulting discrete-time signal using `stem`. It should be easy to see the outline of a sinusoid, because your eyes perform a reconstruction visualizing the envelope of the signal.

b. If necessary, make the plot using `plot`. In this case, the points are connected with straight lines, so the sinusoidal behavior should be obvious. Connecting the signal samples with straight lines is a form of "signal reconstruction" that makes a continuous-time signal from the discrete-time samples. It is not the ideal reconstruction specified by the sampling theorem, but it is good enough to be useful in most situations.

c. Now make a series of plots, just like part (a), but vary the sinusoidal frequency from 100 to 475 Hz, in steps of 125 Hz. Note that the apparent frequency of the sinusoid is *increasing*, as is expected. It might be better to use `subplot` to put four plots on one screen.

d. Make another series of plots, just as in part (c), but vary the sinusoidal frequency from 7525 to 7900 Hz, in steps of 125 Hz. Note that the apparent frequency of the sinusoid is now *decreasing*. Explain this phenomenon.

e. Again make a similar series of plots, but vary the sinusoidal frequency from 32,100 to 32,475 Hz, in steps of 125 Hz. Predict in advance whether the apparent frequency will be increasing or decreasing.

EXERCISE 1.2

Aliasing a Chirp Signal

A linear frequency-modulated signal makes a good test for aliasing, because the frequency moves over a range. This signal is often called a "chirp," due to the audible sound it makes when played through a speaker. The mathematical definition of a chirp is

$$c(t) = \cos(\pi \mu t^2 + 2\pi f_1 t + \psi) \tag{1-3}$$

The instantaneous frequency of this signal can be found by taking the time derivative of the phase (the argument of the cosine). The result is

$$f_i(t) = \mu t + f_1$$

which exhibits a linear variation versus time.

a. Take the parameters of the chirp to be $f_1 = 4$ kHz, $\mu = 600$ kHz/s, and ψ arbitrary. If the total time duration of the chirp is 50 ms, determine the frequency range that is covered by the swept frequency of the chirp.

b. Let the sampling frequency be $f_s = 8$ kHz. Plot the discrete-time samples of the chirp using both `stem` and `plot`. Since the swept bandwidth of the chirp exceeds the sampling frequency, there will be aliasing.

c. Notice that the chirp signal exhibits intervals in time where the apparent frequency gets very low. In fact, the instantaneous frequency is passing through zero at these points. Determine from the plots the times when this happens. Verify that these are the correct times by checking where the aliasing of the swept frequency occurs.

EXERCISE 1.3

Listening to Aliasing

If your computer has the capability for sound output from MATLAB through a D/A converter and a speaker, it will be interesting to listen to the aliased signals created in the previous exercises. To get a reasonable signal, it is necessary to create a much longer signal—perhaps 1 or 2 s in duration. In addition, the signal samples must be created at the natural sampling rate of the D/A converter.

a. For the sampled sinusoid, it makes sense to concatenate several segments, consisting of the sinusoids of slightly different frequency. Each one should be about 200 ms in duration, so putting together 5 to 10 of these will make a signal that can be heard for 1 to 2 s.

b. For the chirp, the duration must be much longer than 50 ms, so the parameter μ must be adjusted to get a swept frequency range that passes through only a few aliases. See if you can pick μ so that a 2-s chirp will pass through exactly 5 aliases. This value for μ will depend on the sampling rate of the D/A converter on your computer system.

PROJECT 2: FREQUENCY-DOMAIN VIEW OF SAMPLING

When a continuous-time signal is sampled, its spectrum shows the aliasing effect because regions of the frequency domain are shifted by an amount equal to the sampling frequency. To show this effect in reality, an oscilloscope is needed. In MATLAB the effect can only be simulated, and that is the goal of this project.

The simulation will consist of a sampling operation, followed by D/A conversion (including a reconstruction filter). This simple system will be driven by sinusoids with different frequencies, and the Fourier transform of the analog signals at the input and output will be compared. The different exercises treat each part of the sampling and reconstruction process. They should be combined into one M-file script that will do the entire simulation.

Hints

To simulate the analog signals, a very high sampling rate will have to be used—at least five times the highest frequency that any analog signal will be allowed to have. Thus there will be two "sampling rates" in the problem—one for the actual sampling under study and the other for simulating the continuous-time signals. A second issue is how to display the Fourier transform of the continuous-time signals. Again, this can only be simulated. The following M-file should be used to plot the analog spectra. Notice that one of its inputs is the `dt` for the simulation.

```
function fmagplot( xa, dt )
%FMAGPLOT
%   fmagplot( xa, dt )
%
%       xa:   the "ANALOG" signal
%       dt:   the sampling interval for
%               the simulation of xa(t)
%
L = length(xa);
Nfft = round( 2 .^ round(log2(5*L)) );   %<-- next power of 2
```

```
Xa = fft(xa,Nfft);
range = 0:(Nfft/4);
ff = range/Nfft/dt;
plot( ff/1000, abs( Xa(1:range) ) )
title('CONT-TIME FOURIER TRANSFORM (MAG)')
xlabel('FREQUENCY (kHz)'), grid
pause
```

EXERCISE 2.1

Signal Generation

To show the aliasing effect we need a simple analog input signal to run through the system. We will use sinusoids, but after you have the simulation working you may want to try other signals. To get started, you must pick a "simulation sampling frequency"; take this to be $f_{\text{sim}} =$ 80 kHz.

a. Generate a simulated analog signal that is a cosine wave with analog frequency $f_\circ$.

$$x(t) = \cos(2\pi f_\circ t + \phi) \qquad 0 \le t \le T$$

Take the phase to be random. Generate samples (at the rate f_{sim}) over a time interval of length T. Choose the signal length T so that you get about 900 to 1000 samples of the simulated analog signal.

b. Plot the time signal with `plot` so that the samples are connected. Make sure that you label the time axis with the true analog time.

c. Plot the Fourier transform of this signal (see `fmagplot` above).

EXERCISE 2.2

A/D Conversion

The A/D converter takes samples spaced by T_s. It is simulated by taking a subset of the samples generated for $x(t)$. To avoid unnecessary complications, the ratio of f_{sim} to the sampling rate of the A/D converter, f_s, should be an integer ℓ. Then every ℓth sample of the $x(t)$ vector can be selected to simulate the A/D conversion.

a. Plot the resulting discrete-time signal when $f_s = 8$ kHz.

b. Compute the DTFT of the discrete-time signal and explain how it is related to the Fourier transform of the analog signal in Exercise 2.1 (c).

EXERCISE 2.3

Design a Reconstruction Filter

The D/A section consists of two parts: a spacing of the discrete-time samples by the sampling time interval T_s, followed by an analog reconstruction filter.

a. The reconstruction filter will, of course, have to be a digital filter to simulate the true analog filter. Use the MATLAB filter design function `cheby2` to design this filter: `[b,a] = cheby2(9,60,fcut)`. This will design a ninth-order filter with 60 dB of stopband attenuation. The analog cutoff frequency has to be at $\frac{1}{2} f_s$. For MATLAB this has to be scaled to `fcut = 2*(fsamp/2)/fsim`.

b. Now use `freqz` to plot the frequency response of the simulated reconstruction filter. To get its true analog cutoff frequency on the plot, you must remember that this is a digital filter, where the frequency $\omega = \pi$ is mapped to $\frac{1}{2} f_{\text{sim}}$.

EXERCISE 2.4

D/A Conversion

The actual D/A conversion phase consists of creating an analog signal $\hat{x}(t)$ from the discrete-time signal $x[n]$ and then filtering with the Chebyshev filter. The MATLAB vector simulating the analog signal $\hat{x}(t)$ is reconstructed from the discrete-time signal vector $x[n]$ by inserting a number of zeros between each sample. The number of zeros depends on the ratio f_{sim}/f_s.

a. Carry out this zero-insert operation on the signal generated in Exercise 2.1 and sampled in Exercise 2.2. Then apply the Chebyshev reconstruction filter to get the smoothed output, $x_r(t)$.

b. Plot the resulting continuous-time output signal $x_r(t)$ and its Fourier transform.

EXERCISE 2.5

Test for Aliasing

All the steps above should be put into one M-file script. Then tests can be run.

a. Take the sampling frequency to be $f_s = 8$ kHz; and let the input signal frequency be $f_\circ = 2$ kHz. Make plots of the input and output Fourier transforms, and compare by plotting them together.

b. Now try a number of different input signal frequencies: $f_\circ = 6$ kHz, 7 kHz, 9 kHz, 10 kHz, and 15 kHz. Since f_{sim} is only 100 kHz, the input frequency should not be taken larger than 20 kHz. Make plots of the input and output Fourier transforms, and compare. Notice where the aliasing starts to occur.

c. To illustrate the aliasing effects on one plot, use `subplot` to put the following four plots together: $x(t)$, $x[n]$, $\hat{x}(t)$, and $x_r(t)$, the analog signal with zeros inserted. Another interesting multiplot would show $x(t)$, $x_r(t)$, and their Fourier transforms together.

d. If possible, try some other signals for which you can predict the result. For example, try to simulate the chirp experiment from Project 1.

PROJECT 3: RECONSTRUCTION OF SIGNALS FROM SAMPLES

Digital signal processing involves, among many other things, the reconstruction of analog signals from digital samples. This project explores various methods that can be used for this reconstruction. Since there are many possible analog signals which can pass through a given set of time samples, the choice of analog signal depends on assumptions made about the properties of the reconstruction.

Consider the case where you are given three samples of an analog signal, $x(t)$, as specified below and shown in Fig. 1.5a:

$$x(0) = 2, \quad x(1) = 1, \quad x(2) = x(t)|_{t=2} = -1 \tag{3-1}$$

No other information is given. To what analog signal do these samples correspond? It is important to realize that there is no one "right answer" to this problem. It depends on the assumptions you make and the reconstruction methods you employ.

For example, one possible analog waveform that corresponds to the samples indicated in Fig. 1.5a is seen in Fig. 1.5b. We have simply drawn an arbitrary curve through the sample points. We do not need to specify where it goes beyond the range shown, and we could have drawn any number of additional arbitrary curves.

To be more concrete, we need to state assumptions and reconstruction methods. For example, we see here three equally spaced samples. We could assume that samples have

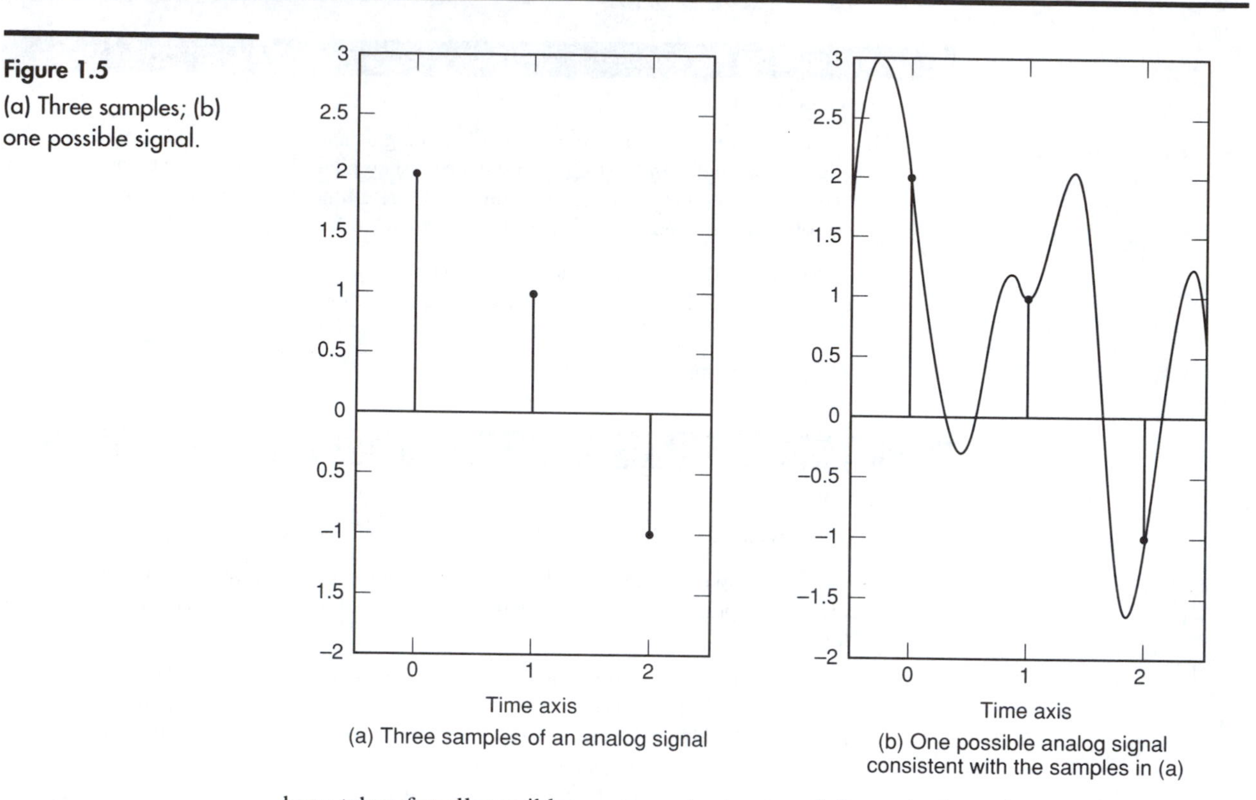

(a) Three samples of an analog signal

(b) One possible analog signal consistent with the samples in (a)

Figure 1.5 (a) Three samples; (b) one possible signal.

been taken for all possible $n = -\infty$ to $+\infty$, and that only these three were found to be nonzero. On the other hand, we could assume that the three samples are a subset of all possible samples for $n = -\infty$ to $+\infty$, but that we were just not given all the other nonzero sample values—only the three shown.

In choosing a reconstruction method, we might decide to fit a polynomial, or fit a sine wave, or use linear interpolation, or use a low-pass filter, or any one of a good number of other methods. For this project, we will be trying to fit the three data points with a sine wave, a polynomial, and then we will try ideal and nonideal low-pass filtering.

EXERCISE 3.1

Fitting a Sine Wave

Assume that the three samples correspond to a sinusoidal waveform of the form

$$x(t) = A\cos(\omega t + \phi) \tag{3-2}$$

You have $x(0)$, $x(1)$, and $x(2)$. Is this enough information to determine A, ω, and ϕ? Can you set up the relevant equations? Can you always solve these equations? If not, give specific numerical values where the process fails.

Can you guess a correct answer? Having found a correct answer, find another answer with a different frequency, ω. Plot the resulting sinusoids on a very fine grid—use a spacing of less than $\Delta t = 0.01$ s.

EXERCISE 3.2

Linear and Polynomial Interpolation

a. Using MATLAB, connect the samples with straight lines. Plot the result on a fine grid with spacing, $\Delta t = 0.01$ s. Explain how `plot` will do this automatically.

b. Convolve the three samples with an impulse response that is triangular, but first insert four zeros between each of them, and use an impulse response 0.2, 0.4, 0.6, 0.8, 1.0, 0.8, 0.6, 0.4, 0.2. Show that this result is identical to linear interpolation if we assume that the samples at $t = -1$ and $t = +3$ are zero.

c. Using MATLAB, fit a second-degree polynomial to the three data points (see `polyfit` and `polyval`). Plot the polynomial on a fine grid for $-5 \leq t \leq 5$. Is this curve realistic in a practical sense? Does it do a good job in extending the signal values beyond the range $0 \leq t \leq 2$?

EXERCISE 3.3

Ideal Low-Pass Filtering

There are no ideal low-pass filters available in reality. However, we can calculate the waveform that would result from an ideal low-pass filter, as follows: An ideal low-pass operation corresponds to a multiplication of the spectrum of a signal by a rectangular function in the frequency domain. This corresponds to a convolution with the inverse Fourier transform, which is a sinc function in the time domain. As applied to point samples, this amounts to sinc interpolation:

$$x_r(t) = \sum_{\ell=-\infty}^{\infty} x(t_\ell) \frac{\sin(\pi(t - \ell T_s)/T_s)}{\pi(t - \ell T_s)/T_s} \tag{3-3}$$

where the samples $x(t_\ell)$ are taken at $t_\ell = \ell T_s$.

a. Write a sinc interpolator based on (3-3). Assume that only a finite number of the signal samples will be nonzero and that the signal need only be reconstructed over a finite time interval.

b. Interpolate a single-point sample of value 1 at $t = 0$. Plot the result from about -5 to $+5$. This should match the sinc function shape.

c. Now interpolate the three-point case given in (3-1) and Fig. 1.5. Compare the result to that obtained from sine-wave fitting.

EXERCISE 3.4

Choice of Assumed Bandwidth

Resolve the following: A signal bandlimited to some frequency f_B can be sampled at $f_s = 2f_B$ and recovered by an ideal low-pass reconstruction filter with cutoff f_B. The same is true for a second signal that is bandlimited to f_b, where f_b is less than f_B, since a signal bandlimited to f_b is also bandlimited to f_B. Also, the signal bandlimited to f_b sampled at f_s can be recovered with an ideal low-pass with cutoff f_b, which has an impulse response (sinc) that is broader than that of the one with cutoff at $f_B = \frac{1}{2}f_s$. Can we interpolate the samples of the signal with bandwidth f_b, sampled at f_s, using the impulse response of the ideal low-pass with cutoff $f_b < \frac{1}{2}f_s$?

ZERO-PHASE IIR FILTERING

OVERVIEW

In many filtering problems it is often desirable to design and implement a filter so that the phase response is exactly or approximately linear. If the filter is restricted to be causal, then exactly linear phase can only be realized with an FIR filter. On the other hand, IIR filters are often preferred for bandpass filtering because they can achieve much sharper transition bands for a given filter order. Unfortunately, the phase response of a causal IIR filter is extremely nonlinear. However, if the IIR filter is implemented as a noncausal operator, its phase response can be made exactly zero, or exactly linear. In these projects we investigate two implementations of noncausal zero-phase IIR filters.

BACKGROUND READING

This method for filtering is discussed in Oppenheim and Schafer (1989), Problem 5.39.

PROJECT 1: ANTICAUSAL FILTERING

Most of the time, we implement recursive difference equations as causal systems for which time runs forward, starting at $n = 0$, then $n = 1, 2, 3, \ldots$. In fact, the MATLAB function `filter` will only do causal filtering. In this project we need a mechanism to perform filtering backward, so that the recursive difference equation is applied at $n = 0$, then $n = -1, -2, \ldots$. In this case, the impulse response of the system is left-sided and the system is *anticausal*.

The bilateral z-transform provides the mathematical framework to describe anticausal systems. If a causal system has a rational z-transform

$$H_c(z) = \frac{B(z)}{A(z)} \qquad \text{ROC} = \{z : |z| > R_{\max}\} \tag{1-1}$$

then the related anticausal system is

$$H_a(z) = \frac{B(1/z)}{A(1/z)} \qquad \text{ROC} = \{z : |z| < 1/R_{\max}\} \tag{1-2}$$

If the radius of the largest root of $A(z)$ satisfies $R_{\max} < 1$, both of these systems are stable, and they will also have exactly the same frequency response magnitude. The region of convergence determines whether the impulse response is right-sided or left-sided. In fact, the two impulse responses are related via a time "flip" (i.e., a time reversal):

$$h_a[n] = h_c[-n]$$

Implementation of the anticausal system, $H_a(z)$, requires a difference equation that will recurse backward. In this project we investigate a method for implementing $H_a(z)$ based on the causal filtering function `filter( )` and time reversals of the signal.

Hints

Since the time base is important when we want to distinguish causal signals from anticausal signals, it will be convenient to adopt a convention for signal representation that involves a pair of vectors. The first vector contains the signal values; the second, the list of time indices. For example, the following code fragment will define an impulse $\delta[n]$ and a step $u[n]$ over the range $-20 \leq n \leq 30$.

```
nn = -20:30;
unit_impulse = (nn==0);
unit_step = (nn>=0);
```

This works because the logical operators == and >= return a vector of 1's and 0's, representing TRUE and FALSE, respectively.

MATLAB has a function in the signal processing toolbox, called `filtfilt`, that will do the zero-phase filtering operation. *It should not be used in these projects.*

EXERCISE 1.1

Group Delay M-File

The group delay is defined as the negative derivative of the phase of the frequency response. However, computation of the group delay is best done without explicitly evaluating the derivative with respect to ω. You may want to consider the projects in the section *Group Delay* for more details about this computation. However, these projects are not required for the following exercises. The M-file below exploits the fact that multiplying by n in the time domain will generate a derivative in the frequency domain. Furthermore, this function is configured for the case where the signal $x[n]$ starts at an index other than $n = 0$, unlike the function `grpdelay( )` in the MATLAB Signal Processing Toolbox. This is accomplished by passing a vector of time indices `[n]` along with the vector of signal values `[x]`.

```
function [gd, w] = gdel(x, n, Lfft)
%GDEL    compute the group delay of x[n] \verb%    usage:
%             [gd, w] = gdel( x, n, Lfft )
%
%     x:    Signal x[n] at the times (n)
%     n:    Vector of time indices
%     Lfft: Length of the FFT used
%     gd:   Group Delay values on [-pi,pi)
%     w:    List of frequencies over [-pi,pi)
%
% NOTE:  group delay of B(z)/A(z) = gdel(B) - gdel(A)
%
X = fft(x, Lfft);
dXdw = fft(n.*x, Lfft);              %--- transform of nx[n]
gd = fftshift(real( dXdw./X ));     %--- when X==0, gd=infinity
w = (2*pi/Lfft)*[0:(Lfft-1)] - pi;
```

Test the group delay function with a shifted unit impulse signal. Define a unit impulse sequence $\delta[n - n_\circ]$ of length 128, over the range $-64 \le n \le 63$. Pick $n_\circ = \pm 5$, and then make a plot of the signal, with the time axis correctly labeled, to show that the impulse is located at $n = n_\circ$. In addition, compute and plot the group delay to verify that the proper value is obtained.

Another simple test signal would be any finite-length signal that is symmetric. In this case, the group delay should be equal to the value of n at the point of symmetry. Try the signal `x = [1 2 3 4 4 3 2 1]` defined over the range $0 \le n \le 7$.

EXERCISE 1.2

Causal First-Order System

Using the MATLAB function `filter`, generate the impulse response of the *causal* system:

$$H_c(z) = \frac{1}{1 - 0.77z^{-1}} \qquad \text{ROC} = \{z : |z| > 0.77\}$$

Plot the impulse response signal over the range $-64 \le n \le 63$. Also calculate and plot the frequency response magnitude and group delay. This can be done in one of two ways: (1) from $H_c(z)$, by directly evaluating an exact formula based on the numerator and denominator polynomials of $H_c(z)$ (i.e by computing with `freqz`); or (2) from a finite section of the impulse response, by computing with the FFT. Implement both computations, and plot them together for comparison. Which one is exact, and which one is an approximation?

Repeat for a pole position closer to the unit circle; try 0.95 instead of 0.77. Explain the significant differences between the two cases.

EXERCISE 1.3

Anticausal First-Order System

For an anticausal filter, the impulse response is zero for $n > 0$. Anticausal filtering can be accomplished in a three-step process: time reverse the input, filter with a causal filter, and then time-reverse the output. The signal can be time-reversed using either `fliplr` or `flipud`. Specifically, the two systems shown in Fig. 1.6 are identical from an input/output point of view. If $H_c(z)$ corresponds to a causal filter, then $H_a(z) = H_c(1/z)$ will correspond to an anticausal filter, and vice versa.

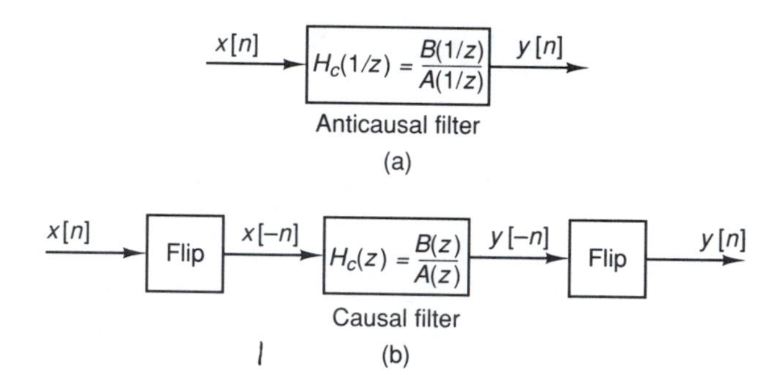

Figure 1.6 Implementation of (a) anticausal filter via (b) flips and causal filtering. The two systems shown are mathematically equivalent.

For the anticausal filter

$$H_a(z) = \frac{1}{1 - 0.95z} = H_c(1/z) \qquad \text{ROC} = \left\{ z : < \frac{1}{0.95} \right\}$$

generate the impulse response over the range $-64 \leq n \leq 63$ by using the method of Fig. 1.6, and plot to verify that the response is left-sided. Then calculate and plot the frequency response magnitude and group delay, by explicitly evaluating the Fourier transform of the finite section of the impulse response. Discuss how the impulse response, frequency response magnitude, and group delay for this filter relate to those for the causal filter in Exercise 1.2. Do the same relationships hold when the pole is at $z = 1/0.77$?

EXERCISE 1.4

Anticausal Filtering Function

Write a general MATLAB function, called `filtrev`, that will implement an anticausal filter whose numerator is specified by the polynomial $B(z)$ and whose denominator is specified by the polynomial $A(z)$. The arguments to `filtrev` should follow the same convention as in the MATLAB function `filter`, except that the vectors `b` and `a` should specify the coefficients of $B(z)$ and $A(z)$ in *increasing* powers of z.

PROJECT 2: FORWARD-BACKWARD FILTERING

In this project, the causal and anticausal implementations will be combined to produce a zero-phase IIR filter. In fact, the overall impulse response will only approximate zero phase, due to constraints on implementing the noncausal part.

Hints

For this project it would be best to have a function that will implement an anticausal filtering operation, such as `filtrev` from Exercise 1.4. This can be used as a building block for the zero-phase filter decompositions.

EXERCISE 2.1

Noncausal Filter as a Cascade

Since a zero-phase filter must have a symmetric impulse response, an IIR zero-phase filter must be noncausal. The basic approach to implementing any noncausal IIR filter is to decompose it into the combination of causal and anticausal subfilters. This can be done in two different ways: as a cascade of the causal and anticausal parts, or as a parallel combination.

a. Consider the noncausal filter

$$H_{nc}(z) = \frac{1}{(1-0.77z^{-1})(1-0.77z)} \qquad \text{ROC} = \left\{z : 0.77 < |z| < \frac{1}{0.77}\right\}$$

Show analytically that the frequency response of this filter is real-valued and that therefore the phase is zero.

b. Generate the impulse response of this filter numerically by treating it as a cascade of the filters implemented in Project 1. Plot the impulse response over the range $-64 \le n \le 63$ and determine whether or not it is symmetric. (It should not matter in which order the two subfilters are applied.)

c. Calculate and plot the frequency response magnitude and group delay, by numerically evaluating the Fourier transform of the finite section of the impulse response.

d. Repeat the implementation, but move the pole location closer to the unit circle—change it from 0.77 to 0.95 and 1/0.77 to 1/0.95. Plot the impulse response and the frequency response (magnitude and group delay). In this case, you are likely to find that the group delay is nonzero, although it should be zero. What is the most likely explanation for this inconsistency? Does this happen in the passband or stopband of the filter's frequency response?

EXERCISE 2.2

Parallel Form for a Noncausal Filter

An alternative implementation of the system in Exercise 2.1 consists of decomposing it as the *sum* of causal and anticausal subfilters. This gives a *parallel* form implementation, involving two subsystems that are flipped versions of one another.

a. To determine the two subfilters from the zero-phase transfer function $H_{nc}(z)$, it is necessary to perform an expansion similar to a partial fraction expansion. Using this approach, $H_{nc}(z)$ in Exercise 2.1 can be expressed as the sum of a causal and an anticausal filter.

$$H_{nc}(z) = \frac{1}{(1-0.77z^{-1})(1-0.77z)} = \frac{\beta+\gamma z^{-1}}{1-0.77z^{-1}} + \frac{\beta+\gamma z}{1-0.77z}$$

Determine the constants β and γ in this decomposition.

b. Generate and plot the corresponding impulse response over the range $-64 \le n \le 63$ by implementing the parallel combination of the causal and anticausal filters.

c. Compute the magnitude and group delay from the finite section of the impulse response generated in part (b). Compare the magnitude response and group delay to the results from Exercise 2.1 and explain any differences.

d. Repeat parts (a)–(c) for the case when the poles are at 0.95 and 1/0.95. Is there any significant difference between the parallel and cascade implementations? Compare in both the passband and stopband.

EXERCISE 2.3

Second-Order Noncausal Filter

Consider the following noncausal filter.

$$H_{nc}(z) = \frac{0.0205z^2 + 0.0034z + 0.0411 + 0.0034z^{-1} + 0.0205z^{-2}}{0.5406z^2 - 1.8583z + 2.7472 - 1.8583z^{-1} + 0.5406z^{-2}}$$

Since $H_{nc}(z)$ is symmetric [i.e., $H_{nc}(z) = H_{nc}(1/z)$], the associated frequency response is purely real, so the filter has zero phase.

a. $H_{nc}(z)$ can be expressed in factored form

$$H_{nc}(z) = \frac{0.1432 + 0.0117z^{-1} + 0.1432z^{-2}}{1 - 1.2062z^{-1} + 0.5406z^{-2}} \cdot \frac{0.1432 + 0.0117z + 0.1432z^2}{1 - 1.2062z + 0.5406z^2}$$
$$= H_c(z)H_c\left(\frac{1}{z}\right)$$

Implement $H_{nc}(z)$ as a cascade of a causal and anticausal filter. Generate and plot the impulse response, frequency response magnitude, and group delay using the range $-64 \leq n \leq 63$.

b. $H_{nc}(z)$ can also be expressed in parallel form as

$$H_{nc}(z) = \frac{0.1149 + 0.0596z^{-1} - 0.0416z^{-2}}{1 - 1.2062z^{-1} + 0.5406z^{-2}} + \frac{0.1149 + 0.0596z - 0.0416z^2}{1 - 1.2062z + 0.5406z^2} \qquad (2\text{-}1)$$

Let $H_1(z)$ denote the causal subfilter and $H_2(z) = H_1(1/z)$ the anticausal one. Implement $H_{nc}(z)$ in this additive form in order to generate its impulse response over the range $-64 \leq n \leq 63$, and plot. Then compute the frequency response magnitude and group delay from this finite section of the impulse response.

c. Determine the pole and zero locations for this system, and derive the mathematical formulas for the group delay of $H_1(z)$ and $H_2(z)$. Compare these exact formulas to the group delay curves computed for these filters from a finite section of their impulse responses.

d. Are there any significant differences between the cascade and parallel implementations of the zero-phase filter? What characteristic of $h_{nc}[n]$ guarantees that both implementations will give an excellent approximation to the zero-phase response?

e. The decomposition of a zero-phase filter into the sum of a causal filter and an anticausal filter can be done in general. It only requires factorization of the denominator and then the solution of simultaneous linear equations to get the numerator coefficients. Write a MATLAB program that will produce this decomposition in the general Nth order case, and verify that the numbers given in equation (2-1) are correct. Note that polynomial multiplication is convolution, and a function such as `convmtx` can be used to set up convolution in the form of linear equations.

EXERCISE 2.4

Zero-Phase Filtering of a Square Wave

The consequences of nonlinear phase versus zero phase can be illustrated in the time domain by processing a pulse-like signal through the two different implementations. For this purpose, construct a signal that is composed of several pulses over the range $-64 \leq n \leq 63$:

$$x[n] = \begin{cases} -2 & \text{for } -22 \leq n \leq -8 \\ 2 & \text{for } -7 \leq n \leq 7 \\ -3 & \text{for } 8 \leq n \leq 22 \\ 0 & \text{elsewhere} \end{cases}$$

a. Process $x[n]$ with the noncausal zero-phase filter in Exercise 2.3. Try both the cascade and parallel implementations, and note whether or not there is any difference.

b. For comparison, process the same input signal through the causal filter from Exercise 2.3(a). For a fair comparison, the signal should be processed by $H_c^2(z)$, so that it is subjected to the same magnitude response. This can be accomplished by running $x[n]$ through the cascade of two $H_c(z)$ filters. Plot the two outputs on the same scale and compare. What is the apparent time delay when comparing these outputs to the input signal $x[n]$? Also, note any differences in the distortion of the pulse shape and in the symmetry of the individual pulses. Explain why the zero-phase filter has certain advantages when viewed in terms of these time-domain characteristics.

chapter 2

DISCRETE FOURIER TRANSFORM

OVERVIEW

The discrete Fourier transform (DFT) is at the heart of digital signal processing, because it is a transform and it is also computable. Although the Fourier, Laplace, and z-transforms are the analytical tools of signal processing as well as many other disciplines, it is the DFT that we must use in a computer program such as MATLAB. Indeed, it was the development of the fast Fourier transform (FFT) algorithm, which efficiently calculates the DFT, that launched modern DSP.

The DFT and MATLAB are perfectly matched, because we can only do computations on finite-length vectors in MATLAB, which is precisely the case handled by the theory of the DFT. An important goal of the projects in this chapter is to develop an understanding of the properties of the DFT and their use in DSP. The relationship of the DFT to Fourier theory is explored in many cases. The purpose of most of the exercises is to develop insight into the transform, so that it can be used for more than just grinding out numbers.

The first set of projects treats basic transform pairs, difficult transforms, and then properties of the DFT. The circular nature of indexing associated with the DFT is emphasized. Properties needed for applications such as computing the FFT of a real-valued sequence are studied. The DFT can be viewed as a matrix operator, so the next set of projects investigates matrix properties such as eigenvalues and eigenvectors of the DFT matrix [1]. This viewpoint is becoming more fashionable with the emergence of programs such as MATLAB, which emphasize the matrix-vector nature of computations. The third set of projects concentrates on the circular convolution property of the DFT. The relation to linear convolution is studied, as well as the extension to block processing and high-speed convolution. The last set of projects treats two transforms that are closely related to the DFT: the discrete cosine transform (DCT) [2] and the discrete Hartley transform (DHT) [3]. Properties of these two transforms and their relation to the FFT are explored.

BACKGROUND READING

All DSP textbooks contain one or more chapters devoted to the DFT, circular convolution, and the FFT algorithm. Since these are major topics, the reader should be able to locate more material on any of these topics in a standard DSP text such as [4], [5], [6], [7], or [8]. For example, in the text by Oppenheim and Schafer, which is used often by the authors, this material is contained in Chapters 8 and 9.

[1] J. H. McClellan and T. W. Parks. Eigenvalue and eigenvector decomposition of the discrete Fourier transform. *IEEE Transactions on Audio and Electroacoustics*, AU-20:66–74, March 1972.

[2] K. R. Rao and P. Yip. *Discrete Cosine Transform: Algorithms, Advantages, Applications*. Academic Press, San Diego, CA, 1990.

[3] R. N. Bracewell. *The Fourier Transform and Its Applications*. McGraw-Hill, New York, second edition, 1986.

[4] L. B. Jackson. *Digital Filters and Signal Processing*. Kluwer Academic Publishers, Norwell, MA, 1989.

[5] A. V. Oppenheim and R. W. Schafer. *Discrete-Time Signal Processing*. Prentice Hall, Englewood Cliffs, NJ, 1989.

[6] R. D. Strum and D. E. Kirk. *First Principles of Discrete Systems and Digital Signal Processing*. Addison-Wesley, Reading, MA, 1988.

[7] R. A. Roberts and C. T. Mullis. *Digital Signal Processing*. Addison-Wesley, Reading, MA, 1987.

[8] J. G. Proakis and D. G. Manolakis. *Digital Signal Processing: Principles, Algorithms and Applications*. Macmillan, New York, second edition, 1992.

[9] C. S. Burrus and T. W. Parks. *DFT/FFT and Convolution Algorithms: Theory and Implementation*. John Wiley & Sons, New York, 1985.

[10] G. H. Golub and C. F. Van Loan. *Matrix Computations*. Johns Hopkins University Press, Baltimore, second edition, 1989.

[11] G. Strang. *Linear Algebra and Its Applications*. Academic Press, New York, 1976.

[12] C. F. Van Loan. *Computational Frameworks for the Fast Fourier Transform*. Society for Industrial and Applied Mathematics, vol. 10, 1992, Philadelphia, PA.

[13] C. Moler. J. N. Little, and S. Bangert. *Matlab User's Guide*, The MathWorks, Inc., South Natick, MA, 1989.

[14] R. N. Bracewell. *The Hartley Transform*. Oxford University Press, New York, 1986.

[15] P. Yip and K. R. Rao. Fast discrete transforms, In D. F. Elliott, editor *Handbook of Digital Signal Processing: Engineering Applications*, chapter 6, pages 481–525. Academic Press, San Diego, CA, 1987.

[16] D. F. Elliott. *Handbook of Digital Signal Processing: Engineering Applications*. Academic Press, San Diego, CA, 1987.

DFT PROPERTIES

OVERVIEW

The properties of the discrete Fourier transform (DFT), while similar to properties of other Fourier transforms, exhibit notable differences due to its finite nature. Because the DFT is our primary calculating tool, we must understand its properties and its relation to the other transforms used in DSP. It is the goal of these projects and exercises to develop familiarity with and insight into the use and properties of the DFT. Features unique to the DFT will be emphasized, especially the circular nature of all indexing in both the time and frequency domains.

The DFT is defined as an operation on an N-point time vector $\{x[0], \ x[1], \dots, x[N-1]\}$:

$$X[k] = \sum_{n=0}^{N-1} x[n] W_N^{nk} \qquad \text{for} \quad k = 0, 1, 2, \dots, N-1 \tag{0-1}$$

where $W_N = e^{-j2\pi/N}$. The operation in (0-1) is a transformation from an N-point vector of time samples $x[n]$ to another N-point vector of frequency-domain samples $X[k]$. The definition (0-1) can also be interpreted as a "frequency sampling" of the DTFT (discrete-time Fourier transform).

A word about the difference between the terms DFT and FFT needs to be made at the outset. The FFT (fast Fourier transform) is just a fast algorithm for computing the DFT; it is not a separate transform. In MATLAB, the function `fft` is always used to compute the DFT; and there is no `dft` function at all. Similarly, the function `ifft` is used to compute the inverse DFT. Therefore, it is usually acceptable to use the terms DFT and FFT interchangeably when referring to the results of computation.

PROJECT 1: EXAMPLES OF THE DFT OF COMMON SIGNALS

In this project we develop the DFT of certain common signals (e.g., pulses, sine waves, aliased sincs, etc.). With MATLAB, you can plot out signals and their transforms easily, so the objective is to visualize a variety of transform pairs. In the process, you should pay attention to the symmetries that might be present.

Hints

All signals used with the DFT are discrete, so they should be displayed using `stem`; similarly, the transform is a vector of discrete values, so it should also be plotted with `stem`. Since the DFT is complex-valued, you will have to plot the real and imaginary parts in most cases. If you want to view simultaneous plots of the real and imaginary parts of both the time-domain and frequency-domain vectors, use the `subplot(22x)` commands prior to each `stem` command to force the four plots to be placed on the same screen, with real and imaginary parts one above the other, as in Fig. 2.1. The program that produces Fig. 2.1 follows:

```
nn = 0:15;
xx = exp(j*nn/3);
XX = fft(xx);
kk = nn;
subplot(221)
stem(kk, real(xx))
title('REAL PART of x[n]'),  xlabel('INDEX (n)')
subplot(223)
stem(kk, imag(xx))
title('IMAG PART of x[n]'),  xlabel('INDEX (n)')
subplot(222)
stem(kk, real(XX))
title('REAL PART of DFT'),  xlabel('INDEX (k)')
subplot(224)
stem(kk, imag(XX))
title('IMAG PART of DFT'),  xlabel('INDEX (k)')
```

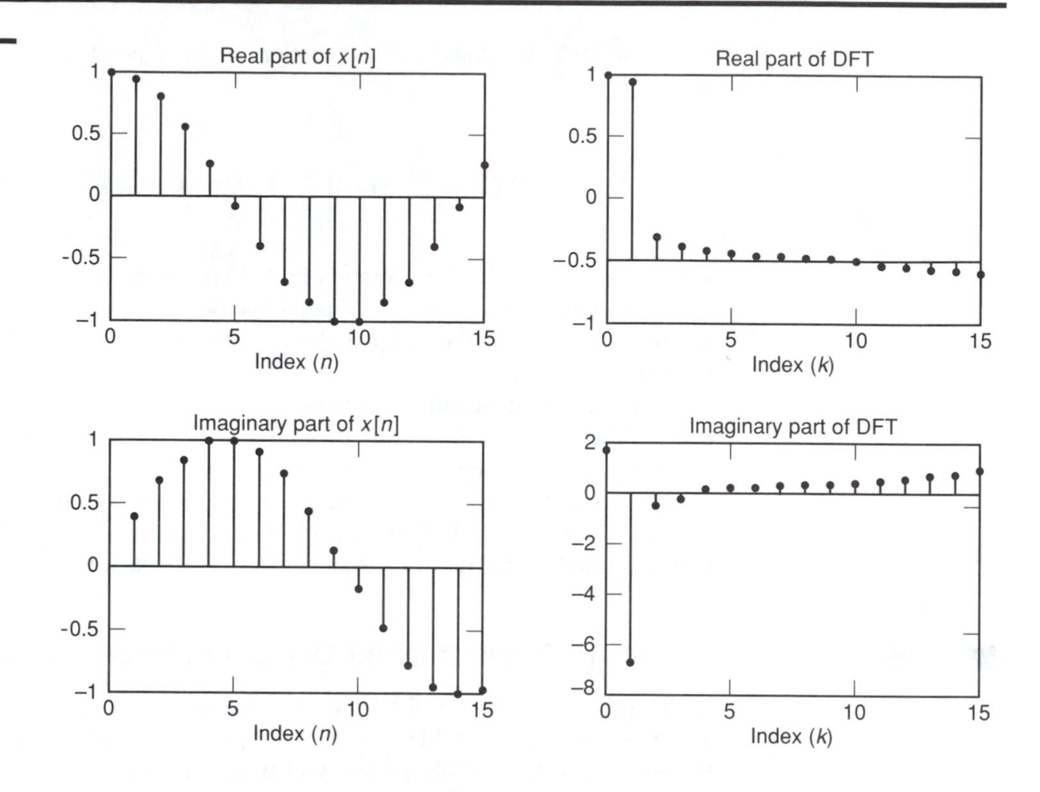

Figure 2.1
Plotting real and imaginary parts of a discrete-time signal and its 16-point DFT with `subplot(22x)` to create a four-panel display. The DFT is on the right and the complex-valued time signal is on the left.

EXERCISE 1.1

Pulses

These are signals containing only ones and zeros. For the following exercises, you can plot the real and imaginary parts of the DFT, but it may be better to plot the magnitude and the phase.

a. *Unit impulse signal*: `xi = [1 0 0 0 0 0 0 0]` corresponds to the mathematical definition:

$$\delta[n] = \begin{cases} 1 & n = 0 \\ 0 & n = 1, 2, \ldots, N-1 \end{cases}$$

For this problem compute an 8-point DFT (i.e., $N = 8$). In general, what is the N-point DFT of $\delta[n]$?

b. *All ones*: `x1 = [1 1 1 1 1 1 1 1]`. Note that this example together with part (a) illustrates the *duality* principle of the DFT.

c. *Shifted impulse*: `xish = [0 0 0 1 0 0 0 0]`. Plot the magnitude of the DFT values. Try other shifts—is there a nonzero shift of `xi` where the DFT is purely real?

d. *Three-point boxcar*: `xb = [1 1 1 0 0 0 0 0]`; try the four-point case also.

e. *Symmetric boxcar*: `xbsy = [1 1 0 0 0 0 0 1]`. Show that this DFT will be purely real. Compare the DFT magnitudes for `xb` and `xbsy`.

EXERCISE 1.2

Sine Waves

A real-valued sinusoid is described by three parameters. The mathematical form of the time signal is

$$s[n] = A\cos(2\pi f_\circ n + \phi) \qquad \text{for } n = 0, 1, 2, \ldots, N-1$$

where N is the signal length, A its amplitude, $f_\circ$ its frequency, and ϕ the relative phase at $n = 0$.

a. Compute the 21-point DFT of a sequence representing exactly one cycle of a cosine wave. Determine the frequency of this sinusoid. Make sure to take exactly one cycle, not one cycle plus one point (i.e., don't repeat the first sample). If done correctly, the answer will be extremely simple.

b. Repeat part (a) for a sine, then for a cosine with a 45° phase shift. Observe carefully the magnitudes of the DFT coefficients (and compare the phases).

c. Repeat part (a) for three cycles of a sinusoid, still using a 21-point DFT. What is the frequency of this sinusoid (in radians per sample)?

d. Try a vector that is 3.1 cycles of a sinusoid. Why is the DFT so different?

e. Experiment with different frequency sinusoids. Show that choosing the frequency to be $f_\circ = k(1/N)$, when k is an integer, gives an N-point DFT that has only two nonzero values.

EXERCISE 1.3

Complex Exponentials

The complex exponential is defined as

$$c[n] = e^{j\omega_\circ n} \qquad \text{for } n = 0, 1, \ldots, N-1 \tag{1-1}$$

The choice of $\omega_\circ$ gives radically different results for the DFT.

a. Choose $\omega_\circ = 6\pi/N$, and compute the $N = 16$-point DFT. How many DFT values are nonzero? Since the complex exponentials are the basis signals of the DFT, the orthogonality property renders their transforms in a simple form.

b. The dual of the complex exponential is the shifted impulse, when $\omega_\circ$ is an integer multiple of $2\pi/N$. Find the 16-point sequence whose DFT is $X[k] = e^{j6\pi k/16}$.

c. Now try $\omega_\circ = 5\pi/N$, and compute the $N = 16$-point DFT. Explain why are there no zeros in the DFT for this case.

d. Euler's formula relates the complex exponential to sine and cosine. Let $\omega_\circ = 6\pi/N$. Show how to construct the DFT of $\sin(\omega_\circ n)$ from the result of part (a).

PROJECT 2: DIFFICULT DFTS

The following DFTs are rather difficult to compute by hand, but the results are not hard to visualize. MATLAB makes it trivial to compute the transform pairs, so the visualization that comes from the duality principle of Fourier analysis is reinforced by the following exercises.

EXERCISE 2.1

Aliased Sinc Sequence

Once mastered, the duality principle is quite powerful. As a good example of its use in computing, consider the DFT of the aliased sinc signal. According to duality, the rectangular pulse and the asinc function are "transform pairs."

a. The DTFT of a rectangular pulse is an aliased sinc function in ω; the N-point DFT just samples the asinc at $\omega = (2\pi/N)k$. For an even-symmetric[1] L-point pulse, the result is

$$R[k] = \text{asinc}(\omega, L)\big|_{\omega=2\pi k/N} = \frac{\sin(\pi k L/N)}{\sin(\pi k/N)} \tag{2-1}$$

Generate a 7-point pulsewidth `boxcar(7)`, and compute its 16-point DFT. Verify that the correct transform values were obtained. Repeat for a 21-point DFT. Explain why the 21-point DFT has so many values equal to zero.

[1] See Project 3 on DFT symmetries for more details.

b. In the dual of part (a), the time sequence is taken to be an asinc. Use MATLAB to calculate and plot the N-point DFT of the following sampled asinc sequence:

$$a_0[n] = \frac{\sin(9\pi n/N)}{\sin(\pi n/N)} \qquad \text{for } n = 0, 1, \ldots, N-1$$

Assume that N is greater than 9, say $N = 16$, or $N = 21$.

c. Find the N-point DFT of the following shifted asinc sequence:

$$a_1[n] = \frac{\sin(9\pi(n+1)/N)}{\sin(\pi(n+1)/N)} \qquad \text{for } n = 0, 1, \ldots, N-1$$

Since the asinc function is periodic, this shift is a circular shift.

d. Note that the factor in the numerator of $a_0[n]$ or $a_1[n]$ must be odd! Try replacing the 9 with 10 in either of the previous parts, and compute the DFT. Why is there so much difference between the even and odd cases?

EXERCISE 2.2

Impulse Train

Perform the following experiment:

a. Generate an impulse train containing 207 samples. In between impulses there should be 22 zeros. The height of each impulse should be a constant. Call this signal $p[n]$.

$$p[n] = \sum_{\ell=0}^{8} \delta[n - \ell M_\circ] \qquad \text{with } M_\circ = 23$$

b. Compute the 207-point DFT of $p[n]$. Observe that in the DFT domain, $P[k]$ also takes on only the values of zero and a constant. Determine the spacing between impulses in the k domain.

c. The period of the input signal, $M_\circ$, is a divisor of the length of the FFT. Use this fact to explain the mathematical form of $P[k]$. Generalize this result. In particular, predict the DFT if $p[n]$ contained 23 impulses separated by 9; then verify with MATLAB.

d. Change the DFT length slightly and compute a 200-point DFT. Since the last 22 points of $p[n]$ are zero, we only need to drop off the zeros at the end. Explain why the transform values are so different, but notice that the DFT still has a number of large regularly spaced peaks.

e. When the length of the DFT is doubled, there should still be zeros in the transform. Compute the 414-point DFT of $p[n]$ and plot the magnitude. Determine which DFT points are exactly equal to zero, and explain why. What would happen for a 621-point DFT? Explain.

f. Compute the 1024-point DFT of $p[n]$, again with zero padding prior to the FFT. Note that the transform has many peaks, and they seem to be at a regular spacing, at least approximately. Measure the spacing and count the peaks. State the general relationship between the period of the input signal, $p[n]$, the length of the DFT, and the regular spacing of peaks in the DFT.

EXERCISE 2.3

A Gaussian

An often quoted result of Fourier theory is that the "Fourier transform of a Gaussian is a Gaussian." This statement is exactly true for the case of the continuous-time Fourier transform, but only approximately true for the DFT and the DFT.[2]

[2]These signals for which $\mathcal{DFT}\{v[n]\} \rightarrow v[k]$ are called eigenvectors of the DFT (refer to the next set of projects, *DFT as a Matrix*).

a. Generate a real-valued Gaussian signal:

$$g[n] = e^{-\alpha n^2} \qquad -L \leq n \leq L$$

The signal is truncated in a symmetric fashion about the origin ($n = 0$). Choose L so that the Gaussian is sampled well out onto its tails. The exact choice of L will depend on α; perhaps the largest exponent αL^2 should be restricted to be less than 100. If we take $\alpha = \frac{1}{2}$, this is a special case where the continuous-time Fourier transform yields a transform that is the same form, $X(\omega) = \exp(-\frac{1}{2}\omega^2)$.

b. Form an N-point vector from the samples of $g[n]$. Note that N will be equal to $2L+1$. Place the samples into the vector so as to preserve the even symmetry. This can be accomplished by rotating the largest sample $g[0]$ to the beginning of the vector.

c. Compute the N-point DFT of $g[n]$. Verify that the result is purely real. If it is not, the time vector was not constructed symmetrically.

d. Plot the real part of the DFT and compare to a Gaussian. It may be necessary to rotate the DFT vector to see that it looks like a Gaussian (see `fftshift`).

e. Experiment with different values of α. Keep the transform length constant. Try to make the width of the Gaussian the same in both the time and frequency domains. Notice that when the width of the Gaussian decreases in the time domain, it increases in the DFT domain. This is a demonstration of the uncertainty principle of Fourier analysis: "The product of time width and frequency width is always greater than a fixed constant." Thus shrinking the time width will necessarily increase the frequency width.

EXERCISE 2.4

Real Exponential

Another common signal is the real, decaying exponential. The z-transform of this signal consists of a single pole, so it is very simple to evaluate. However, it is wrong to think that the DFT is merely a sampled version of the z-transform.

a. Generate a finite portion of an exponential signal: $x[n] = (0.9)^n u[n]$, for $0 \leq n < N$. Take a small number of samples, say $N = 32$.

b. Compute the N-point DFT of $x[n]$, and plot the magnitude of the DFT $|X[k]|$.

c. Compare $|X[k]|$ to samples of $\left|Y(e^{j\omega})\right|$, the magnitude of the DTFT of $y[n] = (0.9)^n u[n]$, an infinitely long exponential.

$$\left|Y(e^{j\omega})\right| = \left|\frac{1}{1 - 0.9e^{-j\omega}}\right| \tag{2-2}$$

Plot the magnitudes on the same graph—explain the difference in terms of windowing.

d. Another related signal can be created by sampling the DTFT of $a^n u[n]$. Create a DFT by sampling the formula (2-2)

$$V[k] = Y(e^{j\omega})\Big|_{\omega=(2\pi/N)k}$$

Take the N-point IDFT of $V[k]$ to obtain $v[n]$. Experiment with different transform lengths for N, because as $N \to \infty$ the result should get very close to $a^n u[n]$.

Since the DFT was formed by sampling in the frequency domain, there should be aliasing in the time domain. Derive a formula for $v[n]$ in terms of $y[n]$, based on this idea of time aliasing.

PROJECT 3: SYMMETRIES IN THE DFT

This project reviews different signal attributes such as even, odd, purely real, and purely imaginary, and explores their implication for a special structure of the resulting DFT. The attribute of conjugate symmetry is also introduced, because it is the important dual of purely real: "A purely real-time signal has a conjugate symmetric DFT, and vice versa." In many cases, these symmetries can be used to simplify computations, especially in different variations of the FFT algorithm.

Hints

The indexing of an N-point vector in MATLAB runs from 1 to N. However, many of the DFT symmetries are expressed in terms of flips with respect to the origin (e.g., the constraint $x[n] = x[-n \bmod N]$ defines an even signal). This presents problems when working in MATLAB. A simple solution would be to create a new M-file to perform the circular flip. This would be based on modulo-N arithmetic, which is implemented via the M-file `mod` described in Appendix A. Starting from `cflip`, you can then write additional M-files for extracting the even and odd parts of a signal vector, as well as its conjugate-symmetric and conjugate-antisymmetric parts.

EXERCISE 3.1

Symmetries Are Circular

All operations with the DFT are done over an indexing domain that is circular. Since all symmetry properties boil down to only *two* basic operations, conjugation and flipping, it is essential to have functions for each. MATLAB provides a built-in conjugate function: `conj`. However, the "flip" operation is tricky. There are built-in MATLAB functions called `fliplr` and `flipud`, for flip-left-right on rows, and flip-up-down on columns. Neither of these is what we want for the DFT, because these flips do the following:

$$y_{\text{out}}[n] = x[N-1-n] \qquad \text{for } n = 0, 1, \ldots, N-1$$

Thus $x[0]$ is exchanged with $x[N-1]$, $x[1]$ with $x[N-2]$, and so on. On the other hand, the circular flip needed for the DFT would satisfy

$$y_{\text{cir}}[n] = x[-n \bmod N] = x[N-n] \qquad \text{for } n = 0, 1, \ldots, N-1$$

In this case, $x[0]$ *stays put*, while $x[1]$ is exchanged with $x[N-1]$, $x[2]$ with $x[N-2]$, and so on.

a. Write an M-file for the circular-flip operation; call it `cflip`. For an N-point row vector `cflip` is simply `[ x(1), x(N:-1:2) ]`, but you should make it work for rows, columns, and matrices (i.e., c-flip each column).

b. Verify the DFT property: "A c-flip in time gives a c-flip in frequency." Use simple test vectors, but make them complex-valued to test this property completely.

c. Verify the DFT property: "A conjugate in time gives a conjugate plus a c-flip in frequency." State the dual of this property; then verify it with a complex-valued test signal.

EXERCISE 3.2

Even and Odd Parts

The primary symmetries are based on evenness and oddness. Of course, these must be defined "circularly."

a. For use in the rest of this exercise, write a MATLAB function that will extract the even part of a vector; do the same for the odd part. These functions should call the `cflip` function written previously.

b. Generate a real-valued test signal $v[n]$ using `rand`. Pick a relatively short length, say $N = 15$ or $N = 16$. Compute the DFT of $v[n]$ to get $V[k]$, and then try the following for both even and odd lengths.

c. Calculate the even and odd parts of $v[n]$.

$$v_e[n] = \tfrac{1}{2}(v[n] + v[-n \bmod N])$$
$$v_o[n] = \tfrac{1}{2}(v[n] - v[-n \bmod N])$$

Then compute the DFTs of these two signals for the next part.

d. For the DFT computed in part (a), extract its real and imaginary parts and make the following comparison:

$$\mathrm{DFT}\{v_e[n]\} \quad \text{vs.} \quad \mathrm{Re}\{V[k]\}$$
$$\mathrm{DFT}\{v_o[n]\} \quad \text{vs.} \quad \mathrm{Im}\{V[k]\}$$

If $v[n]$ is complex, show that these same relations do not hold.

e. The notions of even and odd can be extended to the complex case by defining two attributes called *conjugate symmetric* and *conjugate antisymmetric*. Generate a random complex-valued test signal, $v[n]$. Compute its *conjugate-symmetric* and *conjugate-antisymmetric* parts via

$$\begin{aligned} v_{\text{csy}}[n] &= \tfrac{1}{2}(v[n] + v^*[-n \bmod N]) \\ v_{\text{cas}}[n] &= \tfrac{1}{2}(v[n] - v^*[-n \bmod N]) \end{aligned} \tag{3-1}$$

Write MATLAB functions that will extract these parts from a complex-valued signal vector.

f. Show that the real part of $v_{\text{csy}}[n]$ is always even and that the imaginary part is odd. Verify these facts in MATLAB for the conjugate-symmetric part of a random test sequence. State a similar relation for $v_{\text{cas}}[n]$; verify it.

g. Verify that the DFT of a conjugate-symmetric signal is purely real. What about the DFT of a conjugate-antisymmetric signal?

EXERCISE 3.3

DFT of a Real Sequence

By duality, the DFT of a purely real sequence should be conjugate symmetric. Start with a real N-point sequence $v[n]$, which is neither even nor odd.

a. Calculate its DFT: $V[k] = \mathrm{DFT}\{v[n]\}$.

b. Display $\mathrm{Re}\{V[k]\}$ and $\mathrm{Im}\{V[k]\}$. In addition, display the magnitude and phase of $V[k]$ and note any obvious symmetries.

c. Extract the conjugate-antisymmetric part of $V[k]$, which should be zero.

d. Show that $\mathrm{Re}\{V[k]\}$ is even, by computing the odd part of $\mathrm{Re}\{V[k]\}$, which ought to be zero. Show that $\mathrm{Im}\{V[k]\}$ is odd.

Thus we can conclude that the DFT of a real sequence is conjugate symmetric, which is the expected dual property. What would be the result for a purely imaginary input vector?

EXERCISE 3.4

All Possible Symmetries

Any complex-valued signal can be decomposed into four subsignals, each of which exhibits a certain symmetry. Specifically, the complex signal, $v[n]$, can always be written as

$$v[n] = v_{\text{r.e}}[n] + v_{\text{r.o}}[n] + j(v_{\text{i.e}}[n] + v_{\text{i.o}}[n]) \tag{3-2}$$

where the subscripts denote real (r), imaginary (i), even (e), and odd (o). Thus $v_{\text{i.o}}[n]$ is the odd part of the imaginary part of $v[n]$. The same sort of decomposition can be done for the DFT, $V[k]$.

$$V[k] = V_{\text{r.e}}[k] + V_{\text{r.o}}[k] + j(V_{\text{i.e}}[k] + V_{\text{i.o}}[k])$$

However, it is wrong to assume that the DFT of one of the subsignals matches the corresponding subsequence of the DFT.

a. Write a MATLAB function that will decompose any vector into its four parts, as defined in (3-2).

b. The DFT symmetries can be summed up in a diagram that shows the correspondence of the symmetry in the transform domain to one of the four subsignals in the time domain. Complete the following diagram by connecting each time-domain subsignal • to the appropriate frequency-domain subsequence ○.

DFT SYMMETRIES

$v_{\text{r.e}}[n]$	•	○	$V_{\text{r.e}}[k]$	real and even
$v_{\text{r.o}}[n]$	•	○	$V_{\text{r.o}}[k]$	real and odd
$v_{\text{i.e}}[n]$	•	○	$V_{\text{i.e}}[k]$	imaginary and even
$v_{\text{i.o}}[n]$	•	○	$V_{\text{i.o}}[k]$	imaginary and odd

c. Give numerical examples to show that each connection in the diagram is correct.

PROJECT 4: TRICKS FOR THE INVERSE DFT

In practice, special hardware may be built to compute a DFT. In this case, it is advantageous to use the same hardware for the inverse DFT (IDFT). This project shows three different ways that the IDFT can be computed using a forward DFT algorithm. All are based on the fact that the formula for the IDFT is nearly identical to that for the forward DFT, except for a minus sign in the exponent and a factor of $1/N$.

$$x[n] = \frac{1}{N}\sum_{k=0}^{N-1} X[k]W_N^{-nk} \quad \text{for} \quad n = 0, 1, 2, \ldots, N-1 \tag{4-1}$$

Hints

Three different IDFT functions are described in this project. Therefore, three separate M-files should be written and tested. Inside MATLAB the inverse DFT function `ifft` is actually implemented using one of these tricks. Type out the listing of the function `ifft` to see which one.

EXERCISE 4.1

IDFT via Circular Rotations

This exercise deals with computing the inverse DFT using a property of the transform that is known as *duality*. This method is interesting because it emphasizes the circular nature of all DFT indexing. The M-file `cflip` developed in Project 3 will be useful here.

a. Generate a random sequence $x[n]$ using the `rand` function. The length of the sequence can be anything, but it is often chosen as a power of 2; if so, select $N = 16$. Use `fft` to compute the DFT of $x[n]$, and call this $X[k]$. This sequence will be the test sequence for the IDFT. Perform the following steps with MATLAB:

1. Flip the sequence $X[k]$ using a circular flip. Since $X[k]$ is just a vector of 16 complex values, it can be treated as though it were a vector of time samples. So the flip operation defines a new time vector by $y[n] = X[k]|_{k=(-n) \bmod N}$.
2. Since the sequence $y[n]$ is just a vector of 16 complex values, it can be used as the input to a *forward* DFT (i.e., apply `fft` again). Call the result $Y[k]$.
3. Once again, the sequence $Y[k]$ is just a vector of 16 values, so it can be considered as a time vector if k is replaced with n.
4. Compare the numerical values of $x[n]$ and $v[n] = Y[k]|_{k=n}$.

b. Derive the general rule for the relationship between the values of $x[n]$ and $v[n]$, and prove why they are related in such a simple fashion.

c. Program an M-file that implements the IDFT according to this algorithm. Test it on some known DFT pairs.

EXERCISE 4.2

IDFT via Conjugates

This exercise deals with computing the inverse DFT using several well-placed conjugates. Use a test sequence $X[k]$ generated as in Exercise 4.1.

a. We want to show that the following three steps are equivalent to the IDFT of $X[k]$.

1. Conjugate $X[k]$.
2. Compute the *forward* DFT of $X^*[k]$.
   As before, we have an N-point vector, so we can do this.
3. Conjugate the resulting DFT output.
   If the result of the previous DFT were $Y[k]$, we now have $Y^*[k]$.

The conjugate operator merely changes the sign of the imaginary part, as shown in Fig. 2.2.

Figure 2.2
IDFT via conjugate trick.

Re{X[k]}
Im{X[k]}
FFT
−1
−1
N·Re{x[n]}
N·Im{x[n]}

b. Compare the numerical values of $x[n]$ and $v[n] = Y^*[k]|_{k=n}$. Notice that the effect of the conjugate operations has been to change the sign of the complex exponential. Derive the general rule for the relationship between the values of $x[n]$ and $v[n]$, and prove why they are related in such a simple fashion.

c. Create a MATLAB M-file that will implement this algorithm for the IDFT. Compare with the implementation actually used in MATLAB; the listing of the MATLAB `ifft` function can be obtained via `type ifft`.

EXERCISE 4.3

Another IDFT Trick

This exercise shows yet another way to compute the inverse DFT. Its proof also relies on the conjugate property of the DFT. Use the same test sequence as in the preceding two exercises.

a. We want to show that the following three steps are equivalent to the IDFT of $X[k]$ (see Fig. 2.3).

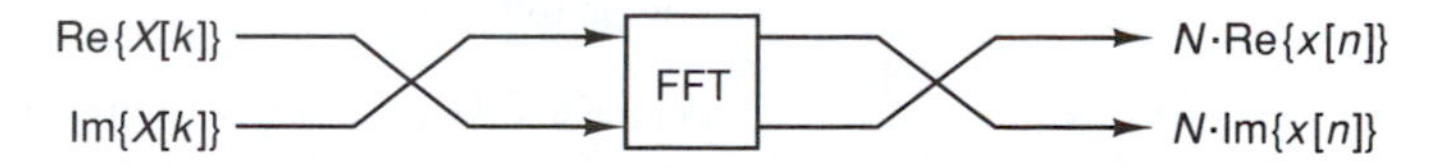

Figure 2.3
IDFT via swap of real and imaginary parts.

1. Swap the real and imaginary parts of $X[k]$ and define the result as a time vector; that is, define $v[n]$ so its real part is the imaginary part of $X[k]$ and its imaginary part is the real part of $X[k]$.
2. Compute the forward DFT of $v[n]$, and call this $V[k]$.
3. Swap the real and imaginary parts of $V[k]$ and define it as a time vector; that is, define $y[n]$ so that its real part is the imaginary part of $V[k]$ and its imaginary part is the real part of $V[k]$.

b. Compare the numerical values of $x[n]$ and $y[n]$. Derive the general rule for the relationship between the values of $x[n]$ and $y[n]$, and prove why they are related in such a simple fashion.

c. Create a MATLAB M-file that will implement this IDFT algorithm.

PROJECT 5: ZERO PADDING AND DECIMATION PROPERTIES

The exercises in this project are concerned with zero padding: at the end of a signal, in the middle of a signal, and between every other sample in the signal. Zero padding with a block of zeros is commonly used to do bandlimited interpolation. Zero padding between sample is a "stretch" operation that is closely related to *decimation*, where every other sample is removed. In all cases, new signals are formed that are either longer or shorter than the original $x[n]$, but only the original data values are present. Relating the DFTs for these new signals to the original $X[k]$ is the point of the following exercises.

EXERCISE 5.1

Pad with Zeros

We can create a long signal by appending zeros to an existing signal. This can be done automatically in the `fft` function by specifying a second argument for the FFT length and making it longer than the signal length.

Create a test signal that is a 16-point sinusoid: $\sin(\omega_0 n)$. Choose $\omega_0 = 2\pi/\sqrt{17}$. Compute the 16-point DFT and call the result $X[k]$. Then compute the DFT for lengths 32, 64, and 256. Show that the same 16 values in $X[k]$ can always be found within each of the longer DFTs. This fact leads to the statement that "zero padding in the time domain gives interpolation in the frequency domain."

EXERCISE 5.2

Zero Padding in the Middle

One difficulty with zero padding is that it destroys symmetry. In other words, if the original signal were such that its DFT were purely real, then after zero padding the resultant DFT is probably not real. There is a method of zero padding that will preserve such symmetries—this is padding "in the middle."

a. Create a real and even-symmetric signal for use as a test signal. Since this is easiest to do when the length is odd, take N to be odd (e.g., $n = 21$). Verify that the DFT of this signal is purely real, and that it is also even-symmetric.

b. Now take the FFT of the same signal, but with the length three times the original (i.e., $N = 63$). Verify that the transform is no longer purely real (unless, of course, your test signal were the impulse).

c. Now create another signal that is also three times longer than the original, but do the padding in the following way:

   1. Put the first $\frac{1}{2}(N+1)$ signal points at the beginning of the output vector. (Remember that N is assumed to be odd.)
   2. Add $2N$ zeros to the vector.
   3. Then tack the last $\frac{1}{2}(N-1)$ signal points at the end of the output.

   Write an M-file that does this operation, and then verify that it preserves symmetry by testing with simple inputs and their DFTs. Also check that the interpolation property still holds.

d. Show how the padding in the middle can be done as the concatenation of three operations: a rotation to the right, zero padding, and then another rotation to the left. Specify the number of samples for each rotation.

e. How would you do this "padding in the middle" when N is even? Write an M-file that exhibits proper adherence to symmetries for the even-length case. The difficulty is $x[N/2]$. If you split it in half, a symmetric signal can be produced after padding, but will this strategy preserve the interpolation through the original DFT values?

EXERCISE 5.3

Stretch—Intersperse Zeros

We can create a longer signal by putting zeros in between the existing samples. For example, if we start with a length N signal, a length ℓN signal can be created as follows:

$$\tilde{x}[=] \begin{cases} x[n/\ell] & \text{when } n \bmod \ell = 0 \\ 0 & \text{when } n \bmod \ell \neq 0 \end{cases} \tag{5-1}$$

Generate a random 10-point signal and compute its DFT. Intersperse three zeros between all the existing samples and then compute the 40-point DFT (i.e., $\ell = 4$). Verify that the longer DFT is just a repetition of the shorter one. Therefore, the duality property is: "Periodicity in one domain goes with interspersed zeros (5-1) in the other domain."

EXERCISE 5.4

Decimation

The dual of the stretch property is decimation. In this case we generate a shorter signal by removing samples. Starting with a length-N signal, we can make a signal of length $M = N/\ell$ by taking every ℓth sample:

$$\hat{x}[n] = x[n\ell] \qquad \text{for } n = 0, 1, \ldots, M-1 \tag{5-2}$$

It is necessary to assume that N contains a factor of ℓ. The M-point DFT of $\hat{x}[n]$ is an "aliased" version of the original DFT, $X[k]$, because sampling is being done in the time domain.

a. Generate a sinusoidal test signal whose length is 60 [i.e., $x[n] = \sin(\omega_0 n)$]. Set the frequency at $\omega_0 = 2\pi/5$. Perform a decimation by $\ell = 2$ and then compute the 30-point DFT. Explain the result in terms of the original 60-point DFT.

b. Redo part (a) with $\ell = 3$. Relate the 20-point DFT to the original 60-point DFT.

c. Try the same experiment with a decimation factor of $\ell = 7$, which is not a divisor of 60. Now there will be nine nonzero samples, so a 9-point DFT should be computed. This result is much different but can be explained by using the fact that the DFT consists of frequency samples of an underlying DTFT.

Project 6: Real Data FFT

There are two cases of interest where the FFT of a real sequence can be computed efficiently. The first is the case of simultaneously transforming two N-point vectors with just one N-point DFT. The second involves the transform of a $2N$-point real vector using a single N-point FFT.

Hints

These algorithms are based on the symmetry properties of the DFT. In Project 3, two M-files should have been written for extracting the conjugate-symmetric and conjugate-antisymmetric parts of a signal vector. These will be needed here [see Exercise 3.2(e)].

EXERCISE 6.1

Computing Two DFTs at Once

Given two real N-point signals, $v_0[n]$, and $v_1[n]$, form a complex signal $v[n] = v_0[n] + jv_1[n]$ and compute its DFT; the result is a complex-valued vector $V[k]$. Since the DFT is a linear operator, it follows that

$$V[k] = \mathcal{DFT}\{v_0[n] + jv_1[n]\} = V_0[k] + jV_1[k]$$

Recognize that $V_0[k]$ is not the real part of $V[k]$, nor is $V_1[k]$ the imaginary part, because both $V_0[k]$ and $V_1[k]$ are complex-valued. Instead, you can use the conjugate symmetry property to show analytically that the following relationships hold:

$$\begin{aligned} V_0[k] &= \text{CSY}\{V[k]\} \qquad \text{(conjugate-symmetric part)} \\ jV_1[k] &= \text{CAS}\{V[k]\} \qquad \text{(conjugate-antisymmetric part)} \end{aligned}$$

See (3-1) for a definition of the conjugate-symmetric and conjugate-antisymmetric operators.

a. Confirm by numerical experiment that $V_0[k]$ and $jV_1[k]$ have the correct symmetry.

b. Write a MATLAB function that will compute the DFT of two real vectors, with only one call to `fft`. Test with known DFT pairs.

EXERCISE 6.2

FFT of Real Data

Given a length-$2N$ real signal $x[n]$, $n = 0, 1, \ldots, 2N-1$, we want to develop a method that will compute the $2N$-point DFT of $x[n]$ using just one N-point FFT. This algorithm relies on one fact from the derivation of the FFT, but it is a relatively easy one to derive, so we develop it here. If we separate the input sequence into two subsequences, one containing the even-indexed members of the original vector, the other the odd-indexed points,

$$\begin{aligned} x_0[n] &= x[2n] \\ x_1[n] &= x[2n+1] \end{aligned} \qquad \text{for } n = 0, 1, 2, \ldots, N-1 \tag{6-1}$$

then we can write the $2N$-point FFT as

$$\begin{aligned} X[k] &= \sum_{n=2\ell} x[2\ell]W_{2N}^{2\ell k} + \sum_{n=2\ell+1} x[2\ell+1]W_{2N}^{(2\ell+1)k} \\ &= \sum_{\ell=0}^{N-1} x_0[\ell]W_N^{\ell k} + W_{2N}^{k} \sum_{\ell=0}^{N-1} x_1[\ell]W_N^{\ell k} \\ &= \mathcal{DFT}\{x_0[\ell]\} + W_{2N}^{k}\mathcal{DFT}\{x_1[\ell]\} \qquad k = 0, 1, \ldots, 2N-1 \end{aligned} \tag{6-2}$$

The DFTs needed here are N-point DFTs. Thus we can get the $2N$-point DFT from two half-length DFTs of real-only data, which is exactly what was done in Exercise 6.1.

a. Form a complex-valued signal $v[n] = x_0[n] + jx_1[n]$, and compute its DFT; the result is a complex-valued vector $V[k]$. Use the conjugate-symmetry property to extract the DFTs of each real sequence:

$$\begin{aligned} X_0[k] &= \text{CSY}\{V[k]\} \qquad \text{(conjugate-symmetric part)} \\ jX_1[k] &= \text{CAS}\{V[k]\} \qquad \text{(conjugate-antisymmetric part)} \end{aligned}$$

b. Apply the result (6-2) from above to get the entire $2N$ points of $X[k]$. Note that each $X_i[k]$ has a period of N, so (6-2) can be simplified for $k \geq N$.

c. Write a MATLAB function that will compute the DFT of a real vector according to this algorithm—just one call to `fft` with length N is permitted.

d. Test your program versus `fft` and show by numerical example that you get the same answer.

e. If you are interested, use `flops` to count the number of floating-point operations used by this method. Verify this count by using the known formula for the number of FLOPs in an FFT, plus the number of operations for the multiplication by W_{2N}^{k}, plus the conjugate-symmetric decomposition. It is likely that the MATLAB count will be higher because it will include additional FLOPs for the computation of the sines and cosines that make up the complex exponential W_{2N}^{k}.

Within MATLAB the `fft` function uses this same trick for the FFT of real data. If you count the number of FLOPs for the FFT of real data and compare to the number of FLOPs for complex data, you will find that the real case takes a little more that 50% of the number for the complex case.

EXERCISE 6.3

IFFT of Conjugate-Symmetric Data

A procedure similar to the real-data FFT can be derived for doing the IDFT of a vector that satisfies the conjugate-symmetric property. This technique is *not* implemented in the MATLAB inverse FFT function, so you should be able to create an improved M-file that is more efficient than the existing `ifft` function.

Given a length-$2N$ conjugate-symmetric vector $X[k]$, $k = 0, 1, \ldots, 2N-1$, we want to develop a method that will compute the $2N$-point IDFT of $X[k]$ using just one N-point IFFT. Again, we need to recall a basic fact from the derivation of the FFT, which we develop below. If we separate the time sequence into two subsequences, one containing the even-indexed members, the other the odd-indexed points, we can write the $2N$-point IFFT as two separate cases:

$$x[2n] = \frac{1}{2N}\sum_{k=0}^{2N-1} X[k]W_{2N}^{-2nk} = \frac{1}{2N}\sum_{k=0}^{N-1}(X[k]+X[k+N])\,W_{N}^{-nk}$$

$$= \tfrac{1}{2}\,\mathcal{IDFT}\{X[k]+X[k+N]\} \qquad n = 0, 1, \ldots, N-1$$

$$x[2n+1] = \frac{1}{2N}\sum_{k=0}^{2N-1} X[k]W_{2N}^{-(2n+1)k} = \frac{1}{2N}\sum_{k=0}^{N-1} W_{2N}^{-k}\,(X[k]-X[k+N])\,W_{N}^{-nk}$$

$$= \tfrac{1}{2}\,\mathcal{IDFT}\{W_{2N}^{-k}\,(X[k]-X[k+N])\} \qquad n = 0, 1, \ldots, N-1 \tag{6-3}$$

The IDFTs needed here are N-point IDFTs, but these two IDFTs can be done simultaneously.

a. Define the two N-point vectors

$$\begin{aligned} X_0[k] &= X[k] + X[k+N] \\ X_1[k] &= W_{2N}^{-k}\,(X[k] - X[k+N]) \end{aligned} \qquad \text{for } k = 0, 1, 2, \ldots, N-1$$

Since $X[k]$ is conjugate-symmetric (mod-$2N$), you can show that $X_0[k]$ is a conjugate-symmetric vector (mod-N); and that $jX_1[k]$ is conjugate-antisymmetric (mod-N). Prove these facts mathematically and then verify by examples in MATLAB.

b. Now we consider how to do two IDFTs at once in this algorithm. Form a complex-valued N-point signal $Q[k] = X_0[k] + jX_1[k]$, and compute its N-point IDFT; the result is a complex-valued vector $q[n]$. Use the conjugate-symmetry property to show that

$$\begin{aligned} x[2n] &= \tfrac{1}{2}\,\mathrm{Re}\{q[n]\} \\ x[2n+1] &= \tfrac{1}{2}\,\mathrm{Im}\{q[n]\} \end{aligned} \qquad \text{for } n = 0, 1, \ldots, N-1 \tag{6-4}$$

The factor of $\frac{1}{2}$ comes from the fact that the IDFT computed will be of length N rather than $2N$.

c. Write a MATLAB function that will compute the IFFT of a conjugate-symmetric vector according to this algorithm—just one call to `ifft` with length N.

d. One difficulty in applying this IFFT is detecting whether or not a signal satisfies the conjugate-symmetric property. Write a MATLAB function that will test whether a vector is conjugate symmetric and return `TRUE` or `FALSE`. This involves a comparison between the vector and its circularly flipped version. Allow for round-off error so that the match does not have to perfect. If the FFT were computed by a real-data FFT, the data will be exactly conjugate symmetric, so the round-off error should not be a problem; otherwise, it will be tricky to set a threshold.

e. Test your program versus MATLAB's `ifft` and show by numerical example that you get the same answer.

f. In MATLAB your M-file should have fewer FLOPs than the built-in MATLAB function `ifft`. Compare the number of floating-point operations used by both methods.

PROJECT 7: DISCRETE FOURIER SERIES

The DFT can be related directly to the discrete Fourier series (DFS) of a periodic signal. If the period of the signal is M, the coefficients produced by an M-point DFT are exactly those needed in a DFS representation of the signal, if they are scaled by $1/N$. In addition, this periodic nature of the signal (from the Fourier series) is equivalent to the circular indexing that is always used in the DFT.

EXERCISE 7.1

Relate DFT to Fourier Series

Consider a 50% duty cycle square wave whose period is $M = 16$.

$$x[n] = \begin{cases} 1 & 0 \le n \bmod M < 8 \\ 0 & 8 \le n \bmod M < 16 \end{cases}$$

a. Compute the 16-point DFT of the sequence
`xx = [ 1 1 1 1 1 1 1 1 0 0 0 0 0 0 0 0 ],`
eight 1's followed by eight 0's.

b. The signal $x[n]$ has a Fourier series that can be written as the following expansion:

$$x[n] = \sum_{k=0}^{M-1} A_k e^{j(2\pi k/M)n} \tag{7-1}$$

Since the vector `xx` is one period of the 50% duty cycle square wave, the DFT result $X[k]$ from part (a) can be used to define the coefficients A_k. Determine the values of A_k from the DFT computed in part (b).

c. Formula (7-1) is correct for all values of n, so it can be used to synthesize a long section of the square wave directly from the DFS representation. Suppose that we would like to produce $x[n]$ over the range $0 \le n < 128$. It is possible to avoid direct evaluation of the sum in (7-1) by using the `ifft` function. Show how to synthesize this section of $x[n]$ by computing a 128-point inverse FFT of a vector produced from the 16 DFS coefficients, A_k.

EXERCISE 7.2

Pulse Train Is a Sum of Cosines

One commonly used periodic signal is the periodic pulse train.

$$p[n] = \delta[n \bmod M] = \begin{cases} 1 & \text{for } n = 0 \bmod M \\ 0 & \text{elsewhere} \end{cases}$$

a. Determine the DFS expansion for this signal, when $M = 21$.

b. Since $p[n]$ is real, and even, its DFS can be written in the following way:

$$p[n] = \sum_k c_k \cos(2\pi kn/N)$$

Determine the values of c_k and the range on k in the summation.

EXERCISE 7.3

Truncation of the Fourier Series

Define a symmetric periodic pulse train as

$$x[n] = \begin{cases} 1 & \text{for } -16 \le n \bmod M \le 16 \\ 0 & \text{elsewhere} \end{cases}$$

The pulse length is 31. If the period is $M = 128$, the signal $x[n]$ can be viewed as a finely sampled version of a continuous-time square wave. In this exercise we consider how the discrete Fourier series is an approximation to the true Fourier series of the continuous-time square wave.

a. Since the period is $M = 128$, the DFS representation would naturally contain 128 coefficients. However, we can truncate the expansion by taking only 21 DFS coefficients (it is best to take an odd number). When doing the truncation, we must maintain symmetry, so we can't take just the first 21. Determine which 21 DFS coefficients to take so that the resynthesized signal (7-1) will be symmetric.

b. The truncation of the Fourier series means that a reconstruction from the 21 DFS coefficients will be bandlimited. Resynthesize two periods of the signal from the 21 coefficients and compare to the original (by overlaying both on the same plot). Use the function `ifft` to do this computation.

c. The bandlimited resynthesis will not match the original at all points. Are there any points where the error between the original and the reconstructed signal is zero?

d. The reconstructed signal should have its worst error near the edge of the pulse—called the Gibbs effect. Measure this worst-case error near the edge of the pulse and compare to the known height of the Gibbs overshoot—approximately 9%.

DFT AS A MATRIX

OVERVIEW

This set of projects will concentrate on properties of the DFT when viewed as a matrix operator. This viewpoint is becoming more fashionable with the emergence of programs such as MATLAB, which emphasize the matrix-vector nature of computations. Very few

DSP textbooks present the DFT in matrix form, but one such presentation can be found in the book by Burrus and Parks [9]. Additional background material on the general subject of linear algebra and matrix computations can be found in [10], [11], and [12].

Consider the N-point DFT as a matrix transformation from the complex vector

$$\mathbf{x} = [\,x[0] \quad x[1] \quad \cdots \quad x[N-1]\,]^T \tag{0-1}$$

to another complex vector, in the frequency domain

$$\mathbf{X} = [\,X[0] \quad X[1] \quad \cdots \quad X[N-1]\,]^T \tag{0-2}$$

In matrix-vector notation, the operation is a matrix multiplication:

$$\mathbf{X} = \mathbf{W}\mathbf{x} \tag{0-3}$$

where the DFT matrix $\mathbf{W}$ has entries that are complex exponentials:

$$\mathbf{W} = \frac{1}{\sqrt{N}}\begin{bmatrix} 1 & 1 & 1 & \cdots & 1 \\ 1 & W & W^2 & \cdots & W^{N-1} \\ 1 & W^2 & W^4 & \cdots & W^{2(N-1)} \\ \vdots & \vdots & \vdots & \cdots & \vdots \\ 1 & W^{N-1} & W^{N-2} & \cdots & W^1 \end{bmatrix} \quad \text{where } W = e^{-j2\pi/N} \tag{0-4}$$

The scaling of $1/\sqrt{N}$ is introduced to make $\mathbf{W}$ a unitary matrix, but this is not the usual definition of the DFT.

PROJECT 1: DFT AS AN ORTHOGONAL MATRIX

The fundamental properties of the DFT rely on the *orthogonality* of the complex exponentials. In terms of the DFT matrix, this property is equivalent to the fact that the column vectors of $\mathbf{W}$ are pairwise orthogonal.

EXERCISE 1.1

Orthogonality

a. Generate an instance of the DFT matrix for a small value of N, say $N = 5$. This can be done without `for` loops by raising $W = e^{-j2\pi/N}$ to different integer powers; in fact, all these powers of W can be computed at once using the pointwise power operator `W .^ M` in MATLAB, where `W` is the complex exponential and `M = [0:(N-1)]' * [0:(N-1)]` is a matrix containing all the integer powers. Remember to divide by $\sqrt{N}$.

b. Another trick for generating the DFT matrix is `W = fft( eye(N) ) / sqrt(N)`. This method works because the `fft` function applied to a matrix will take the DFT of each column vector in the matrix. Write a formula for the ℓth column of `eye(N)` and its DFT to explain why the `fft` function applied to the identity matrix `eye(N)` will generate the complete DFT matrix.

c. Let $\mathbf{w}_j$ denote the jth column of $\mathbf{W}$. Verify that any two columns of $\mathbf{W}$ are orthogonal [i.e., the inner product $\langle \mathbf{w}_i, \mathbf{w}_j \rangle = \mathbf{w}_i^H \mathbf{w}_j = 0$, when $i \neq j$].

d. In fact, the columns of $\mathbf{W}$ are *orthonormal*. Verify that the norm of each column vector is 1:

$$\langle \mathbf{w}_i, \mathbf{w}_i \rangle = \mathbf{w}_i^H \mathbf{w}_i = \|\mathbf{w}_i\|^2 = 1$$

e. The conjugate transpose of **W** is denoted $\mathbf{W}^H$, where the superscript H is called the *Hermitian* operator. In MATLAB, the prime operator does a conjugate transpose. The rows of $\mathbf{W}^H$ are the conjugate of the columns of **W**. Thus, all of the pairwise inner products can be computed simultaneously using the matrix product, $\mathbf{W}^H\mathbf{W}$. Compute this product, and then explain why the result equals the $N \times N$ identity matrix. Since **W** satisfies this property, it is called a *unitary* matrix.

EXERCISE 1.2

Inverse DFT Matrix

The unitary property leads to a trivial definition of the inverse DFT (IDFT). The following three approaches should give the same answer.

a. An inverse DFT matrix can be obtained by computing the inverse $\mathbf{W}^{-1}$ via the MATLAB `inv` function.

b. A second method for computing the IDFT matrix in MATLAB would be analogous to the trick given in Exercise 1.1(a) for **W** [i.e., `Winv = ifft( eye(N) )`]. This corresponds to the definition usually given in textbooks, but it is not unitary and is not equal to $\mathbf{W}^{-1}$ in part (a) because it is off by a scale factor.

c. Due to the unitary property, the inverse of **W** is its Hermitian, $\mathbf{W}^H$. Show that $\mathbf{W}^H = \mathbf{W}^{-1}$ computed in part (a). Determine the scale factor relating $\mathbf{W}^H$ and $\mathbf{W}_{\text{inv}}$ from part (b). For example, look at the result of the division `W' ./ Winv`.

PROJECT 2: EIGENVALUES OF THE DFT MATRIX

The eigenvalues of the DFT matrix follow an amazingly simple pattern [1]. The MATLAB function `eig` makes it easy to compute the eigenvalues and explore these patterns. With two outputs, `eig` will also return the eigenvectors.

EXERCISE 2.1

Eigenvalues

Use MATLAB to find all the eigenvalues of the DFT matrix. Note that the repeated eigenvalues will be shown a number of times equal to their multiplicity.

a. Do this for several consecutive values of N, perhaps $4 \leq N \leq 10$. Make a table of the eigenvalues and their multiplicities. Make sure that the DFT includes the scaling by $1/\sqrt{N}$; otherwise, the eigenvalues will change in magnitude with N.

b. Propose a *general* rule that gives the multiplicity of each eigenvalue as a function of N. Check your rule by finding the eigenvalues for $N = 3$, $N = 17$, and $N = 24$.

EXERCISE 2.2

Characteristic Polynomial

The eigenvalues are also the roots of the *characteristic polynomial* of the matrix. The MATLAB function `poly` will compute the characteristic polynomial satisfied by the matrix [i.e., $p(\mathbf{W}) = \mathbf{0}$]. However, this may not be the minimum order polynomial satisfied by the matrix, which is called the *minimal polynomial.* Since the DFT matrix has repeated eigenvalues, it is necessary to analyze the matrix directly to get the minimal polynomial.

a. Generate an instance of the DFT matrix for a small value of N, say $N = 5$.

b. Suppose that the matrix is applied twice (i.e., take the DFT of the DFT):

$$\mathbf{y} = \mathbf{W}\mathbf{W}\mathbf{x}$$

Define a new matrix $\mathbf{J} = \mathbf{W}^2$ and observe that many entries of $\mathbf{J}$ turn out to be equal to zero. Then it is possible to write a simple expression for $\mathbf{y}$ in terms of $\mathbf{x}$. Why should $\mathbf{J}$ be called a "flip matrix"?

c. Suppose that the matrix $\mathbf{W}$ is applied four times:

$$\mathbf{z} = \mathbf{W}^4\mathbf{x} = \mathbf{J}^2\mathbf{x}$$

What is $\mathbf{z}$ in terms of $\mathbf{x}$?

d. Use the results of the previous two parts to determine the minimal polynomial of $\mathbf{W}$. Compare to the MATLAB function `poly`. Show that both polynomials have the same roots, even though the multiplicities are different. Use `polyvalm` to apply both polynomials to the matrix $\mathbf{W}$ and verify that it does indeed satisfy both.

EXERCISE 2.3

Eigenvectors

The DFT matrix will have a complete set of orthonormal eigenvectors, because it is a unitary matrix.

a. Find all the eigenvectors of the matrix $\mathbf{W}$ when $N = 8$. Associate each eigenvector with one of the four eigenvalues. Each of these four subsets constitutes an eigen-subspace. Verify that the eigenvectors from different eigen-subspaces are pairwise orthogonal.

b. Due to repeated eigenvalues, the set of eigenvectors is not unique. It is possible to normalize and orthogonalize the eigenvector subset belonging to each eigenvalue and, thereby, produce a new set of eigenvectors such that all are pairwise orthonormal. Use the MATLAB function `orth` to perform this task. It is best to apply `orth` to each subspace separately, so that the correspondence between eigenvector index and subspace will not be lost.

c. More than likely, a direct application of `orth` still yields complex-valued eigenvectors. However, it is always possible to find eigenvectors that are purely real. To do this, the first step is to note that the eigenvectors possess symmetry. All the eigenvectors belonging to the real eigenvalues display what would be called *even symmetry*; for the imaginary eigenvalues, the symmetry is *odd*. Even symmetry in these vectors means that the second and last entries are equal, the third and second to last are the same, and so on. In matrix terms, an even symmetric vector is invariant under the operator $\mathbf{J}$ (defined above)

$$\mathbf{J}\mathbf{x} = \mathbf{x} \quad \Longleftrightarrow \quad \mathbf{x} \text{ is even symmetric}$$

Similarly, for odd symmetry we would see a negation due to the flip operator $\mathbf{J}$.

$$\mathbf{J}\mathbf{x} = -\mathbf{x} \quad \Longleftrightarrow \quad \mathbf{x} \text{ is odd symmetric}$$

Verify these symmetries for the eigenvectors.

d. A well-known property of the DFT is that a real even-symmetric input is transformed to a real even-symmetric output. Similarly, an imaginary-even input is transformed to an imaginary-even output. This observation can be used to justify the fact that *either* the real or imaginary part of each eigenvector can be used to construct the orthonormal eigen-subspaces. Justify this procedure and demonstrate that it works by using `orth`, `real`, and `imag` to construct a purely real eigenvector orthonormal basis. The only complication comes when the real or imaginary part is zero, in which case the other part must be used.

EXERCISE 2.4

Orthogonal Expansion

Any matrix that possesses a complete set of eigenvectors can be expanded in terms of those eigenvectors. The expansion takes the following form when the eigenvectors are orthonormal:

$$\mathbf{W} = \sum_{n=1}^{N} \lambda_n \mathbf{w}_n \mathbf{w}_n^H \tag{2-1}$$

In the case of the DFT, there are only four distinct eigen-spaces, so the sum can be grouped according to the different eigenvalues, $\lambda \in \{1, -1, j, -j\}$:

$$\mathbf{W}\mathbf{x} = \sum_{n\in\mathcal{N}_1} \mathbf{w}_n(\mathbf{w}_n^H\mathbf{x}) - \sum_{n\in\mathcal{N}_2} \mathbf{w}_n(\mathbf{w}_n^H\mathbf{x}) + j\left(\sum_{n\in\mathcal{N}_3} \mathbf{w}_n(\mathbf{w}_n^H\mathbf{x}) - \sum_{n\in\mathcal{N}_4} \mathbf{w}_n(\mathbf{w}_n^H\mathbf{x})\right) \tag{2-2}$$

where $\mathcal{N}_1$ is the set of indices for eigenvectors belonging to $\lambda = 1$, $\mathcal{N}_2$ for $\lambda = -1$, and so on. Each term in parentheses is an inner product, requiring N multiplications.

a. Write a MATLAB function that will compute the DFT via this expansion (2-2), specifically for the $N = 16$ case. Verify that the correct DFT will be obtained when the real eigenvectors (determined previously) are used, and compare to the output of `fft`.

b. *Possible computation*: Count the total number of operations (real multiplications and additions) needed to compute the DFT via the orthogonal expansion (2-2). Since the eigenvectors can be chosen to be purely real, the computation of the DFT via the orthogonal expansion will simplify when the input vector is purely real. The real part of the transform will depend only on the first two sums, and the imaginary part on the second two.

EXERCISE 2.5

Gaussian Sum Is the Trace

Consider the following sequence, which has quadratic phase:

$$x_\lambda[n] = e^{-j2\pi\lambda n^2/N} = W_N^{\lambda n^2} \qquad n = 0, 1, \ldots, N-1$$

This signal is a discrete-time chirp because it has linear frequency modulation.

a. When $\lambda = 1$, the sum of $x_\lambda[n]$ from $n = 0$ to $N-1$ is called the *Gaussian sum*. It is also equal to the trace of the DFT matrix and is, therefore, the sum of the eigenvalues. State a general rule, based on the eigenvalues, for the value of the trace as a function of N. This sum is also the dc value of the Fourier transform of $x_\lambda[n]$.

b. Compute the DFT of $x_\lambda[n]$ when $\lambda = 1$. Try several consecutive values of N. Plot the magnitude of the transform. When the length N is even, there are many zeros in the DFT. Explain via a formula why this happens.

   When the length N is odd, the magnitude should be a constant, so it would be interesting to examine the phase. It might be true that the phase versus k is quadratic, just like the phase versus n. To see if this is true, investigate the unwrapped phase of the DFT (see `help unwrap`). Write a formula for the DFT of $x_\lambda[n]$ when N is odd.

c. Compute the N-point DFT of $x_\lambda[n]$ when $\lambda = \frac{1}{2}$. Show that when the length of the DFT is even, the DFT of the linear-FM signal is another N-point linear-FM signal, but with a different value of λ.

d. Derive a general formula for the DFT of a chirp. This is difficult, but one approach is to complete the square in the exponents. This works best for the case when $\lambda = \frac{1}{2}$ and N is even. In this case the magnitude is a constant, so the trace can be used to find the magnitude and phase at dc.

PROJECT 3: DFT DIAGONALIZES CIRCULANT MATRICES

A well-known property of the DFT is its convolution-multiplication property. In terms of matrices, this property is equivalent to the fact that a whole class of matrices will be diagonalized by the DFT matrix. This is the class of circulant matrices, a special kind of Toeplitz matrix. A square circulant matrix has only N distinct elements—it is completely defined by its first column, as shown in the following example for $N = 5$.

$$\mathbf{C} = \begin{bmatrix} 1 & 5 & 4 & 3 & 2 \\ 2 & 1 & 5 & 4 & 3 \\ 3 & 2 & 1 & 5 & 4 \\ 4 & 3 & 2 & 1 & 5 \\ 5 & 4 & 3 & 2 & 1 \end{bmatrix}$$

The circulant matrix is important because the operation of circular convolution can be expressed as a matrix multiplication by a circulant.

Hints

The MATLAB function `diag` can be used to extract the main diagonal, or one of the off-diagonals, from a matrix.

EXERCISE 3.1

Generating a Circulant Matrix

Write a MATLAB function that will generate a circulant matrix. The function should take one argument: the vector that specifies the first column of the matrix. Use the function `toeplitz` as a model. To look at this M-file in MATLAB, do `type toeplitz`.

EXERCISE 3.2

Diagonalization of Cyclic Convolution by the DFT

The circular convolution of $x[n]$ and $h[n]$ can be written as a matrix multiplication if one of the signals is used to generate a circulant matrix. This exercise shows that the convolution-multiplication property is nothing more than a matrix identity about the diagonalization of circulant matrices.

a. Generate a circulant matrix ($\mathbf{C}$) specified by a single column vector. Compute the eigen-decomposition of the circulant. Scale all the eigenvectors so that their first element is equal to $1/\sqrt{N}$.

b. Show that the matrix formed from the eigenvectors is just the DFT matrix, or a permuted version where the columns may be out of order.

c. Show directly that the DFT matrix will diagonalize the circulant. Verify that the similarity transformation $(1/N)\mathbf{WCW}^H$ gives a diagonal matrix. Compare the numbers on the diagonal to the DFT of the first column of the circulant (scale the DFT by $1/\sqrt{N}$).

PROJECT 4: FFT ALGORITHM AS A MATRIX FACTORIZATION

The matrix form of the DFT suggests that the transformation from the n domain to the k domain is a matrix multiply that requires N^2 complex multiplications and $N(N-1)$ complex additions. This is true only when the length N is a prime number, because in many other cases, efficient FFT algorithms have been derived to reduce the number of multiplications and additions. These FFT algorithms can be described in terms of some simple factorizations of the DFT matrix.

EXERCISE 4.1

Stretch and Decimate Matrices

The "stretch" operation involves inserting zeros between the elements of a vector; the "decimate" operation applied to a vector removes all the even-indexed elements, assuming that indexing starts at $n = 1$ (as in MATLAB). Both operations can be represented as matrix multiplications, and both have simple consequences in the frequency domain.

a. In the stretch operation, you start with a length-$N/2$ vector, but the result is a length-N vector, so the stretch matrix must be $N \times N/2$. Call this matrix $\mathbf{S}$, but use $\mathbf{S}(N)$ if it is necessary to specify the length. Since the output vector has zero entries in all even-indexed elements, every corresponding row of $\mathbf{S}$ must be zero. Complete the description of $\mathbf{S}$ by writing a MATLAB function, and give an example for $N = 10$.

b. Similarly, the decimate operation can be represented by a matrix multiply with an $N/2 \times N$ matrix, called $\mathbf{D}(N)$. Describe $\mathbf{D}$ by writing a MATLAB function, and then, for $N = 10$, exhibit all of its entries.

c. Show that the stretch and decimate matrices are related via $\mathbf{S}(N) = \mathbf{D}^T(N)$.

d. Prove that $\mathbf{D}(N)\mathbf{S}(N) = \mathbf{I}_{N/2}$; verify with MATLAB. Give an interpretation of this equation. What is the result of $\mathbf{S}(N)\mathbf{D}(N)$?

EXERCISE 4.2

Stretch Property

The *stretch property* of the DFT states that interspersing zeros between samples in the n domain will cause a periodic repetition in the k frequency domain.

a. Generate a DFT matrix for an even length, say $N = 6$. When applied to the stretched vector, the DFT becomes

$$\mathbf{X} = \mathbf{W}\mathbf{x} = \mathbf{W}\mathbf{S}\hat{\mathbf{x}} \tag{4-1}$$

where $\hat{\mathbf{x}}$ is the $N/2$-point vector prior to zero insertion. Therefore, the matrix product $\hat{\mathbf{W}} = \mathbf{W}\mathbf{S}$ is a reduced matrix whose size is $N \times N/2$. If we let $\mathbf{W}(N)$ denote the N-point DFT matrix, this reduced matrix can be expressed solely in terms of $\mathbf{W}(N/2)$. Derive the form of $\hat{\mathbf{W}}$. Verify with a MATLAB example for $N = 10$.

b. Use the form of $\hat{\mathbf{W}}$ to justify the stretch property; that is, compare the top half and bottom half of $\hat{\mathbf{W}}$ to see the repetition in the vector $\mathbf{X}$. Generalize to the case of stretching by m where $m-1$ zeros lie between each sample.

EXERCISE 4.3

Decimate Property

Repeat the steps in Exercise 4.2 for the *decimate property*: The $N/2$-point DFT of the even-indexed time samples is obtained from the N-point DFT by adding the second half of the DFT vector to the first and dividing by 2. This property is actually aliasing and is a bit harder to prove.

a. In this case, we assume that the N-point DFT is already known:

$$\mathbf{X} = \mathbf{W}(N)\mathbf{x} \qquad \Longleftrightarrow \qquad \mathbf{x} = \mathbf{W}^H(N)\mathbf{X}$$

The objective is to derive the $N/2$-point DFT $\mathbf{Y}$ in terms of $\mathbf{X}$:

$$\mathbf{Y} = \mathbf{W}(N/2)\mathbf{y}$$

where $\mathbf{y} = \mathbf{D}(N)\mathbf{x}$. Use the sampling relationship between $\mathbf{x}$ and $\mathbf{y}$ to expand the equation for $\mathbf{y}$ in terms of $\mathbf{W}(N/2)$ and $\mathbf{X}$. (*Hint*: You must convert the decimate matrix into a stretch matrix).

b. Now finish the derivation by writing $\mathbf{Y}$ in terms of the identity matrix $\mathbf{I}_{N/2}$ and $\mathbf{X}$. Interpret this equation as the decimate property.

c. Verify your derivations with a MATLAB example for $N = 20$.

EXERCISE 4.4

Mixed-Radix Factorization

The matrix notation can be used to illustrate the decomposition needed for the mixed-radix FFT algorithm. When $N = L \times M$, you can find the L-point DFT matrix, and the M-point DFT matrix, inside the N-point one.

a. Generate the DFT matrix for $N = L \times M = 5 \times 4$. Then examine the submatrix `W(1:L:N,1:1:M)` and compare to the M-point DFT matrix. Can you find other submatrices of $\mathbf{W}(N)$ that are equal to either $\mathbf{W}(M)$ or a (complex) scalar multiple of $\mathbf{W}(M)$?

b. The mixed-radix FFT can be written as a six-step process. Consider the specific case of a 20-point DFT that is factored as a 5×4 DFT:

   1. Take the 20-point vector and concatenate it row-wise into a 4×5 matrix. Thus, down one column the entries will consist of every fifth point from the vector.
   2. Compute the four-point DFT of each column.
   3. Multiply *pointwise* by a 4×5 matrix of complex exponentials. This operation is called the "twiddle-factor" multiplication step.
   4. Transpose the result to form a 5×4 matrix.
   5. Compute the five-point DFT of each column.
   6. Reorder the matrix result into a vector. How? That issue is addressed in part (d).

c. Determine the entries for the matrix of twiddle factors and write the MATLAB code that will generate the twiddle-factor matrix.

d. Define the reordering that is needed to build the 20-point k-domain vector from the 5×4 matrix that is the output of the 5-point DFTs. Consider two possibilities: reading out the results one row at a time or one column at a time. Write the MATLAB code for this step. To debug this part, use a test vector that is the IDFT of the vector: `[ 0 : 1 : 19 ]`. Prior to the reordering step, the 5×4 matrix will contain the numbers 0 through 19 in a regular pattern and make the answer obvious.

e. Write a program that does the entire six-step process. Demonstrate that it all works by exhibiting the matrices at intermediate steps. Use either the test vector from part (d), or the signal $x[n] = (-1)^n$.

CONVOLUTION: CIRCULAR AND BLOCK

OVERVIEW

This set of projects will concentrate on the circular convolution property of the discrete Fourier transform (DFT). The relation to linear convolution will be studied, as well as the extension to block processing and high-speed convolution. The operation of N-point circular convolution is defined by the equation

$$y[n] = x[n] \circledN h[n] = \sum_{\ell=0}^{N-1} x[\ell]h[(n-\ell) \bmod N] \qquad \text{(0-1)}$$

Note that circular convolution combines two N-point vectors to give an answer that is also an N-point vector.

In its own right, circular convolution has little or no use. However, it is a by-product of the DFT and is, therefore, easy to compute via the FFT. The reason that the study of circular convolution is an essential part of DSP is that it can be related in a simple manner to normal convolution, $x[n] * h[n]$, which will be called linear convolution here to distinguish it from $x[n]$ Ⓝ $h[n]$. Therefore, the FFT can be used to speed up the computation of a linear convolution. We will study that important connection in detail in several exercises. In MATLAB it is sometimes convenient to express convolution as a matrix-vector multiplication [13], so some background reading in linear algebra and matrix theory would be useful (see [10] and [11]).

PROJECT 1: CIRCULAR INDEXING

Combining N-point vectors according to the circular convolution rule (0-1) becomes easy to visualize with some experience and MATLAB offers the means to do the visualization. Circular convolution and the DFT both require that all indexing be done in a circular (or periodic) fashion. Thus a shifting operation becomes a rotation. In this project we break the circular convolution operation down into its elements: circular shifting and circular flipping.

Hints

Using a simple test signal, such as a ramp, makes it easy to track the circular shifts. Plotting two signals together, via `subplot(21x)`, helps visualize the action done by the shift.

EXERCISE 1.1

Circular Shifts and Rotations

Indexing for the DFT must always be performed in a "circular" fashion. Thus the expression $x[n-1]$, which usually means "shift right by one sample," must be reinterpreted as a rotation by one sample; similarly, the flip operation $x[-n]$, becomes a circular flip $x[(-n) \bmod N]$.

a. To preserve the Fourier property that says "a shift in one domain is multiplication by a complex exponential in the other," we must define the shift using the modulo operator (from number theory).

$$x[n-\ell] \rightarrow x[(n-\ell) \bmod N] = \begin{cases} x[n-\ell] & \text{for } n = \ell, \ell+1, \ldots, N-1 \\ x[n+N-\ell] & \text{for } n = 0, 1, \ldots, \ell-1 \end{cases}$$

This assumes that $0 \leq \ell < N$. The operation is referred to as a circular shift because as the sequence is shifted to the right, indices greater than or equal to N are wrapped back into the smaller indices.

The DFT of a circularly shifted sequence is just the original DFT multiplied by a complex exponential vector, $W_N^{+\ell k}$. Verify this property by using simple test inputs such as shifted impulses, shifted pulses, and shifted sinusoids. For a 16-point DFT, show that a circular shift by 8 is a special case where the transform values will only have their signs modified.

b. Write a function that will compute $n \bmod N$. The `rem` function in MATLAB is not sufficient because it will not handle the case where n is negative. However, a simple modification that uses two calls to `rem` will guarantee an answer in the range $[0, N-1]$. Take advantage of the fact that $(n + N) \bmod N = n \bmod N$.

c. Write a MATLAB function `cirshift` that will shift a vector circularly by ℓ places. The function should work for ℓ greater than N and also handle the case where ℓ is negative by shifting in the opposite direction. Consider how this negative shift could be done by an equivalent positive rotation.

d. Given the sequence `[ 0 1 2 3 4 5 6 7 ]`, rotate this to the new sequence `[ 4 5 6 7 0 1 2 3 ]`. Instead of using `cirshift`, do the computation with only DFT and complex multiply operations. Repeat for a rotation to `[ 2 3 4 5 6 7 0 1 ]`.

EXERCISE 1.2

Circular Flip

The operation of circular flipping was discussed in Project 3 in the section on *DFT Properties*, where a function called `cflip` was developed. When the flip operation $x[-n]$ is interpreted in a circular fashion, the index replacement, $n \to -n$, becomes $(-n) \bmod N$. This is called a circular flip, because the index $n = 1$ is exchanged with $n = N-1$, $n = 2$ with $n = N-2$, and so on. The index $n = 0$ does not move.

Write a function called `cflip()` that will implement this flip operation. Verify that `cflip( [0:1:7] )` returns `[ 0 7 6 5 4 3 2 1 ]`.

EXERCISE 1.3

Flipping and Shifting Signals

In this exercise you should generate the flipped and rotated vectors as found in the circular convolution sum. All indexing within circular convolution is done to stay within the index range $0 \le n < N$. For example, in (0-1) the difference $(n - \ell) \bmod N$ is needed for the signal $h[\cdot]$.

Consider the signal $h[(n - \ell) \bmod N]$ in the circular convolution sum, as a function of ℓ. Starting with the vector for $h[\cdot]$, two steps are needed to construct $h[(n - \ell) \bmod N]$ versus ℓ: a circular flip and a circular shift by n.

a. Use the two functions `cflip` and `cirshift` to produce examples of sequences that are flipped and shifted, circularly of course. In each case, make a two-panel subplot to compare with the original [see `subplot(21x)`].

   1. Start with the 11-point sequence, $x[n] = 2n + 3$.
   2. Plot $x[(\ell - 2) \bmod 11]$ versus $\ell = 0, 1, \ldots, 10$.
   3. Plot $x[(\ell + 3) \bmod 11]$ versus ℓ.
   4. Plot $x[(4 - \ell) \bmod 11]$. Should you flip first and then shift, or vice versa?
   5. Plot $x[(-\ell - 5) \bmod 11]$. If the flip is first, will the shift be by $+5$ or -5?

b. Generate an exponential signal $x[n] = (0.87)^n$ that is 13 points long, and generate the list of 13 indices `nn = 0:12` for n. Perform a circular shift of $x[n]$ to get $y[n] = x[(n-4) \bmod 13]$. Do this by shifting the index vector only, and then plotting $x[n]$ versus n-shifted with `stem`. Which way do you have to rotate `nn`?

c. Repeat for $z[n] = x[(2n + 3) \bmod 13]$. Is this a shift to the "right" or the "left"?

PROJECT 2: CIRCULAR CONVOLUTION

There are several ways to compute a circular convolution besides the sum given in (0-1). Most important among these is calculation using the FFT, which leads to a very efficient implementation.

Hints

Linear convolution can be done in MATLAB with the function `conv`.

EXERCISE 2.1

Function for Circular Convolution

There are two ways to write a function for circular convolution: (1) in the transform domain, or (2) in the time domain. The MATLAB function would need three inputs, the signal vectors `h` and `x`, and `N`, the length of the circular convolution; it would return one output signal `y`.

a. The circular convolution of $x[n]$ and $h[n]$ is equivalent to multiplication of their DFTs, $X[k] \cdot H[k]$. Use this idea to write a circular convolution function that requires three calls to `fft`. Try the following simple tests for $N = 16$ and $N = 21$.

   1. An impulse at $n = a$ convolved with an impulse at $n = b$, where a and b are integers. That is, $x[n] = \delta[(n - a) \bmod N]$ and $h[n] = \delta[(n - b) \bmod N]$. In this case, the output has only one nonzero value, so determine its location. Let $b = -a \bmod N$, for a simple test case.
   2. Two short pulses. Let $x[n]$ be a pulse of length 5 and $h[n]$ a pulse of length 8 starting at $n = 4$. Verify that your function computes the correct output.
   3. Two long pulses such that the output wraps around. Let $x[n]$ be a pulse of length 11 and $h[n]$ a pulse of length 7 starting at $n = 5$. Compute the output and check its correctness versus hand calculation. Explain why the answer is different for the length-16 and length-21 cases.

b. Write a circular convolution function directly from the definition (0-1). This can be done with `for` loops, but good MATLAB programming style demands that vector operations be used instead. Since each output is formed from the inner product of `x` with a circularly flipped and shifted version of `h`, only one loop is needed. Write another circular convolution function based on this idea, and check it on some of the examples above.

c. A refinement of the inner-product approach would be to do the circular convolution as a matrix-vector multiply. Use the circularly shifted versions of $h[n]$ to construct a "circulant" matrix—one whose columns are all just rotations of the first column. A square circulant matrix has only N distinct elements—it is completely defined by its first column (or its first row); see the following example for $N = 5$.

$$\mathbf{C} = \begin{bmatrix} 1 & 5 & 4 & 3 & 2 \\ 2 & 1 & 5 & 4 & 3 \\ 3 & 2 & 1 & 5 & 4 \\ 4 & 3 & 2 & 1 & 5 \\ 5 & 4 & 3 & 2 & 1 \end{bmatrix}$$

Write a function to construct a circulant matrix, and then call that function when computing the circular convolution. Again, check versus the examples from part (a).

EXERCISE 2.2

More Examples of Circular Convolution

The following signals should be combined by circular convolution. Hand calculation for these examples is not difficult, so they can be used to check out the different M-files written in Exercise 2.1. In parts (a)–(c), try both even and odd lengths for N, say $N = 16$ and $N = 21$.

a. Let $x[n] = 1$ for all n, and $h[n] = (-1)^n$. Notice the difference in the even and odd cases. When $N = 16$, can you find other signals $x[n]$ for which the output will be zero? When N is odd, is it possible to find an $x[n]$ for which the output will be zero?

b. Let $x[n]$ be a ramp: $x[n] = n$, and let $h[n] = \delta[n-3] - \delta[n-4]$. Verify that your function computes the correct output.

c. Two periodic pulse trains. Let $x[n]$ be a nonzero constant only when n is a multiple of 3; and let $h[n]$ be nonzero only for n as a multiple of 4. Compute the output and check its correctness. Explain why the answer is different for the length 16 and 21 cases.

d. Generate the signal $x[n] = (-1)^n + \cos(\pi n/2)$, for $n = 0, 1, \ldots, 49$. Generate another signal $h[n]$ as a finite pulse of length 11. Compute the 50-point circular convolution of these two signals; zero-pad $h[n]$ with 39 zeros. Verify your MATLAB answer by checking versus a paper and pencil calculation.

EXERCISE 2.3

Circular Deconvolution

Suppose that $y[n] = x[n] \circledN h[n]$, where the convolution is circular. If $y[n]$ and $h[n]$ are known, recovering $x[n]$ is called the "deconvolution" problem. It is difficult to solve in the time domain, but rather easy in the DFT domain. Consider the following puzzle:

Nine Tech students are seated around a circular table at dinner. One of the students (a DSP expert) challenges the others to a guessing game to demonstrate her "magic" powers. While she leaves the room, each of the other eight students asks those seated to his/her immediate right and left what their IQs are, adds it to his/her own and reports the sum of three IQs. When she returns, these eight sums are given to the DSP student to work her magic. The game is to use just these eight partial sums to determine each student's IQ.

a. The DSP magician is confident that she will be able to solve the puzzle, because it is "just circular deconvolution." Show how to *model* this puzzle as a circular convolution. Call the sums $s_3[n]$, and the IQs $q[n]$, for $n = 0, 1, 2, \ldots, 7$. State an algorithm, based on the DFT, for determining $q[n]$ from $s_3[n]$ and prove that it will work in all cases. Write a simple MATLAB M-file that will solve the puzzle; it should also generate a test case for $s_3[n]$. For example, use randomly chosen IQs.

b. To demonstrate her powers further, the DSP student challenges another student to play the same guessing game but with a slight modification in the rules. This time, each of the remaining eight students will ask only the persons to their left for their ages, and these sums will be reported to the group. Show that, in this case, it might not be possible to get the answer. Use MATLAB to generate a specific counter-example to *prove* that a solution is not generally possible for the case where the sums are taken over two people. Let the age be denoted as $g[n]$, for $n = 0, 1, 2, \ldots, 7$ and the two-person sums as $s_2[n]$.

PROJECT 3: RELATION TO LINEAR CONVOLUTION

Circular convolution is most useful because it can be related to linear convolution, $y[n] = x[n] * h[n]$, which is the normal operator that applies to linear time-invariant systems. The exercises in this project show how circular convolution can be viewed as a "time-aliased" version of linear convolution. Then methods of zero padding are used to make circular convolution give the same answer as linear convolution.

Hints

In MATLAB there is a function called `convmtx` which produces a linear "convolution" matrix. The function `convolm( x, M, '<>' )` in Appendix A is similar, but it also allows for zero padding at one or both ends of the signal `x`.

When computing the length of a signal in MATLAB, it is natural to use the `length` function. However, this function just determines the number of elements in a vector, so it must be applied before zero padding to get a correct signal length.

EXERCISE 3.1

Study the conv Function

In MATLAB, convolution can be performed by the function `conv`. Any two finite-length signals can be convolved.

a. Suppose that two finite-length signals, $x[n]$ and $h[n]$, are convolved to give a result $y[n]$. If the length of $x[n]$ is L_x and the length of $h[n]$ is L_h, determine the length of $y[n]$.

b. Demonstrate the `conv` function by doing the convolution of two rectangular pulses. The expected result is a signal that has a trapezoidal shape. Do this with some examples where the signals have randomly chosen lengths, and verify the length constraint from part (a).

EXERCISE 3.2

Convolution as a Matrix Operation

The operation of convolving two finite-length signals can be represented as a matrix-vector product involving a circulant matrix, which is a special case of a Toeplitz matrix. Interpretation of the convolution sum in this way leads to the "convolution matrix," which is a rectangular $(N \times p)$ Toeplitz matrix whose first row is all zeros except for the first element, and whose first column has zeros in its last $p-1$ elements.

$$\mathbf{H} = \begin{bmatrix} 1 & 0 & 0 & 0 \\ 2 & 1 & 0 & 0 \\ 3 & 2 & 1 & 0 \\ 4 & 3 & 2 & 1 \\ 0 & 4 & 3 & 2 \\ 0 & 0 & 4 & 3 \\ 0 & 0 & 0 & 4 \end{bmatrix}$$

The convolution of $x[n]$ and $h[n]$ can be done by making one of the signals the nonzero part of the first column, and letting the other signal be a vector that multiplies the convolution matrix.

$$\mathbf{y} = \mathbf{H}\mathbf{x} \tag{3-1}$$

a. Do the convolution of a three-point ramp with a seven-point ramp, by constructing the convolution matrix. What are the dimensions of the matrix? Check with the output of `conv`.

b. Suppose that we wanted to perform deconvolution based on the matrix representation of convolution (3-1). Thus we assume that $\mathbf{y}$ and $\mathbf{H}$ are known. Since (3-1) always represents an underdetermined system (i.e., more equations than unknowns, $\mathbf{x}$), the answer will never be unique. In MATLAB the backslash operator is still able to compute a solution for this case. For the case of the two ramps in part (a), let the seven-point ramp be $h[n]$. Apply the backslash operator to see how well the inversion of (3-1) can be done (i.e., compare the result to the expected three-point ramp). This is a noise-free case, so it does not test the robustness of the inversion. To do so, you need to recompute after adding a bit of Gaussian random noise to the $\mathbf{y}$ vector. Use `randn` and scale the standard deviation of the noise to be between 1 and 10% of the maximum signal height (see `help randn`). This will test the robustness of the deconvolution and the sensitivity of the inversion process to small errors in the data.

EXERCISE 3.3

Circular Convolution via Time Aliasing

In this exercise we consider how to modify a linear convolution to obtain a circular convolution. This might seem counterproductive since circular convolution is not a desired operation, but this exercise will allow us to establish the relationship between these two types of convolution.

One way to use `conv` to compute a circular convolution is to take the results of `conv` and perform a time aliasing on the vector.

$$\tilde{y}[n] = \sum_{\ell} y[n+\ell M]$$

The effect of the summation is to produce a value of $\tilde{y}[n]$ by adding up all values of $y[n]$ that are offset by M. The constructed signal $\tilde{y}[n]$ will be periodic with period M. Since all three signals in the circular convolution must have the same length, we would pick M in the time aliasing to be that length.

a. Write a MATLAB function that will perform time aliasing: `y = time_alias( x, M )`. The input parameter M specifies the distance at which aliasing occurs. The output $y[n]$ should have the same length as the input vector but should have a period equal to M.

b. Return to the example of convolving two pulses, and pick both to have a length of 7. Perform a nine-point circular convolution with `conv` followed by time aliasing. This requires that just nine points be taken from the time-aliased output. Check against the results obtained with one of the circular convolution M-files written previously.

c. Repeat for circular convolution lengths of 11, 13, and 16. When does the time aliasing cease to have an effect?

EXERCISE 3.4

Circular Convolution via Periodic Extension

A second way to use `conv` to compute an N-point circular convolution result is via what should be called *periodic convolution*. In this case, three steps are needed:

1. One of the input signals $x[n]$ is extended in a periodic fashion from length N to length N', say $x[n]$. The new signal is called $\tilde{x}[n]$.
2. The `conv` function is applied to compute $\tilde{y}[n] = h[n] * \tilde{x}[n]$. The output length becomes greater than N'.
3. Finally, N points are selected from the output.

a. Determine the minimum length N' needed for the periodic extension.

b. Determine which points to take from $\tilde{y}[n]$ by discarding the endpoints of the linear convolution where $h[n]$ partially overlaps $\tilde{x}[n]$.

c. Write an M-file function that will implement this approach to circular convolution. If the nonzero length of $x[n]$ is less than N, you must zero-pad prior to the periodic extension. Test this M-file on the case where both $x[n]$ and $h[n]$ are pulses, perhaps of different length. This is a simple test case, because the answer is known easily.

EXERCISE 3.5

Zero Padding

As suggested by the periodic-extension method of Exercise 3.4, zero padding can be used to make circular convolution give a correct result for linear convolution. All that is needed is to make the length of the circular convolution long enough so that time aliasing does not come into play.

a. Generate two random signals; the signal $x[n]$ should be length 50, the other signal $h[n]$ length 27. What is the length of $y[n] = x[n] * h[n]$?

b. Compute the N-point circular convolution of these two signals, where N is the length of $y[n]$. Verify that the answer for circular convolution matches the linear convolution.

c. A longer circular convolution will also contain the correct result. Use FFTs to do a circular convolution of length 128. Zero pad both $x[n]$ and $h[n]$. Multiply the DFTs, $X[k]$ and $H[k]$, and then inverse transform to get $\hat{y}[n]$.

d. Verify that $\hat{y}[n]$ contains all the nonzero values of $y[n] = x[n] * h[n]$.

Therefore, we see that any circular convolution longer than the minimum will work.

EXERCISE 3.6

Good Outputs versus Bad Outputs

In the previous examples, the comparison of the circular convolution output to a linear convolution output shows that they are not always the same. Some values may be correct, while others are wrong because of time aliasing.

In the overlap-save method of block convolution, it will be important to identify these good and bad points. So we consider two cases—one with zero padding, one without.

a. Suppose that we convolve (circularly) the signals

$$x[n] = \begin{cases} (0.9)^n & 0 \le n < 13 \\ 0 & \text{elsewhere} \end{cases} \qquad h[n] = \begin{cases} 1 & 0 \le n < 12 \\ 0 & \text{elsewhere} \end{cases}$$

Choose the length of the circular convolution to be $N = 21$. Determine which values of $\hat{y}[n] = x[n] \circledN h[n]$ are the same as those in the linear convolution result $y[n] = x[n]*h[n]$. Give the list of output indices where the values are "bad."

b. Suppose the two signals are defined as

$$x[n] = \begin{cases} (0.9)^n & 0 \le n < 13 \\ 0 & \text{elsewhere} \end{cases} \qquad h[n] = \begin{cases} 1 & 9 \le n < 21 \\ 0 & \text{elsewhere} \end{cases}$$

Where are the good and bad points now that $h[n]$ has zeros at the beginning?

c. Consider the following example, which relates to the overlap-save situation.

$$x[n] = \begin{cases} 1 & 0 \le n < 17 \\ 0 & \text{elsewhere} \end{cases} \qquad h[n] = \begin{cases} \sin(n\pi/13) & 0 \le n < 100 \\ 0 & \text{elsewhere} \end{cases}$$

Suppose that a 100-point circular convolution is performed. (There is no zero padding of $h[n]$.) Determine the indices of the good and bad points in the output.

PROJECT 4: BLOCK PROCESSING

The case of convolving two short signals is not very useful for filtering. In continuous filtering, the input $x[n]$ would be extremely long, at least with respect to the filter's impulse response $h[n]$. Furthermore, the entire input may not be available at one time, so it is not feasible to do the circular convolution with a length that is greater than $L_x + L_h - 1$. Instead, a more reasonable strategy is to chop up the input into blocks and process each one through a circular convolution. This is the basic idea that leads to FFT convolution.

Hints

There are two types of block processing algorithms: overlap-add and overlap-save. Within MATLAB the function `fftfilt` implements the overlap-add method to do long convolutions. However, in the following exercises you should develop your own M-files to implement these two methods of block convolution.

EXERCISE 4.1

Overlap-Add

The overlap-add method works by breaking the long input signal into small nonoverlapping sections. If the length of these sections is M and the length of the impulse response is L_h, a circular convolution length of $N > M+L_h-1$ will avoid all time-aliasing effects (through the use of zero padding). However, each piece of the output is now longer than M. To put the output together it is necessary to add together the overlapping contributions from each segment. Therefore, this method could also be called the "overlap-outputs" method. The description here is brief, so further details must be found in a DSP textbook.

a. The following code was adapted from the MATLAB function `fftfilt`, which implements the overlap-add method. Point out where in this code the overlap-add is taken care of. Notice that the section length is not the same as the FFT length.

```
H = fft(h,Nfft);
M = Nfft - length(h) + 1;        %--- Section Length
%
%******* assume that length(x) is multiple of M *******
%
for ix = 1:M:length(x)
     x_seg = x(ix:ix+M-1);              %--- segment x[n]
     X= fft(x_seg, Nfft);               %--- zero pads
     Y = X.*H;
     y_seg = ifft(Y);
     y(ix:ix+Nfft-1) = y(ix:ix+Nfft-1) + y_seg(1:Nfft);
end
%
%------ check for purely REAL case -----------
if ~any(imag(h)) & ~any(imag(x))
     y = real(y);
end
%-------------------------------------------
```

b. Write an M-file function that implements the overlap-add method of block convolution. One of the inputs to this function should be either the section length M or the FFT length. The circular convolution should be done in the DFT domain. Ultimately, this will provide the fastest running time for your program.

c. Test the function by comparing to a convolution done with `conv`. Use a long vector for `x` and compare the running time of the overlap-add function to `conv` in terms of FLOPs and elapsed time (see `etime`).

EXERCISE 4.2

Overlap-Save

The overlap-save method uses a different strategy to break up the input signal. If the length of the circular convolution is chosen to be N, input segments of length N are taken. The starting location of each input segment is skipped by an amount M, so there is an overlap of $N-M$ points. Thus this method could be called the *overlapped inputs* method.

The filter's impulse response is zero-padded from length L out to length N and an N-point circular convolution is computed using an N-point FFT. Using the idea of good and bad points as in Exercise 3.6, it is possible to identify $M = N-L+1$ good points in the circular convolution result. These M points are then inserted into the output stream. No additions are needed to create the output, so this method is often preferred in practice.

a. Write a MATLAB function to implement the overlap-save method of block convolution. Either the block length M or the circular convolution length N must be specified. It is best to specify N since it should usually be chosen as a power of 2 to exploit the FFT algorithm.

b. Test this function versus `conv` for correctness.

c. Use MATLAB to count the number of floating-point operations via `flops`, and compare to the overlap-add method and to `conv`. Make a table for $L_h = 10, 20, 50$ and $L_x = 1000$; use $N = 128$.

EXERCISE 4.3

Breaking Up Both Sequences

In some rare cases, both sequences, $x[n]$ and $h[n]$, may be too long for the FFT (e.g., if a hardware FFT is being used). Then the circular convolution length N would be fixed by the hard-wired length of the FFT. If both sequences are longer than N, both must be segmented prior to block processing.

One strategy for the segmentation algorithm is to use a loop around either the overlap-add or overlap-save method. In this case, the approach is to break off part of $h[n]$ and convolve it with the entire signal $x[n]$; then break off another piece and convolve. Finally, all the convolutions would be added together.

a. Take the specific case where $N = 32$, $L_x = 68$, and $L_h = 60$. If an overlap-add strategy is used, and the section length is chosen to be 12 for segmenting $h[n]$, determine a section length for $x[n]$, and how many 32-point circular convolutions must be done.

b. Implement this method in an M-file, but try to write the program so that the segment length for each signal is a variable. Experiment with different block lengths.

c. If the performance objective were to minimize the total number of 32-point circular convolutions, determine the best choice of section lengths.

PROJECT 5: HIGH-SPEED CONVOLUTION

Circular convolution implemented in the transform domain together with the FFT algorithm is a powerful combination that yields an extremely fast method for convolution. The best choice for an FFT length is a power of 2, so it is best to choose the block lengths of $x[n]$ and $h[n]$ accordingly. However, should one use the minimum power of 2, or something larger? When would it be better to use direct convolution? A plot of operation counts versus FFT length or versus filter length will give the answer.

Hints

Use `etime` and `clock` to measure elapsed time; or `flops` to count operations.

EXERCISE 5.1

FFT `conv` Function

The `conv` function in MATLAB actually calls `filter` to do its computation. This is efficient when the lengths of the signals to be convolved are small. However, if the signal lengths are long (e.g., > 30), `filter` is quite slow and a convolution based on the FFT should be used. In fact, such an option is available in MATLAB with the `fftfilt` function. However, for this exercise you must write an M-file function that could replace the existing `conv` function. Consider the following requirements for this function:

1. The new convolution function should implement the overlap-save method that calls the FFT to perform circular convolution.

2. The FFT length should be a power of 2, and must be one of the input arguments to the new convolution function.
3. The longer input signal ($x[n]$ or $h[n]$) should be the one segmented in the overlap-save algorithm.
4. The function must return a vector that has the correct length, not one with extra zeros tacked on.

After doing the next exercise, you may want to modify this function so that it uses the FFT only when that would be more efficient; otherwise, it would call `conv`.

EXERCISE 5.2

Crossover Point

The first experiment to run involves a comparison of FFT convolution versus direct convolution, as implemented with `conv`. For extremely short convolutions, the direct convolution is more efficient, but for longer ones the $\log_2 N$ behavior of the FFT makes it much faster. A generally quoted number for the crossover point where the FFT has fewer operations is $N \approx 32$ if you look for the closest power of 2. However, this number is a function of the precise implementation, so we would like to deduce MATLAB's crossover point.

In this exercise we construct the plot by comparing the running time of the FFT convolver from Exercise 5.1 to `conv` (see also Problem 9.34 in [5]).

a. Generate two signals $x[n]$ and $h[n]$, both of length L. The length will be varied over the range $10 \leq L \leq 80$. This range might have to be adjusted if the running time of the FFT on your computer is too slow.

b. Convolve $x[n]$ with $h[n]$ using the `conv` function. Have MATLAB measure the number of floating-point operations. Save these values in a vector for plotting later. Since the FFT can do complex-valued operations, the generated signals should be complex. This will affect the FLOPs counted in `conv`.

c. Now do the convolution by using an FFT that is long enough to contain the entire result of the linear convolution. Use the next higher power of 2. The convolve function written in Exercise 5.1 could be used here. Again, measure the FLOPs and save in a vector.

d. Plot the two vectors of FLOP counts together. Determine the crossover point where the FFT method is faster. Note that there is actually not just one crossover, but it is possible to determine a length beyond which the FFT is almost always better.

e. If you have time, make a true comparison for real-valued signals. This will involve writing a MATLAB function for a conjugate-symmetric IFFT—one that exploits the real-valued nature of the output. The forward FFT in MATLAB already has simplifications for the real-valued input case.

EXERCISE 5.3

Compare with FIR Filtering

In Exercise 5.2 we concentrated on an artificial situation where the lengths of $x[n]$ and $h[n]$ are identical. A more likely case is that of continuous filtering. Here, one of the inputs is indefinitely long—it could be a continuous stream of data. Therefore, the block length for an overlap-add or overlap-save algorithm needs to be chosen. Since the FFT length is arbitrary, there is some flexibility in choosing the block length. Increasing the block length beyond the minimum will make the process more efficient, but only up to a point! As a rule of thumb, it turns out that a block length on the order of 5 to 10 times the filter length is a good choice. To see that there is an optimum choice, we can construct a plot of operation counts versus FFT length.

a. Take the filter length to be $L_h = 40$. To simulate the very long input, make the length of $x[n]$ as large as possible in your version of MATLAB: greater than 20,000 if possible. The two signals can be generated with random numbers (see `rand`).

b. Since the FFT can operate on complex data, we must be fair (to the FFT) and do the comparison for a complex-valued convolution. Otherwise, we should use a modified IFFT algorithm that is simplified for conjugate-symmetric data. Therefore, generate two complex-valued random signals and run the comparison.

c. Use the function from Exercise 5.1, or use `fftfilt`, to do the convolution in sections. One of the input parameters is the section length, so start with $N = 64$ and try successively higher powers of 2. If you put this process in a loop, and print out the length of the FFT each time through the loop, you can see roughly how long it takes each time and gauge which FFT length is best.

d. Measure the FLOPs for each N and collect in a vector. Convert to operations per output point and plot versus $\log_2 N$. For comparison, do the filtering with `conv` and count the FLOPs. Convert to operations per output and plot this number as a horizontal line.

e. Repeat this experiment for longer and shorter filter lengths, L_h. Do enough cases to verify that the rule of thumb stated above is correct. Notice also that several lengths near the optimum yield about the same performance.

RELATED TRANSFORMS

OVERVIEW

This set of projects will introduce two other transforms that are closely related to the DFT: the discrete cosine transform (DCT) [2] and the discrete Hartley transform (DHT) [14]. Properties of these two transforms and their relation to the FFT are explored. Some approaches for computing their values via the FFT are described in the exercises. Readers interested in doing an in-depth study of these transforms should consult [2] and [14] for more details about these alternative transforms.

There are many other discrete orthogonal transforms [15] that could also be studied with MATLAB. For example, `hadamard` in MATLAB will generate a matrix for the Hadamard transform.

PROJECT 1: DISCRETE COSINE TRANSFORM

The discrete cosine transform (DCT) has found widespread use in coding applications. It is now part of the JPEG standard for image coding. Originally, the DCT was developed as an approximation to the optimal Karhunen-Loève transform. Since then, numerous fast algorithms have been developed for its computation. The theory of the DCT is not usually found in an elementary DSP textbook, so the book by Rao and Yip [2], or their chapter [15] in [16], should be consulted for more in-depth information.

There are four types of even DCTs, all of which can be written in the form[3]

$$C_x[k] = \sum_n \phi_{kn}\, x[n] \tag{1-1}$$

[3]There are four additional DCTs, called odd DCTs, because the argument of the cosine function has a factor of $2N - 1$ in the denominator.

where the basis functions, ϕ_{kn} versus n, are defined in terms of cosine functions:

$$\text{DCT-I} \qquad \phi_{kn} = \sqrt{\frac{2}{N}}\left[c_k\, c_n \cos\left(\frac{k\,n\,\pi}{N}\right)\right] \qquad n, k = 0, 1, 2, \ldots, N \tag{1-2}$$

$$\text{DCT-II} \qquad \phi_{kn} = \sqrt{\frac{2}{N}}\left[c_k \cos\left(\frac{k(n+\frac{1}{2})\pi}{N}\right)\right] \qquad n, k = 0, 1, 2, \ldots, N-1 \tag{1-3}$$

$$\text{DCT-III} \qquad \phi_{kn} = \sqrt{\frac{2}{N}}\left[c_n \cos\left(\frac{(k+\frac{1}{2})n\,\pi}{N}\right)\right] \qquad n, k = 0, 1, 2, \ldots, N-1 \tag{1-4}$$

$$\text{DCT-IV} \qquad \phi_{kn} = \sqrt{\frac{2}{N}}\left[\cos\left(\frac{(k+\frac{1}{2})(n+\frac{1}{2})\pi}{N}\right)\right] \qquad n, k = 0, 1, 2, \ldots, N-1 \tag{1-5}$$

and where the factor c_ℓ is

$$c_\ell = \begin{cases} 1/\sqrt{2} & \text{if } \ell = 0 \bmod N \\ 1 & \text{if } \ell \neq 0 \bmod N \end{cases} \tag{1-6}$$

Equation (1-1) defines a matrix transformation, $\boldsymbol{\Phi} = [\phi_{kn}]$, from the time vector with entries $x[n]$ to the transform vector with entries $C_x[k]$. The DCT-I is an $(N+1) \times (N+1)$ transform; the others are $N \times N$.

EXERCISE 1.1

Basic Properties of DCT

The DCT is a purely real transform, unlike the DFT, which requires complex numbers. Some of its properties are the same as those of the DFT, but there are important differences, such as in the circular convolution property.

a. Write an M-file to implement DCT-I. Do the implementation directly as a matrix-vector multiply. In MATLAB, this requires setting up a $\boldsymbol{\Phi}$ matrix defined in (1-1) and (1-2). The factors of c_n and c_k can be implemented as pointwise multiplications of either the input or output vectors; the factor of $\sqrt{2/N}$ must also be applied to the output.

b. Show that the DCT-I is its own inverse and that the DCT-I matrix is orthogonal (its inverse is its transpose). Write an M-file for DCT-IV and show that it is a symmetric orthogonal matrix (i.e., it is also its own inverse).

c. Write M-files to implement the other types of DCTs.

d. Show that the inverse for DCT-II is not DCT-II, but rather, DCT-III. Also show that the DCT-III matrix is the transpose of the DCT-II matrix.

e. Determine whether or not the DCT satisfies the circular convolution property: "The product in the transform domain is equivalent to circular convolution in the signal domain." For DCT-II, let $C_y[k]$ be the DCT of $y[n] = x[n] \circledN h[n]$. Compute the product of the transforms, $C_x[k] \cdot C_h[k]$, and compare to $C_y[k]$. Try the other DCT types also; do any satisfy the circular convolution property?

EXERCISE 1.2

Computing the DCT via an FFT

If we write the basis functions of the DCT-I as

$$\cos\left(\frac{kn\pi}{N}\right) = \cos\left(\frac{2\pi kn}{2N}\right) = \text{Re}\left\{W_{2N}^{nk}\right\}$$

it is easy to see that the DCT can be constructed from a $2N$-point DFT. Let's consider the case of DCT-II in detail. In this case, the basis functions are

$$\phi_{kn} = \left[c_k \cos\left(\frac{k(n+\frac{1}{2})\pi}{N}\right)\right] = c_k \,\text{Re}\left\{ W_{2N}^{k(n+\frac{1}{2})} \right\}$$

If we expand the exponent and plug ϕ_{kn} into the DCT-II definition, we get

$$C_x^{II}[k] = c_k \sqrt{\frac{2}{N}}\,\text{Re}\left\{ W_{2N}^{k/2} \sum_{n=0}^{2N-1} x[n]\, W_{2N}^{nk} \right\} \qquad \text{for } k = 0, 1, \ldots, N-1 \tag{1-7}$$

Therefore, the DCT-II can be computed using the following three steps:

1. Zero-pad $x[n]$ out to a length of $2N$, and compute its FFT. Remember that $x[n]$ starts out as a real-valued length-N sequence, so the FFT output will be conjugate-symmetric.
2. Take the first N outputs of the FFT and multiply them by $W_{4N}^{k} = W_{2N}^{k/2}$.
3. Take the real part and then multiply it by c_k and by $\sqrt{2/N}$.

This approach is preferred in MATLAB because it takes advantage of the built-in `fft` function. Obviously, the same approach can be used for the other three types of DCTs.

a. Implement M-files for all four types of DCTs based on this fast computation strategy.

b. Test these against the direct method of computation implemented in Exercise 1.1.

c. A second approach to computing via the FFT is to create a length-$2N$ input to the FFT so as to eliminate the real-part operator. This is done for the DCT-II case by defining

$$\tilde{x}[n] = \begin{cases} x[n] & \text{for } n = 0, 1, \ldots, N-1 \\ x[2N-1-n] & \text{for } n = N, N+1, \ldots, 2N-1 \end{cases} \tag{1-8}$$

Then the $2N$-point FFT output only needs to be multiplied by $\sqrt{2/N}\, c_k\, W_{4N}^{k}$. Implement this method for DCT-II and verify that it gives the same result as the other two previous implementations. Explain why the signal defined in (1-8) will eliminate the need for the real-part operation in (1-7).

d. The fact that we must use a length-$2N$ FFT is bothersome. A third way to approach the fast computation can reduce the FFT length to N. In this case, the signal $x[n]$ is packed into a new N-point vector as

$$y[n] = \begin{cases} x[2n-1] & \text{for } n = 1, 2, \ldots, N/2 \\ x[N-2n] & \text{for } n = N/2+1, \ldots, N-1 \\ x[0] & \text{for } n = 0 \end{cases} \tag{1-9}$$

Then we compute the N-point FFT and multiply the result by W_{4N}^{k}; call this result $Y[k]$. Now we must extract the real and imaginary parts to create a vector that is almost the DCT.

$$\tilde{Y}[k] = \begin{cases} \text{Re}\{Y[k]\} & \text{for } k = 0, 1, \ldots, N/2 \\ \text{Im}\{Y[N-k]\} & \text{for } k = N/2+1, \ldots, N-1 \end{cases} \tag{1-10}$$

The DCT-II is obtained by multiplying $\tilde{Y}[k]$ by $\sqrt{2/N}\, c_k$. Implement this method and compare the number of FLOPS to the previous computations with the $2N$-point FFT. Explain how the definitions in (1-9) and (1-10) implement a strategy that is similar to computing the FFT of a purely real sequence with a half-length FFT (see Project 6 in the section *DFT Properties*).

EXERCISE 1.3

Discrete Sine Transform

Similar to the DCT, we can define a discrete sine transform (DST) which has eight forms; the four even DSTs are given below. In this case, the basis functions ϕ_{kn} take the following form:

$$\text{DST-I} \qquad \phi_{kn} = \sqrt{\frac{2}{N}}\left[\sin\left(\frac{kn\pi}{N}\right)\right] \qquad n, k = 1, 2, 3, \ldots, N-1 \tag{1-11}$$

$$\text{DST-II} \qquad \phi_{kn} = \sqrt{\frac{2}{N}}\left[\sin\left(\frac{k(n+\frac{1}{2})\pi}{N}\right)\right] \qquad n, k = 0, 1, 2, \ldots, N-1 \tag{1-12}$$

$$\text{DST-III} \qquad \phi_{kn} = \sqrt{\frac{2}{N}}\left[\sin\left(\frac{(k+\frac{1}{2})n\pi}{N}\right)\right] \qquad n, k = 0, 1, 2, \ldots, N-1 \tag{1-13}$$

$$\text{DST-IV} \qquad \phi_{kn} = \sqrt{\frac{2}{N}}\left[\sin\left(\frac{(k+\frac{1}{2})(n+\frac{1}{2})\pi}{N}\right)\right] \qquad n, k = 0, 1, 2, \ldots, N-1 \tag{1-14}$$

The DST-I is an $(N-1) \times (N-1)$ transform; the others are $N \times N$.

a. Write M-files for the four types of DSTs. Use an approach based on the $2N$-point FFT to speed up the computation. In this case, an equation analogous to (1-7) will involve an imaginary part operator to generate the sine terms of the DST.

b. Determine the inverse transform for each type of DST. Generate examples in MATLAB to show that DST-I is its own inverse, then find the inverse of DST-II, DST-III, and DST-IV.

EXERCISE 1.4

Performance of the DCT

The DCT-II finds its main application in coding, because its performance approaches that of the Karhunen-Loève optimal transform [2]. In this exercise we show how to construct a simple experiment to illustrate this fact when the data to be coded come from a first-order Markov process. For such a process the $N \times N$ covariance matrix $\boldsymbol{R}$ has entries $r_{mn} = \rho^{|m-n|}$. The parameter ρ^k gives the degree of correlation between $x[m]$ and $x[m+k]$.

When coding a signal vector with an orthogonal transform, the transform is computed as in (1-1) and then a coding operation is performed in the transform domain. This coding operation consists of keeping a predetermined number of the largest $C_x[k]$'s and zeroing out the rest. The signal can always be reconstructed from its coded representation by doing an inverse transform on the modified (i.e., coded) $C_x[k]$'s. When coding in this fashion, there is a coding error that is equal to the difference between the original $x[n]$ and the signal reconstructed from the coded transform. The most popular measure of the coding error is the total energy in the error signal. Since the signal comes from a random process, the expected value of this energy can be computed from the correlation of the signal.

$$\text{ERROR ENERGY} = \mathcal{E}\{(x[n] - \hat{x}[n])^2\}$$

It may not be immediately obvious, but this error can be computed via the following matrix manipulation (see [15]):

1. Compute $\boldsymbol{\Phi R \Phi}^{-1}$, where $\boldsymbol{\Phi}$ is the $N \times N$ transform matrix.
2. Sort the diagonal elements of the result from largest to smallest.
3. If keeping μ transform coefficients, the error is the sum of the smallest $N - \mu$ diagonal entries.
4. Correspondingly, the sum of the largest μ diagonal entries is the signal energy retained by the coder.

Given this matrix computation of the coding error, it is possible to formulate a design problem to find the optimal coder. The answer to this problem is the Karhunen-Loève transform (KLT), which is optimal in the sense that it packs the most signal energy into the fewest transform coefficients. The KLT in matrix form $\boldsymbol{\Phi}$ is defined by using the eigenvectors of $\boldsymbol{R}$ as the columns of the transform matrix. When coding, the eigenvectors corresponding to the μ largest eigenvalues serve as the basis functions of the KLT coder.

The purpose of this exercise is to compare the performance of DCT-II to that of the KL transform. If the optimal KL method is used, the eigenvalues are a measure of how much energy is kept by the coder, because the matrix $\boldsymbol{R}$ is diagonalized by its eigenvector decomposition.

a. Generate the $N \times N$ covariance matrix $\boldsymbol{R}$ for $\rho = 0.9$; choose $N = 16$.

b. Design the KL transform using `eig`. Make a vector of the eigenvalues in decreasing order, and save for comparison to other methods.

c. Use the DCT-II as the transform matrix $\boldsymbol{\Phi}$. When $\boldsymbol{\Phi R \Phi}^{-1}$ is computed, the result is not a diagonal matrix, but the off-diagonal terms are quite small. Can you explain the meaning of the off-diagonal elements?

d. Again sort the diagonal entries of $\boldsymbol{\Phi R \Phi}^{-1}$ in decreasing order. The total energy in the first μ coefficients can be obtained by adding up the first μ diagonal entries. Do this for $\mu = 1, 2, \ldots, N$ and make a plot in order to compare both methods. Try other values for ρ in the range $0.6 \leq \rho \leq 0.99$. Does the performance of DCT-II versus the optimal KLT depend on the value of the parameter ρ?

e. Now use the DFT matrix for $\boldsymbol{\Phi}$. Explain why the resulting diagonal elements of $\boldsymbol{\Phi R \Phi}^{-1}$ should all be real. Again order these by decreasing size. Make a plot of the running sum of the largest μ terms and compare to the DCT-II and KL results. Notice that the DFT curve lies below the others until $\mu = N$.

f. Implement the other DCTs and DSTs and compare to the DCT-II and KL results.

From this experiment with a first-order Markov process, you can conclude that the DCT, which tracks the KL transform closely, is very close to optimal for this particular coding application. The result, however, is dependent on the nature of the input process, so you might repeat this experiment with a different type of correlation matrix to investigate how well the DCT-II coder would perform in other situations.

PROJECT 2: DISCRETE HARTLEY TRANSFORM

The discrete Hartley transform (DHT) is defined by the equations

$$H_x[k] = \sum_{n=0}^{N-1} x[n]\,\mathrm{cas}(2\pi nk/N) \tag{2-1}$$

$$x[n] = \frac{1}{N}\sum_{k=0}^{N-1} H_x[k]\,\mathrm{cas}(2\pi nk/N) \tag{2-2}$$

where $\mathrm{cas}(\cdot) = \cos(\cdot) + \sin(\cdot)$. It has the advantage of being a purely real transform. More detail about the properties of the DHT can be found in the books by Bracewell [14] and [3].

In our definition, the factor $1/N$ is associated with the inverse transform (2-2) to maintain consistency with the DFT. The definition given in [3] puts the $1/N$ with the forward DHT.

EXERCISE 2.1

Basic Properties of DHT

The first task is to verify the definition of the forward and inverse transforms.

a. Write two M-files to implement the DHT and the inverse DHT. Notice that the inverse transform amounts to doing a forward DHT and then multiplying by the factor $1/N$.

b. Test these functions by computing the DHT followed by its inverse to show that an original vector is obtained.

c. Consider the cas(·) signals. Prove that they are orthogonal:

$$\sum_{n=0}^{N-1} \text{cas}(2\pi nk/N)\,\text{cas}(2\pi n\ell/N) = \begin{cases} N & \text{for } k = \ell \bmod N \\ 0 & \text{for } k \neq \ell \bmod N \end{cases}$$

Demonstrate this fact in MATLAB by constructing some examples of the cas(·) signal.

d. Show that a circular reversal of $x[n]$ gives a circular reversal of its DHT.

e. The shift property is a bit trickier. If this property were analogous to the DFT, a time shift would yield a multiplication of the DHT by a cas(·) function. However, the property involves one additional term. In the DHT domain, the result is a combination of both the DHT and its circular reversal. One term is multiplied by cosine, the other by sine.

f. Show that a "Parseval" relation holds; that is, the energy summed in the frequency domain is N times the energy summed in time domain.

$$\sum_{k=0}^{N-1} H_x^2[k] = N \sum_{n=0}^{N-1} x^2[n]$$

Most of these properties can be proven mathematically, but for these exercises, MATLAB should be used to demonstrate that they are true. In other words, write an M-file to check the property and then test it on many different cases: sines, cosines, rectangular pulses, or random signals.

EXERCISE 2.2

Relation of DHT to FFT

The DHT has a simple relationship to the DFT. Since the kernel of the DFT is

$$W_N^{nk} = e^{-j2\pi nk/N} = \cos(2\pi nk/N) - j\sin(2\pi nk/N)$$

it is obvious that the DHT is just the real part minus the imaginary part of the DFT. Remember that the signal $x[n]$ is real-valued for the DHT.

a. Write an M-file that will compute the DHT with one call to the `fft` function. Verify that it gives the same answer as the DHT functions written in Exercise 2.1.

b. Now try to determine the relationship in the opposite direction. Start with the DHT of a real sequence. What operations have to be performed on $H_x[k]$ to find the FFT of that same sequence? (*Hint*: The answer will involve circular reversals of $H_x[k]$).

c. Once the DHT-to-FFT relationship has been determined, write an M-file and test it on some examples, such as sines, cosines, rectangular pulses, or random signals.

EXERCISE 2.3

Circular Convolution Property

The circular convolution property of the DFT states that multiplication of the transforms gives circular convolution in the time domain. It is one of the most useful properties of the DFT.

a. Demonstrate that the same property does *not* hold for the DHT. Try convolving some shifted impulses (e.g., $\delta[(n-5) \bmod N]$ Ⓝ $\delta[(n-3) \bmod N]$).

b. Show that circular convolution mapped to the DHT domain gives the sum and difference of four products: between the DHTs of the two signals and their reversals. This will require either a mathematical derivation or some experimentation with different combinations in MATLAB.

c. Write an M-file function that does circular convolution via the DHT. Verify that it gives the same answer as `conv` when the signals are zero-padded.

chapter 3

SPECTRUM ANALYSIS

OVERVIEW

In this chapter we present methods for the analysis of signals in the frequency domain. The material here is restricted to the deterministic case, so issues such as windowing and time-frequency analysis are most important. In the analysis of stochastic signals (see Chapter 6), other issues, such as the robustness in the presence of noise, are relatively more important.

The first set of projects treats many of the different kinds of windows that have been proposed for use in spectral analysis and filter design [1]. We study some different ways to measure the performance of a window, beyond the usual measures of mainlobe width and sidelobe height. Finally, we consider the use of windows in spectral analysis, where it is crucial to understand their performance in resolving closely spaced sinusoids.

In the second set of projects, the time-frequency representation of nonstationary signals is studied. The Fourier analysis of such signals needs to be localized and also time-dependent. In this section we concentrate on an elementary time-frequency distribution based on the FFT, called the STFT (short-time Fourier transform), which is easy to compute. When applied to speech, the STFT is called a speech spectrogram, or "voice print." Some computers have hardware for digitizing speech waveforms, so MATLAB can be used to compute and plot a spectrogram to show the changing narrowband features of speech—resonant frequency (called formants) versus time.

The third set of projects is directed at several different situations involving narrowband signals. These include sinusoids, hi-Q filter responses, and periodic pulse trains. One project explores the issue of frequency sampling, which is inherent in the use of the DFT (discrete Fourier transform). Since the DFT is the main computational tool used for numerical Fourier analysis, the frequency sampling relationship between the DFT samples and the DTFT, or between the DFT and the continuous-time Fourier transform, must be understood. This section also treats the synthesis of narrowband pulses and then uses these as test signals to explore the performance of frequency-selective IIR and linear-phase FIR filters. The degradation of pulse shape due to the nonlinear phase of an IIR filter can be demonstrated easily with these test signals.

Another class of narrowband signals are those synthesized from second-order resonators, which can be used to model the formant frequencies in a speech signal, for example. MATLAB provides an easy way to visualize the relationship between the pole locations of the second-order resonator and the time-domain or frequency-domain characteristics of the filter. Another signal found in speech processing is the periodic pulse train, which has a line spectrum. Again, MATLAB provides a simple way to show that when the signal length is finite, the lines are not impulses, but rather, have a finite spectral width inversely proportional to the signal duration.

BACKGROUND READING

The material in this chapter is not necessarily found in just one chapter of a DSP text. Material on the short-time Fourier transform can be found in [2]. In most DSP textbooks the chapter(s) devoted to the DFT, or DTFT, will treat aspects of these projects which are really applications of the DFT. Chapter 11 of [3] discusses some of these applications. In other texts, [3], [4], [5], [6], or [7], some of the material can be found in the DFT chapter(s).

[1] F. J. Harris. *On the use of windows for harmonic analysis with the discrete Fourier transform*. Proceedings of the IEEE, 66:51–83, January 1978.

[2] J. S. Lim and A. V. Oppenheim. *Advanced Topics in Signal Processing*. Prentice Hall, Englewood Cliffs, NJ, 1988.

[3] A. V. Oppenheim and R. W. Schafer. *Discrete-Time Signal Processing*. Prentice Hall, Englewood Cliffs, NJ, 1989.

[4] L. B. Jackson. *Digital Filters and Signal Processing*. Kluwer Academic Publishers, Norwell, MA, 1989.

[5] R. D. Strum and D. E. Kirk. *First Principles of Discrete Systems and Digital Signal Processing*. Addison-Wesley, Reading, MA, 1988.

[6] R. A. Roberts and C. T. Mullis. *Digital Signal Processing*. Addison-Wesley, Reading, MA, 1987.

[7] J. G. Proakis and D. G. Manolakis. *Digital Signal Processing: Principles, Algorithms and Applications*. Macmillan, New York, second edition, 1992.

[8] L. R. Rabiner and R. W. Schafer. *Digital Processing of Speech Signals*. Prentice Hall, Englewood Cliffs, NJ, 1978.

SPECTRAL WINDOWS

In these projects we study a number of different window types. By plotting their frequency response, we can compare their primary characteristics. In addition, we will introduce several metrics to quantify the performance of the different windows. Finally, we will consider the use of windows in spectral analysis where it is crucial to understand their performance in resolving closely spaced sinusoids.

The primary goal of this section is to show that a very large number of windows have been proposed, but there is also a simple way to characterize and compare their performance. A very comprehensive study of many different window types was published in [1]. A presentation of commonly used windows can be found in Chapter 7 of [3].

PROJECT 1: WINDOW TYPES

Many different kinds of windows have been proposed for use in spectral analysis and filter design. In all cases, the window acts in the time domain by truncating the length of a signal:

$$y[n] = x[n] \cdot w[n] \qquad \text{where } w[n] = 0 \quad \text{outside of } 0 \leq n \leq L-1$$

The important properties of the window are usually described in the frequency domain, where the DTFT windowing property states that the windowed signal has a DTFT that is the (periodic) convolution of the true DTFT with that of the window.

$$y[n] = x[n] \cdot w[n] \quad \stackrel{\text{DTFT}}{\longleftrightarrow} \quad Y(e^{j\omega}) = \frac{1}{2\pi} \int_{-\pi}^{\pi} X(e^{j\theta}) W(e^{j[\omega-\theta]})\, d\omega$$

The DTFT of the window is $W(e^{j\omega})$.

In this project we examine many different classes of windows and evaluate their frequency response. In succeeding projects, the performance of these windows is quantified.

Hints

The DTFT of the window is computed by sampling the DTFT $W(e^{j\omega})$ (i.e., by computing a zero-padded FFT).[1] The length of the FFT should be at least 4 to 5 times longer than the window length to get adequate sampling in frequency for plotting. Then the plot of $|W(e^{j\omega})|$ versus ω should be made with `plot`, which will connect the frequency samples and draw a continuous-looking plot. In addition, the sidelobes of $W(e^{j\omega})$ are best compared on a dB plot (see `log10`, `semilogy`, and `db`).[2] In the `db` function, the magnitude of a vector is converted to decibels after the entire vector is scaled to have a maximum amplitude of 1.

In many cases the exercises require that you plot several windows together. Use the MATLAB functions `hold on` and `hold off` to put several plots on one graph, or use the `plot` command with many arguments—one pair for each plot.

EXERCISE 1.1

Rectangular and Triangular Windows

The simplest window is the rectangular window. Anytime a signal is truncated, there is a window; if there is no weighting, the window is, in fact, the rectangular window. In MATLAB this window can be generated by `ones` or `boxcar`.

a. Generate a rectangular window of length 21. Compute its DTFT and plot the magnitude on a dB scale. The plot has regularly spaced zero crossings; explain the location of these in terms of the known transform for the rectangular window.

b. Repeat for different window lengths: 16, 31, and 61. Put all four dB plots on the same graph. Determine a formula that relates the change in the −3 dB point as a function of window length. Also measure the height of the first sidelobe for each case—observe that it remains at about the same level.

c. The triangular-shaped window is also called a Bartlett window. MATLAB has two different functions for this window, and they actually generate windows of different length. Use `bartlett` to generate a length-11 window. Plot the window samples with `stem`. Now use `triang(11)` and redo the plot. What is the actual window length (number of nonzero samples) for each case?

d. Generate the DTFT for triangular windows of length 31 and 61; and plot them on a dB scale. Is it still true that the −3 dB point is halved when the window length is doubled?

---

[1] See the special function `dtft` in Appendix A.

[2] See Appendix A for the db M-file, which thresholds the minimum value in the data to avoid the log of zero.

e. The triangular window can be related to the rectangular window in a simple fashion. Show that an odd-length triangular window (length = L) is the convolution of a length-$(L+1)/2$ rectangular window with itself. Plot the log magnitude of the DTFT of the length-L triangular window together with that of the length-$(L+1)/2$ rectangular window. Explain why the height of the rectangular window sidelobes is exactly twice that of the triangular case (in dB).

f. Determine the minimum length of a triangular window that would have the same mainlobe width (3-dB width) as a length-31 rectangular window.

EXERCISE 1.2

Window Functions in MATLAB

MATLAB has a number of built-in operators for generating windows. These are contained in the signal processing toolbox, and include `hamming`, `hanning`, `blackman`, `chebwin`, and `kaiser`. These are the Hamming window, the Hann[3] window, the Blackman window, the Dolph–Chebyshev window, and the Kaiser window, respectively.[4] The first four of these find most use in spectrum analysis applications. The Kaiser window is important for FIR filter design via the windowing method.

a. For the first three, Hamming, Hann, and Blackman, there is only one parameter to specify—the window length. All three are based on the cosine function. You can type out the functions to see the exact formula (e.g., `type hamming`). For each of the first three windows, generate a window of length $L = 31$ and plot the window coefficients $w[n]$ together on one graph.

b. To illustrate the various mainlobe widths and sidelobe heights for these windows, plot the log magnitude of the DTFT for each. Make one plot with the DTFTs of the Hamming, Hann, and Blackman windows all together. Include the DTFT of the rectangular window for reference, and mark the location $\omega = 2\pi/L$ on the graph. Make this frequency-domain plot as a dB plot, but zoom in on the region from $\omega = 0$ to $\omega = 16\pi/L$. The choice of $16\pi/L$ is arbitrary—just take a sufficient region to see the first few sidelobes. Use the colon operator to select part of the DTFT (see `help :` in MATLAB).

c. For the Dolph–Chebyshev window, there is an additional parameter that must be specified. This window offers control over the sidelobe height, and the second input argument to `chebwin` is the specified sidelobe height. In fact, the resulting sidelobes should all be the same height—called equiripple.[5] Generate three Dolph–Chebyshev windows of length $L = 31$. For the first, specify a sidelobe height of 30 dB, for the second 40 dB, and 50 dB for the third. Plot the window coefficients versus n, and compare to the Hamming window coefficients.

d. Compute the DTFTs of the three Dolph–Chebyshev windows and plot them all together with the Hamming window DTFT. The Hamming is included for comparison because its sidelobe structure is nearly equiripple (at approximately -42 dB). You should observe an inverse trade-off between sidelobe height and mainlobe width for these cases. Quantify this relationship by finding the change in 3-dB width for a drop of 10 dB in sidelobe level.

EXERCISE 1.3

Kaiser Window

The Kaiser window of length L is based on the modified Bessel function $I_0(x)$:

$$w[n] = \frac{I_0(\beta\sqrt{1-(n-M)^2/M^2})}{I_0(\beta)} \qquad \text{for } n = 0, 1, \ldots, L-1$$

---

[3]The term `hanning` is a misnomer, since this window function is attributed to von Hann.

[4]The student version of MATLAB may not have these functions, in which case they will have to be programmed from scratch.

[5]These windows could also be designed with the FIR filter design program `remez`.

The midpoint M is $M = \frac{1}{2}(L-1)$, so for an odd-length window, M is an integer. The parameter β should be chosen between 0 and 10, for useful windows. If the Kaiser window function is not available in your version of MATLAB, it can be programmed from the `besseli`[6] function which is available.

An approximate formula for its frequency response near the mainlobe is

$$W(e^{j\omega}) \approx \frac{2M \sinh(\beta\sqrt{1-(\omega/\omega_\beta)^2})}{\beta I_0(\beta)\sqrt{1-(\omega/\omega_\beta)^2}} \qquad \text{for } \omega \leq \omega_\beta \tag{1-1}$$

$$W(e^{j\omega}) \approx \frac{2M \sin(\beta\sqrt{(\omega/\omega_\beta)^2-1})}{\beta I_0(\beta)\sqrt{(\omega/\omega_\beta)^2-1}} \qquad \text{for } \omega > \omega_\beta \tag{1-2}$$

where $\omega_\beta = \beta/M$. The value ω_β is the approximate width of the mainlobe. Note that for $\omega > \omega_\beta$, the sinh function becomes a sine. Formula (1-2) predicts that the sidelobes will fall off as $1/\omega$, away from the mainlobe.

a. For the Kaiser window, the parameter β offers a trade-off between the sidelobe height and the width of the mainlobe. Generate three Kaiser windows of length $L = 41$. Try different choices of the parameter β: $\beta = 3, 5$, and 8. Plot the window coefficients together on one graph, and compare to the Hamming window coefficients.

b. Compute the DTFTs and plot them all together. Make a zoomed plot in the same way as in part (b) of Exercise 1.2. Note the inverse trade-off between sidelobe height and mainlobe width for these cases.

c. Plot the approximate formulas (1-1) and (1-2), together with the true DTFT of the Kaiser window, for the case $\beta = 5$. Determine how closely the mainlobe and sidelobes follow the approximations. See `sinh` and `besseli` in MATLAB.

d. Show that the first sidelobe in the Kaiser window's DTFT has a height proportional to $\beta/\sinh(\beta)$; determine the constant of proportionality from the approximate formulas (1-1) and (1-2).

The Kaiser window is most useful for filter design, where its frequency-domain convolution with the frequency response of an ideal LPF (or BPF) yields good stopband attenuation. Then the parameter β provides control over the passband and stopband ripples, in a trade-off with the transition width of the filter.

EXERCISE 1.4

Other Windows

There are many other windows that have been proposed. Consult [1] for a rather comprehensive tabulation. In this exercise, a few of these families will be introduced. For each of these, you should write an M-file that will generate the window coefficients for even and odd lengths. In most cases the DTFT should be computed and displayed to judge the quality of the window. Either the Hamming or rectangular window should be included for reference.

a. All these windows are of finite length and have a point of symmetry, so their DTFT has linear phase. It is convenient to give a notation to this point of symmetry: $M = \frac{1}{2}(L-1)$, where L is the window length and M is the point of symmetry (or midpoint).

b. *Cosine series windows*: The Hamming, Hann, and Blackman windows are members of this family, where the window coefficients are given by

$$w[n] = \sum_{\ell=0}^{K} a_\ell \cos\left[\frac{2\pi\ell}{L-1}(n-M)\right] \qquad \text{for } n = 0, 1, \ldots, L-1$$

This class of windows has a simple analytic form for the DTFT, because they are based on linear combinations of cosines. Sometimes the formula is written with $2\pi/L$ in the argument of the cosine. The difference in the resulting frequency response is slight, especially for large L.

---

[6]The function `besseli` does not exist in MATLAB 3.5. Instead use `besseln(0,j*beta)`.

The Hamming and Hann windows require a_0 and a_1. For the Blackman case, there are three coefficients:

$$a_0 = 0.42 \qquad a_1 = 0.50 \qquad a_2 = 0.08$$

If an optimization of these coefficients is done to minimize the maximum sidelobe level, the resulting windows are the Harris–Nuttall windows. For the three- and four-term cases, the coefficients are

	−67 dB	−94 dB
a_0	0.423 23	0.358 75
a_1	0.497 55	0.488 29
a_2	0.079 22	0.141 28
a_3	—	0.011 68

Implement both Harris–Nuttall windows and verify the sidelobe height of their Fourier transforms. Compare their mainlobe width to that of the Hamming and rectangular windows.

c. It is relatively easy to derive the analytic form of the DTFT for the cosine series windows. The result should be a weighted sum of shifted asinc functions.

d. *Parabolic (Parzen) windows*: These are based on a simple polynomial formula,

$$w[n] = 1 - \left[\frac{n-M}{M}\right]^2$$

Plot the window coefficients versus n. Determine the mainlobe width and sidelobe height for this window for the case $L = 41$.

e. *Cauchy window*

$$w[n] = \frac{M^2}{M^2 + \alpha^2(n-M)^2}$$

Make plots of both the time domain and frequency domain for $\alpha = 4$. Experiment with different values of α to see what this parameter controls.

f. *Gaussian window*

$$w[n] = \exp\left[-\tfrac{1}{2}\alpha^2\left(\frac{n-M}{M}\right)^2\right]$$

Do the time- and frequency-domain plots for $\alpha = 3$ and $\alpha = 6$. Determine what α controls.

As you can see, many different functions can be used to create a time window. All have the characteristic taper near the edges, with a peak in the middle. Slight adjustments of the time-domain shape of the window can lead to quite different frequency response attributes, such as sidelobe height.

Project 2: Window Performance

Plots of the window and its frequency response as done in Project 1 indicate that the "performance" of a window can vary dramatically. In this project we examine some different ways to measure the performance of a window. For more detail the reader should consult [1].

Hints

The MATLAB function `find` can be used to locate points on a curve. Suppose that `H` is the frequency response vector and that it has been normalized so that its maximum value is `1.0000`. Then `indx = find( abs(H) > 0.707 )` will list all the indices where $|H(e^{j\omega})| > \frac{1}{2}\sqrt{2}$. From this list of indices we can measure the mainlobe's 3-dB width. If we are examining a window transform that has only one peak, the list of indices will be contiguous, so the first will be the minimum and the last the maximum. Thus the mainlobe width measured in samples is `length(indx)`, which can then be converted to radians.

EXERCISE 2.1

Window Length and Mainlobe Width

The most basic parameter under our control for defining a window is its length. It should always be true that increasing the length of the window will decrease it mainlobe width. Among different types of windows, however, there can be a wide variation in the mainlobe width. To measure mainlobe width, we must establish a common reference point, so we will take the 3-dB width.

a. Verify this fact for the rectangular window by measuring the mainlobe width as the 3-dB width (in amplitude). Take the length to be $L = 10$, 20, 40, 80, and 160; also try some lengths in between. A plot of the mainlobe width versus L should illustrate the inverse relationship.

b. Do the same measurement for the Hamming window and the Hann window. All three of these windows should have a mainlobe width that is γ/L but with different constants of proportionality γ.

c. An alternative definition of the mainlobe width uses a normalization to what is called the *bin width*. For the rectangular window, the first zero crossing in ω is always at $\omega = 2\pi/L$; this number is the bin width for all length-L windows. Thus we can define a normalized width by dividing the 3-dB width by the bin width. This removes the dependence on the length L.

d. Convert the previous 3-dB measurements to a normalized 3-dB width.

e. Verify that a single number will suffice for describing each window's width. In other words, recompute the normalized width for a wide range of L.

f. Determine the normalized width of the Kaiser window as a function of β. Use the approximate formulas (1-1) and (1-2) to discover the precise dependence on β.

EXERCISE 2.2

Sidelobe Height

Control of the sidelobe height is important for filter design applications and for minimizing leakage in neighboring channels of a spectrum analyzer. The rectangular window has the worst sidelobes, and most other windows have been proposed with the objective of providing lower sidelobes. Therefore, we need a measure of sidelobe height. The most common definition is the *maximum sidelobe height*, which often occurs at the first sidelobe.

a. Determine the maximum sidelobe height for the rectangular, triangular, Hamming, Hann, Blackman, and Harris–Nuttall windows. Do this for $L = 41$, but verify that the sidelobe height of these windows cannot be influenced by changing L. Try doubling and halving L to see if there is any change.

b. The advantage of the Kaiser window is that the parameter β can be used to change the sidelobe height. The approximate formulas (1-1) and (1-2) can be used to find the dependence of sidelobe height on β. Run several cases of the Kaiser window to verify the formula which says that sidelobe height is proportional to $\beta/\sinh(\beta)$.

c. Which value of β gives performance that is nearly the same as the Hamming window? For this value of β, plot the windows coefficients of the Kaiser and Hamming windows to compare.

d. Usually, a reduction in the maximum sidelobe height is accompanied by an increase in the mainlobe width. This effect can be shown with the Dolph–Chebyshev window, because its constant sidelobe level can be specified in the design. Run enough cases to make a plot of mainlobe 3-dB width versus sidelobe height, and thus show that a decrease in the sidelobe level is accompanied by a broadening of the mainlobe.

EXERCISE 2.3

Equivalent Noise Bandwidth

Another way to measure the spectral width of the window is to base the measurement on the window's performance as a filter in a spectrum analyzer. If $w[n]e^{j\omega_o n}$ is the impulse response of a filter, and the input to the filter consists of a sinusoid at ω_o plus white noise, the output will have a large peak due to the sinusoid, where the gain is

$$\left|\sum_{n=0}^{L-1} w[n]\right|$$

plus a noise background with variance equal to

$$\sum_{n=0}^{L-1} |w[n]|^2$$

The ratio of the noise variance to the peak amplitude squared tells how well the window performs. It also characterizes the window's bandwidth in the sense that if the mainlobe were narrow and the sidelobes low, the noise contribution would be very low. This ratio is called the *equivalent noise bandwidth*:

$$\text{ENBW} = L\frac{\sum_{n=0}^{L-1} |w[n]|^2}{\left|\sum_{n=0}^{L-1} w[n]\right|^2}$$

a. Prove that ENBW is independent of L for the rectangular window. Demonstrate that for the Hamming window, it is nearly independent of L.

b. Determine ENBW for the rectangular, Hamming, and Harris–Nuttall windows.

c. Compare ENBW to the normalized 3-dB width for the windows in part (b).

EXERCISE 2.4

Scallop Loss

Another measure of the peak and mainlobe performance is again motivated by a spectrum analysis application. In a sliding-window DFT, the FFT length is equal to the window length. Thus samples in frequency lie on a grid at $\omega_k = (2\pi/L)k$. If, on the other hand, the input sinusoid has a frequency halfway between grid points, the output from the FFT will be lower than the true amplitude of the sinusoid, due to the shape of the mainlobe. This reduced gain, called *scallop loss*, is equal to

$$\text{SCALLOP LOSS} = 20\log_{10}\left[\frac{W(e^{j\pi/L})}{W(e^{j0})}\right]$$

a. Determine the scallop loss for some commonly used windows.

b. Now define a composite measure of the scallop loss and the ENBW. The worst case of the ratio of the height of a sinusoid to output noise power is the sum of the scallop loss and the ENBW (in dB). This is called the *worst-case processing loss*. Determine this measure for some of the common windows.

EXERCISE 2.5

Summary of Window Performance

Five measures have been introduced to characterize window performance: mainlobe width, maximum sidelobe height, ENBW, scallop loss, and worst-case processing loss. For as many windows as feasible, make a table that gives all five measures. Make note of any dependence on window parameters, such as β for the Kaiser window.

PROJECT 3: RESOLUTION

The measures of mainlobe width and sidelobe height for different windows are important indicators of how a window will perform in a spectrum analysis situation. Most important, a narrow mainlobe will give good resolution. One of the often misunderstood aspects of windowing in conjunction with the FFT is that of resolution—the length and type *of the window* control resolution, not the length of the FFT. Sometimes these lengths are the same, but often zero padding is used to interpolate in the frequency domain. In this project we study an objective measure of resolution.

EXERCISE 3.1

Definition of Resolution

In spectrum analysis, resolution refers to the ability of the processing system to distinguish between two separate signals whose frequencies are very nearly the same. Thus take the two signals to be

$$x_1[n] = A_1 e^{j(\omega_1 n + \phi_1)} \qquad \text{and} \qquad x_2[n] = A_2 e^{j(\omega_2 n + \phi_2)}$$

The difference between the frequencies $\Delta\omega = |\omega_1 - \omega_2|$ is the parameter that will be varied.

The processing under consideration is the FFT, and the signal to be analyzed is the sum of $x_1[n]$ and $x_2[n]$. For most of this study we will take the amplitudes to be equal, $A_1 = A_2$.

a. Generate a finite portion of the signal $y[n] = x_1[n] + x_2[n]$, for $n = 0, 1, \ldots, L-1$. The amplitudes should be equal and the phases should be constant but chosen at random from the interval $0 \leq \phi < 2\pi$. Take the signal length to be $L = 64$, then compute a 64-point FFT and plot the magnitude. Vary the two frequencies and determine how small you can make $\Delta\omega$ and still see two separate peaks in the DTFT.

b. Repeat the same experiment as in part (a), but use a 128-point FFT and a 256-point FFT with zero padding. The interpolation effect of the longer FFTs permits some increased resolution, but you should verify that the minimum separation is more or less the same when measured in radians. [*Note*: In this case, a formula can be written for the DTFT of $y[n]$, because each sinusoid generates a term like $W(e^{j|\omega-\omega_i|})$, which is a frequency shift of the window's transform.]

EXERCISE 3.2

Peak Finding

To carry out more extensive testing, it is necessary to have an automatic method for peak finding.

a. Write a function that will automatically pick the peaks. This will require a definition of what a peak is. The visual criterion used in Exercise 3.1 will have to be turned into a set of rules for what constitutes a peak.

b. Try to make the program work for different levels of frequency interpolation—ranging from smooth (lots of zero padding) to none (no zero padding).

c. Window the signal $y[n]$ with a rectangular window as you did in Exercise 3.1, and then test your peak-finding function on the DFT of $y[n]$.

d. Now window $y[n]$ with a Hamming window to see if the peak-finding function still works.

EXERCISE 3.3

Measuring Resolution

With the automatic peak-finding algorithm, we are now ready to study different windows. The approach will be to generate a large number of test cases with different separations, collect all the data, and then plot a score that measures successful resolution versus $\Delta\omega$. The score will be the percent of successfully resolved peaks.

a. Make sure that your peak-finding function returns the number of peaks. For the resolution test, a return of 0 or 1 peaks means failure to resolve the peaks, while a return of 2 is success.

b. Write a script that will generate signals, compute their FFTs, and then look for peaks. Generate 10 instances of $y[n]$ at each separation; for each of these, random values of ω_1 should be used. Vary the separation $\Delta\omega$ over a range of values, but work in the neighborhood of $4\pi/L$. When $\Delta\omega < 2\pi/L$, the score should be near 0%, and when $\Delta\omega$ is large, the score approaches 100%. In between you must discover the values of $\Delta\omega$ where the score changes rapidly and try many tests at these points.

c. Collect all the scores (percent correct) and summarize the data in a plot of score versus $\Delta\omega$. Since the expected resolution is inversely proportional to window length, it would be better to make a normalized plot of score versus $\Delta\omega/(2\pi/L)$ (i.e., normalized to the bin width). If feasible, run the test for different lengths L.

d. Apply your resolution test to the Hamming, Hann, and Kaiser ($\beta = 3$, 5, and 8) windows.

Compare these empirically derived values for resolution to the mainlobe width of these windows. Often the 3-dB mainlobe width is quoted as the appropriate value for resolution; do you agree for all windows that you tested?

SLIDING WINDOW DFT

Nonstationary signals are characterized by changing features in their frequency content with respect to time. Therefore, the Fourier analysis of such signals needs to be localized and also time-dependent. The resulting analysis methods are called *time–frequency distributions.* Since it is a function of two variables, a time–frequency distribution is usually presented as an image so that regions of high energy appear as bright spots. Alternatively, a contour plot can be used as shown in Fig. 3.1. These regions indicate important transient events in the signal.[7]

[7]In MATLAB versions 4.0 and 5.0 there is a spectrogram function (`specgram`) as well as the capability to plot the result as an intensity image.

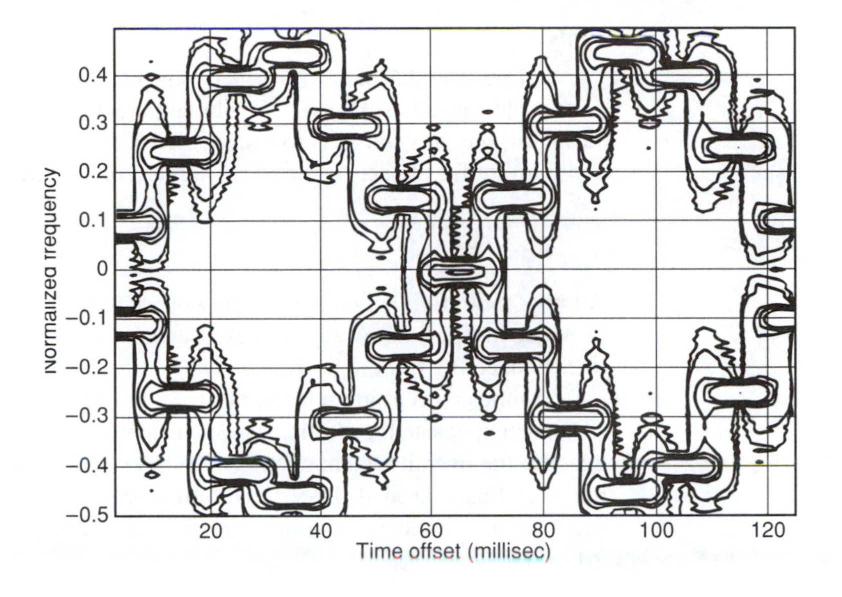

Figure 3.1
Example of a spectrogram plotted with `contour`. The input signal is a stepped-frequency sinusoid whose frequency starts at $\omega = 0.2\pi$ and steps by $\Delta\omega = 0.3\pi$ each time. The plot shows energy concentration in the time–frequency domain at the frequencies contained in the sinusoid. The frequency appear to decrease due to aliasing.

In this set of projects we consider an elementary time–frequency distribution based on the FFT. This function, called the STFT (short-time Fourier transform), is attractive because it is easy and efficient to compute [2, Chap. 4]. When applied to speech, the STFT is called a speech spectrogram [8], or "voice print." Since speech signals are composed of ever-changing narrowband features, this presentation of the data captures these features as formant tracks—resonant frequency versus time. Many other nonstationary signals can be analyzed in more detail if a time–frequency energy distribution is shown. Finally, a considerable amount of DSP research over the last 10 years has been devoted to the study of different types of time–frequency representations. It is not the objective of this set of projects to go into these different distributions, but rather, to concentrate on the one based on the FFT. Specific reading material related to this section can be found in [3, Sec. 11.5].

PROJECT 1: SPECTROGRAM

The functionality needed in a general spectrogram program can be provided rather easily in MATLAB, because the basic functions exist for windowing and the FFT. The spectrogram requires only a few parameters to describe the time–frequency plot desired. It is defined as a short-time Fourier transform of a windowed segment of the signal $x[n]$.

$$X_n(e^{j\omega}) = \sum_{m=n-L+1}^{n} x[m]w[n-m]e^{-jm\omega} \qquad \text{(1-1)}$$

The window $w[\cdot]$ determines how much of the signal will be used in the analysis and controls the frequency resolution of the Fourier analysis. The parameter n denotes the reference position of the window on the signal.

The implementation of (1-1) requires a loop to compute $X_n(e^{j\omega})$ for a succession of window positions n. At each window position, the FFT is computed for an L-point block of the signal. Usually, n is incremented by a fraction (25 to 50%) of the window length L between successive FFTs.

Hints

In version 3.5 of MATLAB it is not possible to make an intensity image, so other plotting formats must be used. Since the time–frequency distribution is a function of two variables, one possibility is to use `mesh` and `contour`. The example in Fig. 3.1 is the contour plot for a spectrogram of the signal $x[n]$ generated in MATLAB via

```
ttt = 2*pi*ones(100,1)*[0.1:0.15:2];
x = cos( ttt(:).*[1:1300]');
```

Another plotting function that might work well for the spectrogram is `waterf`, which is provided in Appendix A. It produces a "waterfall" plot which is similar to `mesh` but without lines connecting front to back.

In most spectrogram plots, there are problems with scaling. First, it may be necessary to compensate for drastic changes in energy from one windowed segment to the next. Second, the overall dynamic range of the plot may be so large that some sort of logarithmic rescaling is needed. In `contour`, this scaling issue can be attacked by specifying the exact contour levels with an optional input argument. The MATLAB code to produce Fig. 3.1 was

```
contour(X,[1 0.9 0.5 0.2 0.1 0.05 0.01 0.005],TT,ff)
```

where the array `X` had already been scaled to have a maximum magnitude of 1, and the vectors `TT` and `ff` contained information to specify the axes of the contour plot. The second argument was the list of nine levels for constructing the contours.

Since the spectrogram is a two-dimensional function, it requires a large array to hold all of its values. Therefore, it may not be possible to do this project with the student versions of MATLAB, which have upper limits on the array size (1k in version 3.5, 8k in version 4, and 16k in version 5). When the maximum array size is 8k or 16k, the choice of frequency resolution and time resolution will have to be made carefully to keep the size of the spectrogram array small enough.

EXERCISE 1.1

Program for Spectrogram

Write an M-file to calculate a spectrogram under rather general conditions. The core of the computation is nothing more than a sliding DFT. The program needs to have five inputs:

1. An input signal, $x[n]$, which is just a long vector of samples.
2. The window. If a vector containing the window values is passed, its length can be determined by MATLAB.
3. The amount to skip from one section to the next. This determines the time sampling of the resultant time–frequency distribution.
4. The frequency range. An upper and lower frequency (in percent of the sampling frequency) could be given so that only a part of the frequency domain would be analyzed.
5. FFT length, including zero padding. This length should probably be taken as a power of 2. It will determine the frequency sampling of the time–frequency distribution.

The output from the spectrogram function should contain three items:

1. The time–frequency distribution, which is complex-valued, even though the magnitude of the spectral values are all that are needed in 99% of the cases.
2. The time axis for use as plot labels.
3. The frequency axis for labels.

The latter two outputs are needed to generate the correct labels on a subsequent plot of the spectrogram. Thus a companion M-file should be written to digest the three outputs from the spectrogram program and produce a correctly labeled plot.

To test the spectrogram program, generate some simple test signals made up of sinusoids whose frequency jumps from one fixed value to another. Verify that the spectrogram plot shows large energy peaks near the known frequencies.

EXERCISE 1.2

Process Frequency-Modulated Signals

The spectrogram function can be used to track frequency variations versus time. In this exercise we consider two common cases involving frequency modulation.

a. Generate a linear-FM (chirp) signal and process it through the spectrogram function. The mathematical form of a continuous-time chirp is

$$\text{LFM CHIRP} = \cos(2\pi\mu t^2)$$

Since the linear-FM chirp has a known functional form for the frequency variation versus time, you can relate that form to a ridge in the time–frequency plot. This function must be sampled to give a discrete-time signal for analysis. Take the sampling frequency to be 10 MHz, and let the chirp rate be $\mu = 2.4 \times 10^{10}$. The signal must have finite duration, so let the time interval of the signal extend from $t = 0$ to $t = 100\,\mu$s.

b. When the frequency modulation is written in the form $x(t) = \cos[2\pi f(t)t]$, it appears that $f(t)$ is the "instantaneous frequency" of $x(t)$. However, the instantaneous frequency of a signal has to be defined as the time derivative of the phase $\phi(t)$:

$$f_i(t) = \frac{1}{2\pi}\frac{d}{dt}\phi(t)$$

For the case of the chirp, determine the instantaneous frequency. Determine whether the slope of the ridge produced by the sliding DFT in the time-frequency plane is an estimate of $f(t)$ or of the instantaneous frequency $f_i(t)$.

c. If the value of μ were changed to $\mu = 3.6 \times 10^{10}$, make a sketch of the spectrogram that would be computed. Verify by running your MATLAB program.

d. Consider a frequency-modulated signal whose instantaneous frequency follows a hyperbolic shape:

$$f_i(t) = \frac{a}{t+b}$$

If the instantaneous frequency equals F_1 at $t = T_1$ and F_2 at $t = T_2$, solve for the parameters a and b.

e. Write an M-file that uses the instantaneous frequency in (d) to generate the phase $\phi(t)$ needed to synthesize a hyperbolic-FM signal:

$$\text{HYPERBOLIC-FM} = \cos(2\pi\phi(t))$$

f. Generate a hyperbolic-FM signal and display its spectrogram. Relate the shape of the ridge in the time–frequency plane to the parameters defining the hyperbolic-FM signal. Take the time and frequency limits to be $[T_1, T_2] = [0, 90\,\mu\text{s}]$, $F_1 = 4.4$ MHz, and $F_2 = 1.1$ MHz. Let the sampling frequency be the same as above, $f_s = 10$ MHz. The very beginning of the signal will not be analyzed unless the signal is zero-padded at the front end. Therefore, it would be convenient to let the time interval start at $t = -10\,\mu$s.

EXERCISE 1.3

Wideband versus Narrowband Speech Spectrogram

In this exercise we study the spectrogram of a voiced speech signal. Chapter 10 has several sections devoted to speech applications, so the ambitious reader is referred to those projects for a more complete investigation of speech processing with MATLAB.

Load in the speech file `vowels.mat`, which is sampled at 8 kHz and contains the English language vowels `E-I-A-I-O` spoken in succession. Since the speech signal for vowels is modeled, to a first approximation, as a pulse train driving a filter, the Fourier transform should exhibit two prominent features [8]. First, the spectral shape of the vocal tract filter should be evident in a few (less than 5) major resonant peaks. Second, the spectrum should have many lines at a regular spacing due to the quasi-periodic nature of the waveform, which has a period of approximately 10 ms.

Experiment with the window length of the sliding FFT to see what sort of spectrogram display you can obtain. The plot will have to be done with `contour` or `mesh`. You should find that the spectrogram analysis can give two different views of the speech waveform:

a. A speech signal has a pulsed behavior when a vowel is spoken. This gives rise to a line spectrum versus frequency. To see these lines on a spectrogram, it is necessary to do a *narrowband* analysis. Determine the frequency resolution parameters needed and pick an appropriate window length. Use your spectrogram program to make this display. Measure the spacing between the lines in the spectrum in hertz and relate this number to the pitch period of the signal (expressed in milliseconds).

b. Another spectral feature of a vowel is that it is made up of a few formants (i.e., four or five resonant peaks in frequency). To see these (and not the line spectrum), it is necessary to do a *wideband* analysis. Determine the appropriate parameters for this analysis, and plot a spectrogram that shows the formant frequencies for several of the vowels.

EXERCISE 1.4

Bandpass Filter Bank

The sliding FFT can be viewed as a channelized spectrum analyzer. If the offset between blocks is taken to be one sample, the processing of a sliding DFT can be interpreted as a filter bank, containing N simultaneous bandpass filters, where N is the FFT length. The frequency response of each filter is determined by the DTFT of the window $W(e^{j\omega})$. The different bandpass filters have a common frequency response, but their center frequencies are offset from one another due to multiplication by a complex exponential.

a. Consider a case where the length of the window is $L = 16$ and a Hamming window is used. Remember that the window is applied to the time signal prior to taking the FFT. If the channels are numbered from $\ell = 0$ to $\ell = 15$, plot the frequency response for channels 0, 3, 6, and 9 on one plot. Then plot channels 6 and 7 together. Since the frequency response of the filters is not ideal, what do these plots say about the "resolution" of the DFT spectrum analyzer?

b. The DFT simultaneously computes N outputs, and each is a single point in the output stream of the N bandpass filters. These outputs can be denoted $y_\ell[n]$, where $\ell = 0, 1, 2, \ldots, 15$ is the channel number. Therefore, we can test the response of the sliding DFT to a sinusoidal input. Generate a sine-wave input at a fixed frequency, $\sin(2\pi(3.4)n/N)$. Make the signal long enough that you can process it with a sliding DFT and get about 128 points in each output $y_\ell[n]$. Take the number of filters to be rather small for this experiment, say $N = 16$. Now compute the average energy in the output of each channel and show that the correct channel (relative to the input frequency) is responding with the most energy. Explain why there is energy in all the channels.

c. Compute the DTFT of two of the channel signals, at $\ell = 3$ and $\ell = 7$. Plot the log magnitude for these two on the same scale and explain what you see.

d. Now consider changing the skip between successive blocks of the FFT. At first, let the skip be 4 samples. Repeat the three steps above: (1) plot all the output signals together, (2) compute the average energy in each channel, and (3) compare the DTFTs of the signals from channels $\ell = 3$ and $\ell = 7$. Explain why the average energy calculation yields the same answer. Then explain why the spectral peak has moved in the DTFTs of channels 3 and 7.

e. Suppose that the only measurement of interest is the average energy from each channel, because we might only be interested in detecting where the input frequency lies. Explain what will happen to the energy calculation if the skip is 2 or 8 samples. What is the largest value that we can use for the skip before the energy calculation gives erroneous results?

PROJECT 2: TONE GENERATOR

This project deals with the implementation of a sliding window DFT spectrum analysis system. To create a scenario that has some element of challenge, imagine a system whose job is to distinguish various tones. An example application might be a touch-tone telephone in which each tone represents a digit. This not how a touch-tone phone actually works, but it makes a reasonable example for the sliding DFT. Since the primary objective of this project is the implementation of a system to estimate frequencies, a data file containing a mystery signal is provided to test your final processing algorithm.

Hints

The mystery signal can be loaded from the file `tonemyst.mat`. In addition, two supporting files that will generate similar mystery signals are provided: `tonegen.m` and `genint.m`. The actual synthesis of `tonemyst.mat` was accomplished in two steps. First, `genint.m` was used to make an interference data file `intfere.mat`. Then `tonegen.m` was used to create a coded sequence and add in the interference (Fig. 3.2). Hopefully, this method of constructing these files will be useful to instructors who wish to create variations of this problem.

For plotting, use axis labels that reflect the natural units in the horizontal and vertical directions; otherwise, it is quite difficult to interpret the 2-D plots. Beware that MATLAB plots 2-D signals in what might be called a *matrix format*, with the origin of a contour plot in the upper left-hand corner. See `rot90` for a function that will rotate the matrix and put the origin in the lower left corner [see also `fliplr()` or `flipud()`].

It is not possible to make a gray-scale *spectrogram* with MATLAB version 3.5, but you can use `contour()` to approximate one. The function `contour()` is a bit tricky to control. If you choose the contour levels correctly, the plot will contour the regions of interest nicely (see Fig. 3.1 and the Hints for Project 1). See `help contour` for information on the optional arguments to `contour`. However, if you use the defaults, there may be so many contour lines that you cannot distinguish any of the important features. You may also need to use a log plot rather than a linear one to see certain features [see `db()` or `log10()`].

In some versions of MATLAB with limited memory, this project will tax the available memory unless you are very careful. If you use big arrays for temporary operations, `clear` them when they are no longer needed.

```
function [y,code] = tonegen(digits, scale, yint)
%TONEGEN    generate "mystery" signal containing tones
%  usage:
%     [Y,C] = tonegen(D,S,Xint)
%       D = vector of digits for a 5-element code
%           if length(D)<5, the function will pick random digits
%       S = scale factor that multiplies the interference signal
%    Xint = interference signal
%       Y = output signal
%       C = output code actually used
%
if( nargin < 3 )
   load intfere.mat
   if ~exist('yint')
      error(' problem loading interference')
   end
end
if( nargin < 2 )
   scale = 1.0;        % add 100% of interference
end
fsamp = 10000;
tones = (250 + 500*[0:9]')/fsamp;
if length(digits) < 5
   digits = mod(fix(clock),10); digits = digits(2:6);
end
code  = mod(fix(digits(1:5)),10);      %--- just 5 digits, must be integers
LL = 50*rand(7,1)-25;     %--- variation in lengths
LL = fix(LL) + [55;175*ones(5,1);95];
Ltot = sum(LL);
if Ltot > length(yint)
   LL = fix(LL*Ltot/length(yint));
end
ttt = [0.5*rand(1);tones(code+1);0.5*rand(1)];    %--- create the tones
for j = 1:7;
   f1 = [ f1; ttt(j)*ones(LL(j),1) ];
end
N = length(f1);    Nm1 = N-1;    nn = [0:Nm1]';
%----------
tau = 0.8;
[ttt,f1i] = filter(1-tau,[1 -tau],f1(1)*ones(99,1));   %--- set init conds.
f1 = filter(1-tau,[1 -tau],f1,0.9*f1i);
y = cos(2*pi*f1.*nn);
y = y + scale*yint(1:N);
```

Figure 3.2 Listing of tonegen.m.

EXERCISE 2.1

Frequency-Coded Input Signal

The input signal, contained in the file `tonemyst.mat`, must be analyzed to determine its frequency content versus time. In this exercise you should become familiar with the characteristics of the mystery signal and the function `tonegen` used to generate it. For example, you can plot the time signal with and without noise and with and without interference to learn more about its nature.

The mystery signal is known to be composed of three different subsignals:

1. The desired signal is a succession of short-duration sinusoids of constant frequency. The duration of each sinusoid is not necessarily the same, but the length of each is usually between 15 and 20 ms. The desired signal encodes a sequence of digits by using different frequencies to represent individual digits, according to the following table:

Digit	0	1	2	3	4	5	6	7	8	9
Freq. (Hz)	250	750	1250	1750	2250	2750	3250	3750	4250	4750

   Thus, the signal changes frequency to one of the values in the table roughly every 15 to 20 ms. Such a scheme bears some resemblance to *frequency shift keying* (FSK) in a communication system, except that the duration is not constant as in FSK. This coded signal has a constant amplitude. For example, the digit sequence 7-3-4-1-9 would consist of a sinusoid of frequency 3750 Hz, followed by one of 1750 Hz, then 2250 Hz, 750 Hz, and finally 4750 Hz. To avoid instantaneous changes in frequency, the step changes in frequency were filtered through a low-pass filter to provide some smoothing of the frequency versus time.

   The coded portion of the signal is not assumed to start at $n = 0$ in the data set, so one part of the problem will be to locate the transition points where the frequency changes. The sampling rate in the system is $f_s = 10$ kHz, and for this problem you can assume that the sampling was carried out perfectly—no aliasing.

2. The second signal component is an interference that is quite a bit stronger than the desired frequency-coded signal. This interference is a narrowband signal (i.e., sinusoid) whose frequency is changing continuously with time. Its peak amplitude is about 10 times stronger than the coded signal. You will have to determine the value of this amplitude and the frequency variation of this signal versus time. However, you are not required to remove this component prior to the sliding-window DFT processor; just estimate its parameters.

3. The third signal component is an additive noise term. This component is zero-mean Gaussian white noise with a variance that is equal to 3% of the amplitude of the *interference* signal. Thus, there is some noise present, but the signal-to-noise ratio is favorable for processing the coded signal. This noise component has been added to make the problem "realistic," but it should not be a major factor in designing your processing scheme.

The function `tonegen(digits)` takes as input a five-element vector and produces an output signal that contains the frequency-coded version of this input vector plus the interference and noise. As a precaution, there is a second output, which is the actual five-element vector used, in case the input was not integers in the range [0,9]. If you supply a second input argument to `tonegen(digits,scale)`, the value of `scale` is interpreted as a multiplier that will scale the additive interference plus noise. Therefore, by using the call `tonegen(digits,0)`, you can experiment with situations in which only the coded signal is present. A third argument (optional) allows a different interference signal to be specified. The coded signal produced by the function `tonegen()` will contain variable-length tones. With just one scalar input argument, `tonegen(0)` will generate a set of "random" digits based on the present clock time and return these in its second output argument.

EXERCISE 2.2

Sliding DFT

The method of analysis to be implemented is the sliding-window DFT, but you must pick the parameters of the system: window length, window type, FFT length, segment skip, and so on. For each design parameter chosen, consider several possibilities.

The primary task in this project is to design and implement the sliding-window DFT system so that it will reliably extract the coded information. For the purposes of implementation, the total signal length can be assumed to be about 100 ms, so that the frequency-coded signal contains 5 digits. Your processing system must be able to identify these 5 digits. This can be done in two ways:

a. Create a spectrogram, and "read" the spectrogram visually to extract the information. A contour plot should be sufficient for this purpose.

b. Design an automatic algorithm based on peak picking such that it will ignore the interference and extract the code automatically. This will be difficult, but try to make it work for any combination of input digits. Explain why some cases are easier than others; take into account the presence of a strong interference.

c. Then examine the complexity of your implementation and justify your choices of window length and segment skip. Is it possible to achieve the same code identification with fewer FFTs and/or with shorter block lengths? Remember that the objective in an efficient system design would be to minimize the amount of computation, because that would reduce cost (e.g., by using less hardware).

EXERCISE 2.3

Demonstration of a Working System

To demonstrate that your system works correctly, you must process the following sets of unscaled data, that is, create the signal using the call `tonegen(digits)`:

a. `digits = [ 9 8 1 2 3 ]`

b. `digits = [ 4 3 4 7 6 ]`

c. `digits` = last five digits of your phone number.

d. Others: If you have implemented an automatic decoder, show that it will work for a random input [e.g., `digits = round( 9.98*rand(5,1) - 0.49 )`].

You should also try to determine the amplitude and duration of the tones from your sliding window DFT processor. Comment on how well or poorly you can make such measurements. Explain how to choose the window length so that your measurement procedure will work as well as possible. Explain the compromises that were made in choosing the window length.

EXERCISE 2.4

Interference Signal

The interference signal also can be estimated. Determine the amplitude of the interference relative to the amplitude to the frequency-coded tones. Also determine a functional form for the frequency variation of the interference signal. The sliding DFT is measuring some sort of "instantaneous frequency," so one possibility is $\cos[2\pi f(t)t]$, but be careful how you relate the ridge in the sliding DFT measurement to $f(t)$. In any event, you should provide a plot of $f(t)$ for the interference, as well as a plot of its instantaneous frequency.

NARROWBAND SIGNALS

The frequency domain is quite useful in describing and processing signals that occupy only a small band in frequency. Such signals, called *narrowband signals*, include sinusoids, high-Q filter responses, and pulse trains. See Fig. 3.3 for an example of a narrowband time signal and its frequency content. The objective in this set of projects is to introduce some notable features of narrowband signals to gain more understanding of the Fourier description of signals in general.

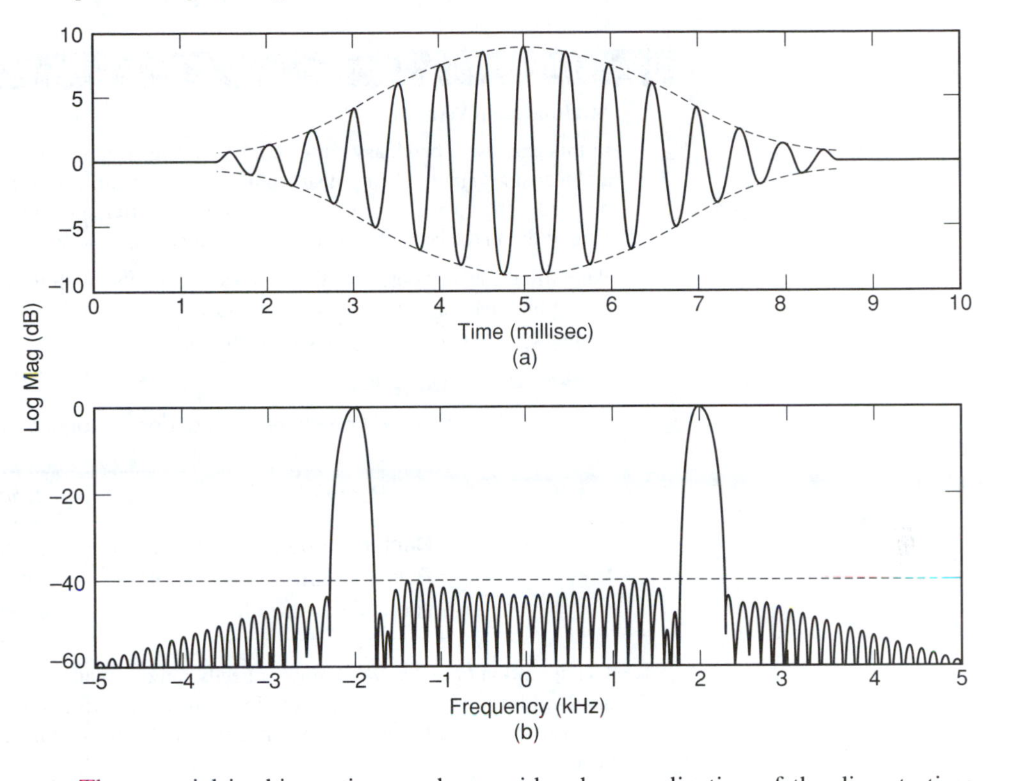

Figure 3.3
(a) Example of a bandpass pulse with a tapered envelope; (b) its Fourier transform magnitude shows the energy concentration in the frequency domain.

The material in this section can be considered an application of the discrete-time Fourier transform (DTFT) and the discrete Fourier transform (DFT). For background reading consult the chapters of your text devoted to the DTFT and DFT (e.g., Chapters 3, 8, and 9 in [3]).

Project 1: Synthesis of a Bandpass Pulse

Whenever computations must be done in both the time and frequency domains, the sampled nature of both domains is an important consideration. This project explores some aspects of frequency sampling, which is inherent in the use of the DFT. Since the DFT is the main computational tool used for numerical Fourier analysis, a central issue of frequency sampling is the relationship between the DFT samples and the DTFT, or between the DFT and the continuous-time Fourier transform if the signal was created via time sampling.

The particular scenario of this project is the synthesis of a class of "bandpass" pulses with prescribed frequency content. The pulses will be synthesized in three different ways: (1) by modulating a window, (2) by truncating an ideal bandlimited pulse, and (3) by imposing specifications directly in the DFT frequency domain. In each case the synthetic pulses can be endowed with perfect linear phase. In the next project we study how FIR and IIR filters affect the phase and amplitude of these narrowband pulses.

Hints

To examine the frequency content of a signal $x[n]$ it is necessary to compute its DTFT $X(e^{j\omega})$. See `freqz` or `fft` in MATLAB or `dtft` in Appendix A. The `find` function can be used to make a list of all frequencies within a certain bandlimit, as in (1-1).

Throughout this project the pulse specifications are given in the continuous-time frequency domain. Therefore, it is necessary to convert such information into the normalized frequency scale of the DTFT and then into the correct frequency indices of the DFT. These conversions are simply linear rescalings of the frequency axes.

EXERCISE 1.1

Modulating a Window

In this exercise a bandpass pulse will be created from the product of a baseband pulse shape and a cosine signal. The baseband pulse must have finite duration, so a window function will be used. The cosine signal will translate the Fourier spectrum from baseband (i.e., centered around dc) to the correct center-frequency location.

We will take as our objective the synthesis of a pulse with the characteristics shown in Table 3-1. Since the pulse should be real-valued, the specifications on the bandwidth and center frequency also apply to negative frequencies.

TABLE 3-1

Desired Parameters for Bandpass Signal Synthesis

Parameter	*Value*	*Units*
Center frequency	$\frac{2}{7}\pi$	radians
Pulse length	41	samples
ssential bandwidth	$\frac{1}{8}\pi$	radians

The definition of *essential bandwidth* needs some explanation. For this project we will define bandwidth from an energy point of view. Thus the essential bandwidth of a pulse centered at $\omega = \omega_o$ will be the frequency amount $\Delta\omega$ such that 99% of the energy in the DTFT lies in the bands

$$|\omega - \omega_o| \leq \Delta\omega \qquad \text{and} \qquad |\omega - (-\omega_o)| \leq \Delta\omega \tag{1-1}$$

Notice that $\Delta\omega$ is a one-way bandwidth because $2\Delta\omega$ is the total bandwidth about the center frequency.

Similarly, we can make a definition of *essential time duration* as the length of that part of the signal containing 99% of the signal energy. This requires that we identify a peak in the signal envelope which is usually possible for symmetric windows such as the Hamming window. Then we evaluate the region centered around the peak for its energy content. Thus we can compute the essential time duration of a symmetric finite-length signal by dropping off samples from both ends until we have thrown away 1% of the energy.

The pulse synthesis will be of poor quality if a rectangular window is used. It is much better to use a shape such as a Gaussian or a Hamming window to create a smooth envelope for the pulse. In effect, these envelopes give the bandpass pulse the appearance of a wavelet—a short burst of high-frequency energy (see Fig. 3.3).

a. Generate a Hamming window of length 41 and compute its DTFT on a very dense grid of frequency samples. Determine the essential bandwidth of this Hamming window, which is a baseband pulse.

b. Determine the essential time duration of the Hamming window pulse.

c. *Modulation*: Any window function $w[n]$ is a baseband signal, so it must be frequency-translated to the desired center frequency, by multiplying by a sinusoid.

$$x_1[n] = w[n]\cos(\omega_o n)$$

Choose ω_o to satisfy Table 3-1, and then create $x_1[n]$ and plot its DTFT magnitude. Show that $|X_1(e^{j\omega})|$ more or less occupies the desired band of frequencies. Compute its essential bandwidth to see if the specifications of Table 3-1 have been met. In the following exercises we address problems in meeting the specs of Table 3-1 exactly.

d. Generate a rectangular window of length 41 and determine its essential bandwidth. Notice that the requirement for concentration of 99% of the energy is so stringent that the rectangular window is unable to meet the specs of Table 3-1 in a reasonable fashion.

EXERCISE 1.2

Truncating an Ideal Pulse

The primary problem with the Hamming pulse for this project is that its essential bandwidth is always inversely proportional to the window length (or essential time duration). Thus the desired specs on bandwidth in Table 3-1 cannot be fulfilled by $x_1[n]$ because the energy in the frequency domain is too concentrated. A different time signal must be used if the bandwidth criterion is to be met exactly. Since the desired DTFT should extend over a band of frequencies, one obvious choice for the time signal would be the inverse DTFT of a pulse in frequency. This is a particularly easy inverse transform to derive analytically, and the result is a *sinc function*:

$$s[n] = \frac{W}{\pi}\,\mathrm{sinc}\left(\frac{Wn}{\pi}\right) = \frac{\sin(Wn)}{\pi n} \qquad \text{for } -\infty < n < \infty$$

where W is the one-way bandwidth. The formula is given for the lowpass case, so cosine modulation will have to be applied to move the transform to the correct center frequency.

One problem with this formula is that it extends over all n, from $-\infty$ to $+\infty$. So, to use it for a bandpass pulse, $s[n]$ must be truncated by a window. Since the main portion of the sinc function lies symmetrically about zero, the window must be placed symmetrically about the $n = 0$ point. For this exercise, keep the window length the same as before (i.e., $L = 41$).

a. Generate the Hamming windowed version of a modulated sinc function, and then compute its DTFT on a dense grid and plot the magnitude response. Visually confirm that the bandwidth looks correct—at least if bandwidth is taken to be the one-way width of the mainlobe.

b. Measure the essential bandwidth of the pulse created in part (a) to verify whether the bandwidth is now correct.

c. In addition, compute the essential time duration of the baseband pulse. Compare this value to $2\pi/W$ and explain any significant difference.

d. Try a rectangular window applied to the sinc function. In the DTFT you should observe higher-frequency-domain sidelobes outside the desired frequency band. Measure the essential bandwidth in this case to see whether or not it gives a reasonable value in meeting the 99% criterion.

EXERCISE 1.3

Synthesis in the Frequency Domain via the DFT

The bandpass pulse can also be created directly from its frequency-domain description. The result should be a pulse similar to that obtained from sinc function synthesis. However, for direct numerical synthesis, the characteristics of the bandpass pulse must be specified in terms of its DFT samples. The DFT samples are set to ones for the in-band frequencies, and zeros for out-of-band. The time-domain pulse is synthesized by taking the inverse FFT and then windowing the result. Choice of the IFFT length is not crucial, but it must be greater than some minimum and should not be so large as to force unnecessary computation. The following steps use the IFFT method to create the *bandpass pulse* signal specified in Table 3-1.

a. First, create the DFT of the signal by setting the appropriate frequency samples to one or zero. Write the MATLAB statements that will define the DTFT samples over both positive and negative frequency. Convert the specifications in Table 3-1, which are given in terms of ω to the DFT index domain k. Recall that the DFT computes samples of the DTFT at $\omega = (2\pi/N)k$. The length of the IFFT to be used is at your discretion. For the moment, take it to be $N = 256$. The method will naturally compute a pulse whose length is N, but later, the pulse length will be shortened by a window applied in the time domain.

b. Use the `ifft()` function to synthesize the signal. Make sure that your output signal is *real-valued*. This involves two issues: (1) make sure that the DTFT is specified with symmetry so that its inverse transform will be purely real, and (2) check the output for tiny values in the imaginary part due to rounding error in the inverse FFT. These would be $\approx 10^{-14}$.

c. Plot the bandpass pulse that was computed. If necessary, rotate the signal within the vector of N samples so that the pulse appears in the middle of the plot (see `help fftshift`). Determine the essential time duration of the synthetic pulse. Explain why this rotation must be done by considering the phase of the DTFT defined in part (a).

d. Window the synthetic pulse to its essential duration (with a rectangular window) and then compute its DTFT with a very long FFT for dense frequency sampling. Plot the magnitude response versus a frequency axis labeled in radians. Note that the transform will not be absolutely confined to the specified frequency region, but there should be very little energy outside the desired region. Also, the in-band spectral content should be nearly flat. Finally, compute the essential bandwidth of the signal.

e. *Alternate way to get real-valued signal*: Show that exactly the same real-valued signal will be generated by using the following steps: (1) define a DTFT that has only positive-frequency components; (2) with no symmetry, the IFFT will yield a complex-valued signal; so (3) take the real part of that signal. Implement this method to verify that it does give the same answer, and then prove why taking the real part generates the appropriate negative-frequency components.

f. The synthesis should be recomputed with longer and shorter FFTs. Generate the same signal with a 512-point FFT, a 128-point FFT, and a 64-point FFT. Explain how the frequency samples would have to be redefined as a function of N. In each case, compute the essential bandwidth of the synthetic pulse to see if it changes very much. Explain any significant difference in the "essential" part of the pulse. What is the minimum-length FFT that could be used?

g. The *uncertainty principle* of Fourier analysis states that the product of time duration and bandwidth is always greater than a universal constant. The exact value of the constant depends on the units used for frequency and on the definition of duration, but it serves as a fixed lower bound for all signals. An example of the uncertainty principle can be seen with the essential bandwidth and time duration of the Hamming window: the essential time duration is $\approx 0.77L$, and the essential bandwidth is approximately $2.65\pi/L$. Thus the time-bandwidth product is 2.04π, which is greater than a lower bound that is around 2π. The obvious consequence of the uncertainty principle is that the specified bandwidth given in Table 3-1 dictates a minimum pulse length. Estimate this minimum "essential" pulse length, and also determine which of the N FFT output points should actually be used. For the parameters given in Table 3-1, express this duration as a total time in number of samples. In effect, the uncertainty principle should justify your choice of the minimum FFT length in part (f).

EXERCISE 1.4

Window the Pulse

If the output of the IFFT is truncated to keep only the essential part, then, in effect, a rectangular window has already been used to create the finite-length bandpass pulse. Better results can be achieved with a Hamming window.

a. Generate an instance of the Hamming window for the essential signal length, L. This length was estimated previously by means of the uncertainty principle. Since the Hamming window tapers to near zero at its ends, its essential time duration is somewhat less than its true length. Therefore, the length of the Hamming window should be taken longer than the rectangular window used in Exercise 1.3(d).

b. Now apply the Hamming window to the main part of the pulse obtained from the IFFT. Take the DTFT of this windowed waveform and plot the log magnitude. Use zero padding to get a dense sampling of the frequency axis. Compare the Hamming windowed case to the previous result for the rectangular window.

EXERCISE 1.5

M-File for Pulse Synthesis

In the next project, these synthetic bandpass pulses will be used as test signals for studying both FIR and IIR filters. Therefore, it will be useful to create an M-file capable of synthesizing a bandpass pulse from arbitrary specifications. Consider the specs listed in Table 3-2, which are given in terms of analog frequency content.

TABLE 3-2

Desired Parameters for Analog Bandpass Signal

Parameter	*Value*	*Units*
Center frequency	6.5	kHz
Pulse length	???	samples
Two-way bandwidth	3	kHz
Sampling period	12.5	μsec

a. Decide which of the three synthesis methods to use. Quite likely, the method based on the DFT samples will offer the best control over the frequency content of the pulse.

b. Write the M-file to synthesize pulses. Its input arguments should be sampling frequency, center frequency, and bandwidth. The output would be the pulse signal values at the specified sampling rate. An optional input would be the window type and the window length. Since the bandpass pulse length can be estimated from the uncertainty principle of the Fourier transform, the window arguments are optional.

c. Use this function to generate the analog bandpass pulse defined in Table 3-2. Show a plot of the pulse versus time (in milliseconds) and a plot of the Fourier magnitude versus analog frequency (in kilohertz).

PROJECT 2: FILTERING THE BANDPASS PULSE

Frequency-selective filters can be used to separate pulses that occupy different frequency bands. In this project we study the action of IIR and FIR filters when processing pulses, such as those synthesized in Project 1. Both types of filters are effective in isolating individual pulses because the magnitude characteristic can be tailored to the frequency band of interest. However, the linear-phase FIR filters introduce no phase distortion to these symmetrical pulses, whereas the IIR filters will cause noticeable changes in the pulse shape, even though all the energy in the pulse is preserved.

The linear-phase bandpass pulses make ideal test signals for frequency-selective digital filters. The processing can be done in three different ways: IIR filtering with a set of second-order bandpass filters, or FIR filtering with linear-phase bandpass filters, or a sliding FFT. The difference between the phase response of FIR and IIR filters can be demonstrated by examining the output of these different processing systems. For background on filter design, consult Chapter 7 of [3].

Hints

In this set of exercises, an unknown signal file is needed. It is contained in the file `b3pulses.mat`. The data file can be loaded into MATLAB via the `load b3pulses.mat` command. This file contains three bandpass pulses: one in the frequency range from 5 to 8 kHz, one between 10.5 and 15.5 kHz, and the third in the band 18 to 20 kHz. These signals are each repeated at a regular interval. The data record is *noisy*. High-pass noise was added to disguise the signals in a plot of the raw data. The signal-to-noise ratio is poor, but the noise is out of band, so frequency-selective filtering ought to work quite well.

To plot very long signals, the function `striplot` from Appendix A can be used to make a plot that extends over several rows of a normal plot. Another useful plotting function is `wp`, which will plot several signals in a one-per-row format. Since the signals are offset from one another, this plotting format makes it easy to measure temporal alignment between two or three signals.

EXERCISE 2.1

IIR Filtering

To isolate the three signals it is necessary to do bandpass filtering. In this part, the bandpass filtering will have to be implemented with three separate IIR filters. Each can be optimized to respond to the known bands occupied by the pulses.

The simplest possible bandpass filter is a second-order section, where the pole locations are chosen to obtain the desired center frequency and 3-dB bandwidth. If we specify a second-order section with poles at $z = re^{\pm j\theta}$ and zeros at $z = \pm 1$, the transfer function is

$$H(z) = \frac{1 - z^{-2}}{1 - 2r\cos\theta\, z^{-1} + r^2 z^{-2}}$$

The frequency response of this system has zeros at $\omega = 0$ and $\omega = \pi$. The poles create a peak at $\omega = \theta$ with a 3-dB bandwidth equal to $2(1 - r)/\sqrt{r}$.

a. Verify the frequency response of the second-order section by choosing θ and r based on the 5- to 8-kHz bandwidth for pulse 1 above. Make a plot of the frequency response versus frequency (in Hertz) to show that the passband extends from 5 to 8 kHz.

b. Use the `filter` function in MATLAB to process the noisy signal. Plot the output as a `striplot` to see if you can detect where the 5- to 8-kHz bandpass pulses lie. Remember that this pulse is repeated several times. Measure the starting time for each repeated pulse.

c. Define the filters for the other two bandwidths: 10.5 to 15.5 kHz and 18 to 20 kHz, and then process the signal. Identify the pulses for each case on a plot of the output signal. Use `wp` to plot the outputs versus t for all three filters together.

d. *Optional*: The bandpass filters can also be designed by standard methods to produce elliptic filters or some other common filter type. If you have already studied this sort of filter design [3, Ch. 7], design three elliptic filters and process the data one more time.

e. After processing, create a table that lists the repetition periods for each signal. Give your answer in milliseconds. Show one or two representative plots to explain how you estimated the time between pulses. You should also count the total number of pulses at each frequency.

These pulses are very easy to find when you have the correct filter bandwidth. Sharp cutoff filters, such as elliptic, are not really necessary.

EXERCISE 2.2

FIR Filtering

A reasonable FIR filter can be created by using what amounts to a matched filter. The suspected finite-length bandpass pulse is synthesized and then used as the impulse response of the filter. The passband of the FIR filter will simply be the frequency range specified for the bandpass pulse.

a. Use the bandpass pulse idea to create the impulse responses for the three bandpass filters needed for processing. Use the M-file developed in Project 1.

b. Compute and plot the frequency response magnitude of all three bandpass filters. Label the horizontal axis with frequency in kilohertz. Verify that the passbands are in the correct location. Define the stopbands of the filters and estimate the attenuation in the stopband for each filter. You should be able to get about 40 dB of rejection if a Hamming window is used.

c. Process the signal through each of the FIR filters and estimate the pulse locations and the interpulse period for each of the three signal components from a plot of the output signal.

d. *Optional*: If you have already studied filter design [3, Chap. 7], it will be straightforward to use the function `remez` to design a better set of FIR bandpass filters. The `remez` function (available in the MATLAB Signal Processing Toolbox) is an implementation of the Parks–McClellan algorithm for FIR filter design. It will provide a filter whose passband ripples and stopband rejection is the best among all FIR filters. Reprocess the data with these filters to see if it makes much of a difference.

EXERCISE 2.3

Comparison of FIR versus IIR Filtering

Compare the output pulse shape for the FIR processing versus the IIR processing. Show that the IIR filter distorts the pulse shape, while the FIR filter causes no distortion at all. Note that both methods also introduce some delay into the output signal—measure the relative difference in delay. Plot the group delay versus frequency for both types of filters to exhibit the nonlinear-phase characteristic of the IIR filter in its passband.

EXERCISE 2.4

Sliding-Window FFT Processing

Since we are searching for events that depend on both frequency and time, a moving FFT can be used to do a time-dependent Fourier analysis (see also the preceding section, *Sliding-Window DFT*). This would be implemented as follows: The first L points of the signal are taken as a section and the FFT is computed. Then the starting point is moved over by M points, and the FFT of L points starting at $n = M$ is computed. This process is repeated as the starting point of the section is moved by M points each time until all the data are exhausted. All of the FFT vectors are put together in one matrix, which can be viewed as a function of two variables: frequency and window position (time). A magnitude plot of this matrix will reveal regions of the time–frequency plane where signal energy is concentrated.

a. The section length L must satisfy two conflicting requirements. It must be short enough to give time localization of rapidly occurring events, but it must be long enough to provide adequate frequency resolution. Each output of the FFT represents the energy in a frequency band whose width is inversely proportional to the number of signal points taken into the FFT. Resolution of the three pulse bandwidths will demand a certain FFT length, but to make a smooth plot in the frequency domain, an even longer FFT might be needed.

   Pick out several representative sections of the signal for experimental processing. Multiply each section by a Hamming window of length L. Compute the FFT (after zero padding) and then plot the magnitude of the FFT. Experiment with the window length to get an acceptable frequency-domain plot. Make sure that the bandpass pulses can be distinguished easily.

b. The skip parameter M for the start of each section should lie between 10 and 50% of the section length when using a Hamming window. You should experiment with the value of M. A smaller skip (10%) will give a smoother contour plot; a larger skip (50%) will minimize the amount of computation. After deciding on the skip percentage, process all the data. Collect all the FFTs together into one large two-dimensional array—one FFT result per column.

c. Make a `contour` plot of the FFT magnitudes to show where the energy peaks lie in the time–frequency plane. Label the contour plot with the correct units of time and frequency. When labeling the time axis, use the middle of the Hamming window as the reference point for the FFT analysis.

d. Identify the three different bandpass pulses on the contour plot. In addition, measure the repetition period for each from the time–frequency plot.

PROJECT 3: RESONANT PEAKS

Another class of narrowband signals consists of those signals synthesized from second-order resonators. These simple signals can be used to model the formant frequencies in a speech signal, for example. In this project we study the relationship between the pole locations of the second-order resonator and the time-domain or frequency-domain characteristics of the filter [3, Sec. 5.3].

Hints

The `contour` function has a number of optional arguments that control things such as axis labeling, the number of contours, and the spacing of contours. Consult `help contour`.

EXERCISE 3.1

Bandwidth of a Resonant Peak

MATLAB can be used to demonstrate the variation in bandwidth for a second-order digital filter:

$$H(z) = \frac{1}{1 - 2r\cos\theta\, z^{-1} + r^2 z^{-2}} \tag{3-1}$$

An informative three-dimensional plot can be produced to illustrate the relationship between the frequency response and the location of the poles (r, θ) with respect to the unit circle in the z-plane. The idea is to show changes in the frequency response, from which the influence of the pole positions can be understood. In this case we are interested primarily in the relationship between the pole radius and the bandwidth of the frequency response.

a. Relate the frequency of the resonant peak in $|H(e^{j\omega})|$ to the pole location (r, θ) in $H(z)$. Give a formula that describes this dependence.

b. An approximate relation between the bandwidth and the pole location is given by the formula

$$\text{3-dB BANDWIDTH} \approx 2\,\frac{1-r}{\sqrt{r}} \qquad \text{(radians)} \tag{3-2}$$

When the pole is very close to the unit circle (i.e., $r \approx 1$), the dependence is nearly linear. Derive this formula (3-2) for the relationship between r and the bandwidth.

c. To see this behavior versus r, a three-dimensional plot of a set of frequency responses can be constructed using the `mesh` command. Write an M-file that will compute the frequency response of a second-order all-pole filter for varying r in the range $0.7 < r < 1$. As the pole location moves closer to the unit circle, the frequency response exhibits a sharp peak, and its bandwidth gets very narrow. Collect all the frequency responses together in a matrix, but include only the frequency range near the peak. Scale each frequency response to have a peak value of one. To make the mesh plot look like a good comparison over the wide range of pole radii being studied, the plot should be constructed on a dB scale.

d. Another view of the set of frequency responses can be provided via the contour plot. This has the advantage of showing the precise location of the 3-dB points in the frequency response. Using an optional argument to `contour`, the contour heights can be specified directly, so 5 contour lines can be drawn very close to the -3 dB level to emphasize that location. In addition, the analytic formula (3-2) for the bandwidth can be superimposed as a dashed line on the contour plot to check its validity.

Thus the mesh plot gives a visual representation of the resonant peak, and the contour plot confirms the approximate formula and shows its range of applicability.

EXERCISE 3.2

Another Pole Movement Example

The pole location of a second-order filter will control the time-domain response of the filter. In fact, the impulse response of a second-order resonator is a decaying sinusoid $y[n]$:

$$y[n] = A\,r^n \cos(\theta n + \phi) \qquad \text{for } n = 0, 1, \ldots, L-1 \tag{3-3}$$

One point of this exercise is to relate the parameters of $y[n]$ to the parameters needed by `filter` for implementation [i.e, the transfer function $Y(z) = H(z) = B(z)/A(z)$].

a. Write an M-file that will synthesize a section of the decaying sinusoid $y[n]$ defined in (3-3). This function should work directly from the signal parameters (r, θ, A, ϕ).

b. Determine the z-transform of $y[n]$, and use this to find a system that will generate $y[n]$ as its impulse response. Then give the correct arguments needed by `filter` to generate $y[n]$.

c. Demonstrate that both techniques will give the same result by generating and plotting the first 50 points of the signal:

$$y[n] = 20\,(0.975)^n \cos\left(\frac{2\pi}{17}n + \frac{\pi}{4}\right) \qquad \text{for } n = 0, 1, \ldots, 49$$

Compare the method based on `filter` to the function that works directly from the signal parameters.

d. Compute and plot the true frequency response by using the `freqz` function. Then compare to the DTFT of the 50-point signal $y[n]$. Compute the DTFT of $y[n]$ for 256 frequency samples via `freqz`. Plot $|Y(e^{j\omega})|$ together with $|H(e^{j\omega})|$ and explain the differences between the transform of the finite segment and the true frequency response of the rational system. Pay special attention to the peak width and sidelobes.

e. For either frequency response, relate the location of the resonant peak to the θ parameter of the signal (3-3). Give a formula that describes this dependence.

f. Generate versions of the same 50-point signal for different values of r and measure the peak width versus r in the range $0.8 < r < 0.99$. Over what range of r is the formula (3-2) given in Exercise 3.1 still valid? Explain why the peak width becomes a constant for r greater than some value $r_{\circ}$. What would happen if the signal length were doubled to $L = 100$?

PROJECT 4: LINE SPECTRA

Periodic signals have line spectra. When the signal length is finite, the lines are not impulses, but rather, have a finite spectral width inversely proportional to the signal duration. If such a finite pulse train is then processed by a filter, the output signal will, in some sense, sample the frequency response of the filter. Thus the output spectrum would exhibit three characteristics: an envelope, spectral lines at a regular spacing, and nonzero width to the spectral lines.

This project is intended to show these three effects. Since voiced speech is a real signal of this type, it is used as an example in the first exercise; the second exercise deals with synthetic signals.

Hints

To load data from a MATLAB file, see help `load`. If your computer has limited memory, use `clear` to remove arrays as soon as they are no longer needed, such as long speech signals.

The function `db` will create a log magnitude vector with a specified range. It will also normalize the maximum value to 0 dB.

EXERCISE 4.1

Fourier Transform of a Speech Vowel

Perform the following experiment with a recorded speech signal:

a. Load the file `bat.mat`, which contains speech data for the word "bat" spoken by a male speaker. After the `load` command the signal in the workspace will be called `sa`. It is about 4000 points in length, so you have to chop out a segment for analysis. Define $x[n]$ to be a 256-point section of the signal somewhere in the range $n = 1400$ to $n = 1800$. If you are using the Student version 3.5 of MATLAB, look for the file `bat1k.mat`, which was derived from `bat.mat` by chopping the speech signal into sections of length 1024. After this file is loaded, four vectors will have been defined: `sa1` through `sa4`. It will then be possible to find the desired 256-point section.

b. Compute the 256-point DFT of $x[n]$, and call it $X_1[k]$; then compute the 1024-point DFT of $x[n]$, and call it $X_2[k]$. Plot the log magnitudes of both on the same scale. One plot should be smoother than the other, but the frequency content should be about the same. The speech data were sampled at $f_s = 8$ kHz. Make the frequency-domain plots so that the horizontal axes are in hertz, labeled with frequencies from 0 up to the Nyquist frequency $(\frac{1}{2}f_s)$.

c. Zoom in on a region of the frequency plot by using the colon notation in MATLAB to select a section of the FFT. Note that the transform has many peaks, but they seem to be at a regular spacing. Measure this interpeak spacing in hertz.

d. The input signal $x[n]$ is nearly periodic—determine its period in milliseconds by measuring directly on a plot of the signal versus time.

e. State the relationship between the period of the speech signal and the interpeak spacing in the Fourier transform. Explain this relationship by giving the underlying Fourier transform pair that exemplifies this behavior.

f. Measure the spectral bandwidth of the lines in the frequency plot. Relate this to the total length of the signal analyzed.

EXERCISE 4.2

Filtering a Pulse Train

In this project we investigate the response of a first-order IIR filter to a pulse train input. This case has some similarity to the speech signal, and it can be analyzed with formulas. Perform the following experiment:

a. Generate an impulse train containing 207 samples. The spacing between impulses should be 23 samples. The height of each impulse can be 1. Call this signal $x[n]$.

$$x[n] = \sum_{\ell=0}^{8} \delta[n - \ell M_\circ] \qquad \text{with } M_\circ = 23$$

b. Derive a formula for the DTFT of $x[n]$. Then compute samples of the DTFT with a 512-point FFT and plot the magnitude. Since $|X(e^{j\omega})|$ is an even function of ω, only the region $0 \le \omega \le \pi$ need be plotted.

c. Filter this input signal with a first-order filter:

$$H(z) = \frac{B(z)}{A(z)} = \frac{1-a}{1-az^{-1}} \qquad \text{with } a = 0.95$$

The numerator value $(1-a)$ is needed to normalize the low-pass filter gain to be 1 at dc. The result of the filtering is an output $y[n]$ that is 207 samples long. Plot $y[n]$ and notice its quasi-periodic nature.

d. Compute the DTFT of $y[n]$ and plot the magnitude response $|Y(e^{j\omega})|$.

e. Compute the frequency response of the first-order filter $H(z)$ and plot its magnitude response.

f. Plot the log magnitudes of both DTFTs on the same graph. One plot should be smooth, while the other contains spectral lines. If you use `db()`, both plots will be normalized. Explain why the smooth one $|H(e^{j\omega})|$ is the envelope of the filtered line spectrum $|Y(e^{j\omega})|$.

g. Zoom in on a region of the log magnitude plot of $Y(e^{j\omega})$ by using the colon notation in MATLAB to select a section of the DTFT. Measure the interpeak spacing, $\Delta\omega$, in radians per second. State the relationship between the period of the input signal, $x[n]$, and $\Delta\omega$.

h. Measure the width of the spectral lines in the zoomed plot and relate this value to the total length of the impulse train signal. Prove that the spectral width is independent of the FFT length used to compute the DTFT samples.

i. Repeat part (f) with a second-order filter (3-1) with a resonant peak:

$$H(z) = \frac{1}{1 - 1.8\cos(6.8\pi/23)z^{-1} + 0.81z^{-2}}$$

Just plot $|H(e^{j\omega})|$ and the DTFT of the filter's output. Notice that the lines in the output spectrum do not necessarily hit the peak of the resonance.

PROJECT 5: FREQUENCY SAMPLING IN THE DTFT

The N-point DFT computes frequency samples of the DTFT located at

$$\omega_k = \frac{2\pi k}{N} \qquad k = 0, 1, \ldots, N-1$$

If the length of the signal $x[n]$ is L points, and L satisfies $L \le N$, the DTFT samples are computed by zero padding the signal $x[n]$ out to a length of N points prior to computing the N-point DFT. This can be accomplished in MATLAB by invoking the `fft` function with two arguments. For example, if the length of the vector `x` were 121 points, then 135 zeros would be appended if `fft( x, 256 )` were executed.

When the signal length is greater than the number of frequency samples N, two possible computations can be done. Either the signal can be truncated to length N, or the correct frequency samples can be computed. In this project we investigate both possibilities.

EXERCISE 5.1

Truncating the Signal

If the call to FFT is `fft(x,N)`, where `N` is less than `length(x)`, no error will be flagged. Instead, MATLAB will *truncate* the vector `x`.

a. Generate a test signal $x[n]$ of length 200, and then compute `fft( x, 128 )`. Show that exactly the same result will be obtained from `fft( x(1:128), 128 )`.

b. The implication of part (a) is that the last 72 samples of $x[n]$ are irrelevant. This causes problems for the frequency-sampling property of the DFT. If $X(e^{j\omega})$ is the DTFT of the 200-point signal $x[n]$, show that the 128 values from `fft( x, 128 )` are not equal to the frequency samples $X(e^{j\omega_k})$ at $\omega_k = (2\pi/128)k$.

EXERCISE 5.2

Fewer DFT Samples Than Data

In `fft( x, N )` when `N` is less than `length(x)`, the computed DFT does not satisfy the frequency-sampling property. In this exercise we develop a formula that modifies the length-L time signal $x[n]$, so that the N frequency samples of its DTFT $X(e^{j\omega})$ can be computed from a single N-point FFT.

a. Generate a test signal $x[n]$ of length 200, and then compute `X = fft( x, 512 )`. Show how the 128 frequency samples of the DTFT at $\omega_k = (2\pi/128)k$ can be obtained from the 512-point vector `X`.

b. The result from part (a) is a 128-point vector containing $X(e^{j\omega_k})$ sampled at $\omega_k = (2\pi/128)k$. If we compute its 128-point IFFT, we obtain a 128-point signal $\hat{x}[n]$ which can be related to the original 200-point signal $x[n]$. Make a plot of $\hat{x}[n]$ and $x[n]$ on a two-panel subplot for comparison. For which indices are the two signals identical? When $\hat{x}[n] \neq x[n]$ determine a simple relationship between the two.

c. In general, the case where $L > N$ should be handled with the following equation:

$$X(e^{j\omega_k}) = \sum_{n=0}^{N-1} \left(\sum_{\ell=0}^{r} x[n + \ell N] \right) e^{-j2\pi nk/N} \tag{5-1}$$

where $r = \lceil L/N \rceil$. The inner summation would be considered time aliasing of $x[n]$. Write a new function that will compute the DFT according to (5-1) when $L > N$. This M-file should compute only one FFT of length N. Test your function by taking the 128-point DFT of a 400-point signal and compare to the correct result, which could be found from a subset of a 512-point zero-padded FFT as in part (a).

chapter 4

MULTIRATE PROCESSING

OVERVIEW

The area of multirate processing is an example of a discipline with no comparable continuous-time signal processing counterpart. In many ways, linear time-invariant digital filters perform the same functions as those of linear time-invariant continuous-time filters. Multirate processing brings the unique advantages of discrete-time signal processing to signal processing. Certain filtering operations can be performed much more efficiently using multirate implementations. For example, a low-pass filtering operation can be split into a cascade of two stages, each of which is a low-pass filter. Since the output of the first low-pass filter has a smaller bandwidth than the input signal, the sampling rate may be reduced, saving computation. A fundamental idea in the area of multirate processing is that one should always perform calculations at the lowest possible rate. If higher rates are required to display a waveform, for example, interpolation can be used.

The area of multirate signal processing is very broad. It includes basic topics such as decimation and interpolation, and more involved topics such as perfect reconstruction filter banks and wavelets. In this chapter, only the basic techniques related to decimation and interpolation are covered.

The chapter begins with a section containing two projects on bandlimited interpolation. Interpolation is accomplished by first upsampling a signal, then filtering the upsampled signal. The first of the two projects compares several types of filters, while the second project develops the design of optimum interpolation filters for bandlimited interpolation.

Interpolation in the frequency domain allows a more detailed examination of the frequency characteristics of a signal. The section *Zoom Transform* considers two different methods for zooming in on the spectrum of a signal. The section *Rate Changing* presents very basic ideas related to the problem of converting from one sampling rate to the other. While decimation and interpolation can be used for changing sampling rates by an integer factor, the problem of rate changing for noninteger factors is more challenging. Combinations of interpolation and decimation are used along with appropriate filters to change rates by rational factors.

BACKGROUND READING

For background reading, a primary source is the book by Crochiere and Rabiner [1] devoted to the subject of multirate filtering. Additional information can be found in standard DSP texts such as Chapter 3 of [2] and Chapter 10 of [3].

[1] R. E. Crochiere and L. R. Rabiner. *Multirate Signal Processing*. Prentice Hall, Englewood Cliffs, NJ, 1983.

[2] A. V. Oppenheim and R. W. Schafer. *Discrete-Time Signal Processing*. Prentice Hall, Englewood Cliffs, NJ, 1989.

[3] J. G. Proakis and D. G. Manolakis. *Digital Signal Processing: Principles, Algorithms and Applications*. Macmillan, New York, second edition, 1992.

[4] M. Golomb and H. F. Weinberger. Optimal approximation and error bounds. In R. E. Langer, editor, *On Numerical Approximation*, chapter 6, pages 117–190. The University of Wisconsin Press, Madison, WI, 1959.

[5] H. W. Schüßler, G. Oetken, and T. W. Parks. New results in the design of digital interpolators. *IEEE Transactions on Acoustics, Speech, and Signal Processing*, ASSP-23(3):301–309, June 1975.

[6] P. P. Vaidyanathan. *Multirate Systems And Filter Banks*. Prentice Hall, Englewood Cliffs, NJ, 1993.

BANDLIMITED INTERPOLATION

OVERVIEW

The process of interpolation essentially corresponds to estimating or reconstructing the values of a signal at locations (times) between the sample values. Figure 4.1 depicts a system for interpolating a signal by a factor of L, where the output of the first system, referred to as a sampling rate expander, is

$$x_e[n] = \begin{cases} x[n/L] & n = 0, \pm L, \pm 2L, \text{etc.} \\ 0 & \text{otherwise} \end{cases}$$

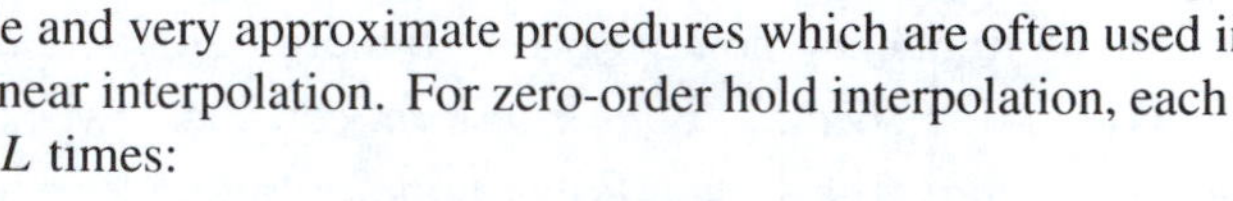

Figure 4.1
General system for interpolation by L.

The low-pass filter interpolates between the nonzero values of $x_e[n]$ to generate the interpolated signal $x_i[n]$. The output $x_i[n]$ essentially corresponds to an upsampled version of $x[n]$. When the low-pass filter is ideal, the interpolation is referred to as bandlimited interpolation.

Accurate bandlimited interpolation requires a carefully designed high-order low-pass filter. Two simple and very approximate procedures which are often used instead are zero-order hold and linear interpolation. For zero-order hold interpolation, each value of $x[n]$ is simply repeated L times:

$$x_i[n] = \begin{cases} x_e[0] & n = 0, 1, \ldots, L-1 \\ x_e[L] & n = L, L+1, \ldots, 2L-1 \\ x_e[2L] & n = 2L, 2L+1, \ldots \\ \vdots & \end{cases} \qquad (0\text{-}1)$$

This can be accomplished by convolving $x_e[n]$ with the impulse response

$$h_{\text{zoh}}[n] = \delta[n] = \delta[n-1] + \cdots + \delta[n-(L-1)] \qquad (0\text{-}2)$$

Zero-order hold interpolation is often used in digital-to-analog converters, resulting in analog "stairstep" waveforms (e.g., each digital sample is converted to a voltage, and that voltage is "held" for the duration of the sampling period).

Linear interpolation can be accomplished using a system with impulse response

$$h_{\text{lin}}[n] = \begin{cases} 1 - |n|/L & |n| \leq L - 1 \\ 0 & \text{otherwise} \end{cases} \tag{0-3}$$

Unlike the zero-order hold interpolator, the linear interpolator is noncausal and has zero group delay. The ideal bandlimited interpolator also has a noncausal impulse response:

$$h_{\text{ideal}}[n] = \begin{cases} \dfrac{\sin(\pi n/L)}{\pi n/L} & n \neq 0 \\ 1 & n = 0 \end{cases} \tag{0-4}$$

For further background reading on the topic of interpolation, a primary source is the book by Crochiere and Rabiner [1] devoted to the subject of multirate filtering. Additional information can be found in standard DSP texts such as Chapter 3 of [2] and Chapter 10 of [3].

Project 1: Interpolation Filter Performance

In this project, the performance of three different interpolation filters is evaluated: the zero-order hold, linear interpolator, and a high-order, sharp-cutoff low-pass filter.

Hints

To perform the sampling rate expansion shown in Fig. 4.1 to generate the signal $x_e[n]$, it will be useful to define the function `srexpand(x,L)`, which takes a sequence $x[n]$ and an integer L as arguments and returns a sequence that is a zero-filled version of $x[n]$, as follows:

```
%srexpand  y = srexpand(x,L) zero fills a sequence X by placing L-1
%          zeros between each sample of the sequence. The resulting
%          sequence has length equal to length(X)*L.
function y = srexpand(x,L)
N = L*length(x);
y = zeros(1,N);
y(1:L:N) = x;
```

The MATLAB function `filter(b,a,x)` implements a difference equation. The output of `filter` will be exactly as long as the input data sequence—the same as convolving the input data sequence with the impulse response and truncating the result to the length of the input sequence. The MATLAB function `conv` performs true convolution. Explain why it is preferable to use the function `filter`, which truncates its output to the same length as its input. Before continuing, type `load BLIdata` at the MATLAB prompt to load the variables for this project.

EXERCISE 1.1

Linear and Zero-Order Hold Interpolation

a. Given $x_e[n]$, the output of the sampling-rate expander in Fig. 4.1, write out the difference equations that correspond to the zero-order hold interpolation filter and the linear interpolation filter for the impulse responses $h_{\text{zoh}}[n]$ and $h_{\text{lin}}[n]$. Note that the impulse response in (0.3) and therefore the difference equation for the linear interpolator is noncausal.

b. Enter the impulse response for the zero-order hold and linear interpolators using the value $L = 5$. The zero-order hold interpolator should satisfy equation (0-2). Note that the linear interpolator in equation (0-3) is a noncausal filter. Since MATLAB does not recognize negative indices, you must enter a causal version, and then "time advance" the result by relabeling the time axis appropriately. Assign the impulse response sequences to the MATLAB variables `hzoh` and `hlin`, respectively. Plot the magnitude of the frequency responses of the zero-order hold and linear interpolation filters on the same graph. Which is a better approximation to the ideal bandlimited interpolator?

c. Using the MATLAB functions `srexpand` and `filter`, implement the interpolation system of Fig. 4.1 with a value of $L = 5$. Use both the zero-order hold and the linear interpolation impulse responses as the interpolation filter. For input to the system use the sequence `data1` that has been provided. Assign the upsampled sequences to the MATLAB variables `xzoh1` and `xlin1`.

d. Using the MATLAB function `stem`, plot the sequences `hzoh`, `hlin`, `xzoh1`, and `xlin1` on a time axis from -10 to $+10$. In establishing the time axis for the plot, be sure to take into account that `hlin` is noncausal. You can check whether or not you have the time axis appropriately lined up for `xzoh1` and `xlin1` by noting that for both the zero-order hold and the linear interpolator, all the values of the original data are preserved exactly on the expanded time axis.

EXERCISE 1.2

Interpolation with a Sharp-Cutoff Filter

a. You have been provided with an interpolation filter which more closely approximates the ideal low-pass filter than either the linear interpolator or the zero-order hold interpolator. The impulse response to a causal version of this filter is located in the MATLAB variable `sharpfilt`. The cascade of the causal version and an ideal time advance of 20 samples is the correct (noncausal) implementation. Analytically express the phase of the frequency response of the *noncausal* filter.

b. Plot the magnitude of the frequency response of the filter `sharpfilt`.

c. Interpolate the sequence `data1` using `sharpfilt`. Assign the resulting sequence to the MATLAB variable `xsf1`.

d. Using the MATLAB function `stem` plot `xsf1` on a time axis from -30 to $+50$.

EXERCISE 1.3

Some Bookkeeping

In Exercises 1.1 and 1.2 we have implemented the system of Fig. 4.1 with $L = 5$ for three different filters. In Exercise 1.4 we want to compare the results with essentially ideal band-limited interpolation. To do that we need to be careful about comparing appropriate segments of $x_i[n]$.

The MATLAB function `filter` filters a finite-length input data vector with a causal filter and returns an output data vector which is truncated to be the same length as the input data vector. Implementing a noncausal interpolation filter requires filtering with a causal version of the impulse response, then time advancing the output. Consequently, after filtering $x_e[n]$ and applying the time advance appropriate to each filter, `xzoh1` `xlin1` and `xsf1` can only be compared on an interval of length less than the length of `data1`.

a. For each of the three filters, with the appropriate time advance incorporated, specify the interval in n over which $x_i[n]$ is available.

b. From your answer in part (a), what is the largest interval common to all three filtered outputs?

c. A second bookkeeping issue is that the MATLAB function `filter` assumes that `data1` is zero for $n < 0$, and consequently, there is a startup transient in filtering until the filter impulse response totally engages the data. The duration of this transient depends on the length of the FIR filter. The three filtered outputs should only be compared after the transient. Taking this into account together with your answer to part (b), `xzoh1, xlin1,` and `xsf1` should only be compared over a common time interval $n_1 \leq n \leq n_2$. Determine n_1 and n_2.

EXERCISE 1.4

Performance

In this exercise we evaluate the performance of the various interpolation filters compared with what would be obtained by essentially bandlimited interpolation. The sequence corresponding to perfect bandlimited interpolation is contained in the MATLAB variable `ideal1`.

A measure of the average interpolation error is

$$e_x = \frac{1}{n_2 - n_1 + 1} \sum_{n=n_1}^{n_2} (x_i[n] - x_{\text{ideal}}[n])^2 \qquad (1\text{-}1)$$

where $x_i[n]$ is the result of using one of the interpolating filters, $x_{\text{ideal}}[n]$ is the ideal bandlimited interpolation, and the parameters n_1 and n_2 are those determined in Exercise 1.3(c). Compute the average interpolation error (1.1) for the three filters, $e_{\text{xzoh1}}, e_{\text{xlin1}}$, and e_{xsf1}. Which interpolation filter has the best performance? Is this what you expected?

EXERCISE 1.5

New Data

Upsample the sequence located in the MATLAB variable `data2` using the interpolators `hlin, hzoh,` and `sharpfilt`. Recompute the interpolation errors by comparing the interpolated sequences with the sequence located in the MATLAB variable `ideal2`. Which filter performs best for this set of data? Why are the interpolation errors so different from those computed above? (*Hint*: Examine the magnitude of the Fourier transforms of the two expanded sequences.)

PROJECT 2: OPTIMUM MIN-MAX INTERPOLATION

In this project, the theory of optimal estimation of signals is applied to the problem of estimating missing samples of a bandlimited signal. For details of the theory, see the chapter by Golomb and Weinberger [4]. The results we need can be summarized as follows. Given signal measurements (linear functionals)

$$F_i(\mathbf{u}) = f_i \qquad i = 1, \ldots, N$$

where the values of the linear functionals F_i are f_i for the signal $\mathbf{u}$, and given that $\mathbf{u}$ belongs to the signal class

$$C = \left\{\mathbf{u} \in \mathcal{H} : \langle \mathbf{u}, \mathbf{u} \rangle \leq r^2,\ F_i(\mathbf{u}) = f_i \qquad i = 1, \ldots, N\right\}$$

the best estimate of the signal, $\hat{\mathbf{u}}$, is a linear combination of the representers ϕ_i of the linear functionals, $F_i(\cdot)$.

$$\hat{\mathbf{u}} = \sum_{i=1}^{N} c_i \phi_i$$

where the coefficients c_i are chosen so that $\hat{\mathbf{u}}$ has the given values f_i of the linear functionals,

$$F_i(\hat{\mathbf{u}}) = f_i \quad i = 1, \ldots, N$$

The signal estimate $\hat{\mathbf{u}}$ is best in the sense that it minimizes the maximum error:

$$\max_{\mathbf{u} \in C} |F(\mathbf{u}) - F(\hat{\mathbf{u}})|$$

In this project, the linear functionals will be the time samples of the signal $\mathbf{u}$, that is,

$$F_i(\mathbf{u}) = u(i)$$

and the class C will be the class of bandlimited, finite-energy ($\sum u(n)^2 < r^2$) signals. A bandlimited signal $\hat{\mathbf{u}}$ with bandwidth B is found which passes through given, equally spaced samples. Samples of this bandlimited signal are computed by filtering a sequence made by inserting zeros in between each of the original known samples, as shown in Fig. 4.2. In the figure, the sampling rate is increased by a factor of $r = 4$ by using a weighted combination of $L = 2$ points on each side of the point to be estimated. The weights on these four points are calculated from the requirement that a bandlimited signal of normalized bandwidth $\alpha/2r$, where $0 < \alpha < 1$, fits the four known samples and has minimum energy.

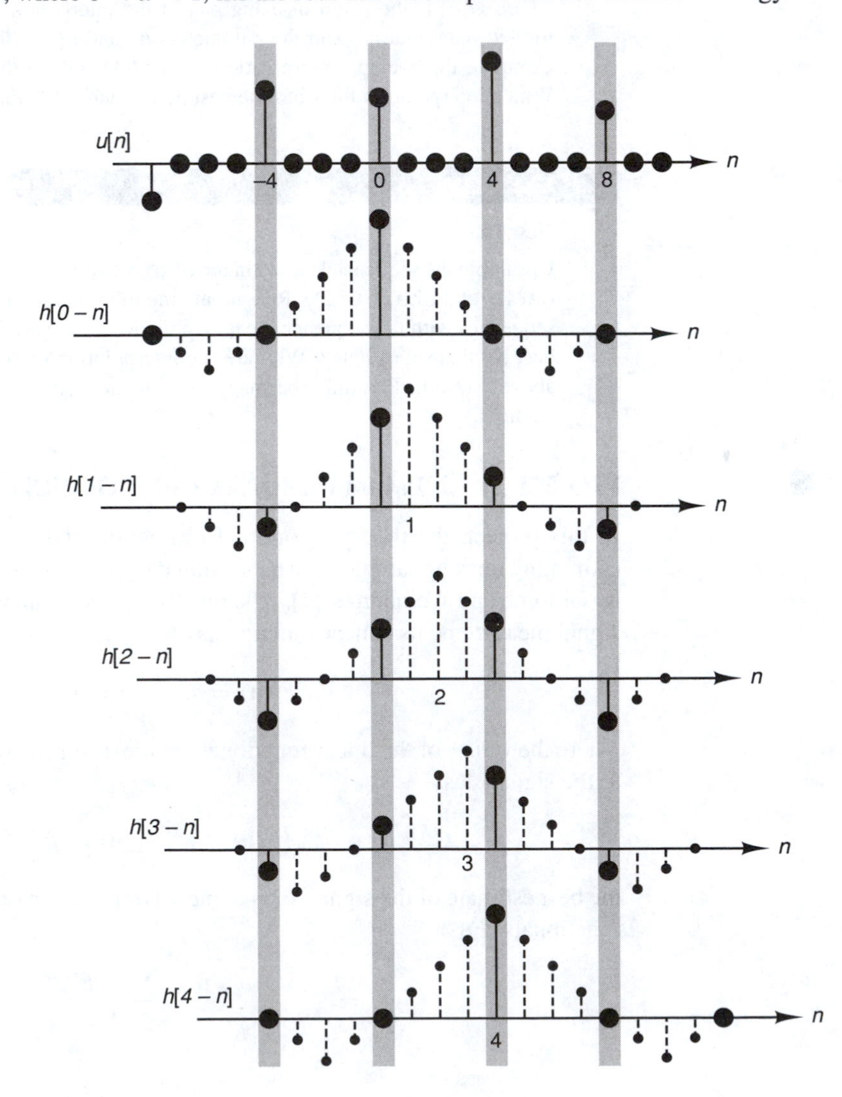

Figure 4.2 Interpolation as a linear filtering process. Each successive line of the figure shows the flipped and shifted version of the impulse response $h[n]$. Note where the zeros of $u[n]$ overlap $h[k-n]$ and where zeros of $h[k-n]$ overlap $u[n]$.

EXERCISE 2.1

Representers

The Riesz representation theorem states that our linear functionals $F_i(\mathbf{u})$ can be represented as inner products of the vector $\mathbf{u}$ and another vector ϕ called the representer of the linear functional F_i. In this exercise you are asked to show that the representers are sinc functions.

Show that if $u[n]$ is a bandlimited discrete-time signal with normalized bandwidth B, then

$$u[n_0] = 2B \sum_{n=-\infty}^{\infty} u[n]\text{sinc}(2B(n_0 - n))$$

where

$$\text{sinc}(x) = \frac{\sin(\pi x)}{\pi x}$$

In other words, the representer of the linear functional $F(\mathbf{u}) = u[n_0]$ is

$$\phi[n] = 2B\ \text{sinc}(2B(n_0 - n))$$

when the inner product is

$$\langle \mathbf{x}, \mathbf{y} \rangle = \sum_{n=-\infty}^{\infty} x[n]y[n]$$

EXERCISE 2.2

Interpolation by a Factor of 4

Assume that a signal $u[n]$ is bandlimited to a bandwidth $B = \frac{1}{8}\alpha$, where $0 < \alpha < 1$. The best linear estimate for $u[k]$, in the range $k = 0, 1, 2, 3$, given $u[-4]$, $u[0]$, $u[4]$, $u[8]$, has the form

$$\hat{u}[k] = \sum_{m=-1}^{2} a_{k.m} u[4m]$$

Show that the coefficients $a_{k.m}$ are given by solution of the following 4×4 system of linear equations:

$$\sum_{m=1}^{4} \text{sinc}(\alpha(m - n))a_{k.m} = \text{sinc}\left(\alpha\left((n-2) - \frac{k}{4}\right)\right) \qquad k = 0, 1, 2, 3 \qquad (2\text{-}1)$$

EXERCISE 2.3

Bandlimited Interpolation Filter

Find the impulse response of a length-15 filter as shown in Fig. 4.2, which when convolved with the zero-filled bandlimited signal

$$u[n] = 0 \qquad n \bmod 4 \neq 0$$

with bandwidth $\frac{1}{8}\alpha$ will give the optimum bandlimited interpolation of the missing samples.

a. Explain how the coefficients $a_{k.m}$ found in (2-1) can be rearranged to form the necessary impulse response (as in Fig. 4.2).

b. Use the function `oetken` shown below to design the desired impulse response (for more details, see [5]).

c. Examine this filter in the frequency domain by plotting the magnitude of its frequency response. How does it compare with the magnitude responses of the zero-order hold interpolation filter and the triangular, linear interpolation filter?

The M-file `oetken` sets up and solves linear equations as in (2-1). The parameter ρ in the M-file corresponds to the variable k in (2-1).

```
function h = oetken( alpha, r, L )
%OETKEN   design interpolation filter via least-squares
%   usage:
%         h = oetken( alpha, r, L )
%    where:
%    alpha/2r = bandwidth of filter (in normalized freq)
%                 0<alpha<1  with alpha=0.9 typical
%       2rL-1 = length of filter
%           h = resultant impulse response
%
% This function designs an FIR filter for minimum norm bandlimited
% interpolation using L points on each side of the point being
% estimated, with an interpolation factor of r.
% Reference:
%     G.Oetken, T.W.Parks, and H.W.Schuessler,
%       "New results in the design of digital interpolators"
%      IEEE Trans. ASSP, vol. ASSP-23, pp.301-309, June 1975.

% 15 May 93    Jim McClellan  (adapted from TW Parks ver 18 July 91)

%----First compute sinc matrix

nm = toeplitz( 0:-1:(1-2*L), 0:(2*L-1) );
S = sinc( alpha*nm );        %--- NOTE: sinc(x) = sin(PI*x)/(PI*x)

%---Next compute RHS, b, for each value of rho=0,...,r-1.

nm = r*[(-L):(L-1)]' * ones(1,r)  + ones(2*L,1)*[0:(r-1)];
b = sinc( (alpha/r) * nm );

%--- Compute the matrix of impulse responses for rho=0,...,r-1.
%---  and then form the impulse response of the interpolator

H = [S \ b]';
h = H( 2:(2*r*L) );
```

The sinc function can be computed for a matrix of values by using the following M-file.

```
function y = sinc( x )
%SINC   compute sin(PI*x)/(PI*x)    (for a matrix x)
%   usage:
%         y = sinc(x)
%
%NOTE: this is the OFFICIAL definition of "sinc"
%       but sometimes the factor of PI is dropped

jkl = abs(x) < 1e-10;
x = pi*x;
y = jkl + (~+jkl).*sin(x)./(x + jkl);
```

ZOOM TRANSFORM

OVERVIEW

The N-point DFT of an N-point sequence represents samples of the DTFT and contains all the information about the DTFT of the sequence. However, some characteristics of the DTFT may not be visually apparent from these samples. Consequently, it is often useful to interpolate the DFT in a frequency band of interest. In this set of exercises we examine two different methods for performing this interpolation, often referred to as the *zoom transform*.

In both approaches, we consider $X_N[k]$ to be the N-point DFT of a finite-length sequence $x[n]$, representing N frequency samples separated in frequency by $2\pi/N$. Given $X_N[k]$, we would like to zoom in on the region between $\omega_c - \Delta\omega$ and $\omega_c + \Delta\omega$. We assume that in this region we want to calculate L equally spaced frequency samples; that is, the result of the zoom transform will be L equally spaced frequency samples in the interval $\omega_c - \Delta\omega$ to $\omega_c + \Delta\omega$, specifically, the frequency samples

$$\omega_k = (\omega_c - \Delta\omega) + \frac{2\Delta\omega}{L}k \qquad k = 0, 1, \ldots, L-1$$

Material specific to this project can be found in [2, Prob. 11.4] and [3, Prob. 10.20].

PROJECT 1: ZOOM TRANSFORM

EXERCISE 1.1

The first method that we consider is shown in Fig. 4.3. Starting from the N-point DFT, $X_N[k]$, $x[n]$ is computed, modulated, and low-pass filtered to form $x_1[n]$, then compressed to form $x_z[n]$. The P-point DFT of $x_z[n]$ contains the L desired zoom transform samples ($P \geq L$). Assume that $h[n]$ is the impulse response of an ideal low-pass filter with frequency response

$$H(e^{j\omega}) = \begin{cases} 0 & -\pi \leq \omega < -\Delta\omega \\ 1 & -\Delta\omega \leq \omega < \Delta\omega \\ 0 & \Delta\omega \leq \omega \leq \pi \end{cases}$$

and that $f[n]$ is the complex exponential sequence

$$f[n] = e^{-j\omega_c n}$$

Depending on the choice of M, the sequence $x_z[n]$ may need be extended with zero values (zero-padded) prior to computing the P-point DFT. Find appropriate values for $\Delta\omega$ and M such that the value of P in Fig. 4.3 can be equal to L, assuming that $\pi/\Delta\omega$ is an integer. With $P = L$, we don't compute more DFT samples than we desire.

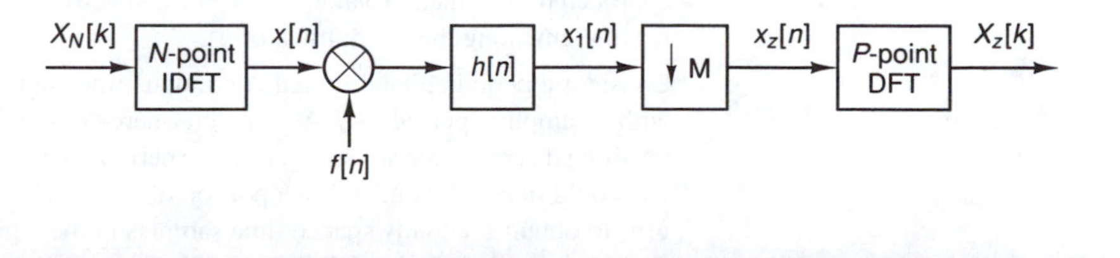

Figure 4.3
Zoom transform, method 1.

EXERCISE 1.2

Consider $x[n]$ with Fourier transform $X(e^{j\omega})$ shown in Fig. 4.4. Sketch the Fourier transforms of the intermediate signals $x_1[n]$ and $x_z[n]$ when $\omega_c = \pi/3$ and $\Delta\omega = \pi/4$, and when M is chosen as in Exercise 1.1.

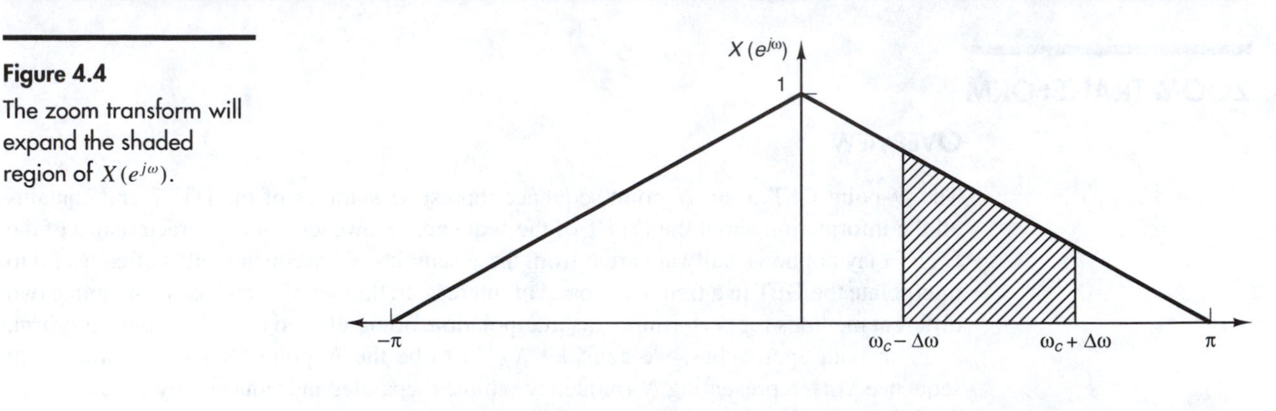

Figure 4.4
The zoom transform will expand the shaded region of $X(e^{j\omega})$.

EXERCISE 1.3

In the system in Fig. 4.3, $h[n]$ is a low-pass filter. If the filter is ideal, $\Delta\omega$ and M can be chosen as in Exercise 1.1. However, since the transition bands of any $H(e^{j\omega})$ we can implement has nonzero width, we will have to choose a smaller value of M to avoid aliasing. For $M = 3$ and $\Delta\omega = \pi/4$ an appropriate set of specifications is:

- Passband edge frequency $\omega_p = 0.25\pi$
- Stopband edge frequency $\omega_s = 0.31\pi$
- Passband tolerance $\delta_1 = 0.01$ [passband varies from $(1 + \delta_1)$ to $(1 - \delta_1)$]
- Stopband tolerance $\delta_2 = 0.01$

Use any appropriate design method to obtain the impulse response of a linear-phase FIR filter meeting these specifications. Be sure to document the method used, and demonstrate that your filter meets the specifications given.

EXERCISE 1.4

Implement the system in Fig. 4.3 with $N = 512$, $L = 384$, $M = 3$, $\Delta\omega = \pi/4$, and $\omega_c = \pi/2$. Choose an appropriate value for P. Test your system on the sequence

$$x[n] = \sin(0.495\pi n) + \sin(0.5\pi n) + \sin(0.505\pi n) \qquad 0 \le n \le 511$$

Turn in a plot of the magnitude of the DFT of $x[n]$ and the magnitude of the zoomed DFT for $\omega_c = \pi/2$ and $\Delta\omega = \pi/4$. Specifically note which points on the zoomed DFT correspond to $\omega_c - \Delta\omega$, ω_c, and $\omega_c + \Delta\omega$.

EXERCISE 1.5

In this exercise we consider the problem of zooming in (in the time domain) on a portion of a discrete-time periodic sequence $\tilde{g}[n]$. In Exercise 1.6 we will then use a somewhat similar approach to obtain an expanded view of the spectrum of a signal (i.e., as an alternative method for implementing the zoom transform).

Consider a periodic, bandlimited continuous-time signal $\tilde{g}_c(t)$ with period $T = 2\pi$, sampled with a sampling period of $T_s = 2\pi/N$ (where N is sufficiently large to avoid aliasing). The resulting discrete-time signal $\tilde{g}[n]$ will be periodic with period N. Given only the samples $\tilde{g}[n]$, we would like to "zoom" in on a portion of $\tilde{g}_c(t)$. This can be accomplished by interpolating $\tilde{g}[n]$ to obtain L equally spaced time samples in the region between $t_c - \Delta t$ and $t_c + \Delta t$. The time samples t_k should satisfy

$$t_k = (t_c - \Delta t) + \frac{2\Delta t}{L}k \qquad k = 0, 1, \ldots, L-1$$

The basic approach is shown in Fig. 4.5.

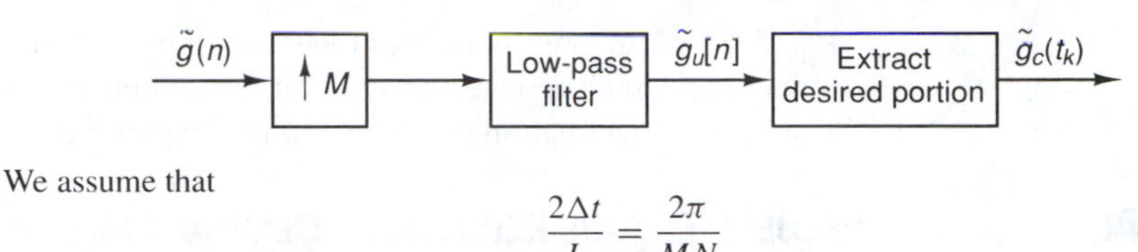

Figure 4.5
Interpolate $\tilde{g}[n]$.

We assume that

$$\frac{2\Delta t}{L} = \frac{2\pi}{MN}$$

If the low-pass filter is a linear-phase filter with integer group delay, the upsampled version of $\tilde{g}[n]$ ($\tilde{g}_u[n]$) will correspond to $\tilde{g}_c(t)$ sampled at integer multiples of T_s/M. If the group delay is not an integer, the samples $\tilde{g}_u[n]$ can be at noninteger multiples of T_s/M.

a. For $M = 3$, determine $H(e^{j\omega})$, the Fourier transform of the low-pass filter in Fig. 4.5, so that the desired output will be obtained.

b. Suppose that you used the following signal as the input to the system:

$$g[n] = \begin{cases} \tilde{g}[n] & 0 \le n < N \\ 0 & \text{otherwise} \end{cases}$$

and the low-pass filter implemented a circular convolution rather than a linear convolution, to obtain a finite-length output $g_u[n]$. Would $g_u[n]$, be equivalent to one period of $\tilde{g}_u[n]$? Explain why or why not.

EXERCISE 1.6

We now want to use an interpolation strategy similar to that in Exercise 1.5 to interpolate the DFT and extract the desired frequency samples. Essentially, the strategy is to upsample $X_N[k]$ as indicated in Fig. 4.6. As before, $X_N[k]$ is the N-point DFT of a sequence $x[n]$, $G_{NM}[k]$ is the NM-point DFT of a sequence $g[n]$, and $H_{NM}[k]$ is the NM-point DFT of a sequence $h[n]$. $\tilde{x}_1[k]$, the output of the "low-time lifter," is the NM-point circular convolution of $G_{NM}[k]$ and $H_{NM}[k]$. The sequence $X_z[k]$ then corresponds to the desired frequency samples.

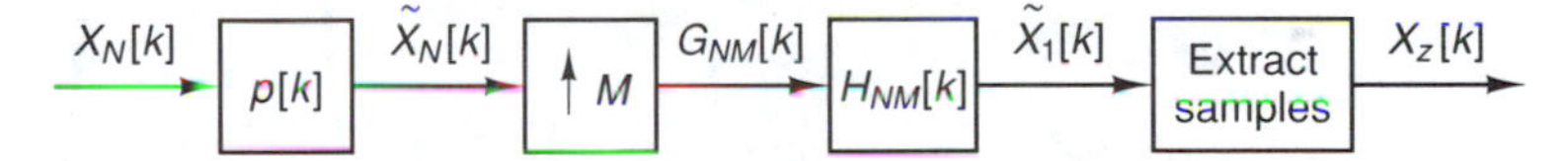

Figure 4.6
Zoom transform, method 2.

a. In terms of N and M, specify $h[n]$ so that $X_1[k]$ will exactly correspond to the NM-point DFT of $x[n]$.

b. One way of approximately implementing the circular convolution in Fig. 4.6 is to replicate $G_{NM}[k]$ several times and implement the linear convolution of this replicated sequence with the finite-length sequence $H_{NM}[k]$. Implement this strategy with an appropriately chosen $H_{NM}[k]$ for the parameters and test signal used in Exercise 1.4. Hand in a plot of the magnitude of the resulting 384 "zoomed" frequency samples. Also, explicitly indicate how many replications of $G_{NM}[k]$ were used to approximate the circular convolution.

c. The desired circular convolution can also be implemented through the use of the DFT (i.e., by multiplying together the IDFTs of the two sequences to be circularly convolved). Implement the system of Fig. 4.6 using this strategy for the circular convolution. Again, hand in a plot of the magnitude of the resulting 384 "zoomed" frequency samples.

RATE CHANGING

OVERVIEW

In this section, we will be looking at the ideas of decimation and rate changing. In particular, we will need to learn how to recognize aliasing in a multicomponent signal, how decimation may cause aliasing, and how a predecimation filter may prevent aliasing (at a cost of some loss of information, however). Rate changing is accomplished with a combination of upsampling and downsamping, or decimation.

A thorough presentation of rate changing can be found in the book by Crochiere and Rabiner [1], which is devoted to the subject of multirate filtering. Background material on the subject of decimation can be found in Chapter 3 of [2] and Chapter 10 of [3].

PROJECT 1: RATE REDUCTION: DECIMATION

This project uses a graphical approach to illustrate the effect of decimation (or rate reduction) on a waveform. To recognize aliasing in the time domain (as plotted by MATLAB), we need to use some type of frequency analyzer. That is, we first need to recognize the individual sinusoidal components of a particular signal and then identify the frequencies of these components. Only then can we determine whether or not these components have been aliased down from higher component frequencies. In the case of a single-frequency component, we only need to look for a single-sinusoidal-output component. This analysis process becomes more difficult when there are more components to identify (and much more difficult with real audio signals such as speech and music).

The eye is quite good at identifying apparent patterns in graphical data. For example, Fig. 4.7 shows a waveform for a signal that consists of a fundamental frequency and a third harmonic. When the signal is plotted as a continuous function, as in Fig. 4.7, we feel confident in identifying the relative frequencies of the two components.

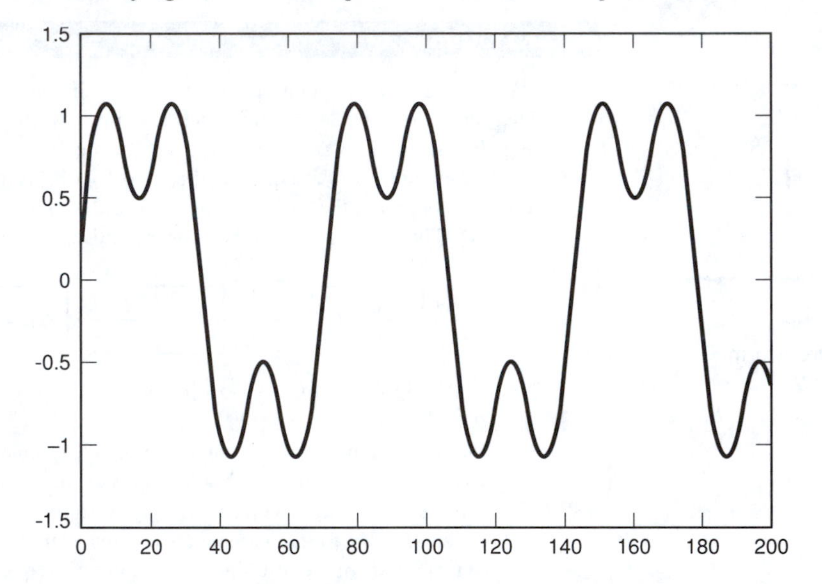

Figure 4.7 Continuous two-component signal.

Similarly, when presented as a dense set of discrete samples (Fig. 4.8), we will usually draw the same conclusions about the frequencies present, assuming that there is no aliasing. What we have seen here is that if we have many samples per cycle at all frequencies of interest, we can get a good idea what frequency components are present by visual inspection of the signal.

Figure 4.9 shows the same signal with 10 times fewer samples than in Fig. 4.8. The signal's frequency content is now clearly less obvious. While the eye still detects a certain periodicity at the lower of the two components, it is not at all clear that there is a higher frequency present (the deviations from the waveform of the lower-frequency component might have been due to noise, for example). If there is a higher frequency, it is not obvious what its frequency might be.

However, if we interpolate the signal of Fig. 4.9 by a factor of 4 (using bandlimited interpolation), we obtain the signal of Fig. 4.10, which again shows strong evidence of the third harmonic. These interpolation methods make the assumption that the signals are bandlimited. In other words, we are attempting to recover samples that would have been obtained if we had simply sampled faster in the first place. This is possible if the sampling theorem was obeyed at the lower (uninterpolated) rate.

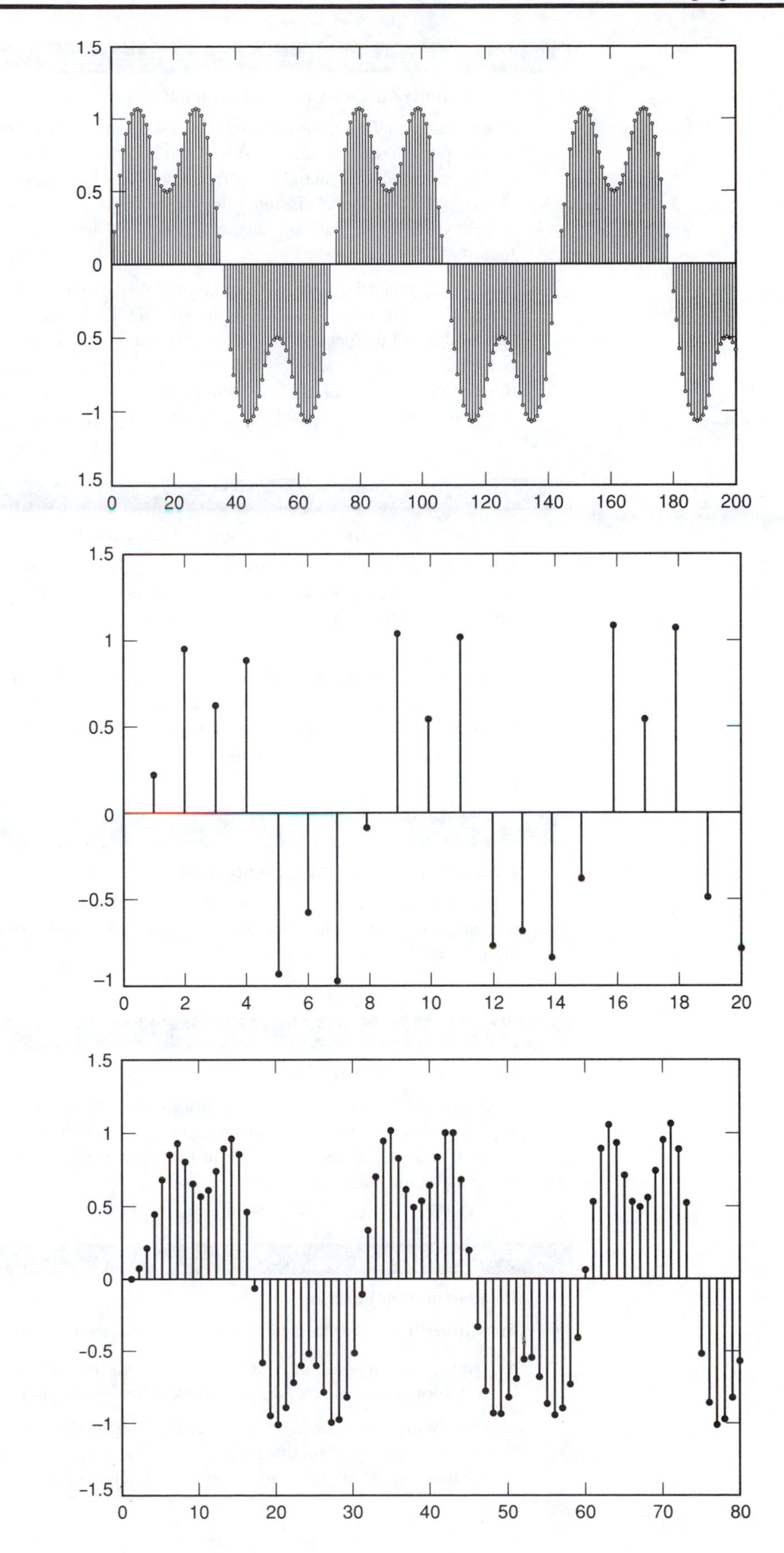

Figure 4.8
Discrete two-component signal.

Figure 4.9
Figure 4.8 at 1/10 the sample rate.

Figure 4.10
Figure 4.9 interpolated by factor of 4.

EXERCISE 1.1

Decimation of a Two-Component Signal

Generate a signal consisting of two sinusoidal components, one with a frequency of 0.02 times the sampling frequency and the other with a frequency of 0.09 times the sampling frequency. Plot about two cycles (at the lower frequency) worth of data. Plot this as a continuous function (MATLAB `plot`) and as discrete samples (MATLAB `stem`) in the manner of Figs. 4.7 and 4.8, respectively. Note that you have more than 10 samples per cycle at the higher frequency. Identify the cycles by eye.

Now we decimate the signal. We will look at this in two ways—with and without an appropriate "predecimation" filter. Whether or not such a filter is needed depends on the signal's original bandwidth and the decimation factor. We can gain considerable insight into this problem by noting that if we take samples and then throw them away, it is the same as not taking them in the first place. Thus, if aliasing would have occurred if the original signal had been sampled at the lower rate, aliasing will occur if we do not use an appropriate predecimation filter.

EXERCISE 1.2

Decimation without Predecimation Filter—Case 1

Decimate the signal obtained above by a 4:1 factor, and try to identify which frequencies are present. However, if we want to identify frequencies visually, we need more samples per cycle! To get around this "Catch 22" we use bandlimited interpolation.

Thus we first decimate by a factor that leaves relatively few samples per cycle at the highest-frequency component. Then we use bandlimited interpolation on the decimated signal. Consult the section *Bandlimited Interpolation* for ways to implement discrete-time interpolation. Implement one of these interpolators to produce samples of the signal at a higher rate and then identify the frequency components.

EXERCISE 1.3

Decimation without Predecimation Filtering—Case 2

Repeat Exercise 1.2, but use a 6:1 decimation factor. Estimate (visually) the frequencies in the sampled signal by interpolating back up to the original rate. Explain why the interpolated signal is different from the original.

EXERCISE 1.4

Use of Pre-Decimation Filter

Decimate 6:1 as in Exercise 1.3, but prior to decimation, low-pass filter the input sequence at slightly below one-half the final sampling rate. This low-pass filter serves as an antialiasing filter. Compare the decimated output to both the original input signal and the output of the antialiasing low-pass filter.

EXERCISE 1.5

Postdecimation Filtering

Sometimes it is possible to carry out a filtering operation either before or after decimation.

a. Apply the low-pass filter of Exercise 1.4 to the input signal and then decimate by 4:1. Interpolate the result so that visual identification of the frequency components is possible.

b. Apply the low-pass filter of Exercise 1.4 to the decimated output signal of Exercise 1.2. Because the low-pass filter was designed with respect to the input sampling rate, it must be used on the interpolated output from Exercise 1.2.

c. Now compare the signals produced in parts (a) and (b). Make sure that both signals are sampled at the same rate. When can you use the predecimation filter as a postdecimation filter with the same results?

PROJECT 2: RATE CHANGING

If you combine rate reduction (decimation by an integer factor q) and rate increase (interpolation by an integer factor p), it is possible to implement a system that will change the sampling rate of a signal by any rational factor p/q. Indeed, in MATLAB the cascade of `decimate` and `interp` (from the Signal Processing Toolbox) would accomplish such a rate change. In this project you are expected to do your own implementation of a rate-changing system, that is, design the appropriate low-pass filters, and implement the correct sequence of compressors, expanders, and filters.

Since a rate-changing system requires interpolation, see the section *Bandlimited Interpolation* for methods of designing interpolation filters. For some background reading about noninteger rate changing, see [2, Sect. 3.6.3].

EXERCISE 2.1

Rate Increase by a Factor of 3/2

Using a signal of your choice, implement a 3/2 sample rate increase. Does the order of interpolation and decimation matter for this case? Consider both the maximum sampling rate involved here and any possible problems with aliasing. Specify the cutoff frequency needed for the low-pass filter(s). Do you need a decimation filter for this case?

Implement your system to process the signal generated in Exercise 1.2. What would happen if input signals at different frequencies were chosen?

EXERCISE 2.2

Rate Decrease by a Factor of 2/3

Using a signal of your choice, implement a 2/3 sample rate decrease. Does the order of interpolation and decimation matter for this case? Do you need a decimation filter for this case? If not, why not? Do you need a separate decimation filter, or does the interpolation filter do the job of both?

Implement the 2/3 rate-changing system for a general input, and then process the signal generated in Exercise 1.2. Then produce an input signal whose normalized frequency is 44% of the sampling frequency and process it through the system. Explain the output that is produced. What is maximum usable frequency range of this particular rate converter?

chapter 5

SYSTEMS AND STRUCTURES

OVERVIEW

In this chapter we deal with linear systems, their possible different descriptions, their division into classes with different properties, and some of the numerous structures available for their implementation. Of special importance are those descriptions that characterize the system completely. MATLAB provides some programs for the transformation of one set of parameters, belonging to one type of description into another. Additional programs of this type are developed in the exercises. It turns out that some of these transformations are simple, while others require rather involved calculations, thus yielding an impression of the feasibility of certain descriptions. Obviously, there is a relation to the identification problem: Starting with an existing system with unknown parameters, measuring procedures are of interest, which provide results to be used for the calculations of a complete description of the system under consideration.

Other characterizations show specific properties of a system only, but they fail if completeness is required. It is one purpose of the corresponding project to show the difference between a complete and an incomplete description of a system. Based on its properties, expressed in either the time or frequency domain, systems can be separated into classes. That includes basic features such as stability and causality, but also distinctions based on the length of the impulse response, which can be of finite or infinite length. Other separations are related to the phase response, yielding the classes of linear-, non-minimum-, and minimum-phase systems. Furthermore, all-passes have to be considered. Besides several practical applications, they are of theoretical interest, since a non-minimum-phase system can always be described as a cascade of a minimum-phase system and an appropriately chosen all-pass. Finally, a distinction between systems based on energy considerations is of interest. Lossless systems, especially, which can be implemented with all-passes as building blocks, found practical applications.

Tests are to be developed yielding a decision about the type of system in terms of these classes, while in other projects, the properties of the different groups are considered. Furthermore, it turns out that the required amount of information for a complete description of a system can be reduced if the class the system belongs to is known.

Considering structures of systems yields a bridge to its actual implementation. There are numerous possibilities for building a system (e.g., when its transfer function is given). Under ideal conditions (i.e., if all wordlength effects are neglected), all different implementations yield the same behavior if the design has been done properly. But the implementations might differ in terms of the required number of arithmetic operations and delay elements. So the few structures considered in this chapter will be compared in terms of their arithmetic complexity, while their differences due to the limited wordlength are one topic in Chapter 7.

BACKGROUND READING

We refer to a large extent to Chapters 5 and 6 in Oppenheim and Schafer [1] and to Chapters 8 and 9 in Roberts and Mullis [2]. Chapter 8 in Lim and Oppenheim [3] contains a discussion of lossless structures and Chapters 2 and 5 in Kailath [4] treat state-variable representations.

[1] A. V. Oppenheim and R. W. Schafer. *Discrete-Time Signal Processing*. Prentice Hall, Englewood Cliffs, NJ, 1989.

[2] R. A. Roberts and C. T. Mullis. *Digital Signal Processing*. Addison-Wesley, Reading, MA, 1987.

[3] J. S. Lim and A. V. Oppenheim. *Advanced Topics in Signal Processing*. Prentice Hall, Englewood Cliffs, NJ, 1988.

[4] T. Kailath. *Linear Systems*. Prentice Hall, Englewood Cliffs, NJ, 1980.

SYSTEMS AND STRUCTURES

OVERVIEW

Linear systems with constant coefficients can be described differently. In the time domain a difference equation, which might be given in state-space notation, or the impulse response yield all information. Correspondingly, the transfer function $H(z)$ in its different representations provides complete descriptions in the z-domain. Its specialization on the unit circle yields the frequency response $H(e^{j\omega})$ and its components. In the first two projects in this chapter we deal with these possibilities. As we shall see, most of these descriptions characterize the system completely, some only partially. If they provide a complete description, all the others can be found by appropriate transformations.

Systems can be divided into classes, characterized by their behavior. These distinct types yield corresponding properties of the transfer function $H(z)$, the frequency response $H(e^{j\omega})$, or its components and the impulse response. They are considered in Project 3.

Concerning structures, Project 4 deals with a few of them. Besides the basic ones, usually known as the direct-, the cascade-, and the parallel structure, the implementation of certain systems in lattice form and of others as a combination of two all-passes will be considered.

PROJECT 1: DESCRIPTION OF SYSTEMS

In this project, seven different descriptions of systems are introduced and their relation considered. Linear systems are described primarily in the time domain either by a difference equation of certain order or by state equations, containing additional information about the internal structure of the system. The difference equation is

$$\sum_{k=0}^{N} a_k y[n-k] = \sum_{\ell=0}^{M} b_\ell v[n-\ell] \qquad a_0 = 1 \tag{1-1}$$

The corresponding state equations are

$$\begin{aligned} \mathbf{x}[n+1] &= \mathbf{A}\mathbf{x}[n] + \mathbf{b}v[n] \\ y[n] &= \mathbf{c}^T\mathbf{x}[n] + dv[n] \end{aligned} \tag{1-2}$$

where $\mathbf{x}[n] = \left[x_1[n],\ x_2[n], \ldots, x_N[n]\right]^T$ is the vector of state variables. If $M = N$, one state representation equivalent to (1-1) is

$$\mathbf{A} = \begin{bmatrix} -a_1 & 1 & 0 & \cdots & 0 \\ -a_2 & 0 & \ddots & & \vdots \\ \vdots & \vdots & \ddots & \ddots & 1 \\ -a_N & 0 & \cdots & \ddots & 0 \end{bmatrix} \qquad \mathbf{b} = \begin{bmatrix} b_1 & - & b_0 a_1 \\ b_2 & - & b_0 a_2 \\ \vdots & & \vdots \\ b_N & - & b_0 a_N \end{bmatrix}$$

$$\mathbf{c}^T = [1, 0, \ldots, 0] \qquad d = b_0 \tag{1-3}$$

Descriptions of a system either by the coefficients a_k and b_ℓ in (1-1) or by the state equations (1-2) are complete, where the particular form of $\mathbf{A}$, $\mathbf{b}$, $\mathbf{c}$, and d in (1-2) provides further information about the particular structure used for the implementation of the system (see Project 4).

Another description of a system in the time domain is given by

$$y[n] = \sum_{k=0}^{\infty} h[k]v[n-k] \tag{1-4}$$

where $h[n]$ is its impulse response, in general a sequence of infinite length. A first closed-form expression can be found as solution of the state equation (1-2) for $v[n]$ being the impulse $\delta[n]$. It is

$$h[n] = \begin{cases} d & n = 0 \\ \mathbf{c}^T\mathbf{A}^{n-1}\mathbf{b} & n \geq 1 \end{cases} \tag{1-5}$$

Particular values of $h[n]$ for $n = 0, 1, 2, \ldots$ can be calculated in MATLAB using the M-file `filter`, with $\delta[n]$ as the input signal, where the system has to be described by the coefficients a_k and b_ℓ. It provides the stepwise solution of (1-1) as

$$h[n] = -\sum_{k=1}^{N} a_k h[n-k] + \sum_{\ell=0}^{M} b_\ell \delta[n-\ell] \tag{1-6}$$

yielding for $n > M$

$$h[n] = -\sum_{k=1}^{N} a_k h[n-k] \tag{1-7}$$

It turns out that only the values $h[n]$, $n = 0 : (N + M)$ are required for a complete characterization of the system. Note that its number is equal to the total number of coefficients a_k, $k = 1 : N$ and b_ℓ, $\ell = 0 : M$.[1]

A description in the z-domain is obtained after applying the z-transform to the difference equation or the state equations. We get the following two equivalent versions of the transfer function:

$$H(z) = \frac{\sum_{\ell=0}^{M} b_\ell z^{-\ell}}{1 + \sum_{k=1}^{N} a_k z^{-k}} \tag{1-8}$$

$$H(z) = \mathbf{c}^T (z\mathbf{E} - \mathbf{A})^{-1}\mathbf{b} + d \tag{1-9}$$

There are other equivalent representations of $H(z)$; besides the form given in (1-8) as a quotient of two polynomials, we can write a product form in terms of poles and zeros:

$$H(z) = b_0 \cdot \frac{\prod_{\ell=1}^{M} (1 - z_\ell z^{-1})}{\prod_{k=1}^{N} (1 - p_k z^{-1})} \tag{1-10}$$

Furthermore, the partial fraction expansion is useful. In case of distinct poles and $M = N$ it is

$$H(z) = B_0 + \sum_{k=1}^{N} \frac{B_k}{1 - p_k z^{-1}} \tag{1-11}$$

The relation to the impulse response $h[n]$ is given by

$$H(z) = Z\{h[n]\} = \sum_{n=0}^{\infty} h[n] z^{-n} \tag{1-12}$$

yielding with the partial fraction expansion (1-11) a further closed-form expression for $h[n]$:

$$h[n] = Z^{-1}\{H(z)\} = B_0 \delta[n] + \sum_{k=1}^{N} B_k p_k^n \tag{1-13}$$

Considering $H(z)$ especially on the unit circle $z = \exp(j\omega)$ yields the frequency response

$$H(e^{j\omega}) = \frac{\sum_{\ell=0}^{M} b_\ell e^{-j\ell\omega}}{1 + \sum_{k=1}^{N} a_k e^{-jk\omega}} \tag{1-14}$$

$H(e^{j\omega})$ is a periodic function in ω with period 2π. Its components and related functions are

$$P(e^{j\omega}) = \operatorname{Re}\{H(e^{j\omega})\}$$
$$Q(e^{j\omega}) = \operatorname{Im}\{H(e^{j\omega})\} \tag{1-15}$$

$$\text{log-magnitude } \ln|H(e^{j\omega})| = \operatorname{Re}\{\ln[H(e^{j\omega})]\} \tag{1-16}$$

$$\text{phase } \varphi(\omega) = \text{phase}\{H(e^{j\omega})\} = \operatorname{Im}\{\ln[H(e^{j\omega})]\} \tag{1-17}$$

$$\text{group delay } \tau_g(\omega) = -\frac{d\varphi}{d\omega} \tag{1-18}$$

[1]The MATLAB notation `0:M` is adopted to indicate a range of integers $\ell = 0, 1, 2, \ldots, M$.

The frequency response $H(e^{j\omega})$ as described by (1-14) can be calculated for distinct points ω_k using `freqz`, yielding after minor manipulations the related functions, as given in (1-16) and (1-17). The M-file `grpdelay` calculates the group delay not as the derivative of the phase as given by (1-18) but according to a description based on the coefficients a_k and b_ℓ (see the section *Group Delay* in Chapter 1).

Note that in (1-1), (1-8), and (1-14) the same coefficients are used. This is called the *transfer function* representation. Obviously, the sets of coefficients a_k and b_ℓ provide a complete description of the system. Other equally sufficient characterizations are given by $[\mathbf{A}, \mathbf{b}, \mathbf{c}^T, d]$, the parameters of the state-space case (1-2), and the representations (1-10) and (1-11), given by the parameters $[p_k, z_\ell, b_0]$ and $[p_k, B_k, B_0]$, respectively. If one of these sets of parameters is given, the others can be calculated.

As has been mentioned, the same holds for a sufficient large number of values $h[n]$, which can be used for calculation of the coefficients a_k and b_ℓ. In MATLAB this can be done by the M-file `prony`. Finally, samples $H(e^{j\omega_k})$, $\omega_k = k \cdot \pi/K$, $k = 0 : K-1$, where $K > (N+M)/2$, corresponding to $2K-1$ real numbers, are sufficient for the description of the system. The M-file `invfreqz` calculates the coefficients a_k and b_ℓ, using these values. In both cases the degrees M and N of the numerator and denominator polynomial must be known (see Exercises 1.2 and 1.3 in this chapter and the section *Design of IIR Filters* in Chapter 8).

For some investigations the autocorrelation sequence $\rho[m]$ of the impulse response is needed (see Exercise 3.2 in this chapter or the section *FFT Spectrum Estimation* in Chapter 6). It is defined as

$$\rho[m] = \sum_{n=0}^{\infty} h[n]h[n+m] = h[m] * h[-m], \qquad \forall m \in \mathbb{Z} \tag{1-19}$$

$$= Z^{-1}\{H(z)H(z^{-1})\} = \frac{1}{2\pi j}\oint H(z)H(z^{-1})z^{m-1}\,dz \tag{1-20}$$

If the partial fraction expansion of $H(z)$ is given as in (1-11), the autocorrelation sequence $\rho[m]$ can be calculated for $m \geq 0$ as

$$\rho[m] = Z^{-1}\{B_0 H(\infty) + \sum_{k=1}^{N} \frac{B_k}{1 - p_k z^{-1}} H(p_k^{-1})\} \tag{1-21}$$

which yields for $m \in \mathbb{Z}$

$$\rho[m] = B_0 H(\infty)\delta[m] + \sum_{k=1}^{N} B_k H(p_k^{-1}) p_k^{|m|} \tag{1-22}$$

In general, the sequence $\rho[m]$ does not give all information about the system. But if it is of minimum phase (see Project 3), the values $\rho[m]$, $m = 0 : N+M$ are sufficient for the calculation of a system, the impulse response $h[n]$ of which belongs to this autocorrelation sequence (see Exercise 3.3). More generally speaking: given $\rho[m]$ of any system, it is possible to calculate the coefficients of a minimum-phase system related to the original system in the sense that both frequency responses have the same magnitude.

Figure 5.1 shows a diagram of all the representations and the relations between them, together with the M-files that transform one representation into another. Most of these M-files are already provided by MATLAB or the Signal Processing Toolbox (`residuez`, `ss2tf`, `tf2ss`, `ss2zp`, `zp2ss`, `filter`, `freqz`, `prony`, `invfreqz`); the others are to be developed in the following exercises.

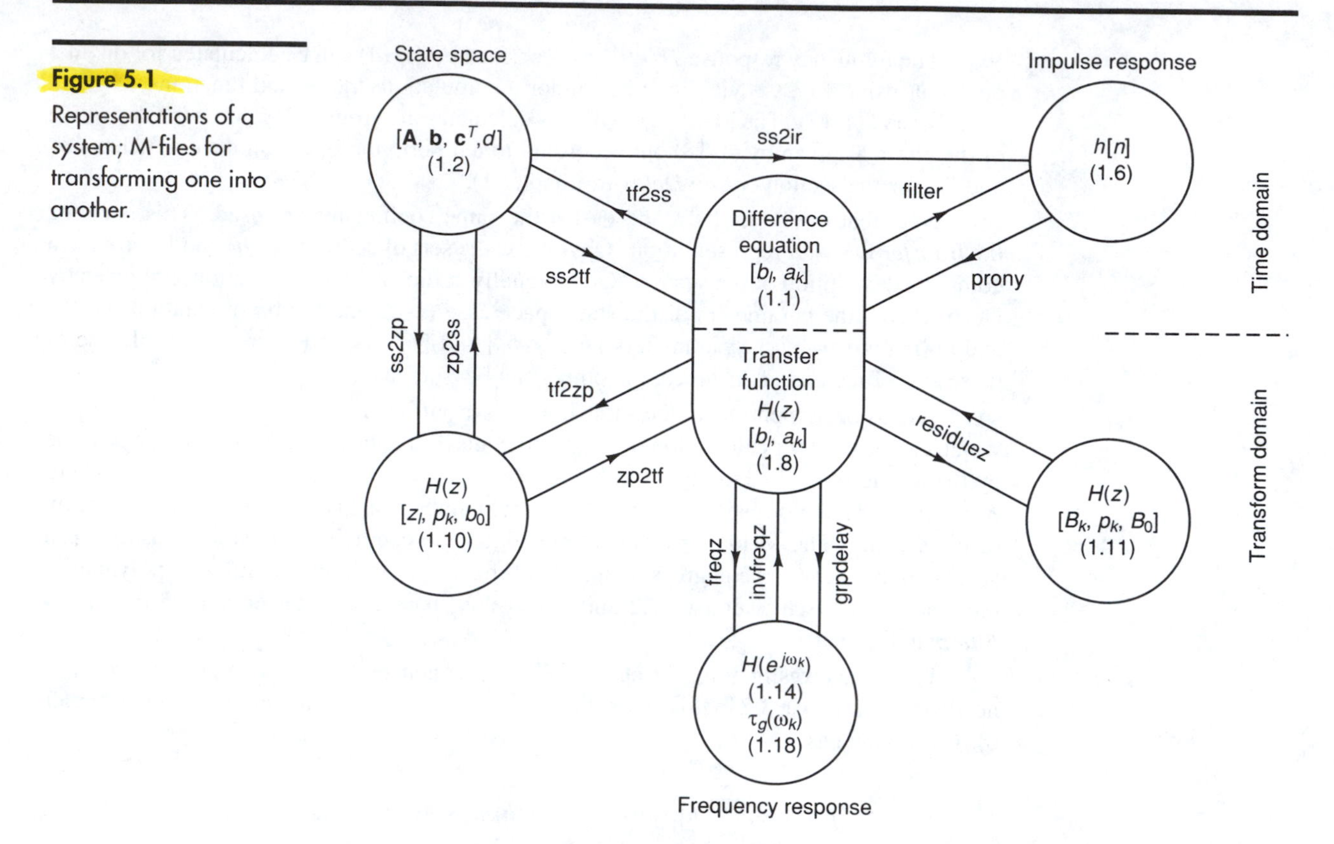

Figure 5.1 Representations of a system; M-files for transforming one into another.

EXERCISE 1.1

Given a Difference Equation

You are given the coefficients a_k and b_ℓ of a difference equation as[2]

$$\mathbf{a} = [1 \quad -0.8741 \quad 0.9217 \quad -0.2672]^T$$
$$\mathbf{b} = [0.1866 \quad 0.2036 \quad 0.2036 \quad 0.1866]^T \qquad (1\text{-}23)$$

a. Calculate the parameters of the following descriptions of the system:

   1. The state-space representation using `tf2ss`
   2. The partial fraction expansion (1-11) using `residuez`

b. Write a function `tf2zp` for calculating the pole–zero representation (1-10) out of the transfer function. Then apply this function to the coefficients given in (1-23). (*Hint*: The factorization of a polynomial can be done with the M-file `roots`.)

c. Calculate 512 samples of the complex frequency response $H(e^{j\omega})$ for $0 \leq \omega < 2\pi$ using `freqz`. Plot the magnitude $|H(e^{j\omega})|$ for $0 \leq \omega < 2\pi$ as well as an amplified version $10H(e^{j\omega})$ for $\pi/2 \leq \omega \leq 3\pi/2$ using `axis('square')` in that case. Note the symmetry of $H(e^{j\omega})$ with respect to the real axis. Plot as well $|H(e^{j\omega})|$ and the phase $\varphi(\omega)$ for $0 \leq \omega \leq \pi$ with `axis('normal')`.

d. Calculate and plot 100 values of the impulse response $h[n]$, first by using `filter` and then based on the closed form (1-13) using the partial fraction coefficients, found in part (a2). Compare your results.

---

[2]Elliptic filter found by `ellip(3,1,20,0.4)`.

EXERCISE 1.2

Given Samples $H(e^{j\omega_k})$ of the Frequency Response

You are given K samples $H(e^{j\omega_k})$ at $\omega_k = k \cdot \pi / K$, $k = 0 : (K-1)$, calculated using `freqz` as in Exercise 1.1(c). Determine the coefficients of the transfer function by using `invfreqz` with $M = N = 3$. See `help invfreqz` for more information. Compare your result with the values given in (1-23), which you started with. Find experimentally the smallest number of samples $H(e^{j\omega_k})$ [i.e., the value $\min(K)$] required for obtaining accurate results.

EXERCISE 1.3

Given Values $h[n]$ of the Impulse Response

You are given L values $h[n]$, $n = 0 : L$, calculated by using `filter` as in Exercise 1.1(d). To find the coefficients with `prony` we need the degree N of the denominator of $H(z)$. If it is not known, as in our example, it can be found by making use of the linear dependencies of the $h[n]$, as expressed in (1-7). As is shown in [4, Sec. 5.1], the rank of a Hankel matrix, built out of $h[n]$, $n \geq 1$ is the desired degree N. So proceed as follows:

a. Generate a Hankel matrix via `S = hankel(h1,h2)`. Let

$$h1 = h[n], \quad n = 1 : L/2$$

$$h2 = h[n], \quad n = L/2 : L$$

b. Compute the rank of the Hankel matrix, `N = rank(S)`, and then proceed with `prony`, using N and $M = N$ for the degrees of the transfer function.

c. Verify your result by comparing it with the values given in (1-23). Find experimentally the smallest number L of values $h[n]$ required for obtaining accurate results.

EXERCISE 1.4

Given Poles and Zeros

You are given a transfer function $H(z)$ as in (1-10) specified by its poles p_k and zeros z_ℓ and the constant b_0.

$$\begin{aligned} p_1 &= 0.9, \quad p_{2,3} = 0.6718 \pm j0.6718 \\ z_1 &= -1, \quad z_{2,3} = \pm j \\ b_0 &= 1/77 \end{aligned} \tag{1-24}$$

a. Write a function `zp2tf` for converting the pole–zero description into the description by the difference equation (1-1) or the transfer function (1-8). (*Hints*: While `roots` yields a factorization of a polynomial, the inverse operation of building a polynomial form out of its roots is accomplished with the function `poly`. Make sure that you obtain polynomials with real coefficients.)

b. Apply your program to the example given above. Transfer your result into the partial fraction expansion. Then compute the impulse response according to (1-13). Compare your result with the result you obtain with `filter`, using the transfer function representation, found before.

c. Calculate the output sequence $y[n]$ for the following input sequences of length 100:

$$v_1[n] = \texttt{ones(1,100)} \quad \text{(yielding the step response)}$$

$$v_2[n] = [1, -1, 1, -1, 1, -1, \ldots]$$

$$v_3[n] = [1, 0, -1, 0, 1, 0, -1, \ldots]$$

Compare and explain why the outputs approach constant values for increasing n. Calculate these values using the transfer function representation of the system and the properties of the input sequences.

d. Find an input sequence $v[n]$ of length 3 such that the corresponding output sequence is proportional to $(0.9)^n$ for $n \geq 3$.

EXERCISE 1.5

Given a State-Space Representation

You are given the description of a system by

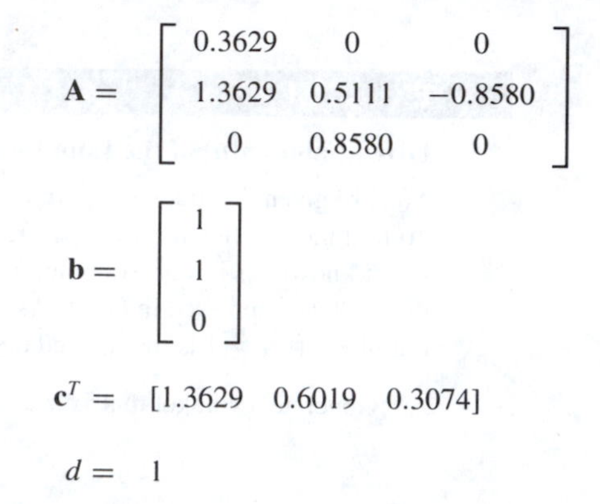

$$\mathbf{A} = \begin{bmatrix} 0.3629 & 0 & 0 \\ 1.3629 & 0.5111 & -0.8580 \\ 0 & 0.8580 & 0 \end{bmatrix}$$

$$\mathbf{b} = \begin{bmatrix} 1 \\ 1 \\ 0 \end{bmatrix}$$

$$\mathbf{c}^T = [1.3629 \quad 0.6019 \quad 0.3074]$$

$$d = 1 \tag{1-25}$$

a. Calculate the transfer function $H(z)$ as expressed by (1-8) using `ss2tf`. Find the poles p_k of the transfer function. Verify that they are equal to the eigenvalues of $\mathbf{A}$.

b. Write a function `ss2ir` for the calculation of a truncated version of the impulse response $h[n]$ by solving the state equations

$$\mathbf{x}[n+1] = \mathbf{A}\mathbf{x}[n] + \mathbf{b}v[n]$$

$$y[n] = \mathbf{c}^T\mathbf{x}[n] + dv[n]$$

such that $\mathbf{x}[n]$ as well as $y[n] = h[n]$ are obtained. Apply your program for $[\mathbf{A}, \mathbf{b}, \mathbf{c}^T, d]$ as given above for $n = 0 : 50$. Plot the components of $\mathbf{x}[n]$ as well as $h[n]$ together using `subplot(22x)`.

c. Compare the result you obtained for $h[n]$ with the one you get with `filter`, using the transfer function, found in part (a).

d. Draw a signal flow graph of the system, described by $\mathbf{A}$, $\mathbf{b}$, $\mathbf{c}^T$, d.

EXERCISE 1.6

Autocorrelation Sequence of an Impulse Response

a. Write a function `acimp` for the calculation of the autocorrelation sequence $\rho[m]$ of an impulse response $h[n]$ according to the closed-form expression in (1-22) starting with the transfer function $H(z)$ given either as in (1-8) by the polynomials a and b or by its poles and zeros and a constant factor as in (1-10). Use `residuez` for calculation of the required partial fraction expansion. Apply your program for the system described by (1-23) by calculating $\rho[m]$ for $m = 0 : 50$. Check your result by calculating $\rho[m]$ approximately either by convolving a truncated version of $h[m]$ with $h[-m]$ using `conv` or as $\text{DFT}^{-1}\{|H(e^{j\omega})|^2\}$.

b. As an alternative, $\rho[m]$ can be calculated for $m \geq 0$ as the impulse response of a system, the transfer function of which is given in partial fraction expansion form as

$$H_{ac}(z) = B_0 H(\infty) + \sum_{k=1}^{N} \frac{B_k}{1 - p_k z^{-1}} H(p_k^{-1}) \tag{1-26}$$

[see (1-20)]. So instead of determining $\rho[m]$ by using (1-22) it can be found with `filter` and the transfer function $H_{ac}(z)$ as the quotient of two polynomials, to be calculated out of (1-26) with `residuez`. Write a correspondingly modified version of your program `acimp`. Check your result by applying it for the example given by (1-23). Compare the resulting sequence $\rho[m]$ with that obtained in part (a).

c. The coefficients of $H_{ac}(z)$ can be calculated as well using samples of $\rho[m]$ by applying `prony`, including the method for determining the degree of the system, as outlined in Exercise 1.3. Use this procedure with the values $\rho[m]$ found in part (a) for the example and compare your resulting $H_{ac}(z)$ with that of part (b).

PROJECT 2: MEASURING THE FREQUENCY RESPONSE FOR $\omega = \omega_k$

Samples of the frequency response $H(e^{j\omega})$ at $\omega = \omega_k$ can be found either by calculation according to (1-14) if the coefficients a_k and b_ℓ are known, or by measurement of the steady-state response for an excitation by $e^{j\omega_k n}$ in the lab or by using the measured impulse response if the system is unknown. In this exercise the accuracy of results found by different measurement schemes and the reason for possible errors are investigated.

To determine the error we use a test system whose parameters a_k and b_ℓ are known. Thus the exact values $H(e^{j\omega_k})$ can be calculated with `freqz` and used for comparison. The first exercise serves as an introduction to the problem.

EXERCISE 2.1

Approaching the Steady State

a. As is well known, a particular value of $H(e^{j\omega_k})$ basically describes the steady-state response of the stable system if excited by $v_k[n] = e^{j\omega_k n}$, $n \geq 0$. In general, this excitation yields an output sequence consisting of two parts:

$$y_k[n] = y_{tk}[n] + H(e^{j\omega_k})e^{j\omega_k n} = y_{tk}[n] + y_{sk}[n]$$

where the transient sequence $y_t[n]$ is decreasing exponentially due to the assumed stability. We want to separate the two parts for four values of ω_k, using the system described by (1-23).

1. Calculate $H(e^{j\omega_k})$ for $\omega_k = 0,\ \pi/4,\ 1.6164,\ 2\pi/3$ using `freqz`.
2. Calculate $y_k[n]$ using `filter` with $v_k[n] = e^{j\omega_k n}$, $n = 0 : 100$.
3. Calculate the steady-state signals

$$y_{sk}[n] = H(e^{j\omega_k}) \cdot v_k[n] = H(e^{j\omega_k})e^{j\omega_k n}$$

4. Calculate the transient signals by subtracting the steady state.

$$y_{tk}[n] = y_k[n] - y_{sk}[n]$$

5. Plot the real parts of $v_k[n]$, $y_k[n]$, $y_{tk}[n]$, and $y_{sk}[n]$ together for each k, using `subplot(22x)`.

Determine by inspection how long it takes until the transient part can be ignored such that

$$y_k[n] \approx H(e^{j\omega_k}) \cdot e^{j\omega_k n}$$

with an error, the magnitude of which is $\leq 1\,\%$. Does the answer depend on $|H(e^{j\omega_k})|$?

b. Show that the measured value for $H(e^{j\omega_k})$ can be obtained as

$$H(e^{j\omega_k}) = u_k[\infty] = \lim_{n\to\infty} y_k[n] \cdot v_k^*[n] = \lim_{n\to\infty} y_k[n] \cdot e^{-j\omega_k n}$$

Calculate $u_k[n] = y_k[n] \cdot v_k^*[n]$ for the given four values ω_k and $n = 0 : 100$. Plot $\mathrm{Re}\{u_k[n]\}$ and $\mathrm{Im}\{u_k[n]\}$. Check the differences between $u_k[100]$ and the true values $u_k(\infty) = H(e^{j\omega_k})$ by calculating $|u_k[100] - H(e^{j\omega_k})|$.

EXERCISE 2.2

Frequency Response Determined with Multitone Periodic Excitation

The operations outlined in Exercise 2.1 basically describe a measurement procedure for use in the lab, where the complex frequency response $H(e^{j\omega})$ is measured point by point. We want to show that this measurement can be done simultaneously at $N/2$ points $\omega_k = k \cdot 2\pi/N$, $k = 0 : N/2 - 1$ as follows:

a. Generate the periodic input sequence

$$v[n] = \frac{1}{2} + \sum_{k=1}^{N/2-1} \cos nk2\pi/N + \frac{1}{2}(-1)^n \qquad n = 0 : \ell N - 1 \qquad (2\text{-}1)$$

where N is chosen to be a power of 2 (e.g., $N = 32$) and ℓ is an integer > 1. (*Hint:* Look for a fast method to generate $v[n]$, taking into account that $v[n]$ is really a Fourier series.) Plot $v[n]$ using `stem`.

b. Then calculate the output sequence $y[n]$ for the example given in (1-23) using `filter`. Note that for sufficiently large n the sequence $y[n]$ is approximately periodic.

c. Select the ℓth part of $y[n]$ as $y_\ell[n] = y[n = (\ell - 1)N : \ell N - 1]$ and calculate its N-point DFT:

$$Y_\ell[k] = \mathrm{DFT}\{y_\ell[n]\}$$

We claim that the following result holds with high accuracy:

$$H(e^{j\omega_k}) \approx \frac{2}{N} Y_\ell[k] \qquad \text{for } \omega_k = (2\pi/N)k, \quad k = 0 : N/2 - 1 \qquad (2\text{-}2)$$

Calculate and plot the magnitude of the error

$$\epsilon_\ell[k] = \left| \frac{2}{N} Y_\ell[k] - H(e^{j\omega_k}) \right|$$

for $\ell = 2 : 4$.

d. Express $y[n]$ in terms of the impulse response $h[n]$, regarding that $v[n]$ as given by (2-1) can be expressed as

$$v[n] = \begin{cases} N/2 & n = \lambda N\,,\ \lambda = 0 : \ell - 1 \\ 0 & n \neq \lambda N \end{cases} \qquad \begin{matrix}\text{Remark:} \\ \text{That is exact,} \\ \text{not an assumption!}\end{matrix} \qquad (2\text{-}3)$$

What is the reason for a possible error? How does it depend on ℓ and N?
Under which condition is statement (2-2) precisely correct?

e. A more general multitone periodic input signal can be used as well. Modify the procedure described above such that it works with

$$v[n] = \sum_{k=0}^{N-1} V[k] e^{jnk2\pi/N} \qquad n = 0 : \ell N - 1 \qquad (2\text{-}4)$$

Here the complex values $V[k]$ can be chosen arbitrarily, except that the two conditions $V[k] \neq 0$ and $V[k] = V^*[N - k]$, such that $v[n]$ is real, must be satisfied. How can the sequence $v[n]$ in (2-4) be calculated efficiently? Which expression has to be used for $H(e^{j\omega_k})$ instead of (2-2)?

EXERCISE 2.3

Frequency Response Determined with a Measured Impulse Response

According to (1-12), samples of the frequency response can be determined as

$$H(e^{j\omega_k}) = \sum_{n=0}^{\infty} h[n]e^{-jnk2\pi/N} \tag{2-5}$$

Using the truncated version $h[n]$, $n = 0 : L-1$ with $L \leq N$, we get

$$H(e^{j\omega_k}) \approx H_{iL}(e^{j\omega_k}) = \sum_{n=0}^{L-1} h[n]e^{-jnk2\pi/N} \tag{2-6}$$

to be executed in MATLAB as `fft(h,N)`.

Use this method to determine the frequency response of the system described by (1-23) again (e.g., with $N = 32$). Calculate and plot the magnitude of the error

$$\epsilon_{iL}[k] = |H_{iL}(e^{j\omega_k}) - H(e^{j\omega_k})|$$

for different values of the length L. Compare your results with those obtained in Exercise 2.2(d).

PROJECT 3: TYPES OF SYSTEMS

Based on possible properties of a system, several classes can be defined, leading to different characteristics of $H(z)$:

- A system is called *real* if its response to any real input signal is real. The consequences for $H(z)$ are: The coefficients in the transfer function (1-8) are real; its poles and zeros are symmetric with respect to the real axis of the z-plane. Furthermore, the frequency response is conjugate symmetric: $H(e^{j\omega}) = H^*(e^{-j\omega})$. (*Remark*: All systems considered in the exercises of this project are real.)
- A system is called *causal* if its impulse response $h[n]$ is zero for $n < 0$. The components $P(e^{j\omega})$ and $Q(e^{j\omega})$ are then related by the Hilbert transform. In case of a real system the following relations hold:

$$\begin{aligned} P(e^{j\omega}) &= \mathrm{Re}\{H(e^{j\omega})\} &&= \sum_{n=0}^{\infty} h[n]\cos n\omega &&= h[0] + \mathcal{H}\{Q(e^{j\omega})\} \\ Q(e^{j\omega}) &= \mathrm{Im}\{H(e^{j\omega})\} &&= -\sum_{n=1}^{\infty} h[n]\sin n\omega &&= -\mathcal{H}\{P(e^{j\omega})\} \end{aligned} \tag{3-1}$$

- A system is called *stable* if its impulse response is absolutely summable. That means for $H(z)$ that all its poles p_k are strictly inside the unit circle, if the system is in addition causal. The stability can be determined either by calculating the roots of the denominator polynomial or by using a stability test procedure such as the Schur–Cohn test.
- A stable system is called *FIR* if its impulse response $h[n]$ is of finite length. That means in case of causality that we have in (1-8) $a_k = 0, k > 0$ (i.e., $N = 0$ and $b_n = h[n], n = 0 : M$). Obviously, all poles p_k in (1-10) are zero. The $M+1$ values of $h[n]$ describe the system completely and immediately. Note that this statement is a special case of the corresponding one for an IIR system with an impulse response of infinite length, as given by (1-6), where $M + N + 1$ values are necessary as a basis for a complete description. But as has been demonstrated in Exercise 1.3, rather lengthy calculations are required in that general case, to get a transfer function characterization of the system.

- A causal and stable system has *linear phase* if it is FIR and if the zeros of its transfer function are on the unit circle or reciprocal to the unit circle. This means that a zero at z_ℓ has a mirror image zero at $1/z_\ell^*$. The numerator of its transfer function is either a mirror-image polynomial if a possible zero at $z = 1$ is of even order, or an anti-mirror-image polynomial if a zero at $z = 1$ is of odd order.
- A causal and stable system is called *minimum phase* if its phase is the smallest possible one of all systems with the same magnitude of the frequency response. There are two types of minimum-phase systems:

  a. Those that are invertible, that is, systems with transfer functions $H(z)$ such that $1/H(z)$ is stable and minimum phase as well. In this case, all zeros of the $H(z)$ are inside the unit circle.
  b. Those that are not invertible. Here all zeros of $H(z)$ are inside or on the unit circle.

We mention that for all invertible minimum-phase systems the group delay satisfies the condition

$$\int_0^\pi \tau_g(\omega)\, d\omega = 0 \tag{3-2}$$

Obviously, (3-2) implies that there are always one or more intervals, where the group delay must be negative.

The real and imaginary parts of $\ln[H(e^{j\omega})]$ are related by the Hilbert transform. If $H(z)$ describes an invertible system, then

$$\ln[H(e^{j\omega})] = \sum_{k=0}^{\infty} c_k e^{-jk\omega} \tag{3-3}$$

where

$$c_k = \frac{1}{2\pi} \int_{-\pi}^{\pi} \ln[H(e^{j\omega})] e^{jk\omega}\, d\omega \tag{3-4}$$

is a causal and real sequence, being the cepstrum of the impulse response $h[n]$ of the system. Here the causality of the sequence c_k is characteristic for a minimum-phase system, while $c_k \in \mathbb{R}$ is only a consequence of the assumption that the system is real. Using the c_k we get

$$\begin{aligned} \operatorname{Re}\{\ln[H(e^{j\omega})]\} &= \ln|H(e^{j\omega})| = \sum_{k=0}^{\infty} c_k \cos k\omega \\ &= c_0 + \mathcal{H}\{\varphi(\omega)\} \end{aligned} \tag{3-5}$$

$$\begin{aligned} \operatorname{Im}\{\ln[H(e^{j\omega})]\} &= \varphi(\omega) = -\sum_{k=1}^{\infty} c_k \sin k\omega \\ &= -\mathcal{H}\{\ln|H(e^{j\omega})|\} \end{aligned} \tag{3-6}$$

In this case the group delay is

$$\tau_g(\omega) = \sum_{k=1}^{\infty} k c_k \cos k\omega \tag{3-7}$$

As an incidental remark we note that a causal and stable system is called *maximum phase* if its phase is the largest possible one of all systems with the same degree and the same magnitude of the frequency response. In this case all zeros of $H(z)$ are outside the unit circle.

- A system with the property $|H(e^{j\omega})| = \text{const.}$ is called an *all-pass*. The zeros of its transfer function are mirror images of its poles with respect to the unit circle. Thus the transfer function of an all-pass is

$$H_A(z) = \frac{b_0 \prod_{k=1}^{N}(1 - p_k^{-1}z^{-1})}{\prod_{k=1}^{N}(1 - p_k z^{-1})} \qquad |H_A(e^{j\omega})| = \left| b_0 \prod_{k=1}^{N} p_k^{-1} \right| \tag{3-8}$$

$$H_A(z) = \frac{z^{-N} + \sum_{k=1}^{N} a_k z^{k-N}}{1 + \sum_{k=1}^{N} a_k z^{-k}} = \frac{z^{-N} D(z^{-1})}{D(z)} \tag{3-9}$$

Obviously, an all-pass is a special case of a maximum-phase system.

A non-minimum-phase system [i.e., a system with a transfer function $H(z)$ having zeros outside the unit circle] can always be described by a cascade of a minimum-phase system and an all-pass with the transfer functions $H_M(z)$ and $H_A(z)$, respectively. Starting with an arbitrary $H(z)$ from (1-10), we can write

$$\begin{aligned} H(z) &= b_0 \frac{\prod_{\ell=1}^{M_1}(1 - z_\ell z^{-1}) \prod_{\lambda=1}^{M_2}(1 - z_\lambda^{-1}z^{-1})}{\prod_{k=1}^{N}(1 - p_k z^{-1})} \frac{\prod_{\lambda=1}^{M_2}(1 - z_\lambda z^{-1})}{\prod_{\lambda=1}^{M_2}(1 - z_\lambda^{-1}z^{-1})} \\ &= H_M(z) H_A(z) \end{aligned} \tag{3-10}$$

where $|z_\ell| \leq 1,\ \ell = 1 : M_1,\ |z_\lambda| > 1,\ \lambda = 1 : M_2,\ M_1 + M_2 = M.$

Remark. The separation of an all-pass described by $H_A(z)$ yields the introduction of M_2 additional poles. A corresponding cascade implementation with the minimum-phase system first would have M_2 uncontrollable natural modes characterized by z_λ^{-n}, $\lambda = 1 : M_2$, while in a cascade with the all-pass first these natural modes are unobservable.

Furthermore, an energy relation can be used to distinguish a general system with $H(z)$ from the corresponding minimum-phase system, described by $H_M(z)$. Let both be excited by the same input sequence $v[n]$, yielding the output sequences $y[n]$ and $y_M[n]$, respectively. We compare the two running energies, defined as

$$w_y[m] = \sum_{n=0}^{m} y^2[n] \tag{3-11}$$

and $w_{y_M}[m]$ correspondingly. It can be shown that

$$w_{y_M}[m] \geq w_y[m]\ \forall m \tag{3-12}$$

That means that the output energy of a non-minimum-phase system never increases faster than that of the corresponding minimum-phase system.

- *Passive* and *lossless* systems: Finally, we mention a class of digital systems with properties corresponding to those of an RLC network. The definition is based on a comparison of the input and output energy of a system. Let $w_v[m]$ be the running energy of the input sequence, defined according to (3-11), and

$$w_v[\infty] = \sum_{n=0}^{\infty} v^2[n] \tag{3-13}$$

its total energy, which is assumed to be finite. The corresponding figures of the output sequence are $w_y[m]$ and $w_y[\infty]$, respectively, to be compared with $w_v[m]$ and $w_v[\infty]$. Now the system is called *passive* if

$$w_y[m] \le w_v[m] \qquad \forall m \tag{3-14}$$

and it is *lossless* if

$$w_y[\infty] = w_v[\infty] \tag{3-15}$$

Using $Y(z) = H(z)V(z)$ and Parseval's equation, the consequences for the transfer function can be derived. It turns out that an all-pass with

$$|H(e^{j\omega})|^2 = 1 \qquad \forall \omega \tag{3-16}$$

is the only lossless system with one input and one output.

Now we consider another system which is more general in the sense that it has one input again but two outputs, yielding the sequences $y_1[n]$ and $y_2[n]$. The total output energy with $\mathbf{y}[n] = [y_1[n],\ y_2[n]]^T$ is now defined as

$$w_y[\infty] = \sum_{n=0}^{\infty} \mathbf{y}^T[n]\mathbf{y}[n] \tag{3-17}$$

If $H_1(z)$ and $H_2(z)$ are the two corresponding transfer functions, the relation

$$|H_1(e^{j\omega})|^2 + |H_2(e^{j\omega})|^2 = 1 \qquad \forall \omega \tag{3-18}$$

is found to be the condition for a lossless system. The two transfer functions are complementary (see the related Fig. 5.2), and they both satisfy the condition

$$|H_{1,2}(e^{j\omega})|^2 \le 1 \tag{3-19}$$

Consequences in terms of sensitivity are considered in Project 3 of Chapter 7.

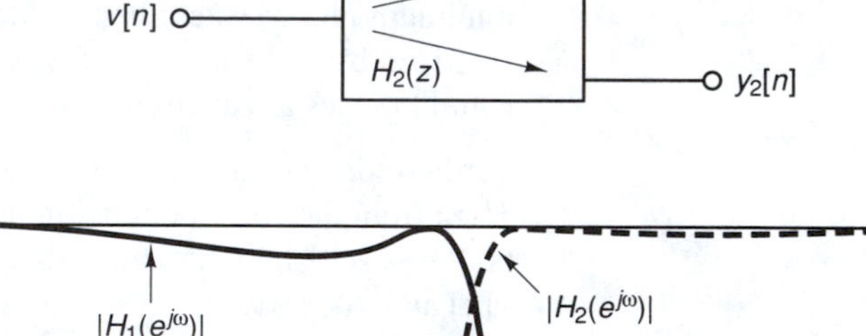

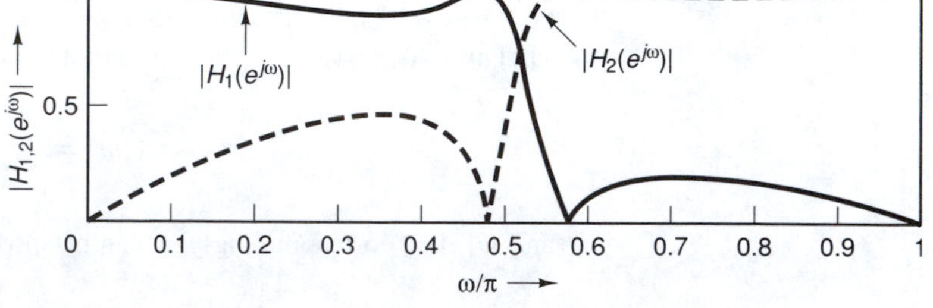

Figure 5.2
Lossless system with two complementary transfer functions.

EXERCISE 3.1

Stability

Suppose that the zeros of the denominator of a real system are given partly by

```
p = rand(1,5).*exp(j*pi*rand(1,5))
```

where `rand` generates a uniform distribution. Get further information with `help rand`.

a. Find the coefficients a_k of the complete denominator polynomial $A(z)$ such that all the resulting coefficients are real.

b. Now the stability of the polynomial has to be checked. Using your knowledge about the generation of the polynomial $A(z)$, do you expect stability?

c. Calculate the roots of $A(z)$; check whether they are inside the unit circle.

d. An alternative possibility is the Schur–Cohn stability test, which is described briefly as follows. Starting with the given polynomial of degree N,

$$A(z) = A_N(z) = 1 + \sum_{\ell=1}^{N} a_\ell^{(N)} z^{-\ell}$$

a sequence of polynomials $A_i(z)$, $i = N : -1 : 0$ are calculated recursively according to

$$\begin{aligned} A_{i-1}(z) = & \frac{1}{1-k_i^2}\left[A_i(z) - k_i z^{-i} A_i(z^{-1})\right] \\ = & 1 + \sum_{\ell=1}^{i-1} a_\ell^{(i-1)} z^{-\ell} \end{aligned}$$

where $k_i = a_i^{(i)}$. Note that $z^{-i}A_i(z^{-1})$ is a flipped version of $A_i(z)$. According to Schur–Cohn, the zeros of the denominator are inside the unit circle iff

$$|k_i| < 1 \qquad i = N : -1 : 1 \tag{3-20}$$

The values k_i are called the reflection coefficients (see Project 4, section on the lattice structure). The Schur–Cohn test can be executed with the following program:

```
function k = atok(a)
%ATOK      implements the Inverse Levinson Recursion
%  usage:
%     K = atok(A) converts AR polynomial representation
%                to reflection coefficients
%         where  A = vector of polynomial coefficients
%           and  K  "        "    reflection coefficients
%

a = a(:);    % make sure we are dealing with a column vector
N = length(a);
k = zeros(N-1,1);     % # refl. coeffs = # of roots-1 = degree-1
for i=(N-1):-1:1
   k(i) = a(i+1);
   b = flipud(a);
   a = ( a - k(i)*b ) / ( 1 - k(i).*k(i) );
   a(i+1)=[];
end
```

e. Make up a polynomial with some zeros on or outside the unit circle and show that the Schur–Cohn test detects the instability.

EXERCISE 3.2

Minimum and Non-Minimum-Phase Systems

The transfer function of a system is partly given by the poles

$$p_1 = 0.9 \cdot e^{j\pi/4} \quad p_2 = 0.8$$

and the zeros

$$z_1 = 1.5 \cdot e^{j\pi/8} \quad z_2 = 0.5$$

a. Find additional poles and zeros as well as a constant gain factor such that the transfer function $H_1(z)$ will describe a real, stable system of degree 3 with $H_1(1) = 1$. Is the system minimum phase?

b. Change the appropriate parameters of the system described by $H_1(z)$ such that the resulting transfer functions $H_\lambda(z)$ of degree 3 have the property $|H_\lambda(e^{j\omega})| = |H_1(e^{j\omega})|$. Find all functions $H_\lambda(z)$, $\lambda = 2, \ldots$ of stable real systems with this property. To check your result, calculate and plot $|H_\lambda(e^{j\omega})|$ for all these systems. Use `zplane` to show the pole–zero locations of the different systems. Which of these systems is minimum phase, and which is maximum phase?

c. Calculate for the resulting transfer functions $H_\lambda(z)$, $\lambda = 1, \ldots$ the phases $\varphi_\lambda(\omega)$, the group delays $\tau_{g\lambda}(\omega)$, the impulse responses $h_\lambda[n]$, and the energy sequences of the impulse responses $w_{h\lambda}[m] = \sum_{n=0}^{m} h_\lambda^2[n]$. Choose $n = 0 : 50$. Plot the comparable functions in one diagram. Use `subplot` to get a complete description of these systems with four subpictures. Verify that the minimum-phase system has the smallest unwrapped phase over the interval $[0, \pi]$. [*Hint*: Use `grpdelay(.)` for the calculation of $\tau_{g\lambda}(\omega)$.]

d. Verify the following statement: Let $\tau_{gM}(\omega)$ be the group delay of a minimum-phase system and $\tau_{g\lambda}(\omega)$ the group delays of the non-minimum-phase systems with the same $|H(e^{j\omega})|$; then

$$\tau_{gM}(\omega) < \tau_{g\lambda}(\omega) \qquad \forall \omega, \lambda \tag{3-21}$$

e. Calculate $\int_0^\pi \tau_{g\lambda}(\omega)\, d\omega$ numerically for all cases. Verify that the values are $\ell\pi/2$, where ℓ depends on the number of zeros outside the unit circle. [*Hint*: Calculate $\tau_{g\lambda}(\omega)$ at $\omega_k = k \cdot \pi/N$, $k = 0 : N-1$. Use `sum(.)` for an approximate integration according to the rectangular rule. A value $N = 512$ is recommended.]

f. Excite the minimum-phase system as well as one of the others with any input sequence, [e.g., with `v=rand(1,100)`]. Calculate the running energies $w_y[m]$ and $w_{yM}[m]$ of the corresponding output sequences $y[n]$ and $y_M[n]$ and verify the relation (3-12).

g. Let $H_M(z)$ be the transfer function of the minimum-phase system found in part (b). Find the corresponding inverse system described by $H_{Mi}(z) = 1/H(z)$. Calculate its group delay $\tau_{gMi}(\omega)$ and compare it with $\tau_{gM}(\omega)$.

EXERCISE 3.3

All-Pass

The transfer function of a real, stable all-pass is partly known by the poles and zeros

$$p_1 = 0.9 \cdot e^{j\pi/4} \qquad p_2 = 0.8$$
$$z_1 = 1.5 \cdot e^{j\pi/2} \qquad z_2 = 1.25$$

a. Find additional poles and zeros as well as a constant factor to complete the transfer function such that $|H_A(e^{j\omega})| = 1$ and $H_A(z)$ has the minimum possible order. Check your result by calculating and plotting the magnitude of the frequency response. Calculate and plot the group delay $\tau_g(\omega)$ and the impulse response $h[n]$. Determine the minimum value of the group delay. Prove that the group delay of a stable all-pass is never negative.

b. Find a closed-form expression for $\int_0^\pi \tau_g(\omega)\, d\omega$ in terms of the number of poles and zeros of the all-pass. Calculate the integral numerically for the example given above to verify your result.

EXERCISE 3.4

Autocorrelation Sequence of Impulse Responses

a. Calculate the autocorrelation sequences $\rho_\lambda[m]$ of the impulse responses $h_\lambda[m]$ of at least two of the systems described by $H_\lambda(z)$ found in Exercise 3.2(b) and for the all-pass of Exercise 3.3. Use the program `acimp` developed in Exercise 1.6(a). Compare your results with those to be expected.

b. Given the autocorrelation sequence $\rho[m]$ of the impulse response $h[n]$ of any system, develop a program for the calculation of the transfer function $H_M(z)$ of the corresponding minimum-phase system using the following basic relations [3, Sec. 8.1]: The autocorrelation sequence can be expressed as

$$\rho[m] = g[m] + g[-m] \tag{3-22}$$

where

$$g[m] = \begin{cases} 0.5\rho[0] & m = 0 \\ \rho[m] & m > 0 \\ 0 & m < 0 \end{cases} \tag{3-23}$$

The z-transform of (3-22) yields

$$R(z) = H(z)H(z^{-1}) = G(z) + G(z^{-1}) \tag{3-24}$$

The procedure works as follows:

1. Design a system with the transfer function $G(z) = Z^{-1}\{g[m]\}$ using Prony's method according to Exercise 1.3.
2. Calculate the poles and zeros of the rational function $R(z)$ as given in (3-24).
3. The transfer function $H_M(z)$ of the minimum-phase system is determined uniquely by the poles p_k of $R(z)$ with $|p_k| < 1$ [being the poles of $G(z)$] and the zeros z_ℓ with $|z_\ell| \leq 1$, where in case of a double zero on the unit circle only one has to be used for $H_M(z)$.
4. Calculate the coefficients a_k and b_ℓ of $H_M(z)$ using the selected poles p_k and zeros z_ℓ (use `zp2tf`, as developed in Project 1). The required constant factor b_0 can be determined using the condition

$$|H_M(1)| = \sqrt{R(1)} = \sqrt{2G(1)} \tag{3-25}$$

Use your program for one of the non-minimum-phase systems of Exercise 3.2(b). Compare your result with the transfer function of the minimum-phase system of that exercise.

Remark. Selecting roots z_ℓ partly or completely outside the unit circle yields a variety of non-minimum-phase systems, all with the same $\rho[m]$ and thus the same $|H(e^{j\omega})|^2$.

EXERCISE 3.5

Linear-Phase Systems

We are given two of the zeros of an FIR system

$$z_1 = 0.9e^{j\pi/5} \quad z_2 = e^{-j\pi/2}$$

a. Find additional zeros such that the resulting transfer function of minimum degree describes a system with the following frequency response:

1. $H_a(e^{j\omega}) = e^{-j\omega\tau_a} H_{0a}(e^{j\omega})$, where τ_a is constant and $H_{0a}(e^{j\omega})$ is real
2. $H_b(e^{j\omega}) = e^{-j\omega\tau_b} H_{0b}(e^{j\omega})$, where τ_b is constant and $H_{0b}(e^{j\omega})$ is imaginary

b. Determine τ_a and τ_b. Calculate and plot the two corresponding impulse responses and the functions $H_{0a}(e^{j\omega})$ and $H_{0b}(e^{j\omega})$.

EXERCISE 3.6

Separation of a Non-Minimum-Phase System

The transfer function $H(z)$ of a linear phase system is partly described by the zeros

$$z_1 = 0.9e^{j\pi/6} \quad z_2 = 0.8e^{j\pi/4}$$

a. Complete the description such that the first value of the impulse response is $h[0] = 1$.

b. Separate $H(z)$ according to

$$H(z) = H_A(z)H_M(z)$$

where $H_A(z)$ is the transfer function of an all-pass and $H_M(z)$ that of a minimum-phase FIR system [see (3-10)].

c. Let $h_A[n]$ and $h_M[n]$ be the impulse responses of the two subsystems and $h[n]$ the overall impulse response, to be expressed as

$$h[n] = h_A[n] * h_M[n] = h_M[n] * h_A[n]$$

What are the lengths of these three impulse responses?

d. Calculate and plot $h_A[n]$, $h_M[n]$, and $h[n]$ using `stem` and `subplot(22x)`. Comment on the results in terms of observability and controllability, taking the different ordering of the subsystems in consideration.

EXERCISE 3.7

Relation between the Real and Imaginary Parts of $H(e^{j\omega})$

a. Given N samples $P(e^{j\omega_k}) = \text{Re}\{H(e^{j\omega_k})\}$, $\omega_k = k \cdot 2\pi/N$, $k = 0 : N-1$, where $H(e^{j\omega_k})$ is the frequency response of a real causal system, find $Q(e^{j\omega_k}) = \text{Im}\{H(e^{j\omega_k})\}$ and the impulse response $h[n]$. Use, for example, with N being even (e.g., $N = 32$)

```
P = [P1 P2]
```

where

```
P1 = randn(1,N/2);
P2 = [randn(1,1) P1(N/2:-1:2)].
```

Is the solution unique? [*Hint*: Calculate `p = ifft(P,N)`. Find the relation between the even sequence $p[n]$ of length N and the finite-length impulse response $h[n]$ of the system.]

b. Given N samples $Q(e^{j\omega_k}) = \text{Im}\{H(e^{j\omega_k})\}$, where the ω_k are defined as in part (a), again belonging to a real causal system,

1. Generate a sequence of N numbers using `randn(1,.)` again, being appropriate samples of $Q(e^{j\omega})$.
2. Find $P(e^{j\omega_k})$ and the impulse response $h[n]$. Is the solution unique?
3. In case you decided that the solution is not unique, modify your result such that $|H(1)| = 1$.

EXERCISE 3.8

Checking the Minimum-Phase Property

Given the impulse response $h[n]$ of a length-N FIR system, we want to develop methods to check whether or not the system has minimum phase. Furthermore, if the system is found to be non-minimum-phase, we want to construct a minimum-phase system with the same magnitude $|H(e^{j\omega})|$.

a. Develop an M-file `minph` based on the roots of

$$H(z) = \sum_{n=0}^{N-1} h[n]z^{-n}$$

executing the following steps:

1. Calculate the zeros of $H(z)$ with

```
z = roots (h)
```

2. Check `abs(z)` > 1.
3. If `abs(z(i))` > 1, introduce `z(i) = 1/conj(z(i))`.
4. Scale the transfer function such that the resulting minimum-phase system satisfies

$$|H_M(e^{j\omega})| = |H(e^{j\omega})|$$

Apply your program with

```
h=randn(1,10);
```

b. Develop an M-file `minphcep` based on the cepstrum c_k as defined in (3-4), which executes the following steps:

1. Calculate the c_k approximately as

```
c = ifft(log(fft(h,M)))
```

where $M \gg N$.

2. To check the causality of the sequence, rearrange it according to

```
cr = [c(M/2+2:M) c(1:M/2+1)]
```

and plot it for $k = -M/2 + 1 : M/2$.

3. If you find the sequence not to be causal and thus the system to be non-minimum phase, determine the corresponding minimum-phase system described by $H_M(z)$ with the following steps. Calculate approximately the coefficients of the Fourier-series expansion of $\ln|H(e^{j\omega})|$ as

```
a = ifft(log(abs(fft(h,M))))
```

Now $M/2 + 1$ samples of the cepstrum c_{mk} of the minimum-phase system are

```
cmk = [a(1) 2*a(2:M/2) a(M/2+1)]
```

Its frequency response is obtained as

```
Hm = exp(fft(cmk,M))
```

its impulse response with

```
hm = ifft(Hm)
```

Check your program starting with the example considered in part (a).

EXERCISE 3.9

Testing the Losslessness Property

a. You are given poles and zeros of a minimum-phase system, described by $H_1(z)$.

$$p_{1,2} = 0.6e^{\pm j\pi/8} \qquad p_3 = 0.5$$
$$z_{1,2} = 0.9e^{\pm j\pi/8} \qquad z_3 = 0.75$$

Furthermore, we use the all-pass, described by $|H_A(e^{j\omega})| = 1$, having the same poles.

1. Determine the corresponding two transfer functions $H_1(z)$ and $H_A(z)$ such that $H_1(1) = H_A(1) = 1$.
2. To learn about the properties of these systems, calculate and plot $|H_1(e^{j\omega})|$ as well as the two group delays τ_{g1} and τ_{gA}.

3. Calculate the two output sequences $y_1[n]$ and $y_A[n]$ for an input sequence

```
v = [rand(1,30),zeros(1,60)]
```

using `filter`. Plot these sequences using `subplot(21x)` and `stem`.

4. Calculate the running energies

$$w_x[m] = \sum_{n=0}^{m} x^2[n] \tag{3-26}$$

for the sequences $x[n] = v[n]$, $y_1[n]$, and $y_A[n]$ and plot them in one diagram.

5. Comment on your results in terms of passivity. Are the systems lossless?

b. In this part we construct a system with two outputs according to Fig. 5.2 in order to study the relations given in equations (3-17), (3-14), and (3-18). You are given the poles of two all-passes of first and second order.[3]

$$AP1: \quad p_1 = 0.1587$$
$$AP2: \quad p_{2,3} = -0.0176 \pm j0.8820$$

1. Determine the two all-pass transfer functions $H_{A1}(z)$ and $H_{A2}(z)$.
2. Excite the two all-passes with

```
v=[randn(1,30), zeros(1,60)]
```

and calculate its output sequences $u_1[n]$ and $u_2[n]$ with `filter`. Furthermore, calculate

$$y_1[n] = 0.5(u_1[n] + u_2[n])$$
$$y_2[n] = 0.5(u_1[n] - u_2[n])$$

3. Calculate the running energies for $v[n]$, $y_1[n]$, and $y_2[n]$ according to (3-26) and plot them in one diagram.
4. Calculate and plot as well the differences

$$d[m] = w_v[m] - (w_{y_1}[m] + w_{y_2}[m])$$

Is the total system with the input signal $v[n]$ and two output signals $y_1[n]$ and $y_2[n]$ passive? Is it lossless?

5. Determine the two transfer functions

$$H_1(z) = 0.5[H_{A1}(z) + H_{A2}(z)]$$
$$H_2(z) = 0.5[H_{A1}(z) - H_{A2}(z)]$$

Calculate and plot in one diagram $|H_1(e^{j\omega})|^2$ and $|H_2(e^{j\omega})|^2$.

6. Compare the numerator of $H_1(z)$ with the numerator b of the elliptic filter, only the denominator of which was used in this exercise. Can you explain or generalize this remarkable result?

Remark. The topic is picked up again in Project 4, Exercise 4.4.

PROJECT 4: STRUCTURES

In this project, different possible structures for the implementation of a desired system will be considered. They will be compared in terms of their complexity or the required number of arithmetic operations, if implemented with MATLAB.

---

[3] p_1, p_2, p_3 are the poles of $H(z)$, the transfer function of an elliptic filter found with `ellip(3, 1, 15, 0.5)` and described by

$$b = [0.3098 \quad 0.4530 \quad 0.4530 \quad 0.3098]$$
$$a = [1.0000 \quad -0.1235 \quad 0.7726 \quad -0.1235]$$

Direct Structure

As has been mentioned in the description of Project 1, there are different algebraic expressions for the transfer function $H(z)$, which yield different elementary structures. The representation of $H(z)$ as quotient of two polynomials [see (1-8)] yields the so-called *direct structures*, in which the coefficients a_k and b_k are parameters. Essentially two different direct forms are available (see [1, Chap. 6]). Direct form II in its transposed version is implemented in the MATLAB function `filter`. Obviously, it can be used for an FIR filter as well (i.e., if $a_k = 0, k = 1 : N$).

Cascade Structure

Based on the representation (1-10) of the transfer function with the poles p_k and zeros z_ℓ, the numerator and denominator polynomials of $H(z)$ can be written as products of polynomials of first or second degree with real coefficients. That yields the representation

$$H(z) = \prod_{\lambda=1}^{L} H_\lambda(z) \tag{4-1}$$

the description of the *cascade structure*. Its subsystems are characterized by

$$H_\lambda^{(1)}(z) = \frac{b_{0\lambda} + b_{1\lambda} z^{-1}}{1 + a_{1\lambda} z^{-1}} \quad \text{and} \quad H_\lambda^{(2)}(z) = \frac{b_{0\lambda} + b_{1\lambda} z^{-1} + b_{2\lambda} z^{-2}}{1 + a_{1\lambda} z^{-1} + a_{2\lambda} z^{-2}} \tag{4-2}$$

where $a_{k\lambda}, b_{k\lambda} \in \mathbb{R}$. Since a different pairing of the zeros z_ℓ and poles p_k of $H(z)$ yields different transfer functions $H_\lambda(z)$ and since the ordering of these subsystems can be changed, there are many different cascade structures besides the fact that quite a few different implementations can be used for realization of the subsystems themselves.

Parallel Structure

The partial fraction expansion as given in (1-11) for the case of distinct poles leads to

$$H(z) = B_0 + \sum_{\lambda=1}^{L} H_\lambda(z) \tag{4-3}$$

describing the parallel structure, a parallel connection of subsystems, the transfer functions of which are now

$$H_\lambda^{(1)}(z) = \frac{b_{0\lambda}}{1 + a_{1\lambda} z^{-1}} \qquad H_\lambda^{(2)}(z) = \frac{b_{0\lambda} + b_{1\lambda} z^{-1}}{1 + a_{1\lambda} z^{-1} + a_{2\lambda} z^{-2}} \tag{4-4}$$

Here we do not have the choice between different implementations other than the possibility of choosing different structures for the subsystems.

State Equations, Normal System

An appropriate description of all structures is possible using state equations. An equivalent system with the same transfer function but different inside structure can be obtained by introducing a new state vector $\mathbf{q}[n]$ by $\mathbf{x}[n] = \mathbf{T}\,\mathbf{q}[n]$, where $\mathbf{T}$ is a nonsingular transformation matrix. The procedure will be shown using a system of second order as an example. Furthermore, we choose $\mathbf{T}$ such that a so-called *normal system* is obtained, described by a normal matrix $\mathbf{A}_N$, characterized by the property $\mathbf{A}_N^T \mathbf{A}_N = \mathbf{A}_N \mathbf{A}_N^T$.

We start with an implementation of the system in the second direct form (1-3) for the case $N = 2$:

$$\mathbf{A} = \begin{bmatrix} -a_1 & 1 \\ -a_2 & 0 \end{bmatrix} \quad \mathbf{b} = \begin{bmatrix} b_1 - b_0 a_1 \\ b_2 - b_0 a_2 \end{bmatrix} \quad \mathbf{c}^T = [1 \quad 0] \quad d = b_0 \tag{4-5}$$

If p and p^* are the eigenvalues of $\mathbf{A}$, the transformation matrix

$$\mathbf{T} = \frac{1}{\mathrm{Im}\{p\}} \begin{bmatrix} 0 & -1 \\ \mathrm{Im}\{p\} & \mathrm{Re}\{p\} \end{bmatrix} \tag{4-6}$$

leads to the normal matrix

$$\mathbf{A}_N = \mathbf{T}^{-1}\mathbf{A}\mathbf{T} = \begin{bmatrix} \mathrm{Re}\{p\} & \mathrm{Im}\{p\} \\ -\mathrm{Im}\{p\} & \mathrm{Re}\{p\} \end{bmatrix} \tag{4-7}$$

and

$$\mathbf{b}_N = \mathbf{T}^{-1}\mathbf{b} \quad \mathbf{c}_N^T = \mathbf{c}^T\mathbf{T} \quad d_N = d \tag{4-8}$$

An interesting property of the structure will be considered in Exercise 4.2. Furthermore, it will be used in Chapter 7.

Lattice Structures

Another group of configurations to be considered are *lattice structures*, here limited to the implementations of an invertible minimum-phase FIR system and an all-pole IIR system (see [1, Sec. 6.6]), the transfer function of which is the reciprocal of that of the corresponding FIR system. It turns out that in the FIR case the corresponding maximum-phase system is obtained as well at another output, while the IIR structure provides simultaneously the all-pass with the same denominator polynomial. The two structures are shown in Figs. 5.3a and b, respectively. Using the notations of Fig. 5.3a, the analysis of one section in the FIR case yields

$$\begin{bmatrix} Y_i(z) \\ U_i(z) \end{bmatrix} = \begin{bmatrix} 1 & k_i z^{-1} \\ k_i & z^{-1} \end{bmatrix} \begin{bmatrix} Y_{i-1}(z) \\ U_{i-1}(z) \end{bmatrix} \tag{4-9}$$

As can be shown, the subtransfer functions $H_i(z) = Y_i(z)/V(z)$ and $G_i(z) = U_i(z)/V(z)$ are related by

$$\begin{aligned} H_i(z) &= H_{i-1}(z) + k_i z^{-i} H_{i-1}(z^{-1}) \\ G_i(z) &= z^{-i} H_i(z^{-1}) \end{aligned} \tag{4-10}$$

yielding for $i = N$ the relation

$$G_N(z) = z^{-N} H_N(z^{-1}) \tag{4-11}$$

So if $H_N(z)$ is the transfer function of a minimum-phase FIR system with all zeros inside the unit circle of the z-plane, $G_N(z)$ describes the corresponding maximum-phase system. Its zeros are mirror images of those of $H_N(z)$.

In the synthesis procedure the coefficients k_i are determined recursively, starting with the given transfer function

$$H_N(z) = 1 + \sum_{n=1}^{N} h^{(N)}[n] z^{-n}$$

The k_i are calculated such that the sequence of transfer functions

$$H_i(z) = 1 + \sum_{n=1}^{i} h^{(i)}[n] z^{-n} \qquad i = N : -1 : 1$$

is generated according to

$$H_{i-1}(z) = \frac{1}{1 - k_i^2}\left[H_i(z) - k_i z^{-i} H_i(z^{-1})\right] \tag{4-12}$$

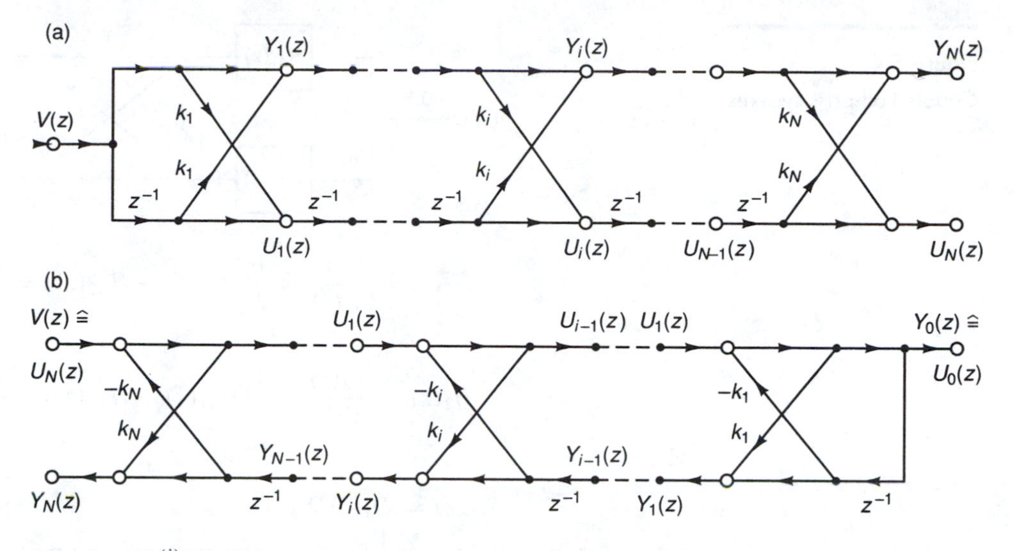

Figure 5.3
Lattice structures: (a) minimum-phase and maximum-phase FIR systems; (b) all-pole and all-pass IIR systems.

where $k_i = h^{(i)}[i]$. This procedure corresponds precisely to the Schur–Cohn stability test, outlined in Project 3. So the program `atok(.)` presented there can be used immediately for calculation of the k_i.

The lattice implementation of an all-pole IIR system with the transfer function

$$H(z) = \frac{1}{A(z)} = \frac{1}{1 + \sum_{m=1}^{N} a_m^{(N)} z^{-m}} \tag{4-13}$$

requires the same procedure, applied to the denominator polynomial $A(z)$. Figure 5.3b shows the corresponding structure, where

$$H(z) = \frac{Y_0(z)}{U_N(z)} = \frac{Y_0(z)}{V(z)} \tag{4-14}$$

Furthermore, the analysis of one section in Fig. 5.3b leads to the relation

$$G_i(z) = \frac{Y_i(z)}{U_i(z)} = \frac{k_i + z^{-1}G_{i-1}(z)}{1 + z^{-1}k_i G_{i-1}(z)} \tag{4-15}$$

$G_i(z)$ is the transfer function of an all-pass of ith order if $G_{i-1}(z)$ describes an all-pass of order $i-1$ and if $|k_i| < 1$. Since $G_0(z) = 1$ and since the k_i are determined as explained above, these conditions are satisfied, yielding finally for $i = N$

$$G_N(z) = \frac{Y_N(z)}{U_N(z)} = \frac{Y_N(z)}{V(z)} = \frac{z^{-N}A(z^{-1})}{A(z)} \tag{4-16}$$

the transfer function of an all-pass, having the same denominator as $H(z)$.

Coupled All-Passes

In Project 3 and Exercise 3.9(b) a lossless system has been presented consisting of two coupled all-passes. The structure is considered again here [3, Sec. 8.6]. Given two all-passes with the transfer functions

$$H_{Ai}(z) = \frac{z^{-N_i}D_i(z^{-1})}{D_i(z)} \qquad i = 1, 2 \tag{4-17}$$

where the $D_i(z)$ are polynomials of degree N_i. As shown in Fig. 5.4, their combination yields a system with one input and two outputs and the transfer functions

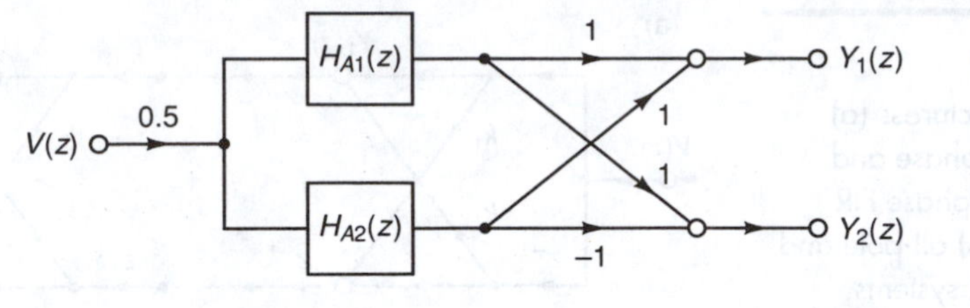

Figure 5.4
Coupled all-pass systems.

$$H_1(z) = \frac{Y_1(z)}{V(z)} = \frac{1}{2}[H_{A1}(z) + H_{A2}(z)] = \frac{B_1(z)}{A(z)} \tag{4-18}$$

$$H_2(z) = \frac{Y_2(z)}{V(z)} = \frac{1}{2}[H_{A1}(z) - H_{A2}(z)] = \frac{B_2(z)}{A(z)} \tag{4-19}$$

Here

$$A(z) = D_1(z)D_2(z) \tag{4-20}$$

$$B_{1,2}(z) = \frac{1}{2}\left[z^{-N_1}D_1(z^{-1})D_2(z) \pm z^{-N_2}D_2(z^{-1})D_1(z)\right] \tag{4-21}$$

As can readily be confirmed, the numerator polynomials have the properties

$$z^{-(N_1+N_2)}B_1(z^{-1}) = B_1(z) \tag{4-22}$$

$$z^{-(N_1+N_2)}B_2(z^{-1}) = -B_2(z) \tag{4-23}$$

Here $B_1(z)$ is a mirror-image polynomial, while $B_2(z)$ is an anti-mirror-image polynomial.

The interdependence of $H_1(z)$ and $H_2(z)$ are found with (4-18) and (4-19) as

$$H_1(z)H_1(z^{-1}) + H_2(z)H_2(z^{-1}) = 1\ \forall z \tag{4-24}$$

yielding for $z = e^{j\omega}$

$$|H_1(e^{j\omega})|^2 + |H_2(e^{j\omega})|^2 = 1\ \forall\omega \tag{4-25}$$

and thus

$$|H_i(e^{j\omega})| \le 1 \qquad \forall\omega \qquad i = 1, 2 \tag{4-26}$$

Furthermore, we get

$$H_1(z) \pm H_2(z) = H_{Ai}(z) \qquad i = 1, 2 \tag{4-27}$$

and on the unit circle

$$|H_1(e^{j\omega}) \pm H_2(e^{j\omega})| = 1 \tag{4-28}$$

In the sense of (4-25), (4-28) both transfer functions are doubly complementary.

Besides the property $|H_i(e^{j\omega})| \le 1$ there are further restrictions due to the required properties of the numerator polynomials $B_1(z)$ and $B_2(z)$, as given in (4-22) and (4-23). For example, only low-passes and high-passes of odd degree can be implemented with this structure. In general, the design starts with a given transfer function. We explain it using an elliptic low-pass of Nth degree as an example, where N is odd. Its transfer function $H_1(z) = B_1(z)/A(z)$ has the required properties: $|H_1(e^{j\omega})| \le 1$, $H_1(1) = 1$, $H_1(-1) = = 0$, $B_1(z)$ being a mirror-image polynomial with zeros on the unit circle only.

To find the transfer functions of the two all-passes, the corresponding transfer function $H_2(z) = B_2(z)/A(z)$ of the complementary high-pass has to be calculated such that (4-24) is satisfied and $B_2(z)$ is an anti-mirror-image polynomial. We show one possibility, based on (4-24). With (4-18) and (4-19) we get

$$z^{-N}B_2(z)B_2(z^{-1}) = z^{-N}A(z)A(z^{-1}) - z^{-N}B_1(z)B_1(z^{-1}) \tag{4-29}$$

a mirror-image polynomial of order $2N$, the roots of which are in this case on the unit circle, each of even order. One of them is located at $z = 1$. Selecting half of these roots appropriately and using the condition $|H_2(-1)| = 1$ yields all information for the calculation of $B_2(z)$. Finally, the two all-passes can be found with (4-27).

The implementation of the all-passes can be done using the direct structure, with a numerator polynomial being the flipped version of the denominator or with the lattice structure as shown in Fig. 5.3b. Its coefficients have to be calculated as shown above. As a modification a cascade of blocks of first and second order can be used, each implemented in lattice form as shown in Fig. 5.5. Their transfer functions are

$$H_{A\lambda}^{(1)}(z) = \frac{Y_\lambda^{(1)}(z)}{V_\lambda(z)} = \frac{k_{1\lambda} + z^{-1}}{1 + k_{1\lambda}z^{-1}}$$

$$H_{A\lambda}^{(2)}(z) = \frac{Y_\lambda^{(2)}(z)}{V_\lambda(z)} = \frac{k_{1\lambda} + k_{2\lambda}(1 + k_{1\lambda})z^{-1} + z^{-2}}{1 + k_{2\lambda}(1 + k_{1\lambda})z^{-1} + k_{1\lambda}z^{-2}}$$

Obviously, its reflection coefficients are simply related to the poles of the individual transfer functions. Find these relations.

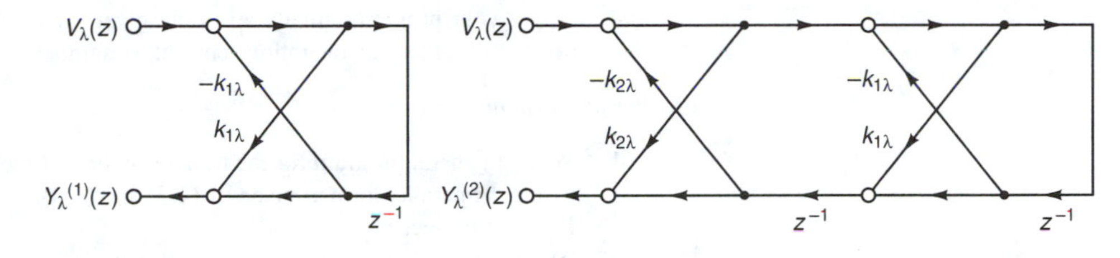

Figure 5.5
All-pass blocks of first and second order, implemented as lattices.

The use of lattices yields so-called structural all-passes. They keep their all-pass property as long as $|k_i| < 1$, showing a low sensitivity against coefficient errors due to quantization (see Chapter 7, Project 3).

EXERCISE 4.1

Complexity of Different Implementations

You are given the coefficients of a seventh-order low-pass[4]

$$\begin{aligned} \mathbf{a} &= [1\ \ -0.5919\ \ 2.1696\ \ -1.1015\ \ 1.5081\ \ -0.5823\ \ 0.3275\ \ -0.0682] \\ \mathbf{b} &= [0.0790\ \ 0.2191\ \ 0.4421\ \ 0.5905\ \ 0.5905\ \ 0.4421\ \ 0.2192\ \ 0.0790] \end{aligned} \tag{4-30}$$

The properties of three different implementations in terms of the required number of arithmetic operations are to be determined.

a. *Direct structure*: The function `filter` yields a solution of

$$y[n] = \sum_{\ell=1}^{M+1} b(\ell)v[n+1-\ell] - \sum_{\ell=1}^{N} a(\ell+1)y[n-\ell]$$

Determine the required number of arithmetic operations (multiplications and additions) for one output value $y[n]$ as a function of M and N. Verify your result for the example given above with $M = N = 7$ and $v[n] = \delta[n]$, $n = 0:99$. [*Hint*: The number of operations can be determined experimentally by setting the counter to zero with `flops(0)` before starting, while at the end `op = flops/length(y)` yields the desired result.]

[4] Elliptic filter found by `ellip(7,0.1,40,0.5)`.

b. *Cascade structure*: We want to develop a program for the transformation of the direct form, given by **a** and **b**, into the cascade form described by (4-1) and (4-2).

1. How many different implementations in cascade form are possible in case of a system of seventh order as in the example if you count all possible pairings of poles and zeros to blocks of second order and all different orderings of the subsystems? The subsystems themselves are to be implemented in direct form.
2. Write the desired general program using the following pairing and ordering scheme [1. Sec. 6.9]:
   A. Start with the pole p_k having the largest magnitude.
   B. Pick the nearest zero z_ℓ.
   C. Calculate the real parameters a_{i1} and b_{i1} of the second-order block characterized by p_k and z_ℓ.
   D. Take this block as the first one.
   E. Proceed similarly for the next block with the remaining poles and zeros.
3. Write a program for execution of the cascade form where the subsystems are real and implemented by `filter`. (Make sure that the coefficients of your subsystems are actually real.)
4. Apply your programs for the example given by (4-30). Determine the required number of arithmetic operations per output sample.

c. *Parallel structure*

1. Write a general program for the transformation of the direct form, given by **a** and **b**, into the parallel form as given by (4-3) and (4-4). Use `residuez` as in Exercise 1.1(b) of Project 1.
2. Write a program for execution of the parallel form, where the subsystems are real and implemented by `filter`.
3. Apply and check the performance of your program in terms of complexity using the example given by (4-30).

d. Summarize your results from parts (a), (b), and (c).

Remark. The number of arithmetic operations is only one criterion, and not the most important. Others related to wordlength effects are considered in Chapter 7.

EXERCISE 4.2

Properties of a Normal System

You are given the description of a system by

$$\mathbf{A} = \begin{bmatrix} 1.9 & 1 \\ -0.95 & 0 \end{bmatrix} \quad \mathbf{b} = \begin{bmatrix} 1 \\ 1 \end{bmatrix} \quad \mathbf{c}^T = [1 \quad 1] \; d = 0$$

a. Let the input sequence be $v[n] = 0$ and the initial state vector $\mathbf{x}_i[0] = [1 \quad 1]^T$.

1. Calculate the state vector $\mathbf{x}_i[n] = [x_{i1}[n] \; x_{i2}[n]]^T$ and the output sequence $y_i[n]$ for $n = 0 : 100$. Plot $\mathbf{x}_i[n]$ using `axis('square')` with

```
plot(x1,x2,'o')
```

   Plot $y_i[n]$ using `stem`.
2. Calculate and plot the stored energy of the system, defined as

$$w[n] = \mathbf{x}^T[n] \cdot \mathbf{x}[n]$$

We mention that a nonexited system is called passive if

$$\Delta w[n+1] = w[n+1] - w[n] \leq 0 \;\forall n \geq 0$$

Is the system under test passive in that sense?

b. Now we assume the initial state to be zero. Calculate the state vector $\mathbf{x}_s[n]$ and the output sequence $y_s[n]$ for an excitation by the unit-step sequence $u[n]$. Compare the resulting final values $\mathbf{x}_s[100]$ and $y_s[100]$ with those for $n \to \infty$, to be calculated easily out of the state equations. Is there a relation to the transfer function $H(z)$ at a specific point of z?

c. Transform the system into a normal one according to (4-6)–(4-8) and repeat part (a). Do not forget to transform the initial state. Note that the state vector $\mathbf{q}[n]$ differs completely from $\mathbf{x}[n]$, while the output sequences in both cases are the same. Compare the two systems in terms of passivity as defined in part (a), point 2.

d. Repeat part (b) for the transformed system. What are the final values $\mathbf{q}_s[100]$ and $y_s[100]$ in this case? Compare them with those to be expected.

EXERCISE 4.3

Lattice Structures

a. Find the state-space descriptions of the two lattice structures as given in Fig. 5.3 for one input and two outputs. For simplicity use $N = 4$. Check your results with

$$\mathbf{a} = [1.0 \;\; -2.1944 \;\; 2.6150 \;\; -1.6245 \;\; 0.5184]$$

having the corresponding reflection coefficients

$$\mathbf{k} = [-0.6707 \;\; 0.8819 \;\; -0.6658 \;\; 0.5184]$$

(*Hint*: The checking should be done by transforming the state-space descriptions into the transfer function representation using `ss2tf` and comparing the results with the given vector **a**.)

b. Given the vector **k** of reflection coefficients, develop two M-files for calculation of the pairs of output sequences of the two lattice structures for an excitation with the input sequence $v[n]$. Call them

`[yN,uN] = firlat(k,v)` and `[u0,yN] = iirlat(k,v)`

With the notation of Fig. 5.3, $y_N[n]$ and $u_N[n]$ are the output sequences of the minimum- and maximum-phase FIR system, respectively, while $u_0[n]$ and $y_N[n]$ belong to the all-pole and the all-pass IIR system. Check your programs by applying them to the example given in part (a) for an excitation by $\delta[n]$. Which results do you expect in the FIR cases? For comparison, generate the impulse response in the IIR case differently by using `filter`.

c. Let $u_0[n]$ be the impulse response of the all-pole system. Calculate the reactions $y_N[n]$ and $u_N[n]$ of the FIR system after excitation with $u_0[n]$. Explain the results.

EXERCISE 4.4

Elliptic Filters Implemented with Coupled All-Passes

You are given the coefficients of the transfer function of a low-pass as[5]

$$\mathbf{a} = [1 \;\; -0.1235 \;\; 0.7726 \;\; -0.1235]$$
$$\mathbf{b} = [0.3098 \;\; 0.4530 \;\; 0.4530 \;\; 0.3098]$$

---

[5]For **a**, see Exercise 3.9.

a. The coefficients of $H_2(z)$, the transfer function of the corresponding complementary high-pass, are to be calculated. Determine the polynomial on the right-hand side of (4-29) and calculate its roots. Select N of them appropriately and calculate the required anti-mirror-image polynomial $B_2(z)$.

b. The coefficients of the two all-pass transfer functions $H_{A1}(z)$ and $H_{A2}(z)$ are to be determined. First, we get, according to (4-27),

$$H_{Ai}(z) = \frac{B_1(z) \pm B_2(z)}{A(z)} \qquad i = 1, 2$$

Since the degrees of the all-passes are N_1 and N_2, respectively, while $A(z)$ has the degree $N = N_1 + N_2$ [see (4-17) and (4-20)], the two numerators and the denominator $A(z)$ must have common roots. So find the denominator polynomials $D_i(z)$ of the $H_{Ai}(z)$ by deleting those roots of $A(z)$, which it has in common with $B_1(z) + B_2(z)$ or $B_1(z) - B_2(z)$, respectively.

The design of the system is continued by calculating the reflection coefficients for lattice implementations of the two all-passes, either according to Fig. 5.3b or as a cascade of blocks of first or second order, as shown in Fig. 5.5. Use your program `irrlat(.)` or write another one for the cascade which calculates the output sequence $y_{N_i}[n]$ of an all-pass having the degree N_i. The implementation is completed as in Exercise 3.9 with

$$y_1[n] = 0.5(y_{N1}[n] + y_{N2}[n])$$
$$y_2[n] = 0.5(y_{N1}[n] - y_{N2}[n])$$

c. We want to explain the property of the structure somewhat differently.

1. Calculate the unwrapped phases $\varphi_1(\omega)$ and $\varphi_2(\omega)$ of the two all-passes for $0 \leq \omega \leq \pi$. Plot

$$\Delta\varphi(\omega) = \varphi_2(\omega) - \varphi_1(\omega)$$

Explain the properties of the structure in terms of this phase difference.

2. Calculate and plot in one figure

$$\left|G_1(e^{j\omega})\right| = \frac{1}{2}\left|1 + e^{-j\Delta\varphi(\omega)}\right| = \left|\cos\frac{\Delta\varphi(\omega)}{2}\right|$$

$$\left|G_2(e^{j\omega})\right| = \frac{1}{2}\left|1 - e^{-j\Delta\varphi(\omega)}\right| = \left|\sin\frac{\Delta\varphi(\omega)}{2}\right|$$

How are $G_{1,2}(e^{j\omega})$ related to $H_{1,2}(e^{j\omega})$?

d. The required all-passes can be found differently as follows:

1. Starting with $H_1(z) = B_1(z)/A(z)$, calculate the roots of $A(z)$ as `p = roots(a)`.
2. Calculate the phase of these roots by `ph = angle(p)` and number them starting with `min(ph)`.
3. Pick all poles having an odd number for the polynomial $D_1(z)$ of $H_{A1}(z)$ and all poles with an even number as zeros of the polynomial $D_2(z)$ of $H_{A2}(z)$.

Repeat the design of the all-passes for the example above with this procedure.

e. Finally, we determine the complexity of this structure, using the filter described by (4-30) as an example.

1. Design the two all-passes using the procedure explained in part (d).
2. To design the lattice implementations of the two all-passes, calculate the two sets of reflection coefficients with `atok`.

3. The output sequence of the low-pass is

$$y[n] = 0.5\,(y_{N1}[n] + y_{N2}[n])$$

where $y_{N1}[n]$ and $y_{N2}[n]$ are the output sequences of the two all-passes, to be calculated with your program `iirlat(k,v)`. As an alternative, use a cascade of all-passes as given in Fig. 5.5. Use, for example, the unit-step sequence `ones(1,100)` as the input signal $v[n]$. Determine the required number of numerical operations as described before.

chapter 6

Stochastic Signals

OVERVIEW

The study of random signals and methods of power spectrum estimation is a rather challenging discipline because the level of mathematics needed for a rigorous presentation of stochastic processes is high. On the other hand, many of the algorithms used to perform spectrum estimation are relatively easy to implement in a language such as MATLAB, with its vector and matrix operations and its random number generator. Furthermore, an understanding of these algorithms and their performance is often gained through simulation. The signals involved are random, so their mathematical description is based on probability distributions. However, in actual processing only one member signal out of the stochastic process is manipulated. Therefore, a simulation should capture the essential behavior of the algorithm through the processing of a few signals. A thorough simulation must be a Monte Carlo simulation, which demands many runs of the algorithm on different random signals from the same underlying process.

The objective of this chapter is to present the essential information about parameter estimation for a stochastic process and relate the theoretical formulas for quantities such as bias and variance. Initially, we consider the estimation of simple quantities such as the mean, variance, and pdf of a random signal. We also examine, by means of simulation, properties of the random signal, such as ergodicity and stationarity. The rest of this chapter is devoted to power spectrum estimation. First, methods based on the FFT are covered in detail. This leads to an understanding of the Welch–Bartlett method, which is the most widely used technique based on the FFT. Then the maximum entropy method (MEM) and other techniques of "modern" spectrum estimation are treated.

In the study of spectrum estimation, we concentrate on the implementation of several representative methods, and then on ways to characterize their performance. To show their merits as well as their deficiencies, we use as an example a random process with known power density spectrum, generated by passing white noise through a linear system with known transfer function. Thus, in our experiments, we can compare to the correct results and can express the observed deviations in terms of this known spectrum. In the section *FFT Spectrum Estimation*, the data window introduces a fundamental resolution limit due to the uncertainty principle of Fourier analysis. One project considers the problem of resolving two closely spaced sinusoids with different windows and at different SNRs.

Obviously, methods that circumvent the resolution limit of Fourier analysis are of interest. A number of such methods have been developed and popularized over the past 20 years, MEM being among the best known. In the section *Modern Spectrum Estimation,* the projects will treat MEM and some related methods based on linear prediction. The important role of the all-pole spectrum, and of pole–zero models in general, will be stressed. Finally, the Pisarenko harmonic decomposition is introduced, along with practical estimation methods based on eigenvalues and eigenvectors (e.g., MUSIC). A key concept in this area is the idea of a signal-noise subspace representation through eigenvectors. A complete study of these methods and their application to signal analysis and spectrum estimation would easily take another entire volume, so we are content to introduce some of the well-known methods. Then we can use MATLAB to implement each one and to do a simulation to gauge their relative performance.

BACKGROUND READING

There are quite a few books on random processes and random variables, beside the classic one by Papoulis [1] (e.g., those by Leon-Garcia [2] and Scharf [3]). A brief introduction to random signals is also given as an appendix in both [4] and [5], as well as Section 2.10 in [4].

Some books are devoted to spectral estimation, especially the books by Marple [6] and Kay [7]. Others have rather comprehensive chapters on the subject: Chapter 13 in [1], Chapters 11 and 12 of [8], Chapter 2 in [9] and Sections 11.5–11.7 in [4]. Chapters 4, 5, and 8 in [5], as well as Chapters 1, 2, and 6 in [9], also address a variety of topics related to spectrum estimation and modeling of random signals. A useful algorithm for estimating the autocorrelation sequence is described in [10]. Finally, two reprint collections have been published by IEEE Press on the topic of modern spectrum estimation [11, 12].

[1] A. Papoulis. *Probability, Random Variables, and Stochastic Processes*. McGraw-Hill, New York, third edition, 1991.

[2] A. Leon-Garcia. *Probability and Random Processes for Electrical Engineering*. Addison-Wesley, Reading, MA, 1989.

[3] L. L. Scharf. *Statistical Signal Processing, Detection, Estimation and Time Series Analysis*. Addison-Wesley, Reading, MA, 1991.

[4] A. V. Oppenheim and R. W. Schafer. *Discrete-Time Signal Processing*. Prentice Hall, Englewood Cliffs, NJ, 1989.

[5] J. G. Proakis, C. M. Rader, F. Ling, and C. L. Nikias. *Advanced Digital Signal Processing*. Macmillan, New York, 1992.

[6] S. L. Marple. *Digital Spectral Analysis with Applications*. Prentice Hall, Englewood Cliffs, NJ, 1987.

[7] S. M. Kay. *Modern Spectral Estimation: Theory and Application*. Prentice Hall, Englewood Cliffs, NJ, 1988.

[8] J. G. Proakis and D. G. Manolakis. *Digital Signal Processing: Principles, Algorithms and Applications*. Macmillan, New York, second edition, 1992.

[9] J. S. Lim and A. V. Oppenheim. *Advanced Topics in Signal Processing* . Prentice Hall, Englewood Cliffs, NJ, 1988.

[10] C. M. Rader. An improved algorithm for high speed autocorrelation with applications to spectral estimation. *IEEE Transactions on Audio and Electroacoustics*, AU-18:439–441, December, 1970.

[11] D. G. Childers, editor. *Modern Spectrum Estimation*. IEEE Press, New York, 1978.

[12] S. B. Kesler, editor. *Modern Spectrum Estimation II*. IEEE Press, New York, 1986.

[13] V. F. Pisarenko. The retrieval of harmonics from a covariance function. *Geophysical Journal of the Royal Astronomical Society*, 33:247–266, 1972.

[14] J. Makhoul. Linear Prediction: A Tutorial Review. *Proceedings of the IEEE*, 63(4):561–580, April 1975.

[15] R. Schmidt. Multiple Emitter Location and Signal Parameter Estimation. *Proceedings of the RADC Spectral Estimation Workshop*, pages 243–258, 1979.

[16] R. Kumaresan and D. W. Tufts. Estimating the Angles of Arrival of Multiple Plane Waves. *IEEE Transactions on Aerospace and Electronic Systems*, AES-19(1):134–139, January 1983.

[17] J. P. Burg. Maximum entropy spectral analysis. *Proceedings of the 37th Meeting of the Society of Exploration Geophysicists*, 1967.

[18] C. L. Lawson and R. J. Hanson. *Solving Least Squares Problems*. Prentice Hall, Englewood Cliffs, NJ, 1974.

[19] D. W. Tufts and R. Kumaresan. Singular value decomposition and improved frequency estimation using linear prediction. *IEEE Transactions on Acoustics, Speech, and Signal Processing*, ASSP-30:671–675, August 1982.

[20] G. H. Golub and C. F. Van Loan. *Matrix Computations*. Johns Hopkins University Press, Baltimore, 1989.

STOCHASTIC SIGNALS

OVERVIEW

The signals we have to deal with in practice are, in most cases, not deterministic. A speech signal, for example, cannot be described by an equation. Nonetheless, it has certain characteristics which distinguish it from, say, a television signal. In fact, almost all signals that we have to handle in communications, and in many other fields of engineering and science, are of a stochastic nature (also called random).

A stochastic signal has two facets: at a fixed time instant its value is a random variable, and as a function of time the random variables might be interrelated. The definition of the random signal is done via its statistical properties: probability density function, joint density function, mean, autocorrelation, and so on. In a theoretical problem, these quantitative descriptions apply to the ensemble of all realizations of the particular random process. They are deterministic functions, well behaved in the mathematical sense. However, in a practical problem they must be estimated, using measurements on a finite set of data taken from observations of the random process. Since these estimates are actually formed from random variables, they are themselves random variables. Thus we can only make probabilistic statements about the closeness of an estimated value to the true values (e.g., 95% confidence intervals).

This set of projects deals with the description and processing of stochastic signals, mainly under the assumptions of stationarity and ergodicity. Furthermore, we investigate how the ensemble averages are influenced when a stochastic signal is processed through a linear filter or a nonlinear mapping. In most of the projects, estimates are computed in MATLAB via *averaging*—either time averages or ensemble averages. These estimates must then be compared to the true values known from theory.

An understanding of these results should open the way to designing systems that generate stochastic signals with desired properties. For example, a parametric description of a given stochastic signal can be obtained in terms of the coefficients of a fixed linear filter, producing this signal, if it is excited by so-called *white noise*, an uncorrelated stochastic signal. Related to this representation is the problem of prediction and decorrelation, that is, the design of a system whose output sequence is approximately an advanced version of its input, or such that the correlated input sequence is transformed into an output that is a white noise sequence.

As in the case of deterministic signals, a description in both the time and frequency domains is of interest. The direct Fourier transform of the random signal is not a useful quantity, but the transform of the autocorrelation function is—it is called the *power spectrum*. An introduction to methods for spectral estimation is given in the following two sets of projects.

BACKGROUND READING

Many books have been published on the subject of random processes and random variables. A classic text used for many years was written by Papoulis [1]. More recent texts are those by Leon-Garcia [2] and Scharf [3]. A brief introduction to random signals is also given as an appendix in [8], as well as Section 2.10 and Appendix A in [4].

PROJECT 1: RANDOM VARIABLES

In this project, the elementary properties of random variables will be introduced. The exercises concentrate on estimating the mean, variance, and probability density function of the random variables.

A random variable (RV) is described by a *probability density function* (pdf):

$$p_{\mathbf{v}}(v) = \frac{d}{dv} P_{\mathbf{v}}(v) \qquad \text{PROBABILITY DENSITY (pdf)} \tag{1-1}$$

where

$$P_{\mathbf{v}}(v) = \text{Probability}[\mathbf{v} \le v] \qquad \text{PROBABILITY DISTRIBUTION} \tag{1-2}$$

Here $\mathbf{v}$ denotes the random variable and v is a particular value of $\mathbf{v}$. The pdf $p_{\mathbf{v}}(v)$ can (loosely) be interpreted as giving the probability that the RV $\mathbf{v}$ will be equal to the value v.

In many cases, only certain ensemble averages of the RV need be computed—usually the mean and variance:

$$m_{\mathbf{v}} = \mathcal{E}\{\mathbf{v}\} = \int_{-\infty}^{+\infty} v\, p_{\mathbf{v}}(v)\, dv \tag{1-3}$$

$$\sigma_{\mathbf{v}}^2 = \mathcal{E}\{|\mathbf{v} - m_{\mathbf{v}}|^2\} = \int_{-\infty}^{+\infty} (v - m_{\mathbf{v}})^2\, p_{\mathbf{v}}(v)\, dv \tag{1-4}$$

These ensemble averages are *constants*, but they cannot be determined exactly from samples of the RV. In these exercises, samples of the RV will be created by a pseudo-random number generator, the properties of which are known with sufficient accuracy. The pdf, mean, and variance will be *estimated* from a finite number of these samples and then compared with the theoretical values.

Hints

The MATLAB functions `rand(M,N)` and `randn(M,N)` will generate an $M \times N$ matrix of pseudo-random numbers. These functions can produce random numbers with two different pdf's: uniform and Gaussian, respectively.

Warning. Since these exercises require the estimation of ensemble properties via averaging, very large sample sets must be used. This can stress the memory capabilities of MATLAB on some machines and in old student versions. If your machine can accommodate large vectors, use very large lengths (e.g., 8000–10,000) so that your results will be nearly perfect. If you are forced to have short vectors, recognize that the small sample size may cause the answers to deviate noticeably from the corresponding theoretical values.

The MATLAB function `hist(x,nbins)` will compute and plot a histogram (as a bar chart) for a vector `x` of pseudo-random numbers. The default for the number of bins is 10; otherwise, `nbins` can be specified.

EXERCISE 1.1

Uniform pdf

Generate a long vector of samples of a uniform random variable; use at least several thousand samples.

a. Use `hist`, `mean`, and `std` to make estimates of the pdf, $m_{\mathbf{v}}$, and $\sigma_{\mathbf{v}}$, respectively. Note that the histogram must be normalized to have a total area equal to 1, if it is to be a legitimate estimate of the pdf.

b. The MATLAB `rand` function produces a uniform density in the range 0 to 1. Thus it is possible to derive the theoretical values for the mean and variance ($\sigma_{\mathbf{v}}^2$). Determine these theoretical values and compare to the estimated ones.

c. Repeat the numerical experiment of part (a) several times to see that the estimated values are not always the same. However, you should observe that the estimated values hover about the true values.

EXERCISE 1.2

Gaussian pdf

Generate a long vector of samples of a *Gaussian* random variable; use at least several thousand samples.

a. As in Exercise 1.1, compute estimates of the pdf, $m_{\mathbf{v}}$, and $\sigma_{\mathbf{v}}^2$. Compare the mean and variance to the true values; and repeat the experiment several times to observe the variability of these two quantities.

b. The histogram plot should approximate the true pdf—in this case, the bell shape of a Gaussian, $\mathcal{N}(m_{\mathbf{v}}, \sigma_{\mathbf{v}}^2)$. The formula for the Gaussian pdf is known:

$$p_{\mathbf{v}}(v) = \frac{1}{\sigma_{\mathbf{v}}\sqrt{2\pi}} e^{-(v-m_{\mathbf{v}})^2/2\sigma_{\mathbf{v}}^2} \tag{1-5}$$

On a plot of the scaled histogram, superimpose the function for $p_{\mathbf{v}}(v)$ (see `help plot` or `help hold` for ways to put several curves on one plot). Be careful to normalize the histogram and Gaussian pdf to the same vertical scale. Experiment with the number of bins and the length of the vector of random samples to get a reasonably good match.

EXERCISE 1.3

Average 12 Uniform RVs to Create a Gaussian

On a computer it is trivial to generate uniformly distributed random numbers. Other pdf's are usually created by applying nonlinear transformations to the uniform density (see Project 4). For the Gaussian case, there is another simple approach based on averaging. This method relies on the central limit theorem [1], which states (loosely) that averaging independent RVs, irrespective of their pdf, will give a new RV whose pdf tends to a Gaussian (see Exercise 5.2). This theorem involves a limiting process, but a practical example of the theorem can be demonstrated by averaging a relatively small number of uniform RVs; 12, in fact, will be sufficient.

a. Use `rand(12,N)` to create a $12 \times N$ matrix of uniform random variables. Take the average of each column of the matrix; this can be done cleverly by using the `mean` function, which does each column separately. The result is a new random vector of length N.

b. Estimate the pdf of this new random variable from its histogram. Also, estimate its mean and variance.

c. Since the default random number generator in MATLAB is a nonzero mean uniform RV, derive the theoretical values for the mean and variance of the approximate Gaussian.

d. Compare the estimated values from part (c) to the theoretical values. Make a plot of the histogram with the theoretical Gaussian pdf superimposed.

EXERCISE 1.4

Independent RVs

The pseudo-random number generator (`randn`) within MATLAB can be invoked twice to produce samples of two distinct random variables. The interaction of these two RVs is described by their joint pdf, which is a function of two variables. Suppose that we call the two RVs $\mathbf{v}_1$ and $\mathbf{v}_2$. Then the joint pdf at (x, y) is the probability that $\mathbf{v}_1$ equals x *and* $\mathbf{v}_2$ equals y. For example, the two-dimensional Gaussian pdf can be written as

$$p_{\mathbf{v}_1\mathbf{v}_2}(x, y) = \frac{1}{2\pi\sqrt{|\mathbf{C}|}} e^{-\frac{1}{2}(\mathbf{v}-\mathbf{m}_\mathbf{v})^T\mathbf{C}^{-1}(\mathbf{v}-\mathbf{m}_\mathbf{v})} \tag{1-6}$$

where the vector $\mathbf{v}$ is $[x \quad y]^T$, while $\mathbf{m}_\mathbf{v} = [m_{\mathbf{v}_1}, m_{\mathbf{v}_2}]^T$, and $\mathbf{C}$ is the covariance matrix of the two zero-mean RVs, $\tilde{\mathbf{v}}_i = \mathbf{v}_i - m_{\mathbf{v}_i}$:

$$\mathbf{C} = \mathcal{E}\left\{\begin{bmatrix}\tilde{\mathbf{v}}_1\\ \tilde{\mathbf{v}}_2\end{bmatrix}[\tilde{\mathbf{v}}_1 \quad \tilde{\mathbf{v}}_2]\right\} = \begin{bmatrix}\mathcal{E}\{\tilde{\mathbf{v}}_1^2\} & \mathcal{E}\{\tilde{\mathbf{v}}_1\tilde{\mathbf{v}}_2\}\\ \mathcal{E}\{\tilde{\mathbf{v}}_2\tilde{\mathbf{v}}_1\} & \mathcal{E}\{\tilde{\mathbf{v}}_2^2\}\end{bmatrix} \tag{1-7}$$

The covariance matrix is always symmetric and positive semidefinite.

There are two ways to estimate the joint pdf: compute a two-dimensional histogram, or assume that the pdf is Gaussian and estimate the covariance matrix for use in (1-6). To test these methods generate two data vectors containing Gaussian RVs. Generate a few thousand points of each. Make them both zero-mean, but set the variance of the first equal to 1 and the variance of the other to 3.

a. Derive a mathematical formula for the joint pdf, which is a two-dimensional Gaussian. See `help meshdom` to generate the (x, y) domain for computing and plotting this formula. Plot the level curves (via `contour`) of the joint Gaussian pdf, and observe that they are elliptical in shape.

b. Compute an estimate of the covariance matrix (1-7) by taking averages of v_1^2, v_2^2, and $v_1 v_2$. Compare this estimate to the true covariance matrix. Plot the two-dimensional Gaussian function based on this covariance estimate and compare its level curves to those found in part (a).

c. Write a MATLAB function (called `hist2`) to compute the two-dimensional histogram for a pair of vectors. Include an argument to specify the number of bins. Take advantage of the existing one-dimensional histogram M-file `hist` and also the `find` function to compute the output histogram matrix one column (or row) at a time. If your function is too slow when processing long data vectors, analyze the code in the M-file `hist` to see if you can find ways to improve your M-file.

d. Use `hist2` to estimate the pdf directly from the data. Plot this estimated pdf on the same graph with the true pdf. Use a `contour` plot, but plot only a few level curves for each pdf so that it will be easy to compare.

e. Since these two RVs were generated independently, they should be uncorrelated. The definition of *uncorrelated* is that the two-dimensional pdf will factor into the product of two one-dimensional pdf's. This implies that the expected value of the product $\mathbf{v}_1\mathbf{v}_2$ is the product of the expected values, which for the zero-mean case would be zero. Verify that $\mathbf{v}_1$ and $\mathbf{v}_2$ are uncorrelated by doing the following: Estimate the one-dimensional pdf's of $\mathbf{v}_1$ and $\mathbf{v}_2$ separately, and multiply them together to produce another estimate of the two-dimensional pdf. Plot this separable estimate versus the true pdf using a contour plot.

EXERCISE 1.5

Joint pdf for Correlated RVs

In this exercise, the joint pdf of two *correlated* random variables will be computed. Generate two Gaussian random vectors, each containing several thousand samples. Both should be zero-mean RVs with variance equal to 1 and 2, respectively. Form two new RVs by taking the sum and difference of the original RVs; these new RVs will be used for the tests in this exercise.

a. Determine the theoretical form of the joint Gaussian pdf. Find the exact entries in the covariance matrix.

b. Make a mesh plot of this multivariate Gaussian from its functional form. Use `meshdom` to generate the (x, y) domain for the calculation and the plot.

c. Estimate the entries in the two-dimensional covariance matrix from the data vectors, and plot this estimated pdf. Compare the estimated and true pdf's by making a contour plot with a few level curves from each one.

d. Estimate the pdf by computing the two-dimensional histogram. Make a contour plot (and a mesh plot) of the histogram, and compare it to the true pdf. In this case, the two RVs are correlated, so you should verify that the pdf will not factor into the product of two one-dimensional pdf's.

PROJECT 2: NONSTATIONARY, STATIONARY, AND ERGODIC RANDOM PROCESSES

Now we consider a random process $\mathbf{v}_n$ characterized by a particular rule, according to which its different members are generated. A random process, which is also called a stochastic process, is an ensemble of time signals, with the additional characterization that its value v_n at the time instant n is a random variable. In the most general case, the pdf of the signal value might be different for each index n, but the more common case is a *stationary* process where the pdf remains the same for all n. In addition, for a stationary process, the joint pdf between signal values at n and m will depend only on the difference $n - m$.

Since the underlying idea of processing stochastic signals is to learn something about the pdf(s) that define the process, an important problem for stochastic signal processing is how to do that estimation from one member of the stochastic process. If we have only one member from the ensemble of signals, we cannot average across the ensemble as was done in the random variable case. Instead, we must operate under the assumption that time averages along the one recorded signal will be sufficient to learn about the pdf's. The assumption that allows us to take this approach is called *ergodicity*, which states loosely that "time averages will converge to ensemble averages." An ergodic process must be stationary because it would never be possible to estimate a time-varying pdf from just one signal.

A stochastic process $\mathbf{v}_n$ is characterized by its (time-dependent) pdf $p_{\mathbf{v}_n}(v_n, n)$, as well as the joint pdf between RVs at the time instants n and $n + m$, and all higher-order pdf's. If the pdf's do not depend on the definition of the time origin, the process is *stationary*; otherwise, it is called a *nonstationary process*.

Usually, we are most interested in measuring the important *ensemble averages*, which include the mean and variance:

$$m_{\mathbf{v}_n}[n] = \mathcal{E}\{\mathbf{v}_n\} = \int_{-\infty}^{+\infty} v\, p_{\mathbf{v}_n}(v, n)\, dv \qquad \text{(MEAN)} \qquad (2\text{-}1)$$

$$\sigma^2_{\mathbf{v}_n}[n] = \mathcal{E}\{|(\mathbf{v}_n - m_{\mathbf{v}_n})|^2\} \qquad \text{(VARIANCE)} \qquad (2\text{-}2)$$

In case of a *stationary process* the time dependence will disappear, so that the mean $m_{\mathbf{v}}$ and variance $\sigma^2_{\mathbf{v}}$ become constants, as in (1-3) and (1-4).

So far we have considered the whole process $\mathbf{v}_n$, consisting of an infinite number of individual signals $v_\lambda[n]$, all generated according to the same rule. For any one of these sequences, averages in time direction can be defined: for example, the time-averaged mean,

$$\langle v_\lambda \rangle = \lim_{N\to\infty} \frac{1}{2N+1} \sum_{n=-N}^{N} v_\lambda[n] \qquad (2\text{-}3)$$

and the corresponding time-averaged variance,

$$\left\langle |v_\lambda - < v_\lambda > |^2 \right\rangle = \lim_{N\to\infty} \frac{1}{2N+1} \sum_{n=-N}^{+N} |v_\lambda[n] - \langle v_\lambda \rangle|^2 \qquad (2\text{-}4)$$

In general, these time-averaged quantities depend on the individual sequence $v_\lambda[n]$. But if $v_\lambda[n]$ is a member of an ergodic process, the time averages do not depend on λ and are equal to the corresponding ensemble averages:

$$\langle v[n] \rangle = \mathcal{E}\{\mathbf{v}_n\} = m_{\mathbf{v}}, \qquad (2\text{-}5)$$

$$\langle |v[n] - m_v|^2 \rangle = \mathcal{E}\{|\mathbf{v}_n - m_{\mathbf{v}}|^2\} = \sigma^2_{\mathbf{v}} \qquad (2\text{-}6)$$

Thus, for an *ergodic* process, one individual sequence $v_\lambda[n]$ is assumed to be representative of the whole process.

In this project, we first want to show the difference between a nonstationary and a stationary process. A second aim is to give an example of a process that is stationary but nonergodic and contrast it with an ergodic process.

Hints

In the exercises below we examine properties of the signals from the following three random processes. Each should be made into a MATLAB program `rp*.m` that will create a matrix of size $M \times N$ containing random numbers. Make sure that the files are placed on the path where MATLAB can find them.

```
function v = rp1(M,N);         %<<------- RANDOM PROCESS #1
a = 0.02;
b = 5;
```

```
Mc = ones(M,1)*b*sin((1:N)*pi/N);
Ac = a*ones(M,1)*[1:N];
v = (rand(M,N)-0.5).*Mc + Ac;

function v = rp2(M,N);        %<<------- RANDOM PROCESS #2
Ar = rand(M,1)*ones(1,N);
Mr = rand(M,1)*ones(1,N);
v = (rand(M,N)-0.5).*Mr + Ar;

function v = rp3(M,N);        %<<------- RANDOM PROCESS #3
a = 0.5;
m = 3;
v = (rand(M,N)-0.5)*m + a;
```

EXERCISE 2.1

Stationary or Ergodic?

First we can visualize different members of the three processes in the time domain, to get a rough idea about stationarity and ergodicity. For stationarity, the issue is whether or not certain properties change with time; for ergodicity, the issue is whether one member of a stationary process is representative of the entire process.

Generate four members of each process of length 100 ($M = 4$, $N = 100$) and display them with `subplot`. Decide by inspection of the four representative signals whether each process is ergodic and/or stationary.

EXERCISE 2.2

Expected Values via Ensemble Averages

Compute the ensemble mean and standard deviation for each of the three processes, and plot versus time n. This can be done by generating many signals for each process, so let $M = 80$ and $N = 100$. Use the MATLAB functions `mean(·)` and `std(·)` to approximate the mean and standard deviation. From the plot of these quantities versus time, decide (again) about the stationarity of the processes.

Remark. The ensemble average, in the case of `mean`, is a sum across λ for each n:

$$\frac{1}{M}\sum_{\lambda=1}^{M} u_\lambda[n] = \hat{m}_{\mathbf{u}_n}[n] \;\approx\; m_{\mathbf{u}_n}[n] = \mathcal{E}\{\mathbf{u}_n\}$$

Similarly, for the ensemble average of the variance we must compute a sum of squares, after removing the mean.

$$\frac{1}{M}\sum_{\lambda=1}^{M} (u_\lambda[n] - \hat{m}_{\mathbf{u}_n}[n])^2 = \hat{\sigma}^2_{\mathbf{u}_n}[n] \;\approx\; \sigma^2_{\mathbf{u}_n}[n] = \mathcal{E}\{|\mathbf{u}_n - m_{\mathbf{u}_n}[n]|^2\}$$

EXERCISE 2.3

Expected Values via Time Averages

Measure approximately the time averages by calculating the time-averaged means $\langle u_\lambda \rangle$, $\lambda = 1, 2, 3, 4$, for four different members of each process.

$$\frac{1}{N}\sum_{n=0}^{N-1} u_\lambda[n] \approx \langle u_\lambda \rangle$$

$$\frac{1}{N-1}\sum_{n=0}^{N-1} (u_\lambda[n] - \hat{m}_{u\lambda})^2 \approx \left\langle |u_\lambda - \langle u_\lambda \rangle|^2 \right\rangle$$

Use $M = 4$ and $N = 1000$ when generating the signals. Decide (again) which of the processes are ergodic; use the results from Exercise 2.2 if necessary.

Note. Strictly speaking, the time average requires a limit, $N \to \infty$, but the large signal length $N = 1000$ should be sufficient to approach the limiting value. Also, a nonstationary process cannot be ergodic, so the time averages will be useless in that case.

EXERCISE 2.4

Theoretical Calculations

Analyze the MATLAB code in the functions `rp*.m` given above, and write the mathematical description of each stochastic process. If possible, determine a formula for the underlying pdf for each process. Decide on that basis whether the processes are ergodic and/or stationary. Calculate the theoretical mean, $m_{\mathbf{v}_n}[n]$, and variance, $\sigma^2_{\mathbf{v}_n}[n]$, for each. Then compare your results with those obtained by the measurements.

EXERCISE 2.5

Ergodic Process

For an ergodic process, the pdf can be estimated via time averaging.

a. For this purpose write a function M-file `pdf` using the available MATLAB function `hist`. Test your function for the simple case of a stationary process with a Gaussian pdf: `v = randn(1,N);`. Pick the length N to be 100 and 1000.

b. For the processes `rp*.m` that are ergodic, determine their pdf approximately using your `pdf` M-file. Do the measurement for $N = 100$, 1000, and 8000. Plot the three estimated pdf's and the theoretical pdf using `subplot`. Compare the results and comment on the convergence as N gets large.

c. Is it possible to measure the pdf of a nonstationary process with your function `pdf`? Comment on the difficulties that would be encountered in this case.

PROJECT 3: INFLUENCE OF A LINEAR SYSTEM ON A STOCHASTIC PROCESS

In this project we consider the influence of a linear system on the properties of an ergodic random process. Especially, we investigate how the autocorrelation sequence, the mean, and the pdf of a process are changed by filtering. The pdf, $p_{\mathbf{v}_n}(v)$, cannot depend on n, since we are assuming the process to be ergodic and thus stationary.

First, we recall some important definitions. The ergodic random process, $\mathbf{v}_n$, is described by its mean $m_{\mathbf{v}_n} = \mathcal{E}\{\mathbf{v}_n\}$ and its autocorrelation sequence:

$$\phi_{\mathbf{vv}}[m] = \mathcal{E}\{\mathbf{v}_{n+m}\mathbf{v}_n^*\} \tag{3-1}$$

The variance $\sigma^2_{\mathbf{v}_n}$ is contained in the value at $m = 0$, because

$$\phi_{\mathbf{vv}}(0) = \sigma^2_{\mathbf{v}_n} + m^2_{\mathbf{v}_n} \tag{3-2}$$

The Fourier transform of $\phi_{\mathbf{vv}}[m]$ yields a function called the *power density spectrum*,

$$\Phi_{\mathbf{vv}}(e^{j\omega}) = \sum_{m=-\infty}^{+\infty} \phi_{\mathbf{vv}}[m]e^{-j\omega m} \tag{3-3}$$

Every linear time-invariant system can be described by its impulse response $h[n]$ and its transfer function

$$H(z) = \sum_{n=-\infty}^{\infty} h[n]z^{-n} \tag{3-4}$$

When excited by the input $v[n]$, its output sequence $y[n]$ will be a member of a random process $\mathbf{y}_n$, whose mean is multiplied by the dc gain of the filter,

$$m_{\mathbf{y}_n} = m_{\mathbf{v}_n} H(e^{j0}) \tag{3-5}$$

and whose autocorrelation sequence is given by

$$\phi_{\mathbf{yy}}[m] = \phi_{\mathbf{vv}}[m] * \rho[m] = \phi_{\mathbf{vv}}[m] * \left(\sum_{n=-\infty}^{\infty} h[n]h[n+m] \right) \tag{3-6}$$

where $\rho[m] = h[m]*h[-m]$ is the (deterministic) autocorrelation sequence of the impulse response. The power density spectrum of the output process is

$$\Phi_{\mathbf{yy}}(e^{j\omega}) = \Phi_{\mathbf{vv}}(e^{j\omega})|H(e^{j\omega})|^2 \tag{3-7}$$

Furthermore, the *cross-correlation* of the input and output sequences is often used to identify the impulse response. We obtain

$$\phi_{\mathbf{vy}}[m] = \phi_{\mathbf{vv}}[m] * h[m] \tag{3-8}$$

Finally, if we consider the pdf of the output process $\mathbf{y}_n$, a simple answer is possible in two cases:

1. If the input process $\mathbf{v}_n$ is normally distributed, the output process will be normally distributed as well. This fact is independent of the autocorrelation sequence $\phi_{\mathbf{vv}}[m]$ and the impulse response $h[n]$ of the system.
2. If the consecutive values of $\mathbf{v}_n$ are statistically independent (i.e., if its autocorrelation sequence is $\phi_{\mathbf{vv}}[m] = \sigma_{\mathbf{v}_n}^2 \delta[m]$ and if the impulse response is sufficiently long), the output sequence will be approximately normally distributed as well. This result does not depend on the pdf or the impulse response. It is essentially a restatement of the central limit theorem (see Exercises 1.3 and 5.2).

All the relations above are based on the assumption that the output process $\mathbf{y}_n$ is ergodic. But if the system is causal and we start the excitation at $n = 0$ with $v[n]$, being a segment of the ergodic process $\mathbf{v}_n$, the corresponding output sequence $y[n]$ will be

$$y[n] = h[n] * v[n] = \sum_{k=0}^{\infty} h[k]v[n-k] \tag{3-9}$$

This sequence $y[n]$ is obviously causal and thus cannot be a member of a stationary process. However, if we assume that the transient time of the system is approximately of finite length (i.e., $h[n] \approx 0$ for $n > L$), the output process will appear stationary for $n > L$.

To be more specific, we consider the mean and the variance of the output process *during the transient time.*

$$m_{\mathbf{y}_n}[n] = \mathcal{E}\{\mathbf{y}_n\} = \mathcal{E}\left\{\sum_{k=0}^{n} h[k]v[n-k]\right\}$$

$$= m_{\mathbf{v}_n} \sum_{k=0}^{n} h[k] = m_{\mathbf{v}_n} r[n] \tag{3-10}$$

where $r[n]$ is the step response of the system.

A similar result for the variance holds when the input sequence consists of statistically independent values (white noise):

$$\sigma^2_{\mathbf{y}_n}[n] = \sigma^2_{\mathbf{v}_n} \sum_{k=0}^{n} |h[k]|^2 \tag{3-11}$$

Hints

In the following exercises, we study three different filters and their impact on processing random input signals generated via `rand` and `randn`. The linear systems are described by the coefficients of their rational transfer functions.

```
filter 1:        b1 = [ 0.3    0 ];
                 a1 = [ 1   -0.8 ]

filter 2:        b2 = 0.06 * [1 2 1];
                 a2 = [ 1 -1.3 0.845 ]

filter 3:        b3 = [ 0.845 -1.3 1 ]
 (all-pass)      a3 = fliplr(b3);
```

Thus the MATLAB function `filter` can be used for their implementation, and the function `freqz` can be used to compute their frequency responses.

The required estimation of the auto- and cross-correlation sequences can be done with the MATLAB functions `acf` and `ccf`, respectively.[1] These programs apply the method proposed by Rader [10]. According to (3-3), the power density spectrum can be calculated approximately by applying `fft` on the measured autocorrelation sequence. More will be said about power spectrum estimation in the following section, *FFT Spectrum Estimation.*

EXERCISE 3.1

Sample Autocorrelation of White Noise

The `rand` function produces statistically independent samples. Generate segments of the two random input sequences via

```
N = 5000;
v1 = sqrt(12) * (rand(1,N) - 0.5);       %<--- zero mean

v2 = randn(1,N);     %<---- Gaussian: mean = 0, var = 1
```

[1]These special M-files are described in Appendix A.

a. Call these signals $v_1[n]$ and $v_2[n]$. Determine the variance of $v_1[n]$. Compute and plot 64 values of their autocorrelation sequences using `[ phi, lags] = acf(v, 64).` The second output `lags` gives the domain over which the autocorrelation sequence was computed.

b. How closely do they match the true autocorrelation sequence expected from theoretical considerations?

EXERCISE 3.2

Filtered White Noise

The signals $v_1[n]$ and $v_2[n]$ are to be used as inputs to a first-order filter.

a. Suppose that the impulse response of the filter is $h[n] = b\,a^n,\ n \geq 0,\ |a| < 1$. Derive the theoretical autocorrelation function, and then compute the output autocorrelation sequences for both inputs through filter 1, where $a = 0.8$ and $b = 0.3$. Compare both calculated results to the theory and explain any significant differences.

b. Estimate the pdf of the output when the input is uniform, $v_1[n]$. Repeat for the Gaussian input, $v_2[n]$. Explain why both pdf's are nearly the same.

c. Repeat part (a) for a first-order all-pass filter.

$$H(z) = \frac{z^{-1} - a^*}{1 - az^{-1}}$$

Derive the theoretical autocorrelation sequence so that your result is applicable to an all-pass with any number of poles. Is there a difference to be expected for the two different input signals?

EXERCISE 3.3

Measured Autocorrelation

This exercise demonstrates the effect of a filter on the autocorrelation sequence.

a. Excite filter 2 with both white noise signals: $v_1[n]$ and $v_2[n]$. Call the resulting outputs $y_{21}[n]$ and $y_{22}[n]$.

b. Compute and display the histograms of both output sequences, $y_{2i}[n]$. Be careful to use data only from the stationary part of the signals. Explain how the impulse response $h_2[n]$ can be used to estimate the length of the transient.

c. Measure the autocorrelation sequences of the stationary part of the two output sequences $y_{2i}[n]$. Display both input and output correlation sequences using a four-panel `subplot`.

d. Measure the variances of all four signals (two inputs and two outputs) and compare them with the results of theoretical considerations.

e. Excite filter 3 with $v_1[n]$ and $v_2[n]$. Measure the autocorrelation sequences and the histograms of the two output sequences $y_{3i}[n]$. Display your results and explain the similarities as well as the differences.

f. Furthermore, measure the cross-correlation sequence between the input and output sequences for filters 2 and 3. Be careful to use only the stationary part of the output signal. Use the Gaussian input first, but then try the uniform input, $v_1[n]$.

g. From the cross-correlation estimates, compute an estimate of the impulse responses of systems 2 and 3 and then plot with `stem`. Compute $h_2[n]$ and $h_3[n]$, the true impulse responses. Plot the true $h_i[n]$ as a dashed-line envelope to compare with the estimate. Use `subplot(21x)` to create a two-panel display with both $h_2[n]$ and $h_3[n]$. Does the answer depend strongly on whether the input is uniform, $v_1[n]$, or Gaussian, $v_2[n]$?

EXERCISE 3.4

Transients

In this exercise we investigate the transient behavior of a system when excited by a stochastic input signal. We use filter 2 as an example and excite it by signals of length 50 out of the normally distributed process generated by

```
v = alfa* randn(160,50) + beta;
```

a. Choose α and β such that the variance $\sigma_v^2 = 2$ and the mean $m_v = 1$.

b. Calculate and display the output sequences $y_\lambda[n]$ for four different input sequences $v_\lambda[n]$.

c. Generate a 160×50 matrix of output sequences using `filter` inside an appropriate `for` loop. Compute estimates of the mean and variance as functions of n by averaging over the ensemble of $M = 160$ signals.

d. Now calculate the true impulse response $h[n]$, the step response $r[n]$, and the sequence $w[n] = \sigma_{v_n}^2 \sum_{k=0}^{n} h^2[k]$. Display these results for $0 < n \leq 50$, superimposed on the estimates obtained above. Compare the results with those to be expected according to (3-10) and (3-11) and comment on the relationship between $r[n]$, $w[n]$ and the transient nature of the stochastic output signals.

EXERCISE 3.5

Moving Averages

The moving average operator can be used to determine approximately the time-averaged mean and variance of a random sequence out of a stationary process. The N-point moving average operator is

$$y_1[n] = \frac{1}{N} \sum_{m=n-N+1}^{n} v[m]$$

The signal $y_1[n]$ can be used to estimate $\langle v[m] \rangle$; correspondingly, we can form $y_2[n]$ from a moving average of $v^2[m]$ and use it to determine an estimate of $\langle v^2[m] \rangle$.

In this exercise we use an FIR filter whose impulse response is $h[n] = 1/N,\ n = 0, 1, 2, \ldots, N-1$. A reasonable value for N is $N = 100$.

a. Generate an uniformly distributed test signal that is at least five times longer than the FIR filter length, via `v = rand(1, 5*N)`.

b. Calculate and display the two output sequences: $y_1[n]$ for the mean and $y_2[n]$ for the variance. Based on the plot, explain why the random signals $y_i[n]$ are not stationary in the interval $0 \leq n < N$.

c. For $n > N$ the sequences $y_i[n]$ are random variables as well, but now out of stationary processes. It is easy to see that the variances of $y_i[n]$ are smaller than the variance of the input sequence. Measure the means and variances of the signals $y_i[n]$ by time averaging over $N \leq n \leq 5N$. Compare the means and variances with the values expected from theory.

d. Compute and plot the histograms of the two output sequences, using the values for $N \leq n \leq 5N$ again. Mark the mean and variance on this plot.

EXERCISE 3.6

Confidence Interval

The foregoing experiments illustrate that the results of a moving average are random variables as well, with a certain mean and variance. Usually, it is reasonable to assume that these values are normally distributed with a variance proportional to $1/N$ and to the (unknown) variance

σ_v^2 of the input signal. To be more specific, we describe the situation for the measurement of the mean m_v of the random signal $v[n]$.

If we compute a single estimate of the mean by averaging N independent samples of the signal, the estimated value $\hat{m}_v$ comes from a normally distributed RV with variance σ_v^2/N. If the estimate is computed again over another N samples, the value will change somewhat, but we would like to bound the expected change. One way to do this is to compute a "confidence interval" centered on $\hat{m}_v$ within which we predict that all estimates of the mean will lie 95% of the time. This approach leads to the following definition: The unknown mean m_v is located inside a *confidence interval* with the probability

$$S = \text{PROB}\left\{\hat{m}_v - \frac{\sigma_v}{\sqrt{N}}c \le m_v \le \hat{m}_v + \frac{\sigma_v}{\sqrt{N}}c\right\} \tag{3-12}$$

where c is the confidence parameter. Based on the assumption that the variables $\hat{m}_v$ are normally distributed, we get $c = \sqrt{2}\text{erf}^{-1}(S)$, which means, for example, that with a probability of $S = 95\%$, the mean m_v is inside the limits $\hat{m}_v \pm 1.96\sigma_v/\sqrt{N}$.

a. To test the confidence interval calculation, we use a test signal that is white Gaussian noise, as in Exercise 3.5. Compute one estimate of the mean with a 100-point FIR filter. Using this estimate and the *true* variance of the signal, compute the 95% confidence interval for the "unknown" mean. This theoretical calculation can now be tested by computing a large number of mean estimates and checking what percentage lies within the confidence interval. To get a reasonable test, several hundred estimates will have to be checked.

b. In practice, there is another difficulty because we would have to estimate the variance before computing the confidence interval. Explain how your test of the confidence interval, from the previous part, will change if a single estimate of the variance over 100 points is used in place of the true variance.

PROJECT 4: INFLUENCE OF A NONLINEAR MAPPING ON A RANDOM PROCESS

In this project we deal with a nonlinear mapping of a random signal. In particular, we consider how the pdf changes. For special cases we compute the autocorrelation sequence and the power spectrum of the mapped process. These nonlinear mappings have an important application: A new process with a desired pdf can be created out of a given one, usually starting from a uniform distribution.

For convenience we recall some equations (see [1, Chap. 5]). Given a random process $\mathbf{v}_n$ with probability density function $p_\mathbf{v}(v)$, its variables are mapped according to

$$x = g(v)$$

As a simplification we assume that $g(v)$ is a monotonically increasing or decreasing function, which has a unique inverse mapping

$$v = g^{-1}(x)$$

It turns out that the mapped process $\mathbf{x}$ has the new pdf

$$p_\mathbf{x}(x) = p_\mathbf{v}[g^{-1}(x)]\left|\frac{d}{dx}[g^{-1}(x)]\right| \tag{4-1}$$

This analytic result can be used to determine the function $g(v)$, which maps a uniformly distributed process with $p_\mathbf{v}(v) = 1$ for $v \in [0, 1]$ into another one with the desired probability density $p_\mathbf{x}(x)$. The result is

$$x = g(v) = P_{\mathbf{x}}^{-1}(v) \tag{4-2}$$

where $P_{\mathbf{x}}^{-1}(v)$ is the inverse of the distribution function of the desired process.

If the variables of a sequence out of a random process are statistically independent, a memoryless nonlinear mapping will yield independent values again. In the more general case of a process with an arbitrary autocorrelation sequence we confine the considerations to a special case: If $\mathbf{v}_n$ is a normally distributed process with zero mean and the autocorrelation sequence $\phi_{vv}[m]$ and if $x = v^2$, we get for the output process $\mathbf{x}_n$ the mean

$$\mathcal{E}\{\mathbf{x}_n\} = \phi_{\mathbf{vv}}[0] \tag{4-3}$$

and the autocorrelation sequence

$$\mathcal{E}\{\mathbf{x}_{n+m}\mathbf{x}_n\} = \phi_{\mathbf{xx}}[m] = \phi_{\mathbf{vv}}^2[0] + 2\phi_{\mathbf{vv}}^2[m] \tag{4-4}$$

(see [1, Sec. 10.2]). Corresponding to (4-4) the power density spectrum will involve a convolution:

$$\Phi_{\mathbf{xx}}(e^{j\omega}) = \phi_{\mathbf{vv}}^2(0)\delta(\omega) + \frac{1}{\pi}\,\Phi_{\mathbf{vv}}(e^{j\omega}) * \Phi_{\mathbf{vv}}(e^{j\omega}) \tag{4-5}$$

Hints

We use the random signals generated in MATLAB via `rand` and `randn`. When a process with statistically dependent values is needed, we can use filter 2 out of the Project 3.

EXERCISE 4.1

Linear Mapping

You are given a random process $\mathbf{v}_n$ with the probability density function $p_{\mathbf{v}}(v)$ and a linear mapping according to $x = \alpha \cdot v + \beta$, α and β being real numbers. Derive a general expression for $p_{\mathbf{x}}(x)$ in terms of an arbitrary $p_{\mathbf{v}}(v)$. Specialize to the case where the process $\mathbf{v}_n$ is normal, $\mathcal{N}(0, 1)$. Generate a random signal with Gaussian pdf, then transform it via the linear mapping, and finally, plot its histogram to verify the change in the pdf.

EXERCISE 4.2

Nonlinear Mapping

Given a process $\mathbf{v}_n$ uniformly distributed in [0,1].

a. Map this process linearly onto the interval $[-\pi/2, \pi/2]$.

b. Map the resulting process ξ with

$$x = \sin \xi$$

onto the interval $[-1, 1]$. Determine a formula for $p_{\mathbf{x}}(x)$.

c. Perform the mapping with MATLAB for $N = 8000$ values. Measure approximately the probability density function using `hist` with 20 bins. Compare the results with the theoretical formula from part (b).

d. Measure the autocorrelation sequence of $\mathbf{x}_n$ using `acf` and explain the result.

EXERCISE 4.3

Laplacian Noise

Starting with a uniformly distributed random process, generate a process $\mathbf{x}_n$ with a Laplacian distribution and with variance $\sigma_{\mathbf{x}}^2 = 1$:

$$p_{\mathbf{x}}(x) = e^{-\sqrt{2}|x|}/\sqrt{2} \tag{4-6}$$

The nonlinear mapping procedure will be applied to a process $\mathbf{v}_n$, uniformly distributed in [0,1].

a. Verify mathematically that the mapping has to be done with

$$x = g(v) = \begin{cases} \ln(2v)/\sqrt{2} & 0 \le v < 0.5 \\ -\ln(2-2v)/\sqrt{2} & 0.5 \le v \le 1 \end{cases}$$

Plot this function over the range [0, 1].

b. Plot the inverse function, $v = g^{-1}(x)$. In MATLAB, this is as easy as swapping the arguments in the `plot` function.

c. Prepare an M-file `lap(M,N)` for the generation of an $M \times N$ matrix, the rows of which are sample sequences out of a random process with Laplacian distribution, zero mean, and unit variance.

d. Generate a single sequence of length 8000 with your program `lap(M,N)`. Measure approximately the probability density function using a scaled histogram with 100 bins. Display the result on a semilogarithmic scale by using `semilogy` and overlay with a plot of (4-6) to compare with the results expected from theory.

EXERCISE 4.4

Histogram Equalization

In image processing, an effective transformation of a photograph to enhance low-level detail is a nonlinear mapping that stretches out the gray scale of the image. This can be accomplished by transforming the image so that the histogram of its gray levels is nearly uniform. Since the input image does not necessarily have a known pdf (or histogram shape), the process of histogram equalization relies on pdf estimates to synthesize the correct mapping. The key is equation (4-2), which gives the relationship between the mapping $g(v)$ and the probability *distribution* function $P_{\mathbf{x}}(v)$. Since the distribution function is just the running integral of the pdf, it can also be estimated from a histogram of the input data.

a. Generate an input signal that is Gaussian noise. Compute the histogram, and then make a plot of the probability *distribution* function $P_{\mathbf{v}}(v)$ defined in (1-2).

b. Create a nonlinear mapping function that will transform the input signal $v[n]$ into a new random process that is uniformly distributed. This can be done only approximately by a piecewise linear approximation of the distribution function. The exact mapping depends strongly on the number of bins used in computing the underlying histogram.

c. Test your mapping on the Gaussian input signal. Plot the histogram of the output. Generate several results for different number of bins.

d. Write an M-file that will perform histogram equalization on any input signal. Test it on Laplacian noise.

EXERCISE 4.5

Squaring a Random Signal

a. Generate a normally distributed sequence $y[n]$ with zero mean by exciting filter 2 out of Project 3 with a white noise input sequence $v[n]$. Measure its output autocorrelation sequence $\phi_{\mathbf{yy}}[m]$.

b. Perform the mapping $x[n] = y^2[n]$ and measure approximately the pdf $p_{\mathbf{x}}(x)$ with 40 bins and the autocorrelation sequence $\phi_{\mathbf{xx}}[m]$. Compare with the expected theoretical results.

PROJECT 5: COMBINING TWO RANDOM PROCESSES

In this project we consider the joint properties of two or more random processes generated by simple arithmetic operations. The following properties hold (see [1, Chaps. 6 and 7]):

1. If $\mathbf{u}_n$ and $\mathbf{v}_n$ are two random processes, *statistically independent or not*, and $g_1(\cdot)$ and $g_2(\cdot)$ are two mapping functions, the mean of the two mapped random processes is separable:

$$\mathcal{E}\{g_1(\mathbf{u}_n) + g_2(\mathbf{v}_n)\} = \mathcal{E}\{g_1(\mathbf{u}_n)\} + \mathcal{E}\{g_2(\mathbf{v}_n)\} \tag{5-1}$$

2. If $\mathbf{u}_n$ and $\mathbf{v}_n$ are *statistically independent*, that is, if

$$p_{\mathbf{uv}}(u, v, m) = p_{\mathbf{u}}(u)p_{\mathbf{v}}(v) \qquad \forall m \tag{5-2}$$

then the mean of the product is separable:

$$\mathcal{E}\{g_1(\mathbf{u}_n)g_2(\mathbf{v}_n)\} = \mathcal{E}\{g_1(\mathbf{u}_n)\}\mathcal{E}\{g_2(\mathbf{v}_n)\} \tag{5-3}$$

3. In the special case $g_{1,2}(\cdot) = e^{j\chi(\cdot)}$ that yields

$$\mathcal{E}\{g_1(\mathbf{u}_n) \cdot g_2(\mathbf{v}_n)\} = \mathcal{E}\{e^{j\chi(\mathbf{u}_n+\mathbf{v}_n)}\} = C_{\mathbf{x}}(\chi) \tag{5-4}$$

where $C_{\mathbf{x}}(\chi)$ is the *characteristic function* of the process $\mathbf{x}_n = \mathbf{u}_n + \mathbf{v}_n$. Obviously, we get

$$C_{\mathbf{x}}(\chi) = C_{\mathbf{u}}(\chi)C_{\mathbf{v}}(\chi) \tag{5-5}$$

and thus the probability density of the sum of two *independent* processes is the convolution of their pdf's:

$$p_{\mathbf{x}}(x) = p_{\mathbf{u}}(x) * p_{\mathbf{v}}(x) \tag{5-6}$$

4. The autocorrelation sequence for the sum of two *independent* processes is

$$\phi_{\mathbf{xx}}[m] = \phi_{\mathbf{uu}}[m] + \phi_{\mathbf{vv}}[m] + 2m_{\mathbf{u}}m_{\mathbf{v}} \tag{5-7}$$

Finally, the autocorrelation sequence of the product process

$$\mathbf{y}_n = \mathbf{u}_n\mathbf{v}_n$$

where $\mathbf{u}_n$ and $\mathbf{v}_n$ are two *independent* processes, turns out to be

$$\phi_{\mathbf{yy}}[m] = \phi_{\mathbf{uu}}[m]\phi_{\mathbf{vv}}[m] \tag{5-8}$$

All these equations can be extended to the summation or multiplication of more than two random processes. Equations (5-2)–(5-8) holds, if all these processes are *mutually independent*.

EXERCISE 5.1

Sum of Two Random Signals

Generate two independent random processes, $\mathbf{u}_n$ and $\mathbf{v}_n$, both with uniform distribution, the first in the interval [1,2], the other in [0,4].

a. Measure the mean and the variance of the sum $\mathbf{x}_n = \mathbf{u}_n + \mathbf{v}_n$, and compare with the results to be expected theoretically.

b. Measure the mean and the variance of the product $\mathbf{y}_n = \mathbf{u}_n \cdot \mathbf{v}_n$, and compare with the results to be expected theoretically.

c. Measure approximately the pdf of the sum $\mathbf{x}_n = \mathbf{u}_n + \mathbf{v}_n$, and compare with the results to be expected according to (5-6).

Generate now two independent, but not white, random processes $\mathbf{y}_{1n}$ and $\mathbf{y}_{2n}$ by exciting filters 1 and 2 out of Project 3 with $\mathbf{u}_n$ and $\mathbf{v}_n$, respectively.

d. Measure the mean and the autocorrelation sequence of the two individual processes.

e. Measure the mean and the autocorrelation sequence of their sum $\mathbf{y}_n = \mathbf{y}_{1n} + \mathbf{y}_{2n}$ and compare with the results to be expected according to (5-1) and (5-7).

Finally, generate two statistically dependent random processes $\mathbf{w}_{1n}$ and $\mathbf{w}_{2n}$ by exciting filters 1 and 2 out of Project 3 with $\mathbf{u}_n$.

f. Measure the mean, the variance, and the pdf of both processes individually.

g. Measure the mean and the autocorrelation sequence of their sum $\mathbf{w}_n = \mathbf{w}_{n1} + \mathbf{w}_{n2}$ and compare as far as possible with the results to be expected theoretically.

EXERCISE 5.2

Pdf of a Sum

Generate M sequences of length N out of independent random processes $\mathbf{v}_{n\mu}$, all uniformly distributed in the interval $[-1, 1]$, with `rand(M,N)`. Use a large length, say $N = 2000$.

a. Calculate the summation processes $\mathbf{x}_{nM} = \sum_{\mu=1}^{M} \mathbf{v}_{n\mu}$, using `sum` for $M = 2, 3, 4, 5$. Compute and plot their pdf's using `hist` with 20 bins. Display these results all together using a four-panel `subplot`.

b. What type of pdf do you expect for increasing M?

EXERCISE 5.3

Sum and Product of Two Processes

This exercise deals with two independent random processes $\mathbf{u}_n$ and $\mathbf{v}_n$ with zero mean, where $\mathbf{u}_n$ is uniformly distributed in $[-1, 1]$ and $\mathbf{v}_n$ is normally distributed. The autocorrelation sequences are $\phi_{\mathbf{uu}}[m]$ and $\phi_{\mathbf{vv}}[m]$. The sum and the product of these processes are to be investigated.

The process $\mathbf{u}_n$ must be generated in a special way so that its power density spectrum will not be constant. Two steps are involved: Filter a white Gaussian noise input process, and then apply a nonlinear mapping to the output process to change the distribution to uniform. To modify the spectrum we filter a zero-mean normal process; use filter 2 from Project 3, and take the input signal length to be $N = 8000$. The filter's output process will be normally distributed as well, but with a nonconstant power density spectrum. This process can be transformed into a uniformly distributed process by using the mapping technique described in Project 4. It turns out that the "error function" `u = erf(`$v/\sigma_{\mathbf{v}} \cdot \sqrt{2}$`)` maps a normally distributed process $\mathbf{v}_n$ with zero mean and variance $\sigma_{\mathbf{v}}^2$ into a process $\mathbf{u}_n$ that is uniformly distributed on $[-1, 1]$. *Remark*: This nonlinear mapping will change the power density spectrum and thus the autocorrelation sequence, but it will still be nonconstant.

a. Do all the steps necessary to produce the uniformly distributed process $\mathbf{u}_n$ so that it has a nonconstant power spectrum. Measure the pdf to check your result.

b. Measure the autocorrelation sequences of both processes using $N = 8000$ samples of the processes.

c. Measure the mean, variance, and the autocorrelation sequences of the sum and the product of both processes. Compare your results with those expected from theory.

FFT SPECTRUM ESTIMATION

OVERVIEW

This set of projects deals with FFT-based methods for spectrum estimation, the approximate determination of the power density spectrum $\Phi_{\mathbf{yy}}(e^{j\omega})$ of a given real ergodic random process $\mathbf{y}_n$.

$$\Phi_{\mathbf{yy}}(e^{j\omega}) = \sum_{m=-\infty}^{\infty} \phi_{\mathbf{yy}}[m]\, e^{-j\omega m} \tag{0-1}$$

In practice, its measurement cannot be done according to the definition (0-1) because the autocorrelation sequence

$$\begin{aligned}\phi_{\mathbf{yy}}[m] &= \mathcal{E}\{\mathbf{y}_{n+m}\mathbf{y}_n\} \\ &= \lim_{N\to\infty} \frac{1}{2N+1} \sum_{n=-N}^{+N} y_\lambda[n+m] y_\lambda[n]\end{aligned}$$

can only be approximated when a finite segment of length N of one member $y_\lambda[n]$ out of the process is available.

In this section we study the performance of several different methods. To show their merits as well as their deficiencies, we use as an example a random process with known power density spectrum, generated with a linear system with known transfer function, excited by a white Gaussian process $\mathbf{v}_n$ with variance $\sigma_{\mathbf{v}}^2 = 1$. Thus in our experiment we can compare to the correct result and can express the observed deviations in terms of this known spectrum. Since this additional information about the true spectrum is not available when processing an unknown signal, we also consider how to derive confidence interval expressions for our estimates.

BACKGROUND READING

There are books devoted to spectral estimation, especially the books by Marple [6] and Kay [7]. Others have rather comprehensive chapters on the subject: Chapter 13 in the book by Papoulis [1], Chapter 2 in Lim and Oppenheim [9], and Sections 2.10 and 11.5–11.7 in Oppenheim and Schafer [4]. A useful algorithm for estimating the autocorrelation sequence is described in [10].

■ ■ PROJECT 1: PERIODOGRAM

In this project we investigate the periodogram, an estimate that is based on one segment of $y[n]$, extracted by windowing with $w[n]$, a window sequence of length N:

$$x[n] = w[n]y[n] \tag{1-1}$$

The periodogram is defined as

$$I_N(e^{j\omega}) = \frac{1}{N}\left|\sum_{n=0}^{N-1} x[n]e^{-j\omega n}\right|^2 \tag{1-2}$$

Obviously, samples of $I_N(e^{j\omega})$ in (1-2) at the points $\omega = \omega_k = (2\pi/N)k$ can be calculated very efficiently with the FFT. In addition, it can be rewritten as

$$I_N(e^{j\omega}) = \frac{1}{N} \sum_{m=-(N-1)}^{N-1} \rho_{\mathbf{xx}}[m]e^{-j\omega m} \tag{1-3}$$

where the autocorrelation sequence of the finite-length sequence $x[n]$ is computed via

$$\rho_{\mathbf{xx}}[m] = \sum_{n=0}^{N-1-|m|} x[n]\,x[n+|m|] \tag{1-4}$$

The similarity of the definition of the power density spectrum given in the overview (0-1) and Eq. (1-3) is the reason for calling the periodogram the "natural estimate" of the power density spectrum. But as we shall show, it yields rather inaccurate results if used without modification.

A theoretical investigation of the properties of the periodogram yields the following results (see Section 13.2 in [1], Chapter 2 in [9], or Section 11.6 in [4]).

Bias

First, the expected value of $\rho_{\mathbf{xx}}[m]$ in (1-4) reduces to

$$\mathcal{E}\{\rho_{\mathbf{xx}}[m]\} = \rho_{ww}[m]\phi_{\mathbf{yy}}[m] \tag{1-5}$$

where $\phi_{\mathbf{yy}}[m]$ is the true autocorrelation and $\rho_{ww}[m]$ is the autocorrelation sequence of the window:

$$\rho_{ww}[m] = \sum_{n=0}^{N-1-|m|} w[n]w[n+|m|] \tag{1-6}$$

Thus the expected value of the periodogram becomes

$$\mathcal{E}\{I_N(e^{j\omega})\} = \frac{1}{N} \sum_{m=-(N-1)}^{N-1} \mathcal{E}\{\rho_{\mathbf{xx}}[m]\}e^{-j\omega m} \tag{1-7}$$

$$= \frac{1}{2\pi N} \Phi_{\mathbf{yy}}(e^{j\omega}) * \left|W(e^{j\omega})\right|^2 \tag{1-8}$$

Here $W(e^{j\omega}) = \sum_{n=0}^{N-1} w[n]e^{-j\omega n}$ is the DTFT of the window. Finally, the bias, which is the difference between the expected value of the estimate and the true mean, can be written as

$$\text{BIAS} = \mathcal{E}\{I_N(e^{j\omega_k})\} - \Phi_{\mathbf{yy}}(e^{j\omega_k}) \tag{1-9}$$

$$= \frac{1}{2\pi N} \Phi_{\mathbf{yy}}(e^{j\omega}) * \left|W(e^{j\omega})\right|^2 - \Phi_{\mathbf{yy}}(e^{j\omega}) \tag{1-10}$$

■ **Example:** We consider the case of the rectangular window whose autocorrelation sequence is a triangle. The corresponding transform is

$$\left|W(e^{j\omega})\right|^2 = \left[\frac{\sin N\omega/2}{\sin \omega/2}\right]^2 \tag{1-11}$$

According to (1-8), the convolution of this asinc-squared function with $\Phi_{\mathbf{yy}}(e^{j\omega})$ yields the mean of the periodogram, which thus turns out to be a smoothed version of the true power spectrum, $\Phi_{\mathbf{yy}}(e^{j\omega})$. Since the convolution kernel $\left|W(e^{j\omega})\right|^2$ has a mainlobe of width $2\pi/N$, we conclude that

$$\lim_{N\to\infty} \mathcal{E}\left\{I_N(e^{j\omega})\right\} = \Phi_{\mathbf{yy}}(e^{j\omega}) \tag{1-12}$$

at every point of continuity of $\Phi_{\mathbf{yy}}(e^{j\omega})$. It follows that $I_N(e^{j\omega})$ is an unbiased estimator of $\Phi_{\mathbf{yy}}(e^{j\omega})$ in the limit as $N \to \infty$, called *asymptotically unbiased.*

This result (1-12) can be generalized to all windows if two conditions hold:

1. The window is normalized such that $\sum_{n=0}^{N-1} w^2[n] = N$.
2. The width of the mainlobe of the spectrum $\left|W(e^{j\omega})\right|^2$ decreases as $1/N$.

On the other hand, for finite N, (1-12) provides a means to calculate the bias of the result if the true spectrum $\Phi_{\mathbf{yy}}(e^{j\omega})$ is known (see Exercise 1.1).

Variance

Now we consider the variance of $I_N(e^{j\omega})$:

$$\operatorname{var}\{I_N(e^{j\omega})\} = \mathcal{E}\{I_N^2(e^{j\omega})\} - [\mathcal{E}\{I_N(e^{j\omega})\}]^2$$

If the process $\mathbf{y}_n$ is Gaussian, a rather long calculation yields

$$\operatorname{var}\{I_N(e^{j\omega})\} = [\mathcal{E}\{I_N(e^{j\omega})\}]^2 + \left|\frac{1}{2\pi N}\int_{-\pi}^{\pi} \Phi_{\mathbf{yy}}(\eta) W(e^{j(\eta-\omega)}) W^*(e^{j(\eta+\omega)})\, d\eta\right|^2 \tag{1-13}$$

Since the first term in (1-13) does not go to zero as $N \to \infty$, the periodogram is not a consistent estimator.

■ **Example:** In the special case of a rectangular window, we get

$$\operatorname{var}\{I_N(e^{j\omega})\} = \left[\mathcal{E}\{I_N(e^{j\omega})\}\right]^2 + \left[\sum_{m=-(N-1)}^{N-1} \frac{\sin\omega(N-|m|)}{N\sin\omega}\,\phi_{\mathbf{yy}}[m]\right]^2 \tag{1-14}$$

If $\mathbf{y}_n$ has no significant correlation beyond lag m_0 (i.e., $\phi_{\mathbf{yy}}[m] \approx 0$ for $|m| > m_0$), then for $N \gg m_0$ this expression yields

$$\operatorname{var}\{I_N(e^{j\omega})\} \approx \Phi_{\mathbf{yy}}^2(e^{j\omega})\left[1 + \left(\frac{\sin\omega N}{N\sin\omega}\right)^2\right] \tag{1-15}$$

Using the fact that the mean is asymptotically unbiased, we get for $N \to \infty$

$$\operatorname{var}\{I_N(e^{j\omega})\} = \begin{cases} 2\,\Phi_{\mathbf{yy}}^2(e^{j\omega}) & \omega = 0, \pi \\ \Phi_{\mathbf{yy}}^2(e^{j\omega}) & \text{elsewhere} \end{cases} \tag{1-16}$$

The essential point of this example is that even for very long windows the standard deviation of the periodogram estimate is as large as the mean, the quantity to be estimated.

Hints

For the exercises below we need a test signal consisting of filtered noise. We start with a signal that is a normally distributed white random sequence $v[n]$ with zero mean and variance $\sigma_v^2 = 1$, generated with the MATLAB function `randn`. The test signal will be created by passing $v[n]$ through a known filter $H(e^{j\omega})$. The MATLAB function `filter` is used to produce the random signal. However, its output sequence will be stationary only after the transient is over; this transient length can be estimated by checking the length of the impulse response.

The function `freqz` can be used to calculate $|H(e^{j\omega})|^2$, which is also the correct power density spectrum $\Phi_{yy}(e^{j\omega})$. Thus our estimates, which are just approximations, can be compared to this standard. In addition, the true autocorrelation sequence, $\phi_{yy}[m]$, which is the inverse DTFT of $\{\Phi_{yy}(e^{j\omega})\}$, can be approximated by calculating the inverse FFT of the sampled power spectrum (see Exercise 1.6 in chapter 5).

In several exercises, the periodograms will be calculated with different values of N, yielding a different number of samples of the estimate of the power density spectrum. For a better comparison with the theoretical function $\Phi_{yy}(e^{j\omega}) = |H(e^{j\omega})|^2$, they should all be interpolated to the same grid. This can be accomplished simply by padding the windowed sequence $x[n]$ by $M - N$ zeros prior to taking the FFT needed for the periodogram (a convenient choice would be $M = 512$ or 256).

EXERCISE 1.1

Formulas for Bias and Variance

In this exercise we verify the mathematical expressions for the true values of the bias and variance. For the filter applied to the WGN, we use the following coefficients:

```
b = 0.06 * [ 1  2  1 ]
a = [ 1   -1.3   0.845 ]
```

The transient portion of the output must be skipped. The poles, which lie at a radius of approximately 0.92, can be used to estimate the duration of the transient.

a. Calculate the true power spectrum, $\Phi_{yy}(e^{j\omega}) = |H(e^{j\omega})|^2$, and plot versus ω.

b. To compute the bias of the periodogram, it is easier to start from the autocorrelation domain, as in (1-5) and (1-7). First, we need a segment of the true autocorrelation sequence, $\phi_{yy}[m]$. Compute an approximation to $\phi_{yy}[m]$ using the `ifft` function. Determine the length needed in the inverse FFT so that the approximation error is negligible. Write an M-file that will return a specified section of $\phi_{yy}[m]$. Then plot $\phi_{yy}[m]$ for $-100 \le m \le 100$. If possible, derive a formula for $\phi_{yy}[m]$ and use this formula to calculate the values.

c. Now we can find the expected value of the periodogram as in (1-5). First, calculate the expected value of the autocorrelation function of the windowed signal $\mathcal{E}\{\rho_{xx}[m]\}$ for the case of a rectangular window of length N. Then use the DTFT to compute $\mathcal{E}\{I_N(e^{j\omega_k})\}$, as suggested in (1-7). Plot the expected value of the periodogram, and overlay with the true value of the power spectrum as a dashed line. Repeat for four different cases, $N = 32$, 64, 128, and 256. Display the results together in a four-panel `subplot`.

d. Now determine the bias of the periodogram (1-10), and plot it versus ω. Make the plots for all four values of N and justify that the bias will go to zero as $N \to \infty$. Explain why the bias is greatest at the peak of the spectrum.

e. Calculate and plot the variance of the periodogram according to (1-14); again for $N = 32$, 64, 128, and 256. Verify that it is *not* decreasing with N.

f. To check the accuracy of the approximation in (1-15), calculate and plot the difference of the two functions given in (1-14) and (1-15) for $N = 128$ and 256.

EXERCISE 1.2

Measuring the Periodogram with a Rectangular Window

Now we compute the periodogram of $y[n]$ directly from the data, using a rectangular window.

a. Calculate four periodograms with $N = 32$, 64, 128, and 256. Plot all four together using a four-panel `subplot`. Overlay with the true power spectrum as a dashed line for comparison. Describe features that improve with increasing N, and relate these observations to the fact that the bias is decreasing.

b. Calculate and plot together 10 periodograms, all for the case where the window length is $N = 64$. To get independent results, nonoverlapping sections must be used and the transient must be skipped. Repeat for $N = 256$. Describe the nature of these plots in view of the fact that the variance is *not* decreasing as $N \to \infty$. For instance, measure all the peak heights to see whether they are less variable for larger N.

EXERCISE 1.3

Measuring the Periodogram with a Tapered Window

We now repeat Exercise 1.2, but with the Hamming window $w_1[n]$ or the von Hann (`hanning`) window $w_2[n]$:

$$w_1[n] = \gamma_1 [0.46 - 0.54 \cos(2\pi n/N)] \qquad \text{for } 0, 1, \ldots, N-1 \tag{1-17}$$

$$w_2[n] = \gamma_2 [\tfrac{1}{2} - \tfrac{1}{2} \cos(2\pi n/N)] \qquad \text{for } 0, 1, \ldots, N-1 \tag{1-18}$$

Pick one of these windows to use throughout this exercise.

a. Determine the gain factor γ_i needed to normalize the energy in the window so that the windowed periodogram estimate is asymptotically unbiased.

b. To investigate the properties of this window, calculate and plot its autocorrelation sequence and the corresponding transform; overlay with a plot of asinc-squared to compare with the rectangular window case. Use $N = 64$ and the MATLAB function `xcorr` for the calculation of $\rho_{ww}[m]$. In the plot of the transform, it would be wise to use a semilogarithmic representation to see details in the sidelobes, and a linear plot to view the mainlobe. Calculate approximately the mainlobe width of the tapered window as a function of N. Will the windowed periodogram estimate be asymptotically unbiased?

c. Compute and plot the bias for the tapered window, as in Exercise 1.1(b)–(d).

d. Repeat Exercise 1.2(b) with the tapered window. Compare your results with those obtained for the rectangular window and with the theoretical function $\Phi_{\mathbf{yy}}(e^{j\omega})$. In which frequency range do you get an improvement, and why?

PROJECT 2: PERIODOGRAM AVERAGING

In Project 1 we found the periodogram to be an easily calculable but biased and inconsistent estimate of the power density spectrum. Here we consider methods for improving the performance of the periodogram. Quite a few modifications have been introduced to improve the results without losing the advantage of an efficient calculation. The essential points of the changes are:

- *Averaging* over a set of periodograms of nearly independent segments
- *Windowing* applied to the segments
- *Overlapping* the windowed segments for more averaging

If we are given M data points from a signal $y[n]$ out of an ergodic process $\mathbf{y}_n$, this data block can be divided into $K = M/N$ segments of length N, where for convenience we assume K to be an integer. According to Bartlett's procedure (see Section 11.6.3 in [4]) we calculate an estimate of the power density spectrum as the average of K periodograms:

$$\Phi_B(e^{j\omega}) = \frac{1}{K} \sum_{r=0}^{K-1} I_N^{(r)}(e^{j\omega}) \tag{2-1}$$

where the periodogram of the rth segment is

$$I_N^{(r)}(e^{j\omega}) = \frac{1}{N}\left|\sum_{n=0}^{N-1} x_r[n]e^{-j\omega n}\right|^2 \tag{2-2}$$

This segment was created by windowing $y[n]$:

$$x_r[n] = w[n]y[n + r(N - N_o)] \tag{2-3}$$

Strictly speaking, (2-3) describes Bartlett's procedure only if $w[n]$ is the rectangular window and the overlap N_o is zero. The generalization to other windows, especially to the case of overlapping windows, where $N_o > 0$, is called *Welch's procedure*.

This averaging process improves the variance. First, we calculate the expected value

$$\mathcal{E}\{\Phi_B(e^{j\omega})\} = \frac{1}{K}\sum_{r=0}^{K-1} \mathcal{E}\{I_N^{(r)}(e^{j\omega})\} = \mathcal{E}\{I_N(e^{j\omega})\} \tag{2-4}$$

Since the periodogram is a biased estimate of the desired power spectrum for a finite-length N (see Project 1), the averaging does not yield an improvement. The bias is only a function of the window length and window type. In fact, if only a finite-length signal is available, segmentation will increase the bias, since it forces the window to be shorter.

For the variance, however, we get

$$\mathrm{var}\{\Phi_B(e^{j\omega})\} = \frac{1}{K_{\mathrm{eff}}}\mathrm{var}\{I_N^{(r)}(e^{j\omega})\} \tag{2-5}$$

If the segments were truly independent, then $K_{\mathrm{eff}} = K$. However, in general, there is correlation between the segments, or the segments are overlapped, so that $K_{\mathrm{eff}} < K$. The number K_{eff} indicates the effective number of averages being accomplished.

Hints

We use again the random sequence $y[n]$, generated as described in Project 1. For an easy comparison with the theoretical function $\Phi_{yy}(e^{j\omega})$, the power spectra should all be interpolated to the same grid—probably 512 frequency samples would be best.

The determination of empirical results for the expected values in Exercise 2.2 could be done by calculating a matrix of estimated power density spectra and using the MATLAB functions `mean` and `std` for calculating these values per column.

EXERCISE 2.1

Program for the Welch–Bartlett Procedure

Write an M-file called `wbpsd` for the Welch–Bartlett spectral estimation technique. It should be possible to specify some or all of the following input parameters:

- M length of the input sequence
- N_o size of the overlapping part of the sections
- K number of segments
- N length of the periodogram (a power of 2)
- $w[n]$ window (so different types can be tried)
- N_{fft} number of frequency samples after interpolation (default $N_{\mathrm{fft}} = 512$)

To make sure that a full-size segment is always used for calculating the averaged periodogram according to (2-2), the number of segments $K = (M - N_o)/(N - N_o)$ must be an integer. If the selected parameters M, N, and N_o do not satisfy this condition, M should be reduced accordingly by ignoring an appropriate number of data at the end of the given sequence.

EXERCISE 2.2

Averaged Periodograms

Use the wbpsd M-file from Exercise 2.1 to process the test signal. Use a data length of $M = 512$ with $N = 64$ and the following window types, overlaps, and number of sections:

Rectangular window:	$N_o = 0$	$K = 8$
Hann window:	$N_o = 0$	$K = 8$
Hann window:	$N_o = 32$	$K = 15$

a. Calculate and plot together 10 Welch–Bartlett estimated power spectral densities for these three cases, using independent length-M sequences each time.

b. Calculate 100 Welch–Bartlett estimated power spectral densities for the foregoing parameters with 100 independent sequences of length M. Use these results for the calculation of the empirical bias and variance as a function of ω. Plot the empirical bias together with the true bias. Is there good agreement between theory and the empirical result, or should the experiment be redone for more than 100 trials?

c. Use the spectra from part (b) to calculate an empirical variance for the Welch–Bartlett estimate. Compare your result to the theoretical formula studied in Project 1. Determine an estimate for K_{eff}, the effective number of averages using (2-5). Explain why the value of K_{eff} is always less than K; in addition, determine which case has the most variance reduction.

d. Show empirical evidence that the variance will converge to 0, as $M, K \rightarrow \infty$, keeping N fixed. This may require that you generate one more test case with a much longer data sequence ($M \gg 512$).

EXERCISE 2.3

MATLAB M-File `Spectrum`

The M-file `spectrum` is MATLAB's realization of the Welch–Bartlett spectral estimator. It is contained in the Signal Processing Toolbox along with a companion function `specplot` that will plot its outputs. The data window `hanning` is hard-coded into `spectrum`, although it could be changed by editing. The length N of the averaged periodogram and the size N_o of the overlapping part of the sections can be specified as input arguments.

It turns out that `spectrum` introduces a *scaling* of the power spectral density estimate, $\Phi_{\mathbf{yy}}^{(s)}(e^{j\omega}) = (1/\alpha) \cdot \Phi_{\mathbf{yy}}(e^{j\omega})$. This scaling can be a problem when experimental data must be calibrated to known units.

a. Use `spectrum` to estimate a *scaled* power spectral density from a length $M = 512$ data sequence. Choose the overlap to be zero ($N_o = 0$); and try three different section lengths, $N = 64$, 128, and 256. Repeat the same estimates with the M-file `wbpsd` written in Exercise 2.1. Determine the scaling factor $\alpha(N)$ by comparing the two estimates. Comment on the possible cause of the different scalings.

b. Determine whether or not the scaling α is dependent on the frequency ω.

EXERCISE 2.4

Empirical Study of Confidence Intervals

The empirical measurement of the variance of a spectral estimate requires that the estimation process is repeated many times. However, in practice one wants to know whether a single estimate is any good; and whether or not a repetition of the experiment is likely to yield an estimate that is close to the first. From one estimate the procedure is to compute a "confidence interval" that marks the upper and lower limits on further estimates. The M-file `spectrum` offers an estimate of the 95% confidence interval, based on a Gaussian assumption for the PSD estimates. In theory, 19 out of every 20 estimates should lie within the confidence interval bounds.

a. Use `spectrum` to calculate both a PSD estimate and its confidence interval from a length $M = 512$ data sequence for $N = 64$ and $N_o = 32$. Plot the results using the M-file `specplot` to display the confidence interval as dashed lines above and below the estimated spectrum.

b. To test the 95% factor, redo the spectrum estimate 20 times and plot all the estimates on the same graph with the confidence interval. Visually compare the estimates to the upper and lower bounds. Does the confidence estimate seem to be realistic? Which frequencies seem correct, and which are definitely wrong?

c. It is possible to count the number of times the estimates lie outside the confidence bounds. This can be done for all 20 estimates and for all the frequencies in the estimate. From this count determine what percentage of the estimates lie inside the confidence limits and compare your empirical result to the projected 95% figure.

d. The confidence limit assumes that the bias is zero, so that one estimate is sufficient to characterize the mean. However, this is not true for the finite sample case. Redo the plot of the confidence interval using the true mean and true variance of the PSD estimate. Compare to the confidence intervals created by `spectrum` and explain any significant differences. Repeat the empirical counting process in part (c) to see whether or not these confidence limits are more realistic.

e. The confidence limit in `spectrum` is based on a Gaussian assumption, where the mean and variance have to be approximated. For the PSD estimate, the mean is taken to be the computed value of $I(e^{j\omega})$, and the variance is set equal to $(1/K_{\text{eff}})$ times the mean squared. The 95% point on a Gaussian pdf can be computed using the function `erfinv`.[2] Use these ideas to write an M-file that will add confidence intervals to the function `wbpsd` written in Exercise 2.1.

The Gaussian assumption is only an approximation, and it is a poor one when the ratio of the variance to the mean is large. In reality, the distribution is chi-squared.

■ ■ PROJECT 3: NARROWBAND SIGNALS

Quite often spectral estimates are desired for systems whose frequency response contains narrow peaks (e.g., passive sonar). In this case the resolution limit imposed by the data window is the main concern—in fact, it causes a bias in the spectral estimate. On the other hand, the approximation to the variance (1-15) no longer holds. In this project we investigate a system with two narrowband peaks and the estimation of its power spectrum via the Welch–Bartlett procedure.

Hints

Long data sequences may be needed to carry out the estimation procedures needed in this project. Even on computers with limited memory, the filtered output can be produced for very long sequences by using the initial and final state-saving capability of `filter`.

EXERCISE 3.1

Resolution of Peaks

Construct a transfer function $H(z)$ with four poles at

$$\text{POLES} = \{0.99\angle \pm 88^\circ, 0.99\angle \pm 92^\circ\} \tag{3-1}$$

The numerator should be a constant, so the system function $H(z)$ is all-pole.

---

[2]In MATLAB version 3.5 this function was named `inverf`.

a. Plot the magnitude squared of the transfer function $|H(e^{j\omega})|^2$, which is also the true power spectrum.

b. From the plot of the true power spectrum, find the frequency separation of the two peaks. Then determine the minimum window length needed to resolve the two peaks when using a rectangular window; also determine the window length needed for a Hamming window.

c. Generate the filter's output signal, $y[n]$. Plot a section of the *random* signal, and comment on its structure. In this case the transient will be rather long, so make sure that you ignore enough points at the beginning of the signal.

d. Apply the Welch–Bartlett method with the two different windows and with averaging over eight segments. You should verify whether or not the window length is sufficient to resolve the peaks. If not, increase the window length until you get resolution.

e. Consider the spectrum estimates generated for different numbers of segments: $K = 4, 8, 16$, and 32. From these results, determine whether or not variance is a significant problem with estimating the narrowband spectrum. In particular, observe whether more averaging improves the estimate.

f. Consider the bias of the spectrum estimate in the following way: Does the estimated power spectrum have the correct shape near the peaks? It will be necessary to "blow up" the plot near the peak locations and make a plot with the estimate superimposed on the true spectrum to answer this question. If they are different, this is a manifestation of bias. Assuming that they are different, how much more would the length of the window have to increased to remove this bias? (*Hint*: Compare the bandwidth of the true peak to the mainlobe width of the window.)

g. *Optional*: Change the system slightly by placing zeros in the transfer function at $z = \pm j$. Redo the experiment for the Hamming window, to see whether or not the minimum window size drops.

EXERCISE 3.2

Different Windows

The mainlobe of the window function is the primary factor in resolution. However, we cannot ignore the sidelobe structure. In Exercise 3.1 the spectrum contained two equal-amplitude peaks. Now we consider another system in which the peak heights are quite different. It is easy to produce an example where some windows will always fail due to their sidelobe structure.

a. Use the following system to generate a random output signal with known power spectrum:

$$H(z) = \frac{2.2 - 2.713z^{-1} + 1.9793z^{-2} - 0.1784z^{-3}}{1 - 1.3233z^{-1} + 1.8926z^{-2} - 1.2631z^{-3} + 0.8129z^{-4}}$$

Plot the true power spectrum for this system.

b. Process the signal with the Welch–Bartlett method using the rectangular window. Use eight sections of data and a window length of 64, which should easily resolve the two peaks.

c. Redo the Welch–Bartlett processing with a length-64 Hamming window. Are the two peaks now visible in the plot of the spectrum estimate?

d. Change the length of the rectangular window and reprocess the data in an attempt to get resolution of both peaks. Compare to the processing with a longer Hamming window. Why is the Hamming window capable of resolving the peaks while the rectangular window cannot? If the length of the rectangular window is made extremely large, will it eventually be possible to "see" the two peaks when producing the spectral estimate with rectangular windowing?

EXERCISE 3.3

Sine Wave Plus Noise

A related case is that of sine waves in noise. This problem arises in radar, for example, when processing Doppler returns. Once again the resolution issue is the main one, but now the signal is deterministic in a background of noise. The test signal for this exercise should contain three sinusoids in a background of White Gaussian Noise (WGN).

1. $\omega_1 = \pi/3$, $A_1 = 50$, $\phi_1 = 45^\circ$
2. $\omega_2 = 0.3\pi$, $A_2 = 10$, $\phi_2 = 60^\circ$
3. $\omega_3 = \pi/4$, $A_3 = 2$, $\phi_3 = 0^\circ$

The noise variance should be set at $\sigma_v^2 = 10$, although other values could be used.

a. Create a signal that has only sinusoid 3 and the noise. If you make a time-domain plot, the sinusoid will not be visible. However, after processing with a long FFT, the peak at the sinusoidal frequency will appear. Try FFT lengths from 32 through 256 to verify this behavior. What is the minimum FFT length needed to pull the sinusoid out of the background of noise? How does this FFT length depend on the SNR?

b. Create the signal containing three sinusoids in a background of WGN, as described above. Use one very long section of the signal (1024 points) to compute an FFT and display the magnitude to verify that the three spectral lines are in the correct place.

c. Based on the known characteristics of the three sinusoids, determine the minimum lengths needed for a Hamming and a rectangular window to resolve the closely spaced sinusoids. Process the data with the Welch–Bartlett method, using eight sections. Make sure that the three peaks are resolved; if not, reprocess with a longer window. Compare the results for the estimates made with the two different window types.

d. Process the data with the Welch–Bartlett method, using $K = 1$ and $K = 8$. Describe the difference in the spectrum estimates for different amount of averaging.

PROJECT 4: CROSS-SPECTRUM

Related to the auto-power spectrum of one signal is the cross-power spectrum between two signals. It is defined as the Fourier transform of the cross-correlation sequence

$$\Phi_{\mathbf{xy}}(e^{j\omega}) = \sum_{m=-\infty}^{\infty} \phi_{\mathbf{xy}}[m] e^{-j\omega m} \tag{4-1}$$

Since the Fourier transform of the cross correlation between two signals is the multiplication of one transform with the conjugate of the other, we can define a cross-spectral estimate directly in the frequency domain:

$$P_{\mathbf{xy}}(e^{j\omega_k}) = X^*[k]Y[k] \tag{4-2}$$

This is analogous to the periodogram definition, so it must be averaged to yield a worthwhile estimate.

EXERCISE 4.1

Function for Cross-Spectrum

Since the cross-spectrum is nearly identical to the auto-spectrum, modify your previous program (`wbpsd`) to accept an additional input, the signal $y[n]$, and then compute the cross-spectrum between $x[n]$ and $y[n]$. Allow for the possibility of sectioning and averaging during the computation.

EXERCISE 4.2

Identification of the Transfer Function

One interesting property of the cross-spectrum is its use in system identification. If the two signals $x[n]$ and $y[n]$ are related by a transfer function $H(z)$ such that $x[n]$ is the input and $y[n]$ is the output, the cross-spectrum is

$$\Phi_{\mathbf{xy}}(e^{j\omega}) = H(e^{j\omega})\Phi_{\mathbf{xx}}(e^{j\omega}) \tag{4-3}$$

a. Equation (4-3) suggests a method of system identification—namely, record the input and output of a system and take the ratio of the cross-spectrum to the auto-spectrum of the input to find the complete frequency response. One of the benefits of this method is that it finds both the magnitude and the phase of $H(e^{j\omega})$. If, on the other hand, you took the ratio of the auto-spectra, the result would be $|H(e^{j\omega})|^2$.

b. Use the following system to test this system identification idea:

```
b = 0.2 * [ 1    0    1 ]
a = [ 1   -1.3   0.845 ]
```

Generate the frequency response of the system and plot both the magnitude and the phase.

c. Generate a long stream of output data, but keep the input data. For now make the input WGN. Compute the cross-spectral estimate by averaging together eight sections when using a Hann window. Plot the result—magnitude and phase.

d. Now compute the auto-spectrum of the input and divide to get the frequency response. Plot the estimated frequency response and the true $H(e^{j\omega})$ on the same graph. Compare the magnitude as well as the phase. Comment on the differences, especially if there are regions where the estimate is very bad.

e. Redo the experiment for an input signal that is not white. For example, use a first-order filter to "color" the noise and show that the division will remove the effect of the input signal. Experiment with low-pass and high-pass filtered noise for the input. Obviously, a flat input spectrum would be the best test signal, but sometimes low-pass or high-pass is all that is available. State a condition on the input spectrum that is necessary for this sort of system identification to work.

MODERN SPECTRUM ESTIMATION

OVERVIEW

The methods of spectrum estimation based on the FFT are all limited by the resolution of Fourier analysis, which is governed by the uncertainty principle. In particular, the resolution of two sinusoids separated by $\Delta\omega$ in frequency requires a data length that is greater than $2\pi/\Delta\omega$. However, in some situations the amount of available data is limited, so it might not be possible to increase the window length. For example, the data may have been recorded in an experiment where the event of interest lasts for only a short time, or the memory in the data acquisition hardware may be limited. In the related problem of beamforming, the amount of data available is equal to the number of sensors in the spatial array, which, in turn, is limited by the spatial aperture of the array.

Therefore, methods that circumvent the resolution limit of Fourier analysis are of interest. A number of such methods have been developed and popularized over the past 20 years. One of the best known is the maximum entropy method (MEM). In this section the projects will be directed at the use of MEM for spectrum estimation and its relation to modeling and prediction.

Most modern spectrum estimation methods are based on pole-zero models, signal modeling, or eigenvector analysis of the signal covariance matrix. We will only consider a few of these: namely, the autocorrelation method of linear prediction (which is also the MEM spectrum estimate for a Gaussian process), the covariance method of linear prediction, and some signal/noise eigenspace methods suggested by the Pisarenko harmonic decomposition [13]. A complete study of these methods and their application to signal analysis and spectrum estimation would easily take another entire volume.

BACKGROUND READING

The methods of modeling for high-resolution spectrum estimation are treated in the books by Marple [6] or Kay [7], as well as Chapters 1, 2, and 6 in [9] and Chapters 4, 5, and 8 in [5]. In addition, two IEEE Press reprint books devoted to modern spectrum estimation are available [11] and [12]. The linear prediction formulation can be found in the tutorial article by Makhoul [14] as well as in many other recent texts. The use of eigenvector methods for decomposing the covariance matrix was introduced by Pisarenko [13]. A practical spectrum estimation method based on this idea is the MUSIC algorithm [15], as well as [16].

PROJECT 1: MAXIMUM ENTROPY METHOD

The maximum entropy method (for a Gaussian process) produces a spectrum estimate in the form of an all-pole spectrum [17]:

$$P_{\text{mem}}(e^{j\omega}) = \frac{G^2}{|A(e^{j\omega})|^2} = \frac{G^2}{|1 + a_1 e^{-j\omega} + a_2 e^{-j2\omega} + \cdots + a_N e^{-jN\omega}|^2} \tag{1-1}$$

where N is the number of poles. The calculation of the parameters $\{G, a_1, a_2, \ldots, a_N\}$ can be done by solving a set of $(N+1) \times (N+1)$ linear equations, which are the normal equations of a least-squares minimization problem [14,18]:

$$\begin{bmatrix} r[0] & r^*[1] & r^*[2] & \cdots & r^*[N] \\ r[1] & r[0] & r^*[1] & r^*[2] & \cdots \\ \vdots & \ddots & \ddots & \ddots & \vdots \\ \vdots & \ddots & \ddots & \ddots & r^*[1] \\ r[N] & \cdots & r[2] & r[1] & r[0] \end{bmatrix} \begin{bmatrix} 1 \\ a_1 \\ a_2 \\ \vdots \\ a_N \end{bmatrix} = \begin{bmatrix} G^2 \\ 0 \\ 0 \\ \vdots \\ 0 \end{bmatrix} \tag{1-2}$$

These equations are based on the autocorrelation sequence $r[\ell]$ of the signal being analyzed.

$$r[\ell] = \frac{1}{L} \sum_{n=0}^{L-1} x^*[n]x[n+\ell] \tag{1-3}$$

The signal $x[n]$ must be extended by padding with zeros whenever the argument $(n+\ell)$ in (1-3) is greater than $L-1$. This method is also referred to as the *autocorrelation method*, for obvious reasons [14]. An alternative method, called the *covariance method*, is treated in Exercise 2.1.

Hints

For computing the autocorrelation when the sequence is very long, the MATLAB function `xcorr` is not suitable, because it computes $r[\ell]$ for all possible lags. It would be better to write a function based on the FFT that will produce only the lags $\ell = 0, 1, 2, \ldots, N$ needed in (1-2). See the M-file `acf` in Appendix A, which was also used in Project 3 of the section *Stochastic Signals*.

To present the results that show variability of the spectral estimates, run the experiment about 10 times and make a scatter plot of the complex roots of the denominator polynomial $A(z)$. See `help zplane` for plotting in the format of Fig. 6.1. Another way to make the comparison is to plot $P_{\text{mem}}(e^{j\omega})$ for a number of runs, as shown in Fig. 6.2.

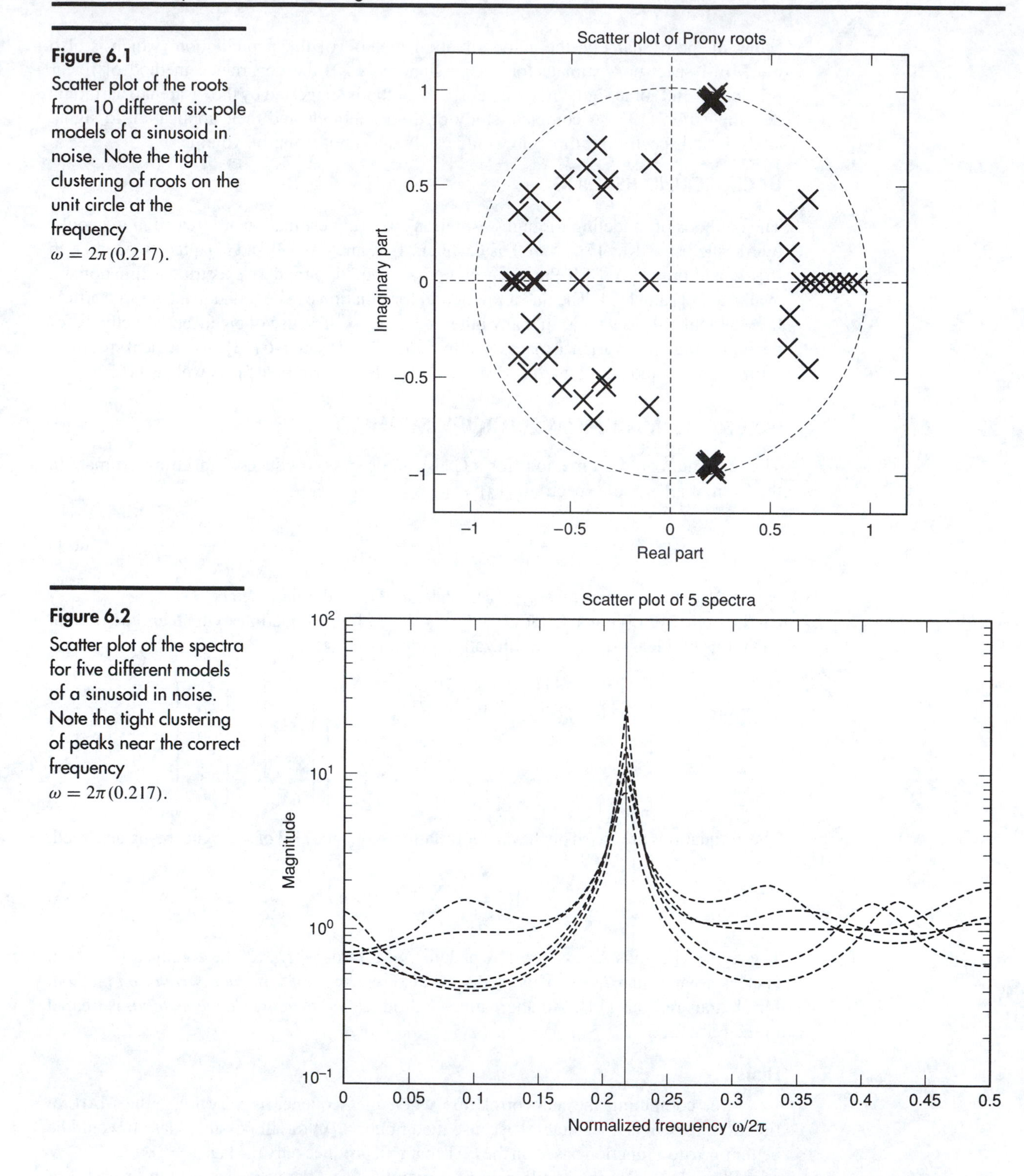

Figure 6.1
Scatter plot of the roots from 10 different six-pole models of a sinusoid in noise. Note the tight clustering of roots on the unit circle at the frequency $\omega = 2\pi(0.217)$.

Figure 6.2
Scatter plot of the spectra for five different models of a sinusoid in noise. Note the tight clustering of peaks near the correct frequency $\omega = 2\pi(0.217)$.

EXERCISE 1.1

MEM Function

Write a MATLAB function that will compute the MEM spectral estimate via (1-1)–(1-3). This consists of two steps:

1. Autocorrelation estimation according to (1-3).
2. Solution of the normal equations (1-2). Be careful, because there are unknowns on both sides of (1-2). The lower N equations must be solved first for the coefficients $\{a_k\}$, then G can be found.

The function should return two outputs: the vector of coefficients $\{a_k\}$ of the denominator polynomial, and the gain term G, which can then be used in (1-1).

EXERCISE 1.2

MEM Estimate

Test your MEM function on the following system, which was also used in the section *FFT Spectrum Estimation*.

```
b = 5

a = [ 1   -1.3   0.845 ]
```

The vectors `a` and `b` define the coefficients of an all-pole filter that will be driven by white Gaussian noise to produce the test signal.

a. Find the frequency response of the filter, and then plot the true power spectrum of the output process. Notice that this is an "all-pole" spectrum.

b. Take a 256-point section of the output signal (after the transient has died out) and apply MEM to compute the all-pole spectrum estimate with a model order of $N = 2$. Since the system has two poles, this choice for N is large enough that MEM might give an exact match to `a`. Compare the estimated coefficients $\{a_k\}$ to the true values.

   To make a fair comparison, this experiment must be run many times to observe the variability in the estimated $\{a_k\}$. To present the results, run the experiment 10 or 20 times and make a scatter plot of a_2 versus a_1 each time. Alternatively, you could generate a scatter plot of one of the complex roots of the denominator polynomial $A(z)$ as in Fig. 6.1. Comment on the variability that you observe in these scatter plots.

c. Another way to make the comparison is to plot $P_{\text{mem}}(e^{j\omega})$ and the true power spectrum on the same graph. As in part (b), this comparison should be repeated many times and then all the estimated spectra plotted together as in Fig. 6.2. Alternatively, you could plot the difference between the true and estimated spectra and observe the maximum deviation between the two. Explain why the largest deviation seems to lie near the peak in the true spectrum.

d. Now try to determine experimentally whether or not the MEM estimate will ever be exactly the same as the true power spectrum. Try section lengths that are hundreds of points long, and also thousands of points long (if feasible on your computer). Compare on the basis of the power spectrum scatter plot, and also by making a scatter plot of the coefficients $\{a_k\}$. From your tests, is it possible to claim that the variability of the estimates (as shown via scatter plots) decreases as L increases?

e. In an attempt to minimize the length of the input data used while still getting a good estimate, the data can be windowed with something other than a rectangular window. Experiment with using a Hamming window on the data prior to computing the autocorrelation estimate (1-3). Rerun the scatter plot of the poles, as in part (b), for $L = 128$ and $L = 256$, in order to compare the Hamming results to those of the boxcar window.

EXERCISE 1.3

MEM with High Model Order

In MEM, the number of poles must be selected a priori. It is an advantage to know the true value of N, but the MEM estimation procedure is robust enough that overestimating the model order N is not disastrous. In fact, with noisy data MEM seems to give better answers when the model order is a bit higher than the minimum. In this exercise we study the performance of MEM when the model order is increased.

a. Generate a signal using the same all-pole filter as in Exercise 1.2. Try higher model orders (e.g., $N = 3, 4, 6, 8, 12, \ldots,$) and for each N, make scatter plots of five spectral estimates to illustrate the variability in the match between the true spectrum and the MEM estimate. Keep the length of the data sequence fixed, but very long, say $300 < L < 1000$. Does the quality of the estimate improve as N is increased? Does the variability in the estimate change as the model order is increased? The answer may depend on whether you look only near the peak of the true spectrum.

b. Make a scatter plot of the pole locations for these higher-order models and compare to the true pole locations. Explain which poles are the most significant in matching the true spectrum by looking not only at the pole positions, but also at the residues of the poles in a partial fraction expansion of the all-pole transfer function (see `residuez` in MATLAB). Note that the poles (for the autocorrelation method) must always be inside the unit circle.

EXERCISE 1.4

Pole–Zero Case

Now take a test spectrum that is not all-pole.

```
b = [ 1    +1.3    0.845 ]
a = [ 1    -1.3    0.845 ]
```

Redo Exercise 1.3, with a fixed value for L and with a model order of $N = 6$. Comment on the differences caused by the numerator $B(e^{j\omega})$ in the true spectrum.

PROJECT 2: SPECTRUM ESTIMATES BASED ON LINEAR PREDICTION

Some variations on MEM can provide better performance in situations such as determining frequencies for narrowband signals and for sinusoids in noise. In this project we consider three variations of MEM which are all based on the same underlying linear prediction problem. Each of these will produce an all-pole spectral estimate of the form (1-1).

The MEM and its relative, the covariance method, both rely on the solution of a linear prediction problem to find the denominator polynomial $A(z)$ in an all-pole model for the signal. This linear prediction operation can be written as a set of linear equations, which is usually overdetermined (i.e., more equations than unknowns). In matrix form, the prediction error equations are

$$\begin{bmatrix} x[0] & 0 & 0 & \cdots & 0 \\ x[1] & x[0] & 0 & \cdots & 0 \\ x[2] & x[1] & x[0] & \cdots & 0 \\ \vdots & \ddots & \ddots & \ddots & \ddots \\ x[N-1] & x[N-2] & x[N-3] & \cdots & x[0] \\ \vdots & \ddots & \ddots & \ddots & \ddots \\ x[L-2] & x[L-3] & x[L-4] & \cdots & x[L-N-1] \\ \ddots & \ddots & \ddots & \ddots & \vdots \\ 0 & \cdots & x[L-1] & x[L-2] & x[L-3] \\ 0 & \cdots & 0 & x[L-1] & x[L-2] \\ 0 & \cdots & 0 & 0 & x[L-1] \end{bmatrix} \begin{bmatrix} a_1 \\ a_2 \\ a_3 \\ \vdots \\ a_N \end{bmatrix} = - \begin{bmatrix} x[1] \\ x[2] \\ x[3] \\ \vdots \\ x[N] \\ \vdots \\ x[L-1] \\ \vdots \\ 0 \\ 0 \\ 0 \end{bmatrix} \tag{2-1}$$

In a more compact form, we have

$$\mathbf{X}\mathbf{a} = -\mathbf{x} \tag{2-2}$$

which can be solved for **a** by minimizing the squared error between the two sides of (2-2), $\mathbf{e} = \mathbf{X}\mathbf{a} - (-\mathbf{x})$.

$$\min_{\mathbf{a}} \mathbf{e}^H \mathbf{e} = \min_{\mathbf{a}} (\mathbf{X}\mathbf{a} + \mathbf{x})^H (\mathbf{X}\mathbf{a} + \mathbf{x}) \tag{2-3}$$

The minimum error solution in (2-3) is the solution to the normal equations:

$$(\mathbf{X}^H \mathbf{X})\mathbf{a} = -\mathbf{X}^H \mathbf{x} \tag{2-4}$$

assuming that the covariance matrix $\boldsymbol{\Phi} = \mathbf{X}^H \mathbf{X}$ is invertible.

The covariance method differs from the autocorrelation method only in that **X** is formed from a subset of the rows in (2-1). The autocorrelation method uses all the rows; the covariance method uses only those rows of **X** and **x** that contain no zeros (i.e., starting from the row whose first entry is $x[N-1]$ and ending with the row whose first entry is $x[L-2]$). In MATLAB the matrix **X** can be generated with the M-file `convmtx`, or with the M-file `convolm` in Appendix A.

EXERCISE 2.1

Covariance Method

The covariance method computes the denominator $A(e^{j\omega})$ for (1-1) via a slightly different set of normal equations. If we reduce (2-4) to a form similar to (1-2), the (i, j)th entry in the matrix $\boldsymbol{\Phi}$ of the normal equations is

$$\phi(i, j) = \sum_{n=N}^{L-1} x^*[n-i]x[n-j] \qquad i, j = 1, 2, \ldots, N \tag{2-5}$$

In effect, the range of summation is changed so that the same number of nonzero terms $(L-N)$ are always involved. The right-hand side of (2-4) is a vector whose elements are $-\phi(i, 0)$.

The MATLAB function for solving the covariance method is called `prony`, because Prony's method of exponential modeling gives the same answer. See `help prony` from the MATLAB Signal Processing Toolbox for details. Other aspects of Prony's method are treated in Chapter 11 in the section *Linear Prediction*.

a. Use the covariance method on the same example as in Exercise 1.2. Make a scatter plot of the pole locations for the case $N = 2$ and $L = 256$. Compare to the results from the autocorrelation method (MEM) to determine if this estimate has lower variability.

b. Plot several spectra as was done in Exercise 1.2(c) and compare them to the MEM spectrum estimate.

c. Test the covariance estimate for a higher model order (e.g., $N = 8$). Plot the pole locations for 10 different runs and observe whether or not the poles ever move outside the unit circle. In this case, two of the poles should closely model the true poles of the second-order system, while the other six could lie anywhere.

EXERCISE 2.2

Forward-Backward Covariance Method

The covariance method has a small drawback in that it limits the number of prediction error equations used when forming **X** in (2-2). Furthermore, to have more equations than unknowns, we need $L - N \geq N$. Getting more equations L would require more data. If this is not feasible, it is possible to form more equations by using the "forward-backward" method of linear prediction. This strategy is based on the idea that a random signal has the same properties if it is flipped in time. Therefore, backward prediction of the signal can be done with the same prediction operator $A(z)$.

For the covariance method, this works as follows: Augment the matrix $\mathbf{X}$ by appending $L - N$ rows formed by running the predictor over a signal that is the flipped and conjugated version of $x[n]$. In this case, the covariance matrix $\boldsymbol{\Phi}$ generated in (2-4) will have entries

$$\phi_{fb}(i, j) = \sum_{n=N}^{L-1} x^*[n-i]x[n-j] + \sum_{n=N}^{L-1} x_r^*[n-i]x_r[n-j] \qquad i, j = 1, 2, \ldots, N \quad \text{(2-6)}$$

where the autocorrelation of $x_r[n] = x^*[L-1-n]$ is the contribution from the backward prediction. The right-hand side of (2-4) has elements $-\phi_{fb}(i, 0)$.

a. Write an M-file to implement the forward-backward covariance method. The input arguments will be the signal $x[n]$ and the model order N.

b. Test this function on the same case as in Exercise 2.1. Redo parts (a)–(c) of that exercise. Since this method works well for sinusoids in noise, reserve judgment on its performance until testing it in the next project.

EXERCISE 2.3

SVD-Based Solution

Tufts and Kumaresan [19] have proposed a solution to the linear prediction problem that gives more robust behavior in lower SNR cases. It is based on the singular value decomposition (SVD). The solution of (2-4),

$$\mathbf{a} = -[(\mathbf{X}^H\mathbf{X})^{-1}\mathbf{X}^H]\mathbf{x} = -\mathbf{X}^\dagger\mathbf{x}$$

involves a special case of the "pseudo-inverse" of $\mathbf{X}$, denoted by $\mathbf{X}^\dagger$. Computation of the pseudo-inverse $\mathbf{X}^\dagger$ generally relies on the singular value decomposition (SVD) of the matrix $\mathbf{X}$. The book by Golub and Van Loan [20] is an excellent reference for a theoretical background on SVD.

In MATLAB, the M-file `svd` will compute the singular value decomposition, and the M-file `pinv` implements a pseudo-inverse by first computing an SVD and then discarding relatively small singular values prior to forming the inverse. Consult the MATLAB `help` on these functions, as well as [18, 20], for more details on SVD and the pseudo-inverse. The M-file below is a modification of the MATLAB function `pinv` to compute a rank-r pseudo-inverse. For example, if we specify $r = 3$, then the three largest elements along the diagonal of `S` will be used to form the rank-3 inverse. The function `pseudinv` can be used later to solve the prediction error equations.

```
function Ainv = pseudinv(A,r)
%PSEUDINV        Pseudo-inverse of rank r.
% usage:     Ainv = pseudinv(A,r)
%   produces the rank-r inverse of A, from the SVD of A.
%   Only r singular values are retained for the inverse
%   with the exception that singular values less than
%   MAX(SIZE(A)) * S(1) * EPS are discarded.
%      Ainv = pseudinv(A) uses all possible singular values.
% See also  SVD and PINV
```

```
[U,S,V] = svd(A);
S = diag(S);
keep = sum( S > (max(size(A)) * S(1) * eps) );
if (nargin == 1)
   r = keep;
else
   r = min( r, keep );
end
S = diag( ones(r,1) ./ S(1:r) );
Ainv = V(:,1:r) * S * U(:,1:r)';
```

One view of the pseudo-inverse $\mathbf{X}^{\dagger}$ is that it is a low-rank approximation to the inverse of the original matrix. The rank of the inverse can be controlled by the number of singular values that are retained when the pseudo-inverse is constructed. With this view, both the covariance method and the autocorrelation method would be called "full-rank" inverses, because the solution of (2-4) is the same as an SVD inverse in which all singular values are kept.

When applied to linear prediction, the philosophy is the following: The signal matrix $\mathbf{X}$ contains values that are due to both the signal and the noise. If we know a priori that the number of signal components is r, then in the noise-free case the rank of $\mathbf{X}$ would be r. However, the noise usually increases the rank to its maximum. Since we need to "invert" $\mathbf{X}$ to solve for $\mathbf{a}$, it is reasonable that we should design an inverse based solely on the signal. The assumption of the rank-r inversion is that the larger singular values and their singular vectors correspond to the signal components, because they span the same subspace as the signal vectors. Therefore, the rank parameter r is chosen to match the expected number of signal components. For example, if the signal were generated by a second-order filter, we would take $r = 2$.

a. Write an M-file to implement the SVD low-rank covariance method. The input arguments will be the signal $x[n]$ and the rank r, which is the suspected number of signals present. The data matrix $\mathbf{X}$ can be defined by either the forward-backward prediction error, or the forward only. The model order N can be larger, but only r poles should be significant.

b. Test this function on the same case as in Exercise 2.1. Redo parts (a)–(c) of that exercise. Again, this method should work well for sinusoids in noise, so its performance will be tested in the next project.

Project 3: Narrowband Signals

This project is nearly the same as Project 3 in the section *FFT Spectrum Estimation.* However, we now have four different all-pole spectral estimators (from Project 2) to evaluate. For this project we synthesize the test signal using a filter with two narrowband peaks, and compare the estimation of its power spectrum among the four all-pole estimators and to the Welch-Bartlett result done in Project 3 of the section *FFT Spectrum Estimation.* For the most part we investigate the resolution of closely spaced narrow peaks and try to minimize the data length needed. Finally, the case of sinusoids in noise and the influence of SNR on the estimation of frequency are studied.

Hints

Make sure that you have M-files for doing all four all-pole estimates considered in Project 2. As in that project, the spectrum estimates will vary for different realizations of the random process under scrutiny. This variation is of some interest, so a scatter plot of the spectra or of pole locations is the preferred way to present the results.

EXERCISE 3.1

Resolution of Peaks

Construct a transfer function $H(z)$ with four poles at

$$\text{POLES} = \{0.99\angle \pm 88^\circ, 0.99\angle \pm 92^\circ\} \tag{3-1}$$

The numerator should be a constant, so that $H(z)$ is all-pole.

a. Determine the transfer function $H(e^{j\omega})$, and then plot the true power spectrum. From this plot, determine the frequency separation of the two peaks, $\Delta\omega$. From the value of $\Delta\omega$, estimate the FFT length that would be need to resolve the two peaks via Fourier analysis with a Hamming window.

b. Generate the synthetic test signal by exciting $H(z)$ with white Gaussian noise to produce the output signal, $y[n]$. In addition, determine the length of the transient by synthesizing the impulse response of the filter. In this case, the transient will be rather long, so make sure that you ignore enough points at the beginning of the signal when selecting a segment for spectrum estimation.

c. Choose one of the four all-pole estimators and process the signal. The primary objective is to resolve the two peaks in the spectral estimate with the minimum segment length L. Therefore, some experimentation with the window length L, and perhaps with the model order N, or rank r, will be necessary. You might also try different window types. When you have finally settled on specific values of L and N, run 10 cases and plot all the spectra together. This type of scatter plot will show how the resolving capability varies for different parts of the random signal. Finally, since only resolution is of interest, the region near the peaks should be blown up when making the plot.

d. Compare the segment length L determined in part (c) to the FFT length needed in the Welch–Bartlett procedure, using the Hamming window. In addition, compare plots of the spectral estimates for both cases and comment on the differences near the peaks and in the sidelobe regions. This part requires that you refer to Project 3 in the section *FFT Spectrum Estimation*.

e. The bias in the spectrum estimate can be studied by concentrating on the shape of $P_{\text{mem}}(e^{j\omega})$ near the peaks. The shape of the spectrum estimates must be compared to the true shape at the peaks. Thus it will be necessary to zoom in on the plot near the peak locations and make a plot with all the estimates superimposed on the true spectrum. To distinguish bias effects from variance, consider the following two questions: Are the differences you observe consistent over all the spectrum estimates? If the estimated spectra were averaged, would the result be closer to the true spectrum? If feasible, try processing the data with a longer window length to see whether or not the bias can be reduced by increasing L.

f. *Optional*: Change the system slightly by placing zeros in the transfer function at $z = \pm j$, and redo the entire experiment.

EXERCISE 3.2

Small Peaks

In the FFT, the sidelobes presented a problem because small peaks could be obscured by the sidelobes of a large peak. Apparently, an all-pole spectrum estimate such as MEM has no sidelobes, so it may not miss a small peak. Test this idea on the example below (which is also from the section *FFT Spectrum Estimation*).

a. Use the following system to generate a random output signal with known power spectrum.

$$H(z) = \frac{2.2 - 2.713z^{-1} + 1.9793z^{-2} - 0.1784z^{-3}}{1 - 1.3233z^{-1} + 1.8926z^{-2} - 1.2631z^{-3} + 0.8129z^{-4}}$$

Plot the true spectrum, and then evaluate the peak heights, which should be quite different.

b. Process the signal with each of the four all-pole estimators. Try to minimize the window length subject to the constraint that the two peaks be resolved and the relative peak heights be preserved. Recall that using a Hamming window with the FFT method is crucial if the small peak is to be identified at all.

c. *Optional*: Redo the experiment with a new transfer function in which the smaller peak is even closer to the larger one. Choose the poles and residues in a partial fraction expansion to control the true spectrum. See how close you can bring the smaller one to the larger one before the methods fail.

EXERCISE 3.3

Sine Wave Plus Noise

For the case of sine waves in noise, the model-based methods should work well because each sinusoid corresponds to a pole pair in the z-transform. Since the sine-wave component of the signal is deterministic, it should turn out that Prony's method will do an excellent job when the SNR is high. For lower SNR, the SVD-based method should be superior.

a. Create a signal containing three sinusoids in a background of white Gaussian noise

   1. $\omega_1 = \pi/3$, $A_1 = 100$, $\phi_1 = 45^\circ$
   2. $\omega_2 = 0.3\pi$, $A_2 = 10$, $\phi_2 = 60^\circ$
   3. $\omega_3 = \pi/4$, $A_3 = 1$, $\phi_3 = 0^\circ$

   The noise variance should be set at $\sigma_v^2 = 100$, although other values of SNR need to be tried later.

b. First, do the estimation with Prony's method (i.e., the "covariance" method). Show how to extract the frequencies of the sinusoids directly from the roots of the predictor polynomial $A(z)$. Experiment with a model order that may be higher than the minimum needed. Try to do the processing with about 50 data points; then try to minimize the length of data segment used. Obviously, there will be a trade-off among the parameters L, N and the accuracy of the frequency estimates. Run 10 trials of the processing and plot the results together to illustrate the variability of the estimate due to additive noise.

c. Next try MEM, the "autocorrelation" method, with a Hamming window. For the same data length L and model order N, compare the accuracy of the frequency estimation to that of Prony's method. This will necessitate running 10 estimates and presenting the results as a scatter plot for evaluation.

d. Repeat part (b) for the forward-backward covariance method of Exercise 2.2.

e. Repeat part (b) for the SVD-based method from Exercise 2.3.

f. Test for different SNRs, $\sigma_v^2 = 10$ and 1000. Show that an excellent estimate of the frequencies can be obtained with a very short data segment when $\sigma_v^2 = 10$. Determine which methods will succeed when $\sigma_v^2 = 1000$.

g. For these all-pole estimators, devise a calculation that will determine the power in each sinusoid. Consider the fact that the peak height in the all-pole spectrum may not be equal to the power in the sinusoid. In fact, the peaks in the all-pole spectrum cannot be correct if they have finite bandwidth, since the true spectrum for the sinusoidal case consists of impulses in ω.

PROJECT 4: EIGENVECTOR-BASED METHODS

When the spectrum estimation problem is primarily concerned with extracting sinusoids in noise, some recent methods based on the eigen-decomposition of the covariance matrix are among the best for high SNR situations [6, Chap. 13]. These methods have also been popularized for angle-of-arrival (AoA) estimation in array processing. In this case the direction of an arriving plane wave gives a sinusoidal variation across the array, so the measurement of this "spatial" frequency amounts to direction finding.

Hints

See `eig` for computation of the eigenvectors and `svd` for the singular value decomposition [20].

EXERCISE 4.1

Pisarenko's Method

Many of the eigenvector-based methods are descended from Pisarenko's method [13], which is based on an amazing property of the covariance matrix. Starting with any positive-definite Toeplitz covariance matrix, it is possible to derive a representation for the covariance matrix as a sum of sinusoids plus white noise. If $\boldsymbol{\Phi}$ is an $L \times L$ covariance matrix, the representation can be written

$$\mathbf{R} = \sum_{k=1}^{\mu} \alpha_k \mathbf{v}(\omega_k)\mathbf{v}^H(\omega_k) + \sigma^2\mathbf{I} \tag{4-1}$$

where $\mu = L-1$, and $\mathbf{v}(\omega_k)$ is a "steering" vector:

$$\mathbf{v}(\omega_k) = [\,1 \quad e^{j\omega_k} \quad e^{j2\omega_k} \quad \dots \quad e^{j(L-1)\omega_k}\,]^T \tag{4-2}$$

In the array processing problem, the frequency ω_k can be related to direction of arrival θ via

$$\omega_k = \frac{2\pi\,\Delta x}{\lambda}\sin\theta$$

where Δx is the intersensor spacing and λ is the wavelength of the propagating plane wave. In this case it is natural to call $\mathbf{v}(\omega_k)$ a "steering vector."

The algorithm that produces this representation (4-1) must find the frequencies $\{\omega_k\}$, which it does by factoring a polynomial. Amazingly enough, the polynomial of interest is defined by one of the eigenvectors of $\boldsymbol{\Phi}$. The resulting procedure suggests many generalizations.

a. For the experiment in this exercise, create a positive-definite Toeplitz matrix. This can be done with `rand` and `toeplitz`, but some care is needed to satisfy the positive-definite constraint. One shortcut to building a positive-definite matrix is to increase the value of the diagonal elements—eventually, the matrix will become positive definite. After you create the Toeplitz covariance matrix, verify that it is positive definite.

b. Since all the eigenvalues of the covariance matrix will be positive, we can identify the minimum eigenvalue and its corresponding eigenvector—called the *minimum eigenvector*. It turns out that the noise power (σ^2) needed in (4-1) is equal to the minimum eigenvalue of $\boldsymbol{\Phi}$. For your matrix $\boldsymbol{\Phi}$, determine σ^2 and extract the minimum eigenvector for use in part (c). There is a small possibility that your matrix from part (a) might have a minimum eigenvalue ($\lambda_{\min}$) with multiplicity greater than 1. In this case the minimum eigenvector is not unique and the following part will not have the correct property. If so, generate another matrix $\boldsymbol{\Phi}$ so that ($\lambda_{\min}$) has multiplicity 1.

c. The amazing property of $\boldsymbol{\Phi}$ discovered by Pisarenko [13] is that the frequencies (ω_k) are obtained from the roots of a polynomial defined by the minimum eigenvector, $\mathbf{v}_{\min}$. Each eigenvector of $\boldsymbol{\Phi}$ has L elements, so these can be used as the coefficients of a degree $L-1$ polynomial:

$$V_{\min}(z) = \sum_{\ell=0}^{L-1} v_\ell z^{-\ell} \tag{4-3}$$

Verify the following property for your example. All roots of the minimum eigenvector polynomial (4-3) lie on the unit circle (i.e., have magnitude 1). Plot the root locations with `zplane` (Appendix A) to see them all on the unit circle.

d. To complete the verification of the sinusoid plus noise representation of the covariance matrix, it is necessary to find the amplitudes α_k in (4-1). This is relatively easy because the α_k enter the problem linearly. Once the values of ω_k and σ^2 are known, the first column of $\boldsymbol{\Phi}$ can be written in terms of the α_k to get $L-1$ equations in $L-1$ unknowns. Use MATLAB to set up and solve these equations.

e. Now that all the parameters of the representation are known, synthesize $\boldsymbol{\Phi}$ according to the right-hand side of (4-1) and check that the result is equal to $\boldsymbol{\Phi}$ from part (a).

EXERCISE 4.2

Orthogonality of Signal and Noise Subspaces

The eigenvectors of $\boldsymbol{\Phi}$ can be used to decompose $\boldsymbol{\Phi}$ into two orthogonal subspaces, called the noise subspace and the signal subspace. The noise subspace is spanned by the minimum eigenvector; the signal subspace, by the steering vectors. The dimensionality of the noise subspace could be greater than 1 if the matrix $\boldsymbol{\Phi}$ has a repeated minimum eigenvalue.

a. The roots of $V_{\min}(z)$ are all on the unit circle and therefore all of the form $z = e^{j\omega_k}$. Furthermore, the polynomial evaluation $V_{\min}(z)$ at $z = e^{j\omega_k}$ is equivalent to the inner product of $\mathbf{v}_{\min}$ with $\mathbf{v}(\omega_k)$.

$$V_{\min}(e^{j\omega_k}) = \mathbf{v}^H(\omega_k)\mathbf{v}_{\min} \tag{4-4}$$

Each root $z = e^{j\omega_k}$ can be used directly to define a steering vector. Use MATLAB to create the $L-1$ steering vectors, and verify that each is orthogonal to the minimum eigenvector. Equation (4-4) is essential in the generalizations to follow.

b. The algorithm described in Exercise 4.1 becomes a bit more complicated when the minimum eigenvalue is repeated. In this case, fewer than $L-1$ sinusoids are needed in the representation (4-1), and every one of the minimum eigenvector polynomials will have only a subset of roots on the unit circle. Interestingly enough, all of these polynomials will have just this subset of roots in common.

To illustrate this behavior, process the 5×5 positive-definite Toeplitz matrix below, which has a repeated minimum eigenvalue. Apply the procedure as before to derive the representation (4-1). In addition, plot the roots of all the minimum eigenvector polynomials to show which roots are in common.

$$\boldsymbol{\Phi} = \begin{bmatrix} 3 & 1-j2 & -1-j\sqrt{2} & -1 & 1-\sqrt{2} \\ 1+j2 & 3 & 1-j2 & -1-j\sqrt{2} & -1 \\ -1+j\sqrt{2} & 1+j2 & 3 & 1-j2 & -1-j\sqrt{2} \\ -1 & -1+j\sqrt{2} & 1+j2 & 3 & 1-j2 \\ 1-\sqrt{2} & -1 & -1+j\sqrt{2} & 1+j2 & 3 \end{bmatrix}$$

c. Write an M-file that will do the three steps of processing required in Pisarenko's method: Find $\lambda_{\min}$, then factor $V_{\min}(z)$ to get ω_k, and finally, compute α_k. Make your M-file work correctly for the extended case of repeated eigenvalues by using the idea suggested in part (b). Save this file for testing later in Project 5.

EXERCISE 4.3

MUSIC

In the application of Pisarenko's method to actual data, the covariance matrix $\boldsymbol{\Phi}$ must be estimated. This is a significant issue because the representation (4-1) is exact. In practice, a method based on approximation might be preferable. An important algorithm along these lines is MUSIC (MUltiple SIgnal Classifier), introduced by Schmidt [15]. The MUSIC algorithm avoids the factoring step and produces a spectrum estimate as in MEM.

In Pisarenko's method, we assume that the noise subspace can be characterized exactly by the minimum eigenvector. Sometimes the minimum eigenvalue has a multiplicity greater than 1, so the dimensionality of the noise subspace is also greater than 1. In fact, if we knew that the data had only two sinusoids, we would expect the noise subspace to have dimension $L - 2$. But with real data it is unlikely that $\boldsymbol{\Phi}$ would have a minimum eigenvalue that repeats $L - 2$ times. In practice, we would probably observe two large eigenvalues and $L - 2$ small ones. To identify the correct number of significant eigenvalues we need a threshold below which all small eigenvalues are associated with the noise subspace.

The MUSIC estimate is

$$P_{\text{music}}(e^{j\omega}) = \frac{1}{\mathbf{v}^H(\omega)\boldsymbol{\Phi}_n\mathbf{v}(\omega)} \tag{4-5}$$

where $\boldsymbol{\Phi}_n$ is the noise subspace portion of $\boldsymbol{\Phi}$.

$$\boldsymbol{\Phi} = \boldsymbol{\Phi}_s + \boldsymbol{\Phi}_n = \sum_{i=1}^{\mu} \mathbf{v}_i\mathbf{v}_i^H + \sum_{j=\mu+1}^{L} \mathbf{v}_j\mathbf{v}_j^H$$

a. Write an M-file to implement the MUSIC technique. The inputs should be the covariance matrix $\boldsymbol{\Phi}$ and a threshold percentage that determines how many λ_i will be assigned to the noise subspace. If this percentage is called η, then λ_i belongs to the noise subspace when

$$|\lambda_i - \lambda_{\min}| < \eta|\lambda_{\max} - \lambda_{\min}|$$

Finally, the MUSIC function should return two outputs: the power spectrum $P_{\text{music}}(e^{j\omega})$ and the number of signal components found, μ.

b. Test your function by generating a 7×7 covariance matrix $\boldsymbol{\Phi}$ for data that contain two sine waves plus white Gaussian noise with a variance equal to $\sigma_{\mathbf{v}}^2 = 0.01$. Make the amplitudes of the sines equal, and pick the frequencies at random. Take $\boldsymbol{\Phi}$ to be the true covariance in the form (4-1), but perturb the matrix entries slightly. Make sure that the perturbed $\boldsymbol{\Phi}$ is still Toeplitz. Set the threshold $\eta = 0.2$, or higher, so that the MUSIC process finds exactly two complex signal components. Plot the spectrum to see if the estimated frequency is correct.

Extensive testing of the MUSIC algorithm on synthetic data will be undertaken in the next project.

EXERCISE 4.4

Kumaresan–Tufts Method

An improved estimate of the polynomial (4-4) for annihilating the signal subspace can be found using the singular value decomposition [16]. When the signal/noise subspace decomposition is performed, the spectral estimate is constructed by finding steering vectors that are orthogonal to the noise subspace. This is equivalent to finding a polynomial $D(z)$ whose coefficients satisfy

$$\mathbf{E}_s^H\mathbf{d} = \mathbf{0} \tag{4-6}$$

where the $L \times \mu$ matrix $\mathbf{E}_s$ is formed from the signal subspace eigenvectors—one in each column. The vector $\mathbf{d} = [\,1 \quad d_1 \quad d_2 \quad \cdots \quad d_{L-1}\,]$ will be orthogonal to the signal subspace, so it can be used to find steering vectors that satisfy $\mathbf{v}^H(\omega)\mathbf{d} = 0$. Some of the roots of $D(z)$ should lie on the unit circle and give the frequencies of the sinusoidal components in the data. In (4-6) the polynomial coefficients $\mathbf{d}$ are seen to annihilate the members of the signal subspace.

Once **d** is calculated, the spectral estimate is formed in a way similar to the all-pole forms:

$$P_{\text{KT}}(e^{j\omega}) = \frac{1}{|\mathbf{v}^H(\omega)\mathbf{d}|^2} \tag{4-7}$$

Equation (4-6) is underdetermined, but can be solved with a pseudo-inverse computed from the SVD (singular value decomposition). See Exercise 2.3 for a discussion of the pseudo-inverse, including an M-file.

a. Write an M-file to compute the Kumaresan–Tufts estimate (4-7). The inputs should be the covariance matrix and a threshold η for separating the signal and noise subspaces.

b. Repeat the test in Exercise 4.3(b).

c. In addition to showing the spectrum estimate, plot the root locations of $D(z)$ to see where the extra roots go.

PROJECT 5: TESTING WITH SYNTHETIC SIGNALS

To test the various signal–noise subspace methods, we need a method for generating noisy data. The typical test involves estimating the frequencies of a few sinusoids in a background of white noise. Therefore, we need to generate data according to the model given in (4-1). The crucial difference, however, is that the covariance matrix $\boldsymbol{\Phi}$ must be estimated from a finite segment of the signal. Thus $\boldsymbol{\Phi}$ might not be exactly correct for the data being observed. In fact, $\boldsymbol{\Phi}$ might not even have the correct form (i.e., Toeplitz).

In this project we develop an M-file for generating noisy data and we use it to test several of the algorithms discussed in Project 4. The test scenario corresponds to the direction estimation problem. Additional test cases for extracting sinusoids in noise can be found in [16] and [19].

EXERCISE 5.1

Generating Sinusoids in Noise

The model in (4-1) states that the covariance matrix $\boldsymbol{\Phi}$ is composed of two parts: sinusoids plus white noise. Therefore, we can generate a signal according to that rule and see how closely the form of $\boldsymbol{\Phi}$ is matched when $\boldsymbol{\Phi}$ is estimated directly from the data.

The following equation describes the signal generation process:

$$x[n] = \sum_{i=1}^{\mu} A_i e^{j(\omega_i n + \phi_i)} + \frac{\sigma_n}{\sqrt{2}} v[n] \qquad \text{for } n = 0, 1, 2, \ldots, L-1 \tag{5-1}$$

where the phase ϕ_i must be random, and uniformly distributed in the interval $[-\pi, \pi)$; otherwise, the sinusoids will be correlated with one another. The noise $v[n]$ is a complex Gaussian process where both the real and imaginary parts have variance equal to 1. The functions `rand` and `randn` can be used to generate uniformly and Gaussian distributed random values, respectively. The ratio A_i/σ_n determines the SNR; in dB, it is $20\log_{10}(A_i/\sigma_n)$.

a. The simplest case would be one complex exponential plus noise in (5-1). For $L = 3$, generate $x[n]$ when $\omega_1 = 2\pi(0.234)$, $A_1 = 1.0$, and $\sigma_n = 0.1$. Use this signal to generate an estimate of the covariance matrix $\boldsymbol{\Phi}$. Explain why the estimate $\boldsymbol{\Phi}_x$ can be formed by computing the outer product:

$$\boldsymbol{\Phi}_x = \mathbf{x}\mathbf{x}^H = \begin{bmatrix} x[0] \\ x[1] \\ x[2] \end{bmatrix} [x^*[0] \quad x^*[1] \quad x^*[2]] \tag{5-2}$$

In particular, verify that this $\boldsymbol{\Phi}_x$ is Hermitian symmetric. Is $\boldsymbol{\Phi}_x$ a Toeplitz matrix? Compare the estimated $\boldsymbol{\Phi}_x$ in (5-2) to the true value from (4-1).

b. In an actual estimation problem, many "snapshots" of the signal vector $\mathbf{x}_m$ would have to be collected and averaged to get an adequate estimate of Φ from the data. This can be accomplished by averaging a number of outer products (5-2).

$$\Phi_x = \frac{1}{M}\sum_{m=1}^{M} \mathbf{x}_m \mathbf{x}_m^H$$

For the two cases considered in parts (a) and (b), compute Φ_x for 10 snapshots, and then for 100 snapshots or more. Compare these estimates to the true value of Φ. Does the estimate converge to a Toeplitz form?

c. Repeat part (b) for $A_1 = 0$ and $\sigma_n = 1$. This is the noise-only case, so the proper form for Φ is a scaled identity matrix. Is that what you observe?

EXERCISE 5.2

M-files for Simulation

a. Write an M-file to synthesize $x[n]$ given in (5-1). The inputs to this function are the number of sinusoids μ, their amplitudes A_i, the noise power σ_n^2, and the number of signal samples L. Its output is a vector containing the signal values.

b. Write an M-file to estimate the covariance matrix Φ from the signal $x[n]$. This function needs an input to specify the number of snapshots and could then call the M-file for synthesizing $x[n]$.

c. Write an M-file to generate the true covariance matrix Φ given the correct values for ω_i, A_i, and σ_n.

d. Generate a signal containing two sinusoids in noise, and estimate its 4×4 covariance matrix from a large number of snapshots. Let $\omega_1 = 2\pi(0.2)$, $\omega_2 = 2\pi(0.3)$, $A_1 = A_2 = 1.0$, and $\sigma_n^2 = 0.01$. Determine how many shapshots are needed to approximate the true values of Φ within 5%.

EXERCISE 5.3

Testing MUSIC

In this exercise the MUSIC algorithm is tested under different signal-to-noise ratios. Refer to Exercise 4.3 for the M-file that implements MUSIC. The test signal should be formed as follows: $x[n]$ consists of two complex exponentials ($\mu = 2$) in noise. The number of data samples (or sensors) is $L = 7$. The amplitudes are equal $A_1 = A_2 = 1.0$, and the noise power is $\sigma_n^2 = 0.01$. The frequencies are $\omega_1 = 2\pi(0.227)$ and $\omega_2 = 2\pi(0.2723)$ If these frequencies are converted to angle θ for an AoA problem, $\omega = (2\pi\Delta x/\lambda)\sin\theta$, the angles would be $\theta = 27^\circ$ and 33° if the intersensor spacing is $\Delta x = \frac{1}{2}\lambda$. This test signal will be used to evaluate the other algorithms.

a. Run MUSIC once to see what the eigenvalue spread will be. Then pick a threshold η so that MUSIC will use a noise subspace with $L-2$ eigenvectors. Make sure that the two frequencies can be resolved in a plot of $P_{\text{music}}(e^{j\omega})$; otherwise, lower the noise power σ_n^2.

b. Run MUSIC 10 times with the number of snapshots $M = 16$. Plot all the spectra together to see how well the method performs when resolving the two closely spaced spectral peaks. Zoom in on the frequency range to show only $2\pi(0.2) \le \omega \le 2\pi(0.3)$. Furthermore, check whether the correct relative peak heights are obtained.

c. Since this is a frequency estimation problem, one way of summarizing the behavior of the MUSIC estimate would be to pick the frequency estimates as the location of the two largest peaks in ω and then collect these results from all the runs. From these estimates, the mean and variance of the two frequency estimates can be computed.

d. Rerun the experiment with $M = 50$ snapshots and with $M = 5$ snapshots and/or with a different SNR ($\sigma_n^2 = 0.25$ and $\sigma_n^2 = 0.0001$). Comment on the performance for these cases. (*Note*: When the SNR is very high the method should give nearly perfect results (i.e., the variance of the frequency estimates should converge to zero). Similarly, when the number of snapshots becomes very large, the variability of the estimate should decrease.)

EXERCISE 5.4

Roots of Pisarenko's Method

In the application of Pisarenko's method to actual data, the covariance matrix $\boldsymbol{\Phi}$ must be estimated. This is a significant issue because the representation (4-1) is exact. Nonetheless, we can use the test case in Exercise 5.3 to examine the root distribution of the minimum eigenvector polynomial.

a. Generate an estimated covariance matrix $\boldsymbol{\Phi}_x$ and extract its eigenvectors. Plot the root locations for $V_{\min}(z)$.

b. Relate the roots near the unit circle to the true frequencies in the data.

c. Select the $L-2$ smallest eigenvalues and their eigenvectors. Extract the roots of all of these and plot together with `zplane`. Comment on the expected clustering of roots near the true frequencies $z = e^{j\omega_1}$, $e^{j\omega_2}$.

EXERCISE 5.5

Testing the K-T Method

Use the M-file for computing the Kumaresan–Tufts estimate (4-7). Refer to Exercise 5.3 for the test signal needed to study the K-T method.

a. Run the K-T method once with the dimension of the signal subspace set equal to 2. Make sure that the two frequencies can be resolved; otherwise, lower the noise power σ_n^2.

b. Run K-T 10 times with the number of snapshots $M = 16$. Plot all the spectra together to see how well the method performs when resolving the two closely spaced spectral peaks. Furthermore, check whether the correct relative peak heights are obtained.

c. Since this is a frequency estimation problem, pick the frequency estimate as the location of the two largest peaks in ω, and calculate the mean and variance of the two frequency estimates. Compare to the results found in Exercise 5.3(c).

d. Rerun the experiment with more snapshots and with fewer snapshots and/or with a lower SNR. Comment on the performance for these cases.

e. Plot all the root locations of the $D(z)$ polynomial to see where the extra roots go.

f. *Optional*: Run the SVD-based covariance method from Exercise 2.3 on this same test data set. Compare to MUSIC and the K-T method.

chapter 7

WORDLENGTH EFFECTS

OVERVIEW

Under ideal conditions (i.e., if implemented with unlimited wordlength of coefficients and variables), a digital filter behaves as expected if the design has been done properly. In this case the choice of one of the numerous structures influences only the complexity, and thus the achievable speed, if a particular hardware is used. Besides that, the performance will always be the same. But in reality, the situation is different and far more complicated:

Coefficients as well as data can be implemented with finite wordlengths only. Quantized coefficients will lead at least to a more or less erroneous behavior of the system (e.g., a different frequency response). The deviation from the expected performance will depend on the chosen structure (i.e., on its sensitivity). It might even happen that the quantization of coefficients turns a stable system into an unstable one.

Another effect will even change the character of the system: Arithmetic operations done with numbers of limited wordlengths usually yield results of larger wordlengths. The necessarily following reduction down to the wordlength the system can handle is a nonlinear process. The resulting nonlinear system might behave completely differently from the desired linear one. That is especially so if limit cycles occur as a result of an instability.

Of course, it can be expected that the various errors become smaller if the wordlengths of coefficients and data are increased. Therefore, it is an important part of the design to determine the required minimum wordlengths for a satisfying operation of the system to be built, or to check if a satisfying performance with tolerable error can be achieved in a particular case (e.g., if a given integrated signal processor with fixed wordlength is supposed to be used).

For these reasons a rather thorough investigation of the effects explained above is of interest. But since a general analysis of a system working with limited wordlength is rather impossible, a stepwise investigation of the different influences is usually done. That requires, in addition, a model of the system and its signals and thus an approximate description of the real effects. So the influence of quantizing coefficients is studied while ignoring the nonlinearity of the real system. The effect of quantizing the input signal or the result of an arithmetic operation such as multiplication by rounding is described by an additional noise source with certain properties, concerning, for example, probability density, the power density spectrum, and its correlation to the original unquantized signals. Finally, the existence of limit cycles and methods to avoid them have been studied primarily for the rather simple case of a block of second order.

In all cases these separate investigations should not only yield information about the error in particular cases but should also lead to rules for the design of a system such that the desired properties can be achieved with minimum expense. Here two aspects are of importance, the speed of an implementation and the deviation from the idealized performance. Methods for the reduction of the error quite often reduce the speed obtained, whereas increasing the speed might increase the error as well. So it has to be decided which of these properties is more important in a particular implementation.

BACKGROUND READING

Some of the material used here can be found in Sections 6.7–6.10 of Oppenheim and Schafer [1], Chapter 5 of Rabiner and Gold [2], and Chapter 9 of Roberts and Mullis [3]. Special topics are presented in Section 8.6 of Lim and Oppenheim [4] and in [5].

[1] A. V. Oppenheim and R. W. Schafer. *Discrete-Time Signal Processing*. Prentice Hall, Englewood Cliffs, NJ, 1989.

[2] L. R. Rabiner and B. Gold. *Theory and Application of Digital Signal Processing*. Prentice Hall, Englewood Cliffs, NJ, 1975.

[3] R. A. Roberts and C. T. Mullis. *Digital Signal Processing*. Addison-Wesley, Reading, MA, 1987.

[4] J. S. Lim and A. V. Oppenheim. *Advanced Topics in Signal Processing*. Prentice Hall, Englewood Cliffs, NJ, 1988.

[5] H. W. Schüßler and Y. Dong. A new method for measuring the performance of weakly nonlinear systems. *Proceedings of ICASSP-89*, pages 2089–2092, 1989.

WORDLENGTH EFFECTS

OVERVIEW

The investigation of wordlength effects is a very challenging task. It becomes a rather lengthy one due to the large number of possibilities for implementation. That starts with the choice between fixed-point and floating-point arithmetic and continues with the various ways for reducing the wordlength of the results of arithmetic operations. Furthermore, there is a large variety of structures for the implementation, which in general behave differently under limited wordlength conditions.

In this chapter only some of these possibilities can be investigated. One of the reasons is that the required wordlength reduction of arithmetic results has to be simulated under

MATLAB. This rather time-consuming procedure forbids, for example, the investigation of filters of higher order. For the same reason, we restrict the exercises to systems with fixed-point arithmetic. Furthermore, only a few of the large number of different structures will be considered. Nevertheless, in this chapter we provide some insight into the most important quantization effects.

Project 1 deals with the analysis and modeling of signals of limited wordlength. We use the well-known description of the error sequence and check by measurements the validity of these common models. Somewhat related is the investigation and modeling of a real multiplier in Project 2 for different possibilities for the quantization of the product. The effect of quantized coefficients in some structures is the topic of Project 3. For this reason the complex frequency responses of different implementations of a filter will be considered. In particular, we show the favorable sensitivity properties of so-called lossless structures.

The effects due to a limitation of the wordlength after arithmetic operations are considered in Projects 4 and 5. First, an unstable behavior is demonstrated, yielding large-scale and granular limit cycles. Methods are shown to avoid them. Finally, noise-like errors are investigated by measuring the power density spectrum at the output of a system of second order. It will be compared with the result found through the usual analysis. The applied measuring method will be explained in the appendix.

A general remark seems to be appropriate: While restricting the investigations to systems implemented with fixed-point arithmetic, we have to keep in mind that we work under MATLAB with floating-point arithmetic. So all numbers used in our experiments are floating-point representations of data or coefficients, usually with a magnitude smaller or equal to 1, having in fixed point the limited wordlength w, to be specified in the particular cases.

PROJECT 1: QUANTIZED SIGNALS

In this project the properties of an A/D converter are investigated. Included in this study are measurements of its I/O characteristic and the S/N ratio as a function of the noise power as well as an investigation of the noise. We examine especially the validity of the common model for describing the noise by its probability density function, its autocorrelation, and the crosscorrelation with the input signal.

Project description

We consider an A/D converter, the input signal of which is the continuous function $v_0(t)$. Usually, its output sequence is described by

$$v[n]_Q = k[n]Q = v[n] + e[n] \tag{1-1}$$

where

$Q = 2^{-(w-1)}$ is the quantization stepsize in a fixed-point number of wordlength w,

$k[n]$ is an integer with $-2^{w-1} \leq k[n] \leq 2^{w-1} - 1$,

$v[n] = v_0(t = n \cdot T)$, with T being the sampling interval

Figure 7.1 shows how an A/D converter will be simulated in MATLAB as well as the modeling according to (1-1).

The implementation in MATLAB is done with

```
vq = fxquant(v,w,'round','sat'),
```

in the following figures modeled by the lower part of Fig. 7.1.

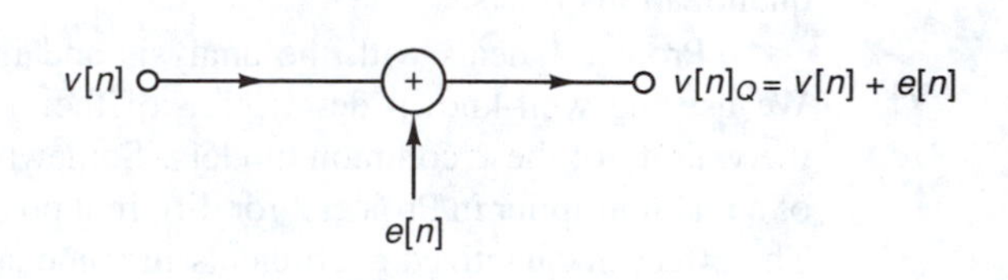

Figure 7.1
Simulating and modeling an A/D converter.

Assuming that $v[n]_Q$ is generated by rounding the samples $v[n]$, the error sequence $e[n]$ is modeled as being a member of a random process $\mathbf{e}_r$ with the following properties:

- It is uniformly distributed in $\left(-\frac{Q}{2}, \frac{Q}{2}\right]$.
- Thus it has a variance $\sigma_{\mathbf{e}}^2 = Q^2/12$.
- It has a constant power density spectrum.
- It is not correlated with $v_0(t)$ or with $v[n]$. (1-2)

A first and obvious condition for these properties of $e[n]$ is that no overflow occurs; that is, we assume that

$$-1 - 0.5Q \leq v_0(t) < 1 - 0.5Q$$

Furthermore, a sufficient condition is that $v_0(t)$ is uniformly distributed in all quantization intervals:

$$[(k - 0.5)Q, \ (k + 0.5)Q] \qquad -2^{w-1} \leq k < 2^{w-1} - 1$$

The objective of this project is the verification of the model, described by (1-1) and (1-2), if appropriate signals $v_0(t)$ are used, but also to show its limitation in other cases.

Furthermore, possibilities for reducing the quantization error by oversampling and error feedback are investigated.

Hints

For the investigation of an A/D converter we use samples of signals $v_0(t)$ generated with floating-point arithmetic. They can be regarded as being unquantized with sufficient accuracy. We shall use two types of random input signals, one uniformly distributed in (-1, 1), the other normally distributed with different variances (see `help rand` and `help randn` for further information). In addition, we shall use sinusoidal sequences of different amplitudes.

The quantization, which simulates the A/D converter, is done using the M-file `fxquant`, written for the simulation of fixed-point arithmetic with specified wordlength w (see Fig. 7.1). The program allows the choice of different types of quantization (rounding, 2's-complement truncation, magnitude truncation) and different modes to handle the overflow. For more information, see `help fxquant`.

For the investigation of the error sequence $e[n]$ the programs `hist`, `spectrum`, `acf`, and `ccf` can be used. Information about these programs can be found with `help` as well. The probability densities of $e[n]$ should be determined for 20 bins. If `hi` is the output of `hist`, the required normalization can be achieved with `p = hi/sum(hi)`. The correlation sequences $\phi_{\mathbf{ee}}[m]$ and $\phi_{\mathbf{ve}}[m]$ should be calculated for $m = -16 : 15$.

EXERCISE 1.1

Quantizer

We start with an investigation of the M-file `fxquant` determining experimentally the non-linear relationship between input and output signal and the characteristics of the error sequence. Choose as the input signal a sequence $v[n]$ linearly increasing between -1.1 and $+1.1$, for example, with stepsize 2^{-6}. Generate the corresponding quantized sequence having a wordlength of $w = 3$ bits using rounding and a saturation overflow characteristic with

```
vq = fxquant(v,3,'round','sat')
```

Calculate `dq = vq - v`; plot separately `vq` versus `v` and `dq` versus `v`. Which deviations from the above-mentioned model do you observe so far?

EXERCISE 1.2

S/N Ratio of an A/D Converter

Now we use a normally distributed input sequence with zero mean. After separating the error sequence

$$e[n] = v[n]_Q - v[n]$$

we measure the ratio of the signal power $S = \sigma_v^2$ to the noise power $N = \sigma_e^2$ under different conditions.

a. Choose $\sigma_v^2 = 0.01$; measure and plot S/N in dB as a function of the wordlength w for $w = 3 : 16$. Use 8000 samples for each point. What is the improvement of the S/N ratio per bit?

b. The following experiment will be done with $w = 8$ bits. Measure and plot S/N in dB as a function of S in dB for S = (–60 : 5 : 20) dB. In a certain range the S/N ratio increases linearly with S. Determine and explain the points where this property is lost. Especially verify experimentally that for $\sigma_v \approx 0.3$ the power of the error due to overflow has the same order of magnitude as the rounding error.

EXERCISE 1.3

Quantization Error

a. Now we investigate the properties of the error sequences $e[n]$ for the following different input signals $v[n]$:

   $v_1[n]$ is a white random signal, uniformly distributed in (-1, 1).

   $v_2[n]$ is a white random signal, normally distributed with variance 1.

   Use, for example, 8000 values of these sequences and generate the error sequences

$$e[n] = v[n]_Q - v[n]$$

   for a wordlength $w = 8$, rounding and saturation overflow characteristic. Determine approximately the following properties of $e[n]$:

   - The mean and the variance using `mean` and `std`
   - The probability density of $e[n]$ with `[hi,x] = hist(e,20)`
   - Its power density spectrum with `spectrum(e,.)`
     (for more information, see `help spectrum`)
   - Its autocorrelation sequence either with `acf` or as `ifft(spectrum(e,.))`

- The cross spectral density of $e[n]$ and $v[n]$ with `spectrum(e,v,.)`
- The crosscorrelation sequence of $e[n]$ and $v[n]$ either with `ccf` or as `ifft(spectrum(e,v,.))`

Comment on the deviations from the theoretical model of the quantization error. Explain especially the reasons for these differences.

b. We consider now the influence of quantization on a sinusoidal sequence:

$$v_3[n] = a \sin[n\omega] \qquad \text{with } \omega = 2\pi/8000\,, \; n = 0 : 8000$$

1. Calculate the error sequences for the two amplitudes $a = 1$ and $a = 1/32$, using the wordlength $w = 8$ and $w = 3$. As before, use rounding and saturation overflow characteristic. Plot the four error sequences using `subplot(22.)`.
2. Determine the six properties of the different $e[n]$'s, as given in part (a). Comment again on the deviations from the theoretical model.

EXERCISE 1.4

Oversampling

In this exercise we investigate the reduction of the quantization noise by oversampling plus an appropriate filtering. Suppose that we are given a continuous signal $v_0(t)$, the spectrum of which is approximately zero for $f \geq f_c$, thus requiring a sampling rate of $f_{s1} \geq 2f_c$. Instead, we use an A/D converter with a much higher sampling rate $f_{s2} = rf_{s1}$, where r is an integer > 1. We want to verify that the following filtering with a low-pass of cutoff frequency f_c combined with a subsampling by a factor r yields a reduction of the noise power by a factor of r or $10 \log_{10} r$ dB in comparison with a direct A/D conversion of $v_0(t)$ with the sampling frequency f_{s1}.

Since we deal with sampled signals in our computer experiments, we have to generate a digital version of a bandlimited signal first (see Fig. 7.2). For this purpose we use a low-pass filter with the transfer function $H_0(z)$ having a stopband cutoff frequency $\omega_c = \pi/r$. It yields the desired sequence $v[n]$, corresponding to $v_0(t)$, but with sampling rate f_{s2}. In the upper branch of the figure it is shown that after subsampling by a factor r, the usual A/D conversion with f_{s1} (corresponding to a quantization in our experiment) yields the output sequence $y_1[n]$, while the extraction of the error sequence $e_1[n]$ can be done as in Exercise 1.3.[1] In the lower branch the conversion by oversampling is done. Another low-pass filter described by $H_1(z)$, having a passband of a width corresponding to the bandwidth of $v_0(t)$, eliminates the noise spectrum for $f > f_c$. A subsampling by the factor r finally generates the desired signal $y_2[n]$ in this case. The separation of the noise sequence $e_2[n]$ requires an identical filter for the generation of a signal out of $v[n]$ for comparison. (Why?)

The following steps have to be taken:

a. Assuming an oversampling factor $r = 4$, use a low-pass, such as an elliptic filter $H_0(z)$ designed with `ellip(7,.1,60,.195)`, having a stopband cutoff frequency $\omega_c \approx \pi/4 = \pi/r$ and an attenuation of 60 dB in the stopband. As an input signal $w[n]$ we apply white noise, uniformly distributed in $(-1, 1)$.

b. Use `fxquant`, for example, with $w = 10$ bits for the quantizers, representing the A/D converters. Later, try other wordlengths.

c. Use FIR filters described by $H_1(z)$ with linear phase, appropriate cutoff frequency, and attenuation of 60 dB in the stopband.

---

[1] Note that the A/D converters are modeled according to their effect by adding the noise sequences $e_1[n]$ and $e[n]$ (see Fig. 7.1).

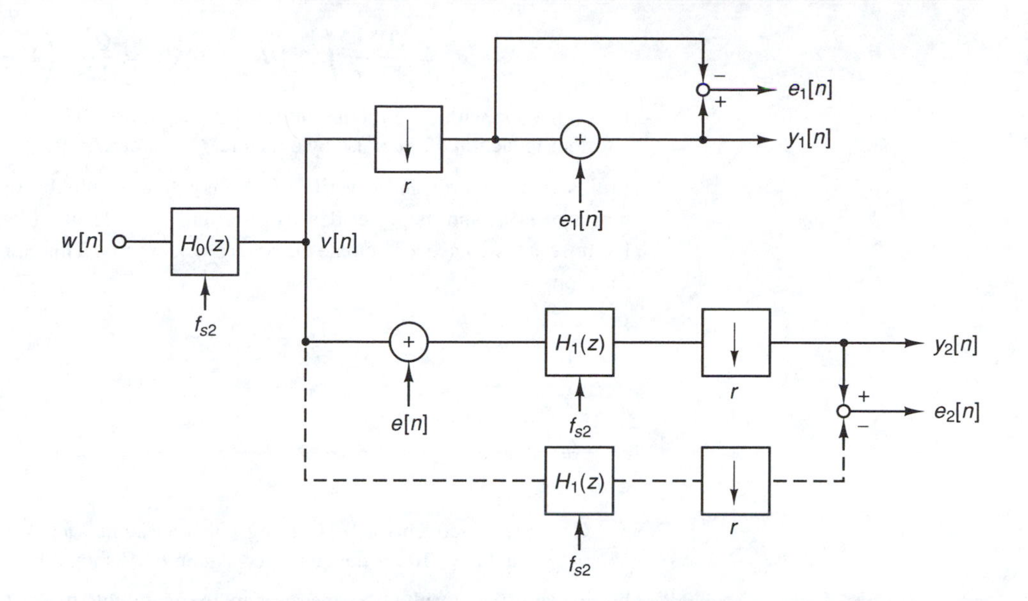

Figure 7.2
Structure for comparative investigation of an oversampling A/D converter.

d. Generate the sequences $e_1[n]$ and $e_2[n]$ working with input signals of length 8000. Measure and compare their variances. Comment on the results.

EXERCISE 1.5

Sigma-Delta A/D Converter

A further reduction of noise is possible with a *sigma-delta A/D converter*, working with error feedback. A discrete model of a simple example is shown in Fig. 7.3.

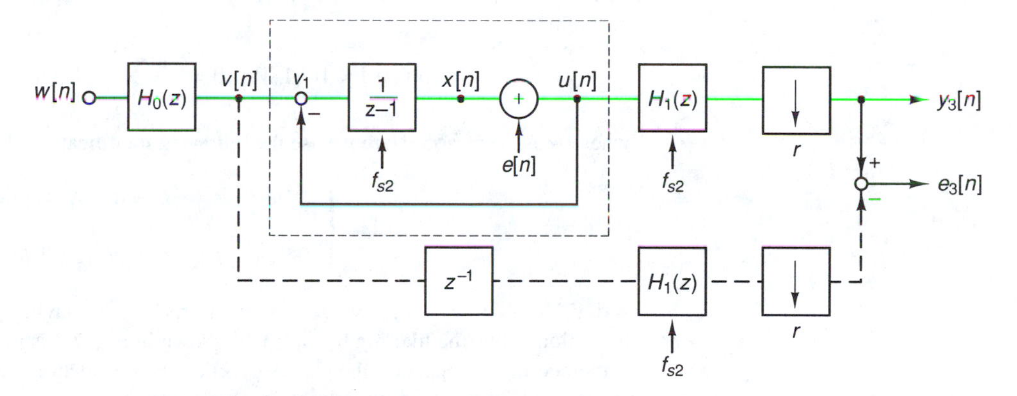

Figure 7.3
Structure for investigating an A/D converter working with oversampling and error feedback.

The analysis of the basic block yields for the Z-transform of its output signal $u[n]$,

$$U(z) = Z\{u[n]\} = \frac{1}{z}V(z) + \frac{z-1}{z}E(z) \tag{1-3}$$

where $E(z)$ is the Z-transform of the noise sequence $e[n]$ generated by the A/D converter. Thus the noise transfer function turns out to be

$$|H_E(e^{j\omega})|^2 = 4\sin^2\omega/2 = 2(1-\cos\omega) \tag{1-4}$$

The low-pass filter $H_1(z)$ eliminates the noise for $\omega > \pi/r$. So the power of the noise sequence $e_3[n]$ at the output becomes in case of an ideal LPF,

$$\sigma_{\mathbf{e}}^2 = \frac{Q^2}{12}\frac{1}{\pi}\int_0^{\pi/r} |H_E(e^{j\omega})|^2 \, d\omega = \frac{Q^2}{12}\frac{2}{r}\left(1 - \frac{\sin \pi/r}{\pi/r}\right) \tag{1-5}$$

In comparison with the structure in Fig. 7.2, the "noise shaping" as described by (1-4) yields a further reduction of the noise by a factor $2\,(1 - \sin(\pi/r)/(\pi/r))$, which $\rightarrow 0$ as $r \rightarrow \infty$.

In this exercise we want to verify these theoretical results by measuring approximately the power of $e_3[n]$ and its power density spectrum. Besides the filters and subsamplers used in Exercise 1.4, we have to simulate the basic block exactly. This can be done using the following for-loop:

```
x = 0,
for n = 1 : length(v),
  u(n) = fxquant(x,w,'round','sat');
  v1 = v(n) - u(n);
  x = v1+x;
end
```

a. Generate the sequence $e_3[n]$ working with input signals $w[n]$ of length 8000 and wordlengths $w = 8$ and $w = 10$ for the quantizer. Again $w[n]$ should be white noise with zero mean.

b. To verify the result (1-5), measure approximately its power (with `std^2`) and its power density spectrum for $\omega < \pi/r$ (using `spectrum`).

EXERCISE 1.6

Different Measuring Method

Furthermore, we use another method for the investigation of the basic block, shown in Fig. 7.3. The procedure is explained in some detail in the appendix to this chapter, where a block diagram and a MATLAB program are given as well. We recommend that this section be read before proceeding. As described by (A-1), the required ensemble of periodic input signals is generated as

$$\tilde{v}_\lambda[n] = \text{IFFT}\{V_\lambda[k]\} = \frac{1}{N}\sum_{k=0}^{N-1} V_\lambda[k] w_N^{-kn} \in \mathbb{R} \tag{1-6}$$

where N is the period. Here we use the following modification: With

$$V_\lambda[k] = \begin{cases} |V| e^{j\varphi_\lambda[k]} & k = 0 : (N/(2r) - 1) \\ 0 & k = N/(2r) : N/2 \end{cases} \tag{1-7}$$

and $V_\lambda[N-k] = V_\lambda^*[k]$, we get a periodic real signal having precisely the desired band-limitation. Thus the filtering by $H_0(z)$, as shown in Fig. 7.3 is not necessary anymore. As is explained in the appendix, the phases $\varphi_\lambda[k]$ are independent random variables, thus yielding an ensemble of periodic but independent sequences $v_\lambda[n]$.

In this experiment we are interested in the estimate $\hat{\Phi}_{\mathbf{ee}}(e^{j\omega})$ of the power density spectrum of the noise. Equation (A-11) gives the desired result at the points $\omega = \omega_k = k \cdot 2\pi/N,\ k = 0 : N/2$. Here we can use the output sequence $u[n]$ of the quantizer directly, the spectrum $U_\lambda[k]$ of which replaces $Y_\lambda[k]$ in (A-11). The variance of the noise in the frequency band up to $\omega_c = \pi/r$ is obtained by a simple summation:

$$\sigma_{\mathbf{e}}^2 = \frac{2\pi}{N}\sum_{k=0}^{N/(2r)-1} \hat{\Phi}_{\mathbf{ee}}(e^{j\omega_k}) \tag{1-8}$$

Neither filtering with $H_1(z)$ nor separation of the noise as explained in Fig. 7.3 is required.

Use this method for measuring $\hat{\Phi}_{\mathbf{ee}}(e^{j\omega_k})$ and $\sigma_{\mathbf{e}}^2$ with $w = 8:2:12$ and $r = 4$. The procedure should be applied with $N = 1024$ and $L = 40$ (the number of experiments to be averaged).

Note that the modification made here requires two changes in the MATLAB program given in the appendix: the definition of the phase as explained above and in the calculation of `SumH`. (Why?)

PROJECT 2: MODELING A MULTIPLIER

In this project we investigate a real multiplier used for the multiplication of a sequence of random numbers $v[n]$ out of a process $\mathbf{v}$ by a constant factor c, which might be a filter coefficient in a practical application. Dealing with fixed-point arithmetic, we assume that the magnitudes of both factors are not larger than 1. The wordlengths are w and w_c, respectively. Thus, in general, the product $p[n] = c \cdot v[n]$ will have a wordlength $w_p = = w + w_c - 1$. Its further processing, for example, in a digital filter requires a quantization of the result down to w bits, the wordlength inside the filter. This project deals with an investigation of the error sequences produced by different methods for the quantization of the product. The goal is a model of a real multiplier, applicable to the noise analysis of a digital system.

Project description

The real multiplier we are interested in is modeled by an ideal multiplier combined with a quantizer. It yields signals

$$p_Q[n] = [cv[n]]_Q = cv[n] + e[n] \tag{2-1}$$

having the required wordlength w. We investigate the properties of the error sequence

$$e[n] = p_Q[n] - cv[n]$$

The following notations will be used:

$Q = 2^{1-w}$ is the quantization step of the input signal $v[n]$ and of $p_Q[n]$.

$Q_c = 2^{1-w_c}$ is the quantization step of the coefficient c.

$Q_p = 2^{1-(w+w_c)}$ is the quantization step of the product $p[n] = cv[n]$.

The quantization will be done by one of the following methods:

- 2's-complement rounding, described in MATLAB notations by

  `pqr = Q * floor(p/Q+.5)` (2-2)

- 2's-complement truncation, described by

  `pqt = Q * floor(p/Q)` (2-3)

- Magnitude truncation, described by

  `pqm = Q * fix(p/Q)=sign(p).*(Q * floor(abs(p/Q)))` (2-4)

The simple and most common models for the description of the error sequences $e[n]$ are in the three cases:

- *2's-complement rounding*: $e[n]$ is a member of a random process $\mathbf{e}_r$, being uniformly distributed in the interval $[-Q/2, Q/2)$. Thus it has the properties

$$m_r = \mathcal{E}\{\mathbf{e}_r\} \approx 0; \quad \sigma_r^2 = Q^2/12 \tag{2-5}$$

Its power density spectrum is constant; it is not correlated with the input process **v**. Note the similarity to the description of an A/D converter.

- *2's-complement truncation*: Now the random process $\mathbf{e}_t$ is uniformly distributed in the interval $(-Q, 0]$. Thus its properties are

$$m_t = \mathcal{E}\{\mathbf{e}_t\} = -Q/2 \quad \mathcal{E}\{\mathbf{e}_t^2\} = Q^2/3 \quad \sigma_t^2 = Q^2/12 \tag{2-6}$$

The power density spectrum of the unbiased version of $\mathbf{e}_t$ is constant and not correlated with the input process **v**.

- *Magnitude truncation*: The random process is uniformly distributed either in the interval $(-Q, 0]$ if $p \geq 0$ or in $[0, Q)$ if p is negative. It can be expressed as

$$e_m[n] = -\frac{Q}{2}\,\text{sign}\,[p] + e'_m[n] \tag{2-7}$$

where $e'_m[n]$ has the same properties as $e_r[n]$, the error sequence in case of rounding. If p is normally distributed with zero mean, we get

$$m_m = \mathcal{E}\{\mathbf{e}_m\} = 0 \quad \sigma_m^2 = Q^2/3 \tag{2-8}$$

Usually, the obvious correlation between $\mathbf{e}_m$ and **v** will be ignored.

The foregoing characteristics of the error sequences have been found to be useful if Q_c is small enough and if $p[n] = cv[n]$, the product to be quantized, is large enough. That implies a dependence on the variance $\sigma_{\mathbf{v}}^2$ of **v** and on the coefficient $c = \lambda_c Q_c$, where λ_c is an integer. Note that the number of possible values of the error sequence is $[Q_c]^{-1}$.

Here one point has to be noted: There might be a big difference between the nominal quantization stepsize Q_c and the effective one Q_{ce} in an actual case. For example, $c = 0.$ x 1 000 000 with x $\in \{0, 1\}$ is a number with nominal $Q_c = 2^{-8}$. But it is actually one out of two possible positive numbers with the effective stepsize $Q_{ce} = 2^{-2}$. Thus the number of possible different values $e[n]$ is now $[Q_{ce}]^{-1}$. A careful investigation of the means and variances of the error sequences yields some modifications of former results. With $c = \lambda_c Q_c$ and $lcd(\lambda_c, Q_c^{-1})$ being the largest common divisor of λ_c and Q_c^{-1}, the *effective quantization stepsize* becomes

$$Q_{ce} = Q_c lcd(\lambda_c, Q_c^{-1}) \tag{2-9}$$

Now we get in the different cases:

- *2's-complement rounding*: Instead of (2-5),

$$m_r = \frac{1}{2} Q Q_{ce} \quad \mathcal{E}\{\mathbf{e}_r^2\} = \frac{Q^2}{12}(1 + 2Q_{ce}^2) \quad \sigma_r^2 = \frac{Q^2}{12}(1 - Q_{ce}^2) \tag{2-10}$$

- *2's-complement truncation*: Instead of (2-6),

$$m_t = -\frac{Q}{2}(1 - Q_{ce}) \quad \mathcal{E}\{\mathbf{e}_t^2\} = \frac{Q^2}{3}\left(1 - \frac{3Q_{ce}}{2} + \frac{Q_{ce}^2}{2}\right) \quad \sigma_t^2 = \frac{Q^2}{12}(1 - Q_{ce}^2) \tag{2-11}$$

- *Magnitude truncation*: With $m_m = 0$ as before, now, instead of (2-8),

$$\sigma_m^2 = \frac{Q^2}{12}\left(1 - \frac{3Q_{ce}}{2} + \frac{Q_{ce}^2}{2}\right) \tag{2-12}$$

Furthermore, in cases of rather large values Q_{ce} but small variances σ_v^2 of the signal $v[n]$, a correlation between $e[n]$ and $v[n]$ will be observed, even in case of rounding. Here the error sequence $e[n]$ includes a term proportional to $v[n]$ but dependent on c and σ_v. Thus we get, instead of (2-1),

$$p_Q[n] = c[1 + \Delta c(c, \sigma_v)]v[n] + e_1[n] \tag{2-13}$$

where now the correlation of $e_1[n]$ and $v[n]$ is rather small. But the result is an incorrect coefficient,

$$c' = c[1 + \Delta c(c, \sigma_v)] \tag{2-14}$$

Hints

In these experiments we use either the expressions (2-2)–(2-4) for doing the different quantization or `fxquant` (see Project 1). Use `help` to get further information about `round`, `floor`, `fix`, and `fxquant`. For measuring the power spectral density of the noise as well as a possible deviation Δc of the coefficient, we use the measuring procedure explained in the appendix. Furthermore, the programs `hist`, `spectrum`, `acf`, and `ccf` can be used.

As input signal $v[n]$ of the multiplier we use quantized versions of either a uniformly or a normally distributed sequence, both with zero mean, generated with `rand` or `randn`, respectively. Their variances will be specified. The quantization down to the wordlength w the multiplier is working with should be done with rounding, according to (2-2).

EXERCISE 2.1

Quantizers

We want to determine the nonlinear I/O characteristics of the three quantizers of the product p, described by (2-2)–(2-4). Determine `pq.` and `e. = pq. - p` as a function of p for $w = 3$ ($Q = 2^{-2}$) and

a. $pa = -0.5 : 2^{-6} : 0.5$ (corresponding to $Q_{ce} = 2^{-4}$)

b. $pb = -1 : 2^{-4} : 1$ (corresponding to $Q_{ce} = 2^{-2}$)

Plot with `subplot(22.)`; `pq./Q` (case a), `pq./Q` (case b), `e./Q` (case a), `e./Q` (case b). Use `plot(p,...,'x')` to show the distinct points.

EXERCISE 2.2

Error Sequences in the Usual Case

For the following measurements we use an unbiased, uniformly distributed input signal $v[n]$ with standard deviation $\sigma_v = 0.25$ and `length(v)` = 8000. Furthermore, we apply $w = 10$.

a. Determine the effective stepsizes Q_{ce} for the coefficients

$$c = 16/128,\ c = 44/128,\ c = 63/128,\ c = 120/128$$

b. Pick a coefficient c with the effective stepsize $Q_{ce} = Q = 2^{-9}$ and calculate the product $p[n] = cv[n]$ and its three quantized versions $p_{Q.}[n]$. Determine the corresponding error sequences

$$e.[n] = p_{Q.}[n] - p[n]$$

and their means and variances. Compare your results with the characteristics of the models, as given in (2-5)–(2-8).

c. Measure approximately the power spectrum densities of the $e_{\cdot}[n]$ and the cross-spectral densities of $e_{\cdot}[n]$ and $v[n]$ using `spectrum`. Plot your results. Alternatively, measure the autocorrelation sequences of the $e_{\cdot}[n]$ with `acf` and the crosscorrelation sequences of $e_{\cdot}[n]$ and $v[n]$ with `ccf`. Comment on your results, especially the differences of the quantizers with regard to the crosscorrelation.

EXERCISE 2.3

Error Sequences in Special Cases

Now we investigate the properties of the different quantizers under other conditions. Especially, we use coefficients c with larger effective quantization step Q_{ce} and in addition smaller variances of the normally distributed input signal $v[n]$. The wordlength of $v[n]$ and $p_{Q\cdot}[n]$ is again $w = 10$.

a. Measure the means and the variances of the error sequences $e[n]$ for the three quantizers using the following parameters:

$\sigma_v = 0.01$ and 0.05 ; $\text{length}(v) = 8000$

$c = \lambda_c Q_c$ with $Q_c = 2^{-7}$ and two values of λ_c such that the effective quantization stepsize becomes $Q_{ce} = 2^{-4}$(e.g., $c = 8/128$ and $72/128$)

Compare your results with those to be expected according to (2-10)–(2-12).

b. Now we want to measure the effective coefficient c', as defined by (2-14) and the power density spectrum $\hat{\Phi}_{\mathbf{ee}}(e^{j\omega})$ of the remaining noise. The measurement should be done with the method outlined in the appendix using $N = 512$ and $L = 50$ trials. The magnitude of the input signal has to be selected such that the desired values as in part (a) are obtained. The result $H(e^{j\omega})$ of the measurement corresponds to c'. (Why?) Furthermore, we get the desired power density spectrum. Comment on your results.

PROJECT 3: SENSITIVITY OF FILTER STRUCTURES

In this project we investigate the properties of different structures for the implementation of digital filters. We are especially interested in a comparison of their performances if they are used to implement systems, described by the same transfer function $H(z)$. Of course, they will have an identical input–output behavior if they work with unlimited wordlength, provided that the design was done properly. But the necessary quantization of coefficients as well as data causes deviations of different size for different structures.

To simplify the problem we investigate first the influence of quantizing the coefficients, disregarding the arithmetic errors. It is essential that such a system is still linear and thus can be described by its frequency response. Its deviation from the ideal frequency response, to be calculated with coefficients of unlimited wordlength, yields a measure of the sensitivity of different structures.

Project description

In Project 4 of Chapter 5 some different structures for the implementation of a certain transfer function $H(z)$ have been described in some detail. We are referring to the corresponding project description. The influence of limiting the wordlength of its coefficients is investigated here by calculating or measuring the frequency response after rounding the parameters to a specified wordlength. The following structures will be investigated:

- Direct form II (in its transposed form implemented in the MATLAB function `filter`). It will be used for the implementation of an FIR and of IIR systems.

- The cascade structure, consisting of blocks of first or second order.
- The parallel structure, again consisting of blocks of first and second order, which differ to some extent from those in the cascade form.
- The lattice structure for the implementation of an all-pole system.
- Coupled all-passes, each implemented using a lattice structure or a cascade of appropriate blocks of first and second order.

We add a few remarks about the last one of these structures by showing that it has a "built-in" insensitivity in the passband [4, Chap. 8]. That results out of its structural property,

$$|H(e^{j\omega})| \leq 1 \tag{3-1}$$

Suppose that the coefficients a_{ki} of the two all-passes are somewhat incorrect, due to the required limitation of their wordlengths. Thus we have $a_{ki} + \Delta a_{ki}$ instead of a_{ki}. Furthermore, we assume that the structure of the all-pass systems is such that their main property $|H_{Ai}(e^{j\omega})| = 1$ for all ω as well as the stability of the system is not affected by the erroneous coefficients, a condition that can be satisfied easily, for example, by using lattices for their implementation. Under these assumptions, equation (3-1) still holds. Now let ω_λ be a point where $|H_i(e^{j\omega_\lambda})| = 1$, with ideal coefficients. Due to (3-1), any change of any coefficients a_{ki} will result in a decrease of $|H_i(e^{j\omega_\lambda})|$, regardless of the sign of Δa_{ki}. Thus the slope of $|H_i(e^{j\omega})|$ as a function of any a_{ki} is precisely zero at $\omega = \omega_\lambda$. So the first-order sensitivity with respect to a_{ki}, defined as

$$S_{ki} = \frac{a_{ki}}{|H_i(e^{j\omega})|} \frac{\partial |H_i(e^{j\omega})|}{\partial a_{ki}} \tag{3-2}$$

is zero at all points ω_λ, where $|H_i(e^{j\omega})|$ is unity. Thus we can expect a small sensitivity of the system in the passband under rather mild conditions.

In the first two projects we dealt with the quantization of data, the magnitude of which was or was supposed to be smaller than 1. Here we cannot apply the same methods if the absolute value of the coefficients is larger than 1. Thus a normalization is required before the quantization step, such that the largest magnitude in a set of implemented coefficients is smaller than 1, as necessary in fixed-point arithmetic. To use completely the wordlength provided, a normalization should be done as well if the largest magnitude in a set of given coefficients is less than 0.5. Thus the normalized version a_{nk} of the elements of a vector $\mathbf{a}$ should satisfy the constraints

$$0.5 \leq \max |a_{nk}| < 1 \tag{3-3}$$

The required denormalization will be obtained by multiplying the output signal by an appropriately chosen factor. This final operation will be reduced to a simple shift if the scaling is done as

$$a_{nk} = \frac{a_k}{n} \tag{3-4}$$

where $n = 2^e$ is the smallest power of 2, being larger or equal to the maximum magnitude of the a_k:

$$e = \lceil \log_2(\max |a_k|) \rceil \tag{3-5}$$

The rounded version $[a_{nk}]_Q$ of the values a_{nk} will be used in our exercises as in an actual fixed-point implementation of the system. The following program yields $a_q = [a_{nk}]_Q$ as well as the normalization factor, called `nfa`.

```
function [aq,nfa] = coefround(a,w);

% This program quantizes a given vector a of coefficients
% by rounding to a desired wordlength w.
f = log(max(abs(a)))/log(2);          % Normalization of the vector a by
n = 2^ceil(f);                        % n, a power of 2, such that
an = a/n;                             % 1>=an>=-1.
aq = fxquant(an,w,'round','sat');     % Quantization of an such that
                                      % 1> aq >= -1;
nfa = n;                              % Normalization factor;
```

Hints

Starting with the coefficients of the various structures obtained during the design step, perform a quantization by rounding using the program `coefround`.

There are two ways to determine the required frequency responses of the resulting filters with quantized coefficients. If programs are available for the implementation of the different structures, the measuring method described in the appendix can be applied. For the first three of the structures listed above, that can be done easily using `filter`. For the implementation of the lattice, the programs `firlat` and `iirlat`[2] can be used.

The other possibility is based on the program `freqz` for the calculation of the frequency response of a system given in the direct form. It can be used as well for the cascade or parallel form, requiring in these cases appropriate combinations of the frequency responses of the subsystems. In the other cases, calculation of the resulting transfer function is required, which describes the lattice with its quantized coefficients. That can be done with `ktoa` [see Exercise 3.2(d)]. After that, its frequency response is obtained with `freqz`.

EXERCISE 3.1

FIR Halfband Filter

We are given the coefficients of an FIR halfband filter with linear phase of $M = 30$th degree.[3]

$$b = [\ -0.0021\quad 0\quad 0.0046\quad 0\quad -0.0094\quad 0\quad 0.0172\ 0\quad -0.0298\quad 0\quad 0.0515$$
$$0\quad -0.0984\quad 0\quad 0.3157\quad 0.5000\quad \ldots]\qquad b[32-\ell] = b[\ell],\ \ell = 1:31$$

The influence of coefficient quantization is to be investigated for two different structures.

a. *Direct structure*:

1. Calculate for later comparison the frequency response
$$H(e^{j\omega}) = \sum_{\ell=0}^{M} b_\ell e^{-j\ell\omega} \qquad M = 30 \tag{3-6}$$
for $\omega = \omega_k = k \cdot \pi/256$, $k = 0:255$ using `freqz(b,1,256)`. Calculate as well the real function
$$H_0(e^{j\omega}) = e^{j\omega M/2} H(e^{j\omega}) \tag{3-7}$$
2. Quantize the coefficients by rounding them to a wordlength of $w = 10$ bits. In this particular example it is $\max |b_\ell| = 0.5$. So the rounding can be done with
```
bq = Q * floor(b/Q+0.5)        (3-8)
```

[2]These functions were written as part of Exercise 4.3 in Chapter 5.
[3]Coefficients found with `remez(30,f,m)`, where `f = [0 0.4 0.6 1], m = [1 1 0 0]`.

Calculate the corresponding frequency response

$$\left[H(e^{j\omega})\right]_Q = \sum_{\ell=0}^{M}[b_\ell]_Q e^{-j\omega\ell} \tag{3-9}$$

as well as

$$\left[H_0(e^{j\omega})\right]_Q = e^{j\omega M/2}\left[H(e^{j\omega})\right]_Q \tag{3-10}$$

Check the influence of the quantization by plotting $|H(e^{j\omega})|$ and $\left|\left[H(e^{j\omega})\right]_Q\right|$ together, once for the passband ($k = 0:104$) and, using a dB scale, for the stopband ($k = 150:255$). Furthermore, calculate and plot the deviation

$$D_0(e^{j\omega}) = \left[H_0(e^{j\omega})\right]_Q - H_0(e^{j\omega}) \qquad k = 0:255 \tag{3-11}$$

3. Let $\Delta b_\ell = [b_\ell]_Q - b_\ell$ with $|\Delta b_\ell| \le Q/2$ be the quantization error of the coefficient $[b_\ell]_Q$. Show that

$$D_0(e^{j\omega}) = e^{j\omega M/2}\sum_{\ell=0}^{M}\Delta b_\ell e^{-j\omega\ell} \tag{3-12}$$

What is the upper limit of $|D_0(e^{j\omega})|$ expressed in terms of $\max|\Delta b_\ell|$? Verify your result using the example given.

4. Does the quantization of the coefficients destroy the linear-phase property? Give reasons for your answer.
5. The halfband property of the system can be described as follows: If $H_0(e^{j\omega})$ as defined by (3-7) is the real frequency response of the low-pass and

$$H_1(e^{j\omega}) = 1 - H_0(e^{j\omega}) \tag{3-13}$$

that of the corresponding high-pass, the relation

$$H_1(e^{j(\pi-\omega)}) = H_0(e^{j\omega}) \tag{3-14}$$

characterizes the halfband filter. Does the system with the quantized coefficients and the real low-pass frequency response $[H_0(e^{j\omega})]_Q$ still satisfy (3-13) and (3-14)?

b. *Cascade structure*: The FIR system can be implemented as well by a cascade of blocks of second degree. To find their coefficients we need the zeros of the transfer function $H(z)$, the numerator polynomial of which is given by the vector **b**. Perform the following steps:

- Calculating `z = roots (b)` yields two real values z_1, $z_2 = 1/z_1$ and $M/2 - 1$ pairs z_λ, z_λ^*, $\lambda = 2:M/2$.
- Pick z_1, z_2 as well as the pairs z_λ, z_λ^* and calculate the coefficients of the corresponding polynomials $\mathbf{b}_\lambda$ of second degree using `real(poly(.))`. That yields the preliminary representation

$$H_{cp}(z) = \prod_{\lambda=1}^{M/2} H_\lambda(z) \tag{3-15}$$

with

$$H_\lambda(z) = 1 + b_{\lambda 1}z^{-1} + b_{\lambda 2}z^{-2} \tag{3-16}$$

Note that in some cases the coefficients are larger than 1.

- With the scaling coefficient

$$\hat{b}_0 = \frac{H(1)}{H_{cp}(1)}$$

we get finally

$$H_c(z) = \hat{b}_0 H_{cp}(z) = H(z)$$

as required.

1. To check your design, calculate the frequency response as

$$H_c(e^{j\omega}) = \hat{b}_0 \prod_{\lambda=1}^{M/2} H_\lambda(e^{j\omega}) \tag{3-17}$$

using `freqz(b..,1,256)` for the subsystems and compare your result with that found in part (a).

2. Since the coefficients of the different subsystems have values with magnitudes ≥ 1, we have to do the rounding combined with a normalization as

```
[b...q,n...] = coefround(b...,w)
```

Determine the quantized values $[b_\ell]_Q$, $[\hat{b}_0]_Q$ and the normalization factors n_ℓ for $w = 10$ bits. Calculate the frequency response

$$\left[H_c(e^{j\omega})\right]_Q = \hat{b}_n \prod_{\lambda=1}^{M/2} \left[H_\lambda(e^{j\omega})\right]_Q \tag{3-18}$$

where $\hat{b}_n = [\hat{b}_0]_Q \prod_{\ell=1}^{M/2} n_\ell$ is the overall normalization factor. Does $\left[H_c(e^{j\omega})\right]_Q$ have precisely linear phase? If not, which changes have to be made in some of the transfer functions $H_\lambda(z)$ as given in (3-17) to yield linear phase? If this property is achieved, calculate the real frequency response

$$\left[H_{c0}(e^{j\omega})\right]_Q = e^{j\omega M/2} \left[H_c(e^{j\omega})\right]_Q \tag{3-19}$$

3. Check the influence of the quantization on the performance of the cascade structure by plotting $|H(e^{j\omega})|$ and $\left|\left[H_c(e^{j\omega})\right]_Q\right|$ in one diagram for the passband and in one for the stopband, as described in part (2). Furthermore, calculate and plot the difference

$$D_{c0}(e^{j\omega}) = \left[H_{c0}(e^{j\omega})\right]_Q - H_0(e^{j\omega}) \tag{3-20}$$

Compare your results with those obtained for the direct structure.

4. Check whether the quantization of the coefficients yields a system with the halfband property as described by (3-14) and (3-15). Look for an explanation if that is not the case precisely.

Remark. So far the design of the cascade system did not take into account other effects besides the quantization of coefficients. Scaling of the different blocks and ordering them in an appropriate sequence was not done. Both are required in practice to avoid an overflow and to minimize the noise.

EXERCISE 3.2

Elliptic Filter

The design of an elliptic filter of seventh degree according to the specifications

Passband: $0 \geq 20\log_{10}|H(e^{j\omega})| \geq -0.1$ dB for $0 \leq |\omega| \leq \omega_p = 0.4\pi$

Stopband: $20\log_{10}|H(e^{j\omega})| \leq -40$ dB

done with `[b,a] = ellip(7,.1,40,.4)` yields the coefficients

$$\mathbf{b} = [0.0463\ 0.0462\ 0.1251\ 0.1274\ 0.1274\ 0.1251\ 0.0462\ 0.0463]^T$$
$$\mathbf{a} = [1.0\ -2.2818\ 4.0692\ -4.2947\ 3.5055\ -1.8711\ 0.6857\ -0.1228]^T$$

It turns out that the stopband of the low-pass starts at $\omega = 0.429\pi$.

Four different structures for its implementation will be investigated concerning the influence of quantizing its coefficients. We have to realize that any quantization yields a deviation of the equiripple behavior of $|H(e^{j\omega})|$ in both bands for any structure. While working with fixed-point arithmetic and limited wordlength, we always have to accept changes of the tolerance scheme. So any implementation with wordlengths smaller than that of the computer during the design yields in this example the following characteristics:

Passband: $\delta_{p1} \geq 20\log_{10}\left|\left[H(e^{j\omega})\right]_Q\right| \geq \delta_{p2}$, where δ_{p1} might be positive and $\delta_{p2} < -0.1$ dB.

Stopband: $20\log_{10}\left|\left[H(e^{j\omega})\right]_Q\right| \leq \delta_S$, with $\delta_S > -40$ dB.

Furthermore, the cutoff frequencies of both bands will change.

In this experiment we determine the values δ_{p1}, δ_{p2}, and δ_S for four structures and different wordlength w. Of higher practical importance is the determination of the minimum required wordlength such that, for example, $|\delta_{p1} - \delta_{p2}| \leq 0.2$ dB *and* $\delta_S \leq -38$ dB.

a. *Direct structure*: First calculate `H = freqz(b,a,512)` as a reference for later comparisons.

   1. The system to be implemented has at least to be stable. Determine the required wordlength w_0 of the quantized coefficients $[a]_Q$ for stability by calculating the magnitudes of the roots of the different polynomials defined by these coefficients. Do the quantization with `aq = coefround(a,w)`.
      Using this w_0, calculate

```
[bq, nb] = coefround(b, w0)
[aq, na] = coefround(a, w0)
```
(3-21)

      and

```
Hq = nb/na*freqz(bq,aq,512)
```
(3-22)

      Plot $|H(e^{j\omega})|_Q$ and determine the characteristic values δ_{p1}, δ_{p2}, and δ_S of the resulting tolerance scheme.

   2. Calculate the rounded versions of **b** and **a** and the normalization factors `nb` and `na` for $w = 10$, 12, 13, and 14 bits using `coefround(..,..)` again. Calculate the corresponding frequency responses according to (3-22).

   Plot the frequency responses `20*log10(abs(Hq(1:208)))` of the passbands for these four cases and determine experimentally the values δ_{p1} and δ_{p2}. Check separately the performances in the stopband by plotting `20*log10(abs(Hq(215:512)))`. Which values $\delta_S = \max\left[20\log_{10}\left|\left[H(e^{j\omega})\right]_Q\right|\right]$, $\omega \geq 0.429\pi$ do you find? Use `subplot(22.)`. What is the minimum required wordlength if $|\delta_{p1} - \delta_{p2}| \leq 0.2$ dB *and* $\delta_S \leq -38$ dB are prescribed?

b. *Cascade structure*:

   1. Design a cascade implementation of the given filter using the procedure described in Exercise 4.1 of Chapter 5:

- Calculate the poles and zeros of the transfer function as `p = roots(a)` and `z = roots(b)`.
- Pick the pole p_k with the largest magnitude and the nearest zero z_ℓ. Choose them and their complex conjugate values for the first block of second order in the sequence. Calculate its denominator and numerator polynomials using `real(poly(..))`.
- Proceed similarly for the next block with the remaining poles and zeros. Since the given example is a system of seventh degree, the last block will be of first order.
- Calculate the common factor $\hat{b}_0$ as described in Exercise 3.1 for the cascade implementation of an FIR filter.

We end up with the representation

$$H_c(z) = \hat{b}_0 \prod_{\lambda=1}^{L} H_\lambda(z) \tag{3-23}$$

where either

$$H_\lambda(z) = \frac{1 + b_{\lambda 1} z^{-1} + b_{\lambda 2} z^{-2}}{1 + a_{\lambda 1} z^{-1} + a_{\lambda 2} z^{-2}}$$

or

$$H_\lambda(z) = \frac{1 + b_{\lambda 1} z^{-1}}{1 + a_{\lambda 1} z^{-1}} \tag{3-24}$$

(*Note*: In our particular example we get $b_{\lambda 2} = 1$ and for the subsystem of first order $b_{\lambda 1} = 1$. Explain why that is the case.)

To check your design, calculate the frequency response

$$H_c(e^{j\omega}) = \hat{b}_0 \prod_{\lambda=1}^{L} H_\lambda(e^{j\omega}) \tag{3-25}$$

using `freqz(b..,a..,512)` for the subsystems and compare the result with $H(e^{j\omega})$ found in part (a) for the direct structure.

2. Recall the stability conditions for a polynomial in z of second order, described by $A(z) = 1 + a_1 z^{-1} + a_2 z^{-2}$. Calculate and plot the zeros of all polynomials with the coefficients $a_{1\ell} = \ell_1 Q$ and $a_{2\ell} = \ell_2 Q$ for $Q = 2^{-4}$ while choosing the integers ℓ_1 and ℓ_2 such that these zeros are inside the unit circle.
   Referring to the example, use the subsystem that has the largest coefficient $a_{\lambda 2}$ to test whether rounding can yield an unstable system. If that is the case, determine the required wordlength for stability. The zeros of the transfer function are located on the unit circle. Can this property get lost by rounding the coefficients $b_{k\lambda}$ of the numerator polynomials of the various subsystems?
3. Quantize the coefficients $\mathbf{b}_\lambda$ and $\mathbf{a}_\lambda$ of the subsystems by rounding to wordlengths $w = 6, 8, 10$, and 12 according to (3-21). Calculate the corresponding frequency responses $[H(e^{j\omega})]_Q$ using (3-25) with the $[H_\lambda(e^{j\omega})]_Q$ of the subsystems, to be calculated as in (3-22). Compare their magnitudes in the passband and stopband separately, as outlined in part (a) for the direct structure. Determine the values δ_{p1}, δ_{p2}, and δ_S as well as the minimum required wordlength under the same constraints as in part (a). (*Hint*: To simplify the work, write a program for the calculation of the $[H(e^{j\omega})]_Q$ for the different wordlengths.)
4. Notice that the cascade structure is rather insensitive in the stopband. Can you give reasons for this observation?

c. *Parallel structure*:

1. Design an implementation of the filter according to the representation

$$H_p(z) = B_0 + \sum_{\lambda=1}^{L} H_\lambda(z) \tag{3-26}$$

with

$$H_\lambda^{(1)}(z) = \frac{b_{0\lambda}}{1 + a_{\lambda 1} z^{-1}} \qquad H_\lambda^{(2)}(z) = \frac{b_{0\lambda} + b_{1\lambda} z^{-1}}{1 + a_{\lambda 1} z^{-1} + a_{\lambda 2} z^{-2}} \tag{3-27}$$

as a parallel connection of corresponding blocks of first and second order with real coefficients $b_{\ldots\lambda}$ and $a_{\lambda\ldots}$. [*Hint*: To get subsystems with transfer functions as given by (3-27), do a partial fraction expansion of $H(z)/z$. That can be done using `residue(b,am)`, where the coefficients of the modified denominator polynomial a_m are `am = [a 0]`. Proceeding this way, you get, for example, B_0 as the residue of the added pole at $z = 0$. While the coefficients $a_{\lambda 1}$, $a_{\lambda 2}$ are the same as in the cascade structure, the $b_{0\lambda}$, $b_{1\lambda}$ are to be calculated as `2*real(r(i)*poly(p(i+1)))`. Here p_i and $p_{i+1} = p_i^*$ are a pair of poles, r_i and $r_{i+1} = r_i^*$ the corresponding residues. Use `help residue` for more information]. Check your design by calculating the frequency response

$$H_p(e^{j\omega}) = B_0 + \sum_{\lambda=1}^{L} H_\lambda(e^{j\omega})$$

and comparing it with $H(e^{j\omega})$ found in part (a).

2. Quantize the coefficients of the transfer functions $H_\lambda(e^{j\omega})$ as you did in the other cases, now with $w = 8$, 9, 10, and 12 bits. Calculate the resulting frequency responses $[H_p(e^{j\omega})]_Q$ and compare them in the passband and stopband separately as you did before. Again determine the limits δ_{p1}, δ_{p2}, and δ_S as well as the minimum required wordlength under the same constraints as above.

3. If you compare the performance of the parallel structure in the stopband with that of the cascade structure (e.g., for $w = 8$ bits), you notice a fundamental difference. Give reasons for this observation.

d. *Coupled all-passes in lattice structure*

1. Design an implementation of the low-pass by determining two all-pass transfer functions $H_{A1}(z)$ and $H_{A2}(z)$ such that

$$H_1(z) = 0.5[H_{A1}(z) + H_{A2}(z)] = H(z) \tag{3-28}$$

where $H(z)$ is again the transfer function of our low-pass example. For more information see Project 4 in Chapter 5, especially Exercise 4.4. The design can be done with the following steps:

- Calculate the poles of the transfer function $H_1(z) = H(z)$ with `p = roots(a)`.
- Determine its phase using `ph = angle(p)`, and number the poles according to these phases, starting with its minimum value.
- Pick all the poles with odd number as poles p_1 of $H_{A1}(z)$ and those with even number as poles p_2 of $H_{A2}(z)$. Calculate the corresponding sets a_1 and a_2 of coefficients of the denominator polynomials as `a = real(poly(p...))`.
- Calculate the reflection coefficients k_1 and k_2 for lattice implementations of the all-passes as `k... = atok(a...)`.

As pointed out in the general description of this project, there are essentially two ways to calculate the frequency response of the resulting structure and thus to check whether the design was done properly:

- Implement the all-passes by applying the program `iirlat(k...)`[4] with its all-pass option. Determine the frequency responses $H_{A1}(e^{j\omega})$ and $H_{A2}(e^{j\omega})$ of the all-passes using one of the methods described in Project 2 of Chapter 5, Exercises 2.2 and 2.3. Then calculate

$$H_{ca}(e^{j\omega}) = 0.5[H_{A1}(e^{j\omega}) + H_{A2}(e^{j\omega})] \tag{3-29}$$

and compare it with $H(e^{j\omega})$ found in part (a).

[4]This function was written as part of Exercise 4.3 in Chapter 5.

- We can calculate the transfer function representations of the all-passes by transforming the reflection coefficients k_i back into a_k, $i = 1, 2$ of the corresponding polynomials with `a... = ktoa(k...)`. The numerator coefficients are found as `b... = fliplr(a...)`. Then the frequency responses of the all-passes are calculated with `freqz(b...,a...,512)` and used for getting $H_{ca}(e^{j\omega})$, as shown above.

The following program converts the reflection coefficients k into those of the polynomial:

```
function a = ktoa(k);
% a = ktoa(k) converts reflection coefficients,
% given in vector k, into the corresponding
% polynomial a.

N = length(k);
k = k(:);                  % converts k into column vector
a = [1;zeros(N,1)];        % leading coefficient is 1
for i=1:N;
  a(1:i+1) = a(1:i+1) + k(i)*flipud(a(1:i+1));
end;
```

Check your design by calculating the frequency response using one of these methods and compare it with $H(e^{j\omega})$ found in part (a) for the direct structure.

2. Quantize the coefficients k_i of the two all-passes the same way as before, this time with $w = 6, 7, 8$, and 10 bits. Determine 512 values of the resulting frequency responses $[H_{Ai}(e^{j\omega})]_Q$ using one of the ways described above. Calculate $[H_{ca}(e^{j\omega})]_Q$ according to (3-29) and plot its magnitudes for the different wordlengths on a logarithmic scale for passband and stopband separately, as you did before. Which value δ_{p1} do you find in this case independent of w? Determine δ_{p2} and δ_S again in these cases. What is the required wordlength now for satisfying the constraints given above?

EXERCISE 3.3

Minimum-Phase FIR and All-Pole Filter

We are given the coefficients of an invertible minimum-phase FIR system having the transfer function $H_1(z)$.[5] They are, as well, the coefficients of the all-pole system, having the transfer function $H_2(z) = 1/H_1(z)$:

$$\mathbf{b} = [1.000 \quad 0.7542 \quad 0.2217 \quad -0.2037 \quad -0.1171 \quad 0.1029 \quad 0.0680 \quad -0.0602$$
$$-0.0393 \quad 0.0361 \quad 0.0215 \quad -0.0209 \quad -0.0106 \quad 0.0111 \quad 0.0045 \quad -0.0059]$$

For both cases the direct structure and the lattice structure are to be compared in terms of coefficient sensitivity.

a. *Direct structure:*

1. Calculate for later comparison the frequency response

$$H_1(e^{j\omega}) = \sum_{\ell=0}^{15} b_\ell e^{-j\ell\omega} \tag{3-30}$$

and

$$H_2(e^{j\omega}) = 1/H_1(e^{j\omega}) \qquad \text{for } \omega = \omega_k = k \cdot \pi/256,\ k = 0:255$$

[5]The system is related to the FIR halfband filter of Exercise 3.1 as follows: Its coefficient $b(15)$ was changed to 0.6. The 15 zeros of the resulting polynomial located inside the unit circle are used for $H_1(z)$.

2. Round the coefficients to $w = 8$ and 10 bits as `bq = coefround(b,w)` and calculate the corresponding frequency responses $[H_{1d}(e^{j\omega})]_Q$ and $[H_{2d}(e^{j\omega})]_Q = 1/[H_{1d}(e^{j\omega})]_Q$ as well as

$$D_{...d}(e^{j\omega}) = [H_{...d}(e^{j\omega})]_Q - H_{...}(e^{j\omega})$$

Plot the magnitudes of these differences using `subplot(22...)`.

b. *Lattice structure*: Calculate the reflection coefficients k as `k = atok(b)`. Quantize k to $w = 8$ and 10 bits as `kq = coefround(k,w)`. Calculate the coefficients of the corresponding direct form as `blq = ktoa(kq)` and the resulting frequency responses $[H_{1\ell}(e^{j\omega})]_Q$ and $[H_{2\ell}(e^{j\omega})]_Q = 1/[H_{1\ell}(e^{j\omega})]_Q$. Finally, calculate again the differences

$$D_{...\ell}(e^{j\omega}) = [H_{...\ell}(e^{j\omega})]_Q - H_{...}(e^{j\omega})$$

Plot the magnitudes of these $D_{...\ell}(e^{j\omega})$ and compare your results with those obtained in part (a).

Project 4: Limit Cycles

As has been mentioned in the overview, an implemented digital system is no longer linear but is nonlinear, for two reasons:

1. The range of all numbers to be represented in a digital system is limited. If the summation of two or more numbers yields a result beyond this limit, an overflow and thus an error occurs, the value of which depends on the nonlinear characteristic, chosen in the particular implementation, and on the actual signal. This might result in instability of the total system, yielding an oscillation of rather large amplitude despite the fact that no further excitation is applied. It is called a *large-scale* or *overflow limit cycle*.
2. The second nonlinear effect is caused by the required limitation of the wordlength after a multiplication. It has been considered in Project 2. Here we are not interested in a model of a real multiplier, introduced there by using a random error source, but in a possible instability caused by this nonlinearity, yielding an oscillating behavior of the system. Since the resulting limit cycles usually (but not necessarily) have a small amplitude and since their values are expressed as multiples of the quantization step Q, they are called *small-scale* or *granular limit cycles*.

The possibility and the size of limit cycles depend on:

- The type of the nonlinearity, i.e., the overflow characteristic on one hand and the quantization method (rounding or truncation) on the other.
- The location of the poles of the transfer function of the linear system to be implemented; in addition to their influence on the size, they determine the period of the limit cycle.
- The structure of the system.
- The actual excitation. Starting with an initial state $\mathbf{x}(0) \neq \mathbf{0}$, a limit cycle can occur with a zero input signal; it can appear as well with constant or periodic input signals.

A complete investigation of this rather complicated problem is not possible here. The following exercises are restricted to two points:

1. The existence of both types of limit cycles and its dependence on the conditions mentioned above are demonstrated with examples for the special case of systems of first or second order with zero input.
2. Methods of avoiding limit cycles are discussed and demonstrated.

Project description

First we consider large-scale, zero-input limit cycles in a system, described in the state space partly by $\mathbf{A} = \begin{bmatrix} -a_1 & 1 \\ -a_2 & 0 \end{bmatrix}$ (see [3, Chap. 9]). We assume again fixed-point arithmetic and 2's-complement representation of numbers. If the required quantization of a sum $s[n]$ according to

$$-1 \leq f(s[n]) < 1 \tag{4-1}$$

is done using the 2's-complement characteristic directly (see Fig. 7.4a), large-scale limit cycles have to be expected if the coefficients a_1 and a_2 do not satisfy the very restrictive condition

$$|a_1| + |a_2| < 1 \tag{4-2}$$

But for any pair of coefficients belonging to a linearly stable system, an implementation with a characteristic inside the hatched region of Fig. 7.4b will avoid overflow oscillations. A saturation characteristic or a triangular-shaped one can be used for this purpose (see Fig. 7.4c and d).

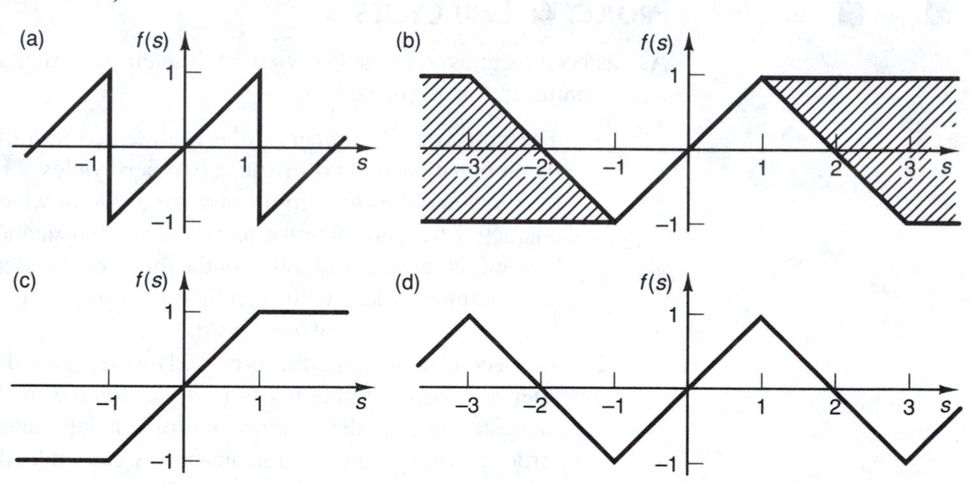

Figure 7.4 Overflow characteristics: (a) 2's complement; (b) constraints for guaranteed stability; (c) saturation; (d) triangular.

Furthermore, no zero-input overflow limit cycles can occur in a normal system (see Project 4 in Chapter 5). Thus if

$$\mathbf{A} = \mathbf{A}_n = \begin{bmatrix} \text{Re}\{p\} & \text{Im}\{p\} \\ -\text{Im}\{p\} & \text{Re}\{p\} \end{bmatrix} \tag{4-3}$$

where p with $|p| < 1$ is the pole of the transfer function of the stable system, there will be no large-scale limit cycles with any overflow characteristic.

Granular limit cycles depend on the location of the poles, the type of quantization (rounding, 2's-complement truncation, or magnitude truncation), and the structure. For a rough description they can be characterized by their period and their extremal value, expressed as

$$\max |y[n]| = LQ \tag{4-4}$$

where Q is the quantization step.

In a first-order system, limit cycles can be observed if rounding is used. It turns out that

$$L \leq \left\lfloor \frac{0.5}{1 - |a_1|} \right\rfloor = \text{floor}\left(\frac{0.5}{1 - |a_1|}\right) \tag{4-5}$$

Thus limit cycles are to be expected if

$$0.5 \leq |a_1| < 1 \tag{4-6}$$

Its period will be 1 if $a_1 < 0$ (deadband effect), and 2 if $a_1 > 0$. There will be no limit cycles if magnitude truncation is applied for the quantization.

For a second-order system the problem is more complicated. Here a simple model yields a rough but in most cases useful description (e.g., [2, Chap. 5]): A limit cycle produced by a nonlinear system can be regarded as output of a linear oscillator, to be described by the *effective coefficients* $\hat{a}_2 = 1$ and $\hat{a}_1 \approx a_1$. That results in an estimate for the extremal value similar to (4-4):

$$L \le \left\lfloor \frac{0.5}{1 - |a_2|} \right\rfloor \tag{4-7}$$

yielding values $L > 0$ for $0.5 \le a_2 < 1$. In this model the estimated frequency results out of $1 + a_1 z^{-1} + z^{-2} = 1 - 2\cos\omega_{0e} z^{-1} + z^{-2}$ as

$$\omega_{0e} = \cos^{-1}(-a_1/2) \tag{4-8}$$

We mention that limit cycles with period 1 can be observed as well. They are characterized by an unstable equilibrium point $\mathbf{x}[\cdot] \neq \mathbf{0}$.

Limit cycles can be suppressed if in the linear case the stored energy w_x cannot increase, that is, if

$$\Delta w_x[n+1] = \mathbf{x}^T[n+1]\mathbf{x}[n+1] - \mathbf{x}^T[n]\mathbf{x}[n] \le 0 \qquad \forall n \tag{4-9}$$

and if the quantization yields

$$\left[\mathbf{x}^T[n]\right]_Q [\mathbf{x}[n]]_Q \le \mathbf{x}^T[n]\mathbf{x}[n] \tag{4-10}$$

The first condition is satisfied if a normal system is used; the second requires magnitude truncation for the quantization of the products.

Two programs are provided, to be used during the following exercises. The program `filtqz` simulates a system with one input and one output, applying `fxquant` for the necessary limitation and the quantization. Since we are interested in the state vector $\mathbf{x}[n]$, a state-space description of the system is applied. Figure 7.5 shows the structure and the location of the quantizers. Note that only some of the numerous possibilities for the implementation of a block of second order can be simulated with this configuration by specifying the coefficients. The program can be used for the investigation of both types of limit cycles.

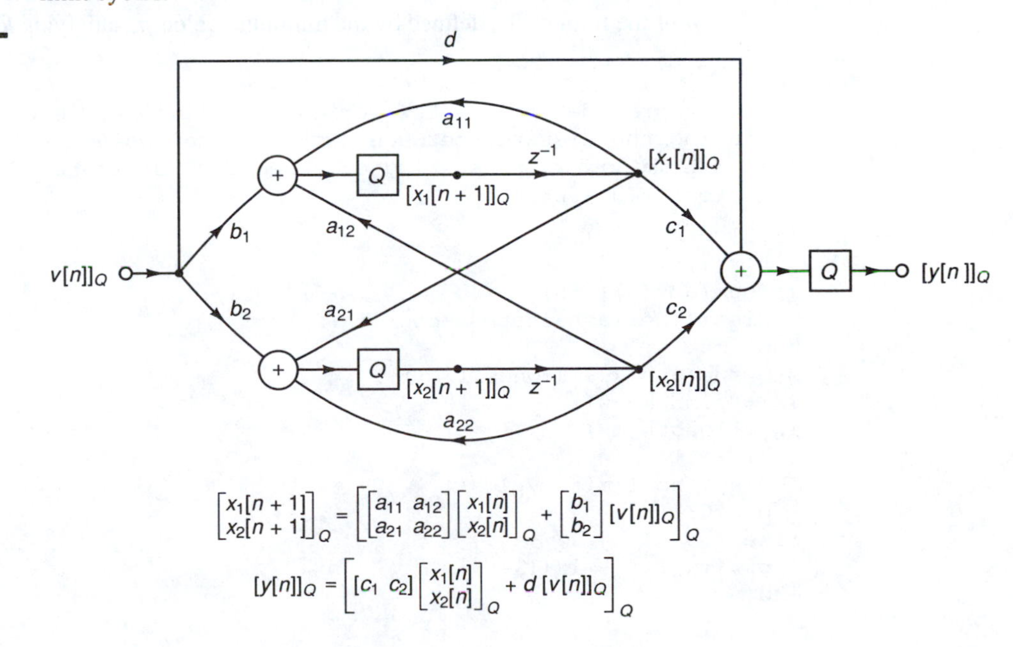

$$\begin{bmatrix} x_1[n+1] \\ x_2[n+1] \end{bmatrix}_Q = \left[\begin{bmatrix} a_{11} & a_{12} \\ a_{21} & a_{22} \end{bmatrix}\begin{bmatrix} x_1[n] \\ x_2[n] \end{bmatrix}_Q + \begin{bmatrix} b_1 \\ b_2 \end{bmatrix}[v[n]]_Q\right]_Q$$

$$[y[n]]_Q = \left[[c_1 \; c_2]\begin{bmatrix} x_1[n] \\ x_2[n] \end{bmatrix}_Q + d\,[v[n]]_Q\right]_Q$$

Figure 7.5
Structure of a block of second order, implemented with `filtqz`.

```
function [Y,X]=filtqz(A,B,C,D,v,bit,rmode,lmode,x0)
% [Y,X]=filtqz(A,B,C,D,v,bit,rmode,lmode,x0)
% recursive filter with one input and one output
% simulation of fixed point wordlength and quantization effects
%
% x(n+1) = [ A*x(n)  + B*v(n)]q         with    [a]q= a quantized
% y(n)   = [ C*x(n)  + D*v(n)]q
%
% A,B,C,D : system matrices with already quantized coefficients
% v       : input vector
% bit     : number of bits for wordlength of coefficients and state variables
% rmode   : rounding mode,  ('round','trunc','magn') see fxquant.m
% lmode   : limiting mode,  ('sat','overfl','none') see fxquant.m
% x0      : initial state,  zero if missing
% Y       : output vector
% X       : state vector, may be missing;

v=fxquant(v,bit,rmode,lmode);   % quantization of input vector
if nargin==9
  x=fxquant(x0,bit,rmode,lmode);
elseif nargin==8
  x=zeros(size(B));                         % start values for state variables = 0
else
  error('wrong number of input arguments')
end
X=zeros(length(x),length(v)+1);
Y=zeros(size(v));                              % define y in full length
for n=1:length(v)               % do filter loop
  Y(n)=C*x+D*v(n);
  X(:,n)=x;
  x=fxquant(A*x+B*v(n),bit,rmode,lmode);          % state quantization
end
X(:,length(v)+1)=x;
Y=fxquant(Y,bit,rmode,lmode);   % output quantization
```

Furthermore, a program `licyp(x)` is provided for the determination of the period p of the limit cycle, defined by the minimum value p, satisfying $\mathbf{x}[n+p] = \mathbf{x}[n]$.

```
function p = licyp(x)

% The program determines the period of a limit cycle by
% finding the distance between two identical columns of
% an NxL matrix x, consisting of a sequence of state
% vectors x[n], n=1:L.

[N,L] = size(x);
l = x(:,L);
if(sum(abs(l))==0)
  disp('The last state vector is zero')
end
s = sum(abs(x - l*ones(1,L)));
ix = find(s==0);
lx = length(ix);
if(lx==1)
  disp('no limit cycles')
  p = 0;
else
  p = ix(lx) - ix(lx-1);
end
```

EXERCISE 4.1

Large-Scale Limit Cycles

a. First we investigate the overflow characteristics, implemented with `fxquant`. Use an input sequence $s = -3 : 0.01 : 3$ and determine with $w = 4$ bits the limited and quantized versions s_q for 2's-complement-, saturation-, and triangular-overflow characteristics. See `help fxquant` for further information. Plot the three functions with `plot(s,sq)` using `subplot(22.)`.

b. 1. The condition (4-2) for the absence of large-scale limit cycles yields a restriction for the possible pole p_k of the corresponding transfer function. Determine the section inside the unit circle for the p_k such that (4-2) is satisfied.

2. You are given

$$\mathbf{A}_1 = \begin{bmatrix} 0.6 & 1 \\ 0.3 & 0 \end{bmatrix} \quad \mathbf{B} = \begin{bmatrix} 0 \\ 0 \end{bmatrix} \quad \mathbf{C} = [1\ 0] \quad \mathbf{D} = 0$$

and the initial condition $\mathbf{x}(0) = \begin{bmatrix} 0.8 \\ 0.4 \end{bmatrix}$

Calculate with `filtqz`, `v=zeros(1,100)`, `'round'`, and $w = 16$ bits the state vector $\mathbf{x}[n]$ and the output sequence $y[n]$. Use any overflow characteristic. Plot the state vector $\mathbf{x}[n]$ with `plot(x(1,:), x(2,:))`, `axis('square')`, and `axis([-1 1 -1 1])` and check if a large-scale limit cycle results.

c. Now you are given

$$\mathbf{A}_2 = \begin{bmatrix} -1.2 & 1 \\ -0.95 & 0 \end{bmatrix} \quad \mathbf{B},\ \mathbf{C},\ \text{and } \mathbf{D} \text{ as before and } \mathbf{x}(0) = \begin{bmatrix} 0.9 \\ 0 \end{bmatrix}$$

1. Check the stability of the system, assuming linearity.
2. Calculate with `v=zeros(1,100)`, `'round'`, and $w = 6$ bits the state vectors for the three overflow characteristics. Furthermore calculate the state vector of the corresponding normal system, described by $\mathbf{A}_{2n}$ as defined by (4-3), using the 2's complement characteristic. Plot the resulting four state vectors in one diagram using `subplot(22.)`, `axis('square')` and `axis([-1 1 -1 1])`. In which case do you get a large-scale limit cycle?
3. In all four cases use `licyp(x...)` to determine the period of a limit cycle if one exists. Can you explain the result? Save the state vectors you obtained with the saturation and triangular characteristic and with the normal system for further investigations.
4. Considering again the system described by $\mathbf{A}_2$, $\mathbf{B}$, $\mathbf{C}$, and $\mathbf{D}$, calculate the state vectors using rounding, 2's-complement overflow characteristic, the same initial condition, and `v=zeros(1,150)`, but now with the further wordlengths $w = 7, 8,$ and 9 bits. Using, in addition, the corresponding result you got before with $w = 6$ bits, plot the state vectors again together in one diagram. Comment on the differences. Determine the periods $p_{...}$ of the resulting limit cycles with `licyp(.)`. Calculate the spectrum $Y_{...}[k]$ of one period of each output sequence $y_{...}[n]$. Plot the magnitudes $|Y_{...}[k]|$ with $k = 0 : (p_{...} - 1)$ using `stem(k,abs(Y...(...)))`. Calculate and compare the actual frequencies of the main spectral lines as

$$\omega_0 = k_{0...} \cdot \frac{2\pi}{p_{...}}$$

where $\max |Y_{...}[k]| = |Y_{...}[k_{0...}]|$, while $p_{...}$ is again the period in the particular case.

EXERCISE 4.2

Granular Limit Cycles

a. *Deadband effect*: We start with an investigation of a block of first order using rounding and for overflow control, the saturation characteristic. You are given

$$A = (1 - 2^{-6}) \qquad B = 0 \qquad C = 1 \qquad D = 0$$

It is $x(0) = 1$. Calculate $y[n]$, $n = 1 : 150$ with the wordlengths $w = 6, 7, 8$, and 9 bits and for comparison with $w = 24$ bits. Plot your results together in one diagram and check them with regard to relation (4-5). Furthermore, use $B = 2^{-6}$, A, C, D as before, $x(0) = 0$, and `v = ones(1,150)`. Calculate $y[n]$ for the same wordlengths as above. Again plot your results together in one diagram and comment on them with regard to (4-5). Now use $A = -0.6$, $B = 0$, $C = 1$, $w = 9$, and $x(0) = 10Q$, where $Q = 2^{(1-w)}$. Calculate $y[n]$ for $n = 1 : 25$. Plot $y[n]/Q$. Do you get a limit cycle? What is its period? How about the relation to (4-5)? For comparison calculate $y[n]$ for one of the examples above, using magnitude truncation (`'magn'`) and the wordlength $w = 9$ bits. Comment on your results.

b. *Limit cycles in second-order systems due to rounding*: In the following exercises we want to determine the characteristic values of the limit cycles, if there are any:

- The period p with `licyp(.)`
- $L = \max |y[n]/Q|$ in one period
- The frequency ω_0 of the main spectral component by finding $|Y[k_0]| = \max |Y[k]|$ with `Y = fft(y,p)` and calculating $\omega_0 = k_0 \cdot 2\pi/p$

Furthermore, we want to visualize the limit cycle by plotting the last $p + 1$ values of the normalized version $\mathbf{x}_r[n]/Q$:

```
l = length(xr); xrn = x(:,l-p:l)/Q;
plot(xrn(1,:), xrn(2,:), '*')
```

1. We consider again the system given in Exercise 4.1(c). Calculate $y_r[n]$ and $\mathbf{x}_r[n]$ for $n = 1 : 75$ using rounding, any overflow characteristic, $w = 16$ bits, and $\mathbf{x}(0) = Q[8\ 8]^T$. Perform the steps outlined above. Check if your results for L and ω_0 satisfy (4-7) and (4-8) at least approximately.
2. Try to eliminate the limit cycle by using 2's-complement or magnitude truncation instead of rounding.
3. Repeat the experiment now with the initial condition $\mathbf{x}(0) = Q[10 \quad 5]^T$.
4. Investigate the properties of the small-scale limit cycles you found in Exercise 4.1 while dealing with overflow limit cycles.

c. *Equilibrium points* $\mathbf{x}(\cdot) \neq \mathbf{0}$: We are given a system with

$$\mathbf{A}_3 = \begin{bmatrix} 1.9 & 1 \\ -0.9525 & 0 \end{bmatrix} \qquad \mathbf{B},\ \mathbf{C},\ \text{and } \mathbf{D} \text{ as before}$$

1. Use the following initial conditions and check if limit cycles will result.

$$\mathbf{x}_1(0) = Q[1 \quad -1]^T \qquad \mathbf{x}_2(0) = Q[1 \quad 1]^T$$
$$\mathbf{x}_3(0) = Q\lambda[-1 \quad 1]^T, \quad \lambda = 1:3 \qquad \mathbf{x}_4(0) = Q[3 \quad 0]^T$$

Calculate $\mathbf{x}[n]$ for $n = 1 : 25$ with $w = 16$ bits. Which one of the initial values given is an equilibrium point?

2. Now use magnitude truncation `'magn'` with the same system and the initial conditions

$$\mathbf{x}_1(0) = Q[1 \quad 1]^T \qquad \mathbf{x}_{2\lambda}(0) = Q[\lambda \quad (1-\lambda)]^T, \quad \lambda = 1:5$$
$$\mathbf{x}_3(0) = Q[-1 \quad -1]^T$$

Does magnitude truncation always eliminate the limit cycle or the equilibrium points in a system, implemented with the second direct form?

d. *Eliminating limit cycles*:

1. We now investigate a system described by

$$\mathbf{A}_4 = \begin{bmatrix} 0 & 1.0 \\ -0.9985 & 1.94 \end{bmatrix} \qquad \mathbf{B}, \mathbf{C}, \text{ and } \mathbf{D} \text{ as before}$$

Calculate $[y, \mathbf{x}]$ for the following cases:

$$\mathbf{x}(0) = Q[-20 \quad 28]^T \qquad \text{rounding}$$
$$\mathbf{x}(0) = Q[-20 \quad 28]^T \qquad \text{2's-complement truncation}$$
$$\mathbf{x}(0) = Q[3 \quad 19]^T \qquad \text{magnitude truncation}$$

Check if limit cycles will occur, and if so, determine their parameters.

2. Summarizing the results you found so far: Did you find a quantization method that eliminated limit cycles in all systems implemented in the first or second direct form?

3. Now we transform the system described by $\mathbf{A}_4$ into the normal one, yielding

$$\mathbf{A}_{4n} = \begin{bmatrix} 0.97 & 0.24 \\ -0.24 & 0.97 \end{bmatrix}$$

Calculate $y[n]$ and $\mathbf{x}[n]$ with rounding, 2's-complement truncation, and magnitude truncation. Check if there are limit cycles by plotting the normalized versions $\mathbf{x}[n]/Q$. Determine and plot in all three cases

$$\Delta w_x[n+1] = \left[\mathbf{x}^T[n+1]\right]_Q [\mathbf{x}[n+1]]_Q - \left[\mathbf{x}^T[n]\right]_Q [\mathbf{x}[n]]_Q$$

Comment on your results with regard to relation (4-9).

Project 5: Quantization Noise in Digital Filters

In addition to the danger of instability, the nonlinearity of an implemented digital system always yields arithmetic errors of random nature. Thus the total performance of a filter has to be described by the signal/noise ratio observed at the output. In general we are interested in a more complete picture of the real system, including the power density spectrum of the error as well as the actual frequency response, which is influenced by the quantization of the coefficients, considered in Project 3. We obtain it with a model consisting of a parallel combination of a linear subsystem, characterized by the transfer function $[H(e^{j\omega})]_Q$ and another system producing the noise error sequence $e[n]$, to be described by its power density spectrum $\Phi_{\mathbf{ee}}(e^{j\omega})$. So if overflow oscillations are avoided by scaling and by using an appropriate characteristic, and if granular limit cycles are either eliminated or small enough to be ignored, a system implemented with a limited wordlength can be described as shown in Fig. 7.6 as a parallel connection of a linear system with a frequency response $[H(e^{j\omega})]_Q$ as in Project 4, and another one, the output sequence $e[n]$, which represents the noise-like error. The separation has to be done such that the two sequences $y_L[n]$ and $e[n]$ are orthogonal to each other. In the appendix, a method is presented for measuring estimated samples of $[H(e^{j\omega})]_Q$ and $\Phi_{\mathbf{ee}}(e^{j\omega})$ at equidistant points ω_k.

The power density spectrum of the noise, considered in this project, depends on the following:

- To a large extent on the filter to be built
- On the structure, used for the implementation
- On the type of quantization after a multiplication
- On the wordlength used

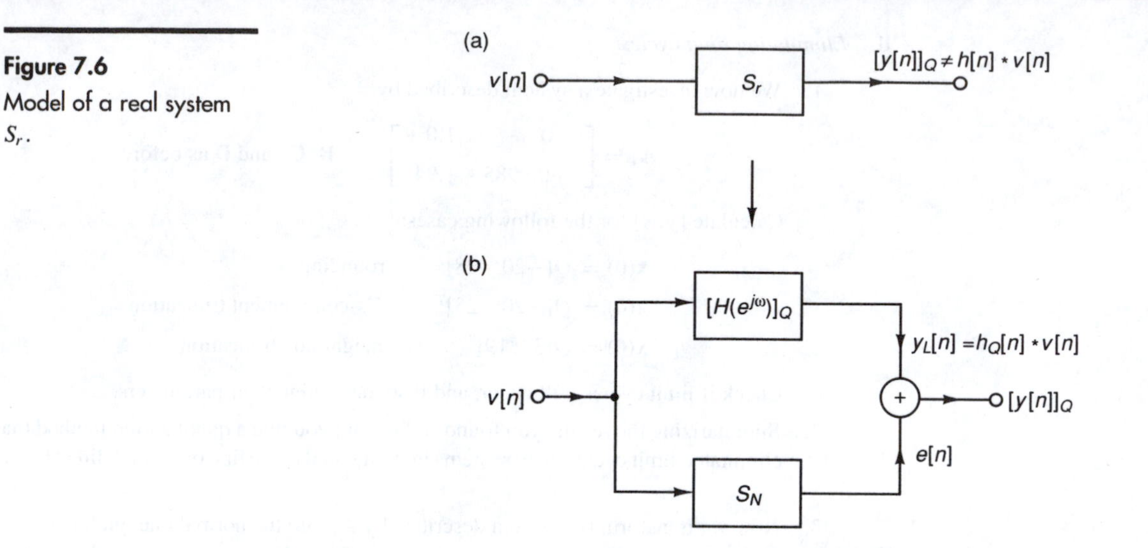

Figure 7.6
Model of a real system S_r.

In a particular case, i.e., for a given transfer function and after decisions have been made concerning the structure, the quantization method, and the wordlength available (i.e., besides others, a decision concerning a particular integrated signal processor to be used), there are two ways to determine the resulting power density spectrum $\Phi_{\mathbf{ee}}(e^{j\omega})$ of the error. They are exemplified in this project:

1. The calculation of $\Phi_{\mathbf{ee}}(e^{j\omega})$, based on models for all quantization points as well as the total system.
2. Its measurement in a simulated implementation, using, for example, the method described in the appendix.

Due to the large number of choices, a complete investigation is beyond the scope of this book. Also, measurement of the power density spectrum is rather time consuming for a higher-order system if the simulation required, including all the quantizations has to be done with MATLAB, as is the case here.

Project description

The investigation of quantization noise is based on a linear model with additional white noise sources at all quantization points. That corresponds to the models for A/D conversion and multipliers, as considered in Projects 1 and 2. Essential is the further assumption that all these noise sources are independent. The overall situation is shown in Fig. 7.7a, assuming that a continuous signal $v_0(t)$ has to be processed by a digital system, working with a quantization step Q_i, while the A/D conversion as well as the final D/A conversion is done with Q. Usually, it is $Q > Q_i$. Figure 7.7b shows the model, assuming that the quantization is done by rounding, such that all noise sources are described by the variance $Q_i^2/12$. In this general model the dependence on the actual multiplier coefficient is ignored, which was considered in Project 2. All noise sources contribute to the output noise over individual linear transmission systems, described by $G_\lambda(e^{j\omega})$. Due to the assumed independence the total power density spectrum of the output noise becomes

$$\Phi_{\mathbf{ee}}(e^{j\omega}) = \frac{Q^2}{12}\left[|H(e^{j\omega})|^2 + 1\right] + \frac{Q_i^2}{12}\sum_{\lambda=1}^{L}|G_\lambda(e^{j\omega})|^2 \tag{5-1}$$

where L is the number of quantization points inside the system and $H(e^{j\omega})$ the transfer function of the overall system. The total noise power becomes

$$\frac{1}{2\pi}\int_{-\pi}^{\pi}\Phi_{\mathbf{ee}}(e^{j\omega})d\omega = \frac{Q^2}{12}\cdot N_{eo} + \frac{Q_i^2}{12}N_{ei} \tag{5-2}$$

where N_{eo} and N_{ei} are the *noise figures* due to the input–output quantization and the inner rounding points, respectively. With $Q_i = Q \cdot 2^{-\Delta w}$ we introduce the required additional wordlength Δw inside the system to be chosen such that the contribution $(Q_i^2/12)N_{ei}$ to the total noise power is equal to the unavoidable part $(Q^2/12) \cdot N_{eo}$. We obtain with

$$\Delta w = \left\lceil \frac{1}{2} \log_2 \frac{N_{ei}}{N_{eo}} \right\rceil \tag{5-3}$$

a measure for the quality of a particular implementation of the system with the transfer function $H(e^{j\omega})$, if used between A/D and D/A converters. Here $\lceil x \rceil$ denotes the smallest integer $\geq x$.

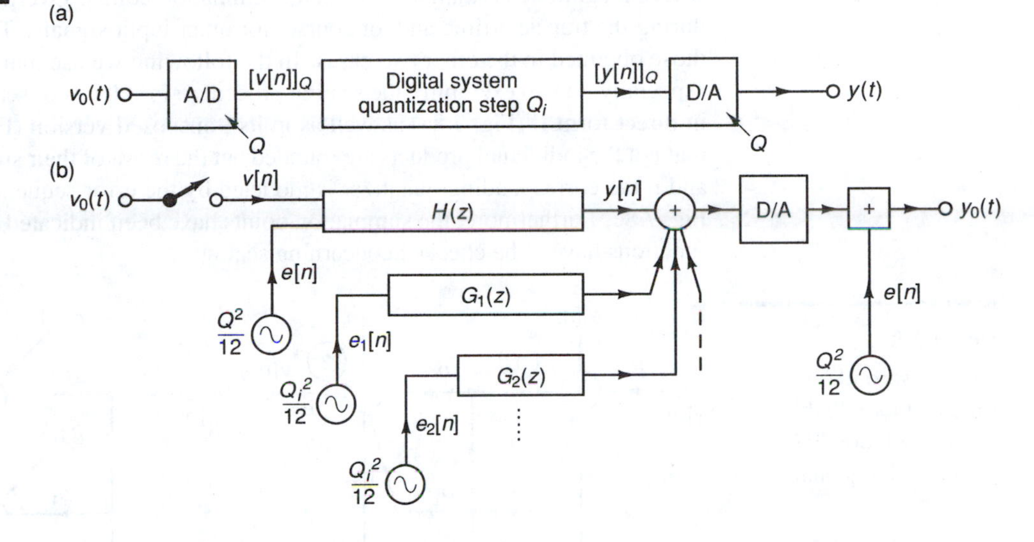

Figure 7.7 Model of a real system with multiple rounding points.

This analysis is based on the assumption that no overflow occurs, neither at the output nor at any summation point inside the filter, a condition that has to be satisfied by proper scaling, based on the frequency responses $F_\lambda(e^{j\omega})$ from the input to all the summation points. Three different norms may be considered for the scaling (e.g., [3, Sec. 9.6]):

$$l_1\text{-norm:} \quad \|f_\lambda\|_1 = \sum_{n=0}^{\infty} |f_\lambda[n]| \tag{5-4}$$

where $f_\lambda[n] = (1/2\pi) \int_{-\pi}^{+\pi} F_\lambda(e^{j\omega}) e^{j\omega n}\, d\omega$ is the impulse response from the input to the λth summation point:

$$l_2 = L_2 \text{ norm:} \quad \|f_\lambda\|_2 = \sqrt{\sum_{n=0}^{\infty} f_\lambda^2[n]} = \sqrt{\frac{1}{2\pi} \int_{-\pi}^{+\pi} |F_\lambda(e^{j\omega})|^2\, d\omega} \tag{5-5}$$

$$L_\infty\text{-norm:} \quad \|F_\lambda\|_\infty = \max_\omega |F_\lambda(e^{j\omega})| \tag{5-6}$$

It can be shown that

$$\|f_\lambda\|_2 \leq \|F_\lambda\|_\infty \leq \|f_\lambda\|_1 \tag{5-7}$$

Scaling according to one of these norms means that the input sequence $v[n]$ is multiplied by a scaling coefficient c_{sk}, $k = 1, 2, \infty$, being the inverse of the corresponding norm, where $|v[n]| \leq 1$ has been assumed. Since only one scaling coefficient can be applied, it has to be determined in each case as

$$c_{sk} = 1/\max_\lambda(\| \bullet_\lambda \|_k) \tag{5-8}$$

If the scaling is done according to the l_1-norm, no overflow will occur. But since in this case $v[n] = \text{sign}\{h[n_0 - n]\}$ and thus $|v[n]| = 1$, $\forall n$ as the worst-case input is required to obtain the value 1 at the summation point, this very conservative scaling yields a rather bad use of available dynamic range for common input signals and thus a bad signal/noise ratio at the output.

It turns out that an l_2 scaling leads to a very good S/N ratio at the expense of a rather high probability of overflows. That can be reduced with a subjectively chosen security factor δ such that

$$c_{s2} = 1/(\delta \max(\|f_\lambda\|_2)) \tag{5-9}$$

The widely used L_∞-norm yields for sinusoidal excitation in the steady state a response whose magnitude is smaller than 1 at all summation points. Overflows have to be expected during the transient time and, of course, for other input signals. The S/N ratio is between those obtained in the other two cases. In the following we use mainly this norm. We use it especially now to exemplify the scaling procedure for blocks of second order implemented in direct form II (Fig. 7.8a) as well as in its transposed version (Fig. 7.8b). It is assumed that not the individual products are rounded but the result of their summations. In Figs. 7.8a and b the corresponding quantizer is modeled by the error sequence $e[n]$, as illustrated in Fig. 7.8c. Furthermore, the summation points have been indicated up to which the transfer functions have to be checked concerning scaling.

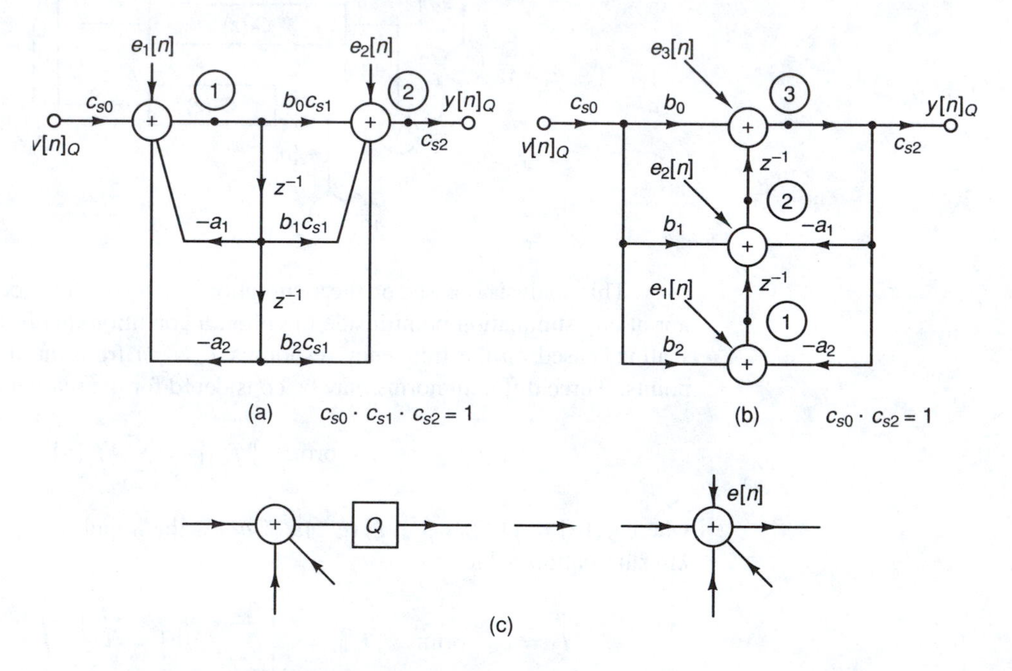

Figure 7.8 Block of second order: (a) direct form II; (b) transposed form II; (c) model of the quantizer.

In the first case the two functions

$$F_1(e^{j\omega}) = \frac{1}{1 + a_1 e^{-j\omega} + a_2 e^{-j2\omega}} \tag{5-10}$$

and

$$F_2(e^{j\omega}) = \frac{b_0 + b_1 e^{-j\omega} + b_2 e^{-j2\omega}}{1 + a_1 e^{-j\omega} + a_2 e^{-j2\omega}} = H(e^{j\omega}) \tag{5-11}$$

have to be checked. Assuming that the overall transfer function $H(e^{j\omega})$ already satisfies the scaling condition, we have to deal with $F_1(e^{j\omega})$ only. Its norm, $\|F_1\|_\infty$ =: `ma`, can be determined approximately with `a = [1 a1 a2]` as

$$\texttt{ma} = \texttt{max(abs(freqz(1, a, 1024,'whole')))} \tag{5-12}$$

To simplify the implementation we often use for scaling a power of 2, thus replacing the multiplication otherwise required by an appropriate shift. In that case the scaling coefficient becomes

$$c_{s0} = 2^{-m} \qquad \text{with } m = \lceil \log_2(\text{ma}) \rceil \tag{5-13}$$

The necessary rescaling at the output can be done by using the coefficients $b_k c_{s1}$ with $c_{s1} = 1/c_{s0}$ if $|b_k c_{s1}| < 1$. In that case the output scaling factor $c_{s2} = 1$ has to be used. In general two steps are required such that $c_{s0} c_{s1} c_{s2} = 1$.

The power density spectrum at the output due to the two noise sources at the summation points becomes

$$\Phi_{\mathbf{ee}}(e^{j\omega}) = \frac{Q_i^2}{12}\left[|H(e^{j\omega})|^2 \cdot c_{s1}^2 + 1\right] c_{s2}^2 \tag{5-14}$$

The noise figure N_{ei} introduced in (5-2) can be calculated approximately in dB with `H = freqz(b,a,1024,'whole')` as

$$\texttt{Nei} \;=\; 10 * \texttt{log10((mean(abs(H).^2)} \;*\; \texttt{cs1^2} + 1) \;*\; \texttt{cs2^2)} \tag{5-15}$$

Equation (5-14) shows that the power density spectrum of the noise at the output is a shifted and multiplied version of the magnitude square of the overall transfer function. That is different in the other case. Again assuming that the overall transfer function $H(z)$ has been scaled appropriately, this time two subfrequency responses have to be checked for scaling:

$$F_1(e^{j\omega}) = \frac{(b_2 - a_2 b_0) + (a_1 b_2 - a_2 b_1)e^{-j\omega}}{1 + a_1 e^{-j\omega} + a_2 e^{-j2\omega}} \tag{5-16}$$

and

$$F_2(e^{j\omega}) = \frac{(b_1 - a_1 b_0) + (b_2 - a_2 b_0)e^{-j\omega}}{1 + a_1 e^{-j\omega} + a_2 e^{-j2\omega}} \tag{5-17}$$

As pointed out above, we now scale according to

$$\max_{\lambda} \|F_\lambda\|_\infty =: \text{ma} = \max[\max|F_1(e^{j\omega})|\,,\ \max|F_2(e^{j\omega})|] \tag{5-18}$$

Here the rescaling can be done at the output only as shown in Fig. 7.8b with $c_{s2} = 1/c_{s0}$. The three noise sequences $e_\lambda[n]$ are transmitted to the output over subsystems described by

$$G_\lambda(e^{j\omega}) = c_{s2}\frac{e^{-j(3-\lambda)\omega}}{1 + a_1 e^{-j\omega} + a_2 e^{-j2\omega}}, \qquad \lambda = 1:3 \tag{5-19}$$

Since these $G_\lambda(e^{j\omega})$ differ by delays only, we get for the total power density spectrum due to rounding inside the system,

$$\Phi_{\mathbf{ee}}(e^{j\omega}) = 3|G(e^{j\omega})|^2 Q_i^2/12 \tag{5-20}$$

Its noise figure can be calculated as described correspondingly by (5-15).

In the following exercises, first the three types of scaling will be exemplified. Then the two structures of Fig. 7.8a and b will be compared in terms of their noise performances. We also try to verify the model shown in Fig. 7.7 by comparing analysis results with those of measurements. Furthermore, the implementations of a low-pass of seventh degree in direct form and as two different cascades of blocks of second order are investigated.

The two programs `fx2filter(b,a,v,cs,w)` and `fxtfilter(b,a,v,cs,w)` can be used for the implementation of a system in direct form II and its transposed version, respectively, with wordlength w for the data paths. The scaling coefficient c_s has to be determined during the design phase. In `fx2filter(.)` it can be introduced as a vector

$c_s = [c_{s0}\ c_{s1}\ c_{s2}]$ with $c_{s1}c_{s2} = 1/c_{s0}$, as described above. If $c_s = c_{s0}$ is given as a scalar, $c_{s1} = 1,\ c_{s2} = 1/c_{s0}$ will be used.

In case of an overflow, the programs use a saturation characteristic. The number of potential overflows is counted. The programs are used in context with the scaling exercise and in combination with the measuring program `nlm`, explained in the appendix.

```
function y = fx2filter(b,a,v,cs,w)
% FX2FILTER  Nth order digital filter in second structure
% done with simulated fixed-point arithmetic
%
% y = fx2filter(b,a,v,cs,w)
%     b    numerator coefficients
%     a    denominator coefficients with a(1) = 1
%     v    input signal
%     cs   scaling coefficient, to be precalculated such that no internal
%          overflow occurs; cs should be a power of 2;
%             cs = [cs0, cs1, cs2]:   input scaling with cs0,
%                                     output compensation with b*cs1 and cs2
%             cs scalar:              cs0 = cs, cs1 = 1, cs2 = 1/cs
%          the number of overflows will be displayed, if any occur
%     w    internal (data-path) wordlength
%     y    output signal

a = a(:).'; b = b(:).';               % row vectors
oc = 0;                               % overflow counter
if length(b) < length(a);
        b = [b, zeros(1,length(a)-length(b))];
elseif length(b) > length(a);
        a = [a, zeros(1,length(b)-length(a))];
end
a = a(2:length(a));
if length(cs) == 3;      b = b * cs(2); cs2 = cs(3);
elseif length(cs) == 1; cs2 = 1/cs;
else        error('cs (scaling coefficient): 3-vector or scalar required.');
end

Plus1 = 2^(w-1);
v   = v *cs(1)*Plus1;
x = zeros(length(a),1);
LEN = length(v);
n=0;
while n < LEN;
        n = n + 1;
        u = round(v(n) - a*x);
        if u+1 > Plus1;                    % saturation nonlinearity
                u = Plus1-1; oc = oc + 1;
        elseif u < -Plus1;
                u = -Plus1; oc = oc + 1;
        end
        y(n) = b * [u;x];
        x = [u;x(1:length(x)-1)];
end
```

```
y = round(y*cs2);                % Quantization point #2
f = find(y > Plus1-1);           % saturation: positive branch
y(f) = Plus1-1 + 0*f;
oc = oc + length(f);
f = find(y < -Plus1);            % saturation: negative branch
y(f) = 0*f - Plus1;
oc = oc + length(f);

y   = y /Plus1;
if oc > 0; disp(['!!! ',int2str(oc),' OVERFLOW(S) ENCOUNTERED !!!']);
end

function y = fxtfilter(b,a,v,cs,w)
% FXtFILTER  Nth order digital filter in the second structure
% in its transposed form, done with simulated fixed-point arithmetic
%
% y = fxtfilter(b,a,v,cs,w)
%    b    numerator coefficients
%    a    denominator coefficients with a(1) = 1;
%    v    input signal
%    cs   scaling coefficient, to be precalculated such that no internal
%         overflow occurs; cs should be a power of 2;
%            cs = [cs0, cs2]:   input scaling with cs0,
%                               output compensation with cs2
%            cs scalar:         cs0 = cs, cs2 = 1/cs
%         the number of overflows will be displayed, if any occur
%    w    internal (data-path) wordlength
%    y    output signal

N   = length(a);
LEN = length(v);
b   = b(:).'; a   = a(:).';      % row vector
oc = 0;                          % overflow counter
if length(b) < length(a);
        b = [b, zeros(1,length(a)-length(b))];
elseif length(b) > length(a);
        a = [a, zeros(1,length(b)-length(a))];
end
Plus1 = 2^(w-1);
v   = cs(1)*Plus1 * v;
x = zeros(1,N);
y = zeros(1,LEN);
n=0;
while n < LEN;
        n = n + 1;
        x = v(n) * b + x;
        u = round(x(1));
        if u+1 > Plus1;                  % saturation nonlinearity
                u = Plus1-1;
                oc = oc + 1;
        elseif u < -Plus1;
                u = -Plus1;
                oc = oc + 1;
        end
```

```
        y(n) = u;
        x = x - u*a;
        x = [round(x(2:N)), 0];
        f = find(x > Plus1-1);          % saturation: positive branch
        x(f) = Plus1-1 + 0*f;
        oc = oc + length(f);
        f = find(x < -Plus1);           % saturation: negative branch
        x(f) = 0*f - Plus1;
        oc = oc + length(f);
end
if length(cs) == 2; cs = cs(2);
else                cs = 1/cs;
end
y  = y *cs;                     % output scaling may cause overflow!
f = find(y > Plus1-1);          % saturation: positive branch
y(f) = Plus1-1 + 0*f;
oc = oc + length(f);
f = find(y < -Plus1);           % saturation: negative branch
y(f) = 0*f - Plus1;
oc = oc + length(f);

y  = y /Plus1;
if oc > 0;
        disp(['!!! ',int2str(oc),' OVERFLOW(S) ENCOUNTERED !!!']);
end
```

EXERCISE 5.1

Types of Scaling

We are given an all-pole system described by

$$\begin{aligned} a &= [1 \quad a_1 \quad a_2] \\ &= [1 \quad -2r\cos\psi \quad r^2], \quad r = 0.975 \qquad \psi = (0:50)\cdot\pi/50 \end{aligned}$$

to be implemented in direct form II as described by Fig. 7.8a. The system is excited with white noise uniformly distributed in $(-1, 1)$ and thus with a variance $\sigma_v^2 = 1/3$. Using this example we investigate the properties of the three scaling methods described above. That will be done by comparing the ratios of the total noise and the signal power at the output as a function of ψ, the angle of the pole $p = re^{j\psi}$. Furthermore, the number of overflows will be checked using the program `fx2filter(.)`.

a. Show that

$$\frac{N_{eik}}{S} = \frac{3}{c_{sk}^2}$$

where c_{sk}, $k = 1, 2, \infty$ is the scaling coefficient in the three cases, while N_{eik} is the corresponding noise figure and S the total signal power at the output, taking $\sigma_v^2 = 1/3$ into account.

b. Write a program for calculation of the scaling coefficients c_{sk} as functions of ψ according to (5-8), using (5-4)–(5-6). Calculate and plot

```
10*log10(3/(csk^2))  =  10*log10(Neik/S)
```

[*Hints*: Calculate the l_1-norm approximately based on 500 values of the impulse response, to be determined with `filter(.)`. The l_2-norm can be calculated in closed form as

$$\|f_1\|_2 = \left[\frac{1+r^2}{(1-r^2)(1+r^4-2r^2\cos 2\psi)}\right]^{1/2} \tag{5-21}$$

while the L_∞-norm has to be determined approximately using (5-12).]

c. Use at the input 500 values of a random sequence, uniformly distributed in $(-1, 1)$, and determine the number of overflows with the program `fx2filter(.)` using $c_s = [c_{sk} \quad 1 \quad 1]$ with c_{sk} as found in part (b) for three different arbitrarily selected pole positions, characterized by their angles ψ. Choose, if required, an appropriate security factor δ as described in (5-9). Comment on your results. What do you expect with a normally distributed input signal of variance $\sigma_v^2 = 2$? Check your expectation experimentally.

EXERCISE 5.2

Verification of the Model of a Real System

You are given the coefficients of a second-order system

$$b = [0.28519 \quad -0.11792 \quad 0.28519]^T$$
$$a = [1.0 \quad -0.5682 \quad 0.9469]^T$$

a. Calculate the required scaling coefficients $c_{s\infty}$ according to (5-6) and (5-8) for implementations of this system with the structures shown in Fig. 8a and b.

b. Calculate for both cases the power density spectrum of the noise $\Phi_{\mathbf{ee}}(e^{j\omega})$ according to (5-14) and (5-20) using $w = 16$ bits. Plot your results in one diagram as $10\log_{10}[\Phi_{\mathbf{ee}}(e^{j\omega})]$, together with the magnitude $|H(e^{j\omega})|$ in dB. Calculate the total power of the noise in both cases.

c. Measure estimates of the power density spectrum $\Phi_{\mathbf{ee}}(e^{j\omega})$ for $\omega = \omega_\kappa = k \cdot 2\pi/N$ with $N = 256$ using the program `nlm(.)` for both structures. Plot your results in comparison to those obtained in part (b). Comment on your results. [*Remarks*: (1) One iteration cycle in `nlm(.)` requires roughly 6 s, if $N = 256$ is used and the measurement is done with a 386-PC having a clock frequency of 25 MHz. With $L = 50$, a usable estimate can be expected. (2) The given program, `nlm`, calls `fxtfilter` for simulation of the transposed structure. Thus a change on line 25 in `nlm` is required (replacing `fxtfilter` by `fx2filter`) if the other structure is to be investigated.]

EXERCISE 5.3

Comparing Two Structures

In this exercise we compare the two structures of Fig. 7.8a and b more thoroughly. We calculate the noise figures as a function of the pole and zero positions for the following groups of transfer functions:

$$H_1(z) = b\frac{1-2\cos\psi_n z^{-1}+z^{-2}}{1-2r\cos\psi_p z^{-1}+r^2z^{-2}} \qquad r = 0.975$$

$$\psi_p = \ell\cdot\pi/50;\ \ell = 0:50$$

$$H_2(z) = \frac{r^2-2r\cos\psi_p z^{-1}+z^{-2}}{1-2r\cos\psi_p z^{-1}+r^2z^{-2}} \qquad \psi_n = \psi_p + \tfrac{\pi}{6}(1-\ell/25)$$

The first transfer function describes a typical subsystem in a cascade implementation of a filter with zeros on the unit circle in a certain distance from the pole location. As a second example we consider an all-pass.

Remark concerning practical implementation: For a certain range of the angles ψ_n and ψ_p the coefficients b_1 and a_1 do not satisfy the condition that their magnitude has to be smaller than 1 (see Project 3). In an actual implementation the problem is solved by a multiplication with $b_1/2$ (or $a_1/2$, respectively) and adding the result twice. We ignore this difficulty here.

a. Write a program for the calculation of b such that $\max |H_1(e^{j\omega})| = 1$ for all the different pole and zero locations.

b. In the case of transposed form II, the noise figure is

$$N_{ei} = 3 \cdot \frac{1}{2\pi} \int_{-\pi}^{+\pi} \left| \frac{1}{1 + a_1 e^{-j\omega} + a_2 e^{-j2\omega}} \right|^2 d\omega$$

$$= 3 \cdot \frac{1}{2\pi j} \oint G(z) G(z^{-1}) \frac{dz}{z}$$

with $G(z) = \dfrac{1}{1 + a_1 z^{-1} + a_2 z^{-2}}$, where $a_1 = -2r \cos \psi_p,\ a_2 = r^2$. Show that

$$N_{ei} = 3 \cdot \frac{1 + r^2}{(1 - r^2)(1 + r^4 - 2r^2 \cos \psi_p)}$$

[see, for comparison, (5-19) and (5-21)].

c. Write two programs for calculation of the scaling factors and the noise figures N_{ei} in dependence of the pole and zero locations for the two structures of Fig. 7.8a and b. Do the scaling according to the L_∞-norm. Apply your programs to both transfer functions. Plot your results together for the two structures but separately for H_1 and H_2. Comment on your results.

EXERCISE 5.4

Elliptic Filter

In this exercise we examine the quantization noise of different implementations of an elliptic filter, the coefficients of which are to be calculated as `[b,a] = ellip(7,.1,40,.4)` (see Exercise 3.2).

a. *Implementation with direct form II*: For simplicity we ignore the fact that there are some denominator coefficients with magnitudes larger than 1.

1. Calculate the scaling coefficient c_{s0} according to the L_∞-norm using (5-12) and (5-13).
2. Calculate the power density spectrum $\Phi_{\mathbf{ee}}(e^{j\omega})$ of the noise with (5-14) using $w = 16$ bits and appropriate values c_{s1} and c_{s2} as well as the noise figure N_{ei} with (5-15). Plot $10\log_{10}[\Phi_{\mathbf{ee}}(e^{j\omega})]$ as well as $20\log_{10}[|H(e^{j\omega})|]$ in one diagram. Using (5-1) and (5-2), calculate the outer noise figure N_{eo}, and with (5-3), the required additional wordlength Δw.
3. If you want to compare your result with that of measurement, you can do that using `nlm(...)` again. But now you have to use roughly $L = 200$ iterations and thus to spend more time. Can you say why more cycles are needed to get a reasonable result?

b. *Implementation as a cascade of subsystems of second order*: We refer to Exercise 3.2(b), where the cascade structure for implementation of the same elliptic filter was investigated concerning its sensitivity. Here we determine the quantization noise at the output, described primarily by the inner noise figure N_{ei} as introduced by (5-2) or (5-15). As will be shown, it depends to a large extent on the pairing of poles and zeros to subsystems of second order and on the ordering of these subsystems.

1. Given a system to be implemented as a cascade of L blocks of second order, show that the number of different cascade structures is $(L!)^2$. [*Remark*: In case of a system of odd order as in our example, the two subpolynomials of first order will be regarded as special cases of quadratic polynomials with one coefficient being zero. So in our example we end up with $L = 4$ and $(4!)^2 = 576$ different cascade implementations.]

2. Given a particular cascade structure described by the transfer function

$$H(z) = \hat{b}_0 \prod_{\lambda=1}^{L} H_\lambda(z)$$

with

$$H_\lambda(z) = \frac{b_0^{(\lambda)} + b_1^{(\lambda)} z^{-1} + b_2^{(\lambda)} z^{-2}}{1 + a_1^{(\lambda)} z^{-1} + a_2^{(\lambda)} z^{-2}}$$

Using direct form II for each subsystem, we get the signal-flow graph shown in Fig. 7.9, where the scaling and rescaling coefficients $c_{s0}^{(\lambda)}$ and $c_{s1}^{(\lambda)}$ are indicated. Determine in general form the transfer functions $F_{1,2}^{(\lambda)}(z)$ required for scaling as well as the $G_{1,2}^{(\lambda)}(z)$ from the rounding points to the output. [*Hint*: We get, for example,

$$c_{s0}^{(1)} F_1^{(1)}(z) = \frac{c_{s0}^{(1)}}{1 + a_1^{(1)} z^{-1} + a_2^{(1)} z^{-2}}$$

$$c_{s0}^{(1)} c_{s0}^{(2)} F_1^{(2)}(z) = H_1(z) c_{s1}^{(1)} \frac{c_{s0}^{(1)} c_{s0}^{(2)}}{1 + a_1^{(2)} z^{-1} + a_2^{(2)} z^{-2}}$$

$$G_1^{(1)}(z) = \frac{1}{c_{s0}^{(1)}} \cdot H(z) \qquad G_2^{(1)}(z) = c_s \prod_{\lambda=2}^{4} c_{s0}^{(\lambda)} c_{s1}^{(\lambda)} H_\lambda(z)]$$

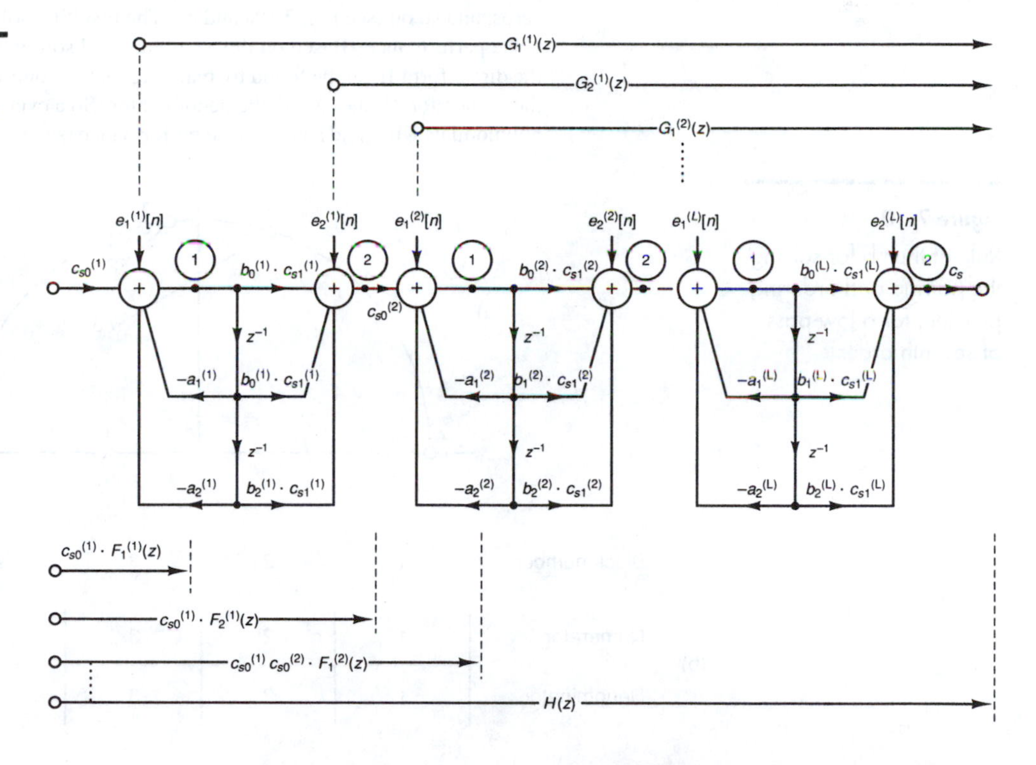

Figure 7.9 Signal-flow graph of the cascade structure, subtransfer functions indicated.

3. You are given a particular cascade structure for implementation of the elliptic filter by the angles of the poles and zeros of the subtransfer functions:

	Block 1	Block 2	Block 3	Block 4
Zero angles	π	± 1.8843	± 1.4685	± 1.3625
Pole angles	± 1.2745	± 1.2196	± 0.9914	0

Do the noise analysis of the system by performing the following steps:

- Calculate the poles and zeros of $H(z)$ as `p = roots(a) ; z = roots(b)` and their angles.
- Identify the poles and zeros for the four subsystems according to the list above, and calculate the coefficients of the subtransfer functions $H_\lambda(z)$ using `poly(.)`.
- Determine the scaling coefficients $c_{s0}^{(\lambda)}$ and $c_{s1}^{(\lambda)}$ as well as c_s based on the L_∞-norm. Use powers of 2 for scaling.
- Calculate the different noise figures $N_{ei1.2}^{(\lambda)}$ approximately according to (5-15), using the corresponding transfer functions $G_{1.2}^{(\lambda)}(z)$.
- Calculate $N_{ei} = \sum N_{ei1.2}^{(\lambda)}$ and the required additional wordlength according to (5-3). Compare your result with that obtained in part (a).
- To check the correct scaling, calculate and plot $|F_{1.2}^{(\lambda)}(e^{j\omega})|\ \lambda = 1:4$ as well as $|H(e^{j\omega})|$. Use `subplot(22.)`.

4. Finding a good cascade structure. In a particular case the best solution for the pairing and ordering problem can be found by a complete search over all $(L!)^2$ possibilities. But that is obviously very time consuming. Instead, we apply a rule of thumb which usually yields a good solution, if not the best one (Sec. 6.9 in [1]). We explain it in two steps for our example. For implementation with the transposed form II, we use for $H_1(z)$ the pole pair with the largest magnitude and the zero pair closest to it. We proceed similarly with $H_2(z)$, with the remaining poles and zeros, and so on (see Fig. 7.10a and b). The resulting solution shows a rather good noise performance. Based on this result, a good solution for implementation with the direct form II can be found by realizing that in contrast to the transposed form, the numerator "comes after" the denominator. So a cyclic shift of the denominator polynomials will yield a good noise performance (see Fig. 7.10c for our example).

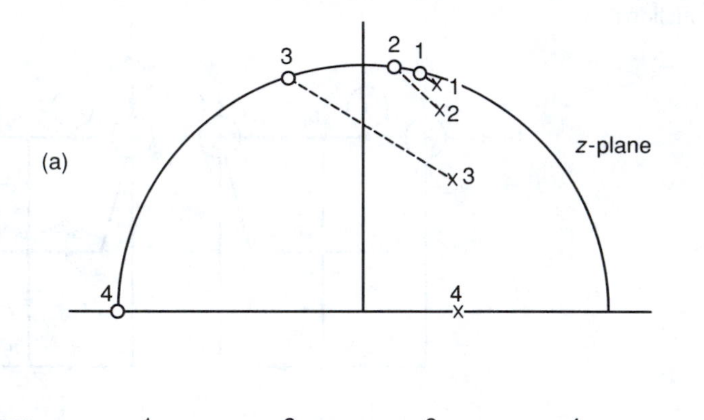

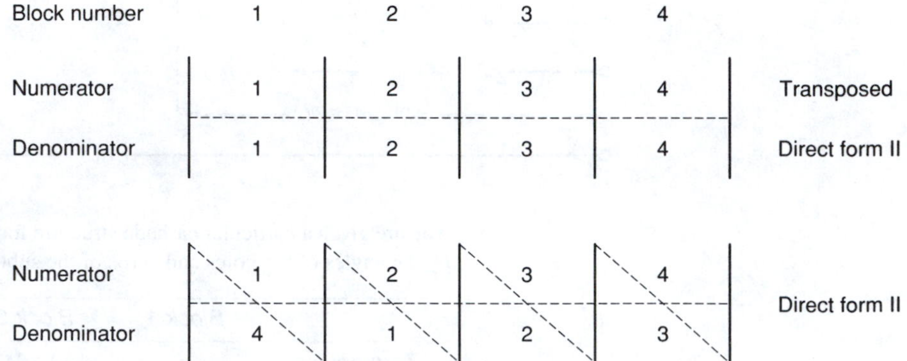

Figure 7.10 Rule of thumb for solving the pairing and ordering problem for a low-pass of seventh order.

- Find the subsystems $H_\lambda(z)$ according to the rule of thumb explained above.
- Do the scaling and calculate the total noise figure N_{ei} as well as $\Delta\omega$ as you did in point 3.

Calculate and plot again the functions $|F_{1,2}^{(\lambda)}(e^{j\omega})|$, $\lambda = 1:4$ and $|H(e^{j\omega})|$. Use `subplot(22.)`. Compare the three implementations in terms of the noise performance and the Δw required.

APPENDIX: METHOD FOR MEASURING THE PERFORMANCE OF AN IMPLEMENTED DIGITAL SYSTEM

The system to be investigated is only approximately linear, due to the effects of limited wordlength inside the filter. It can be modeled by a parallel connection of a linear system, the output of which is $y_L[n]$, and another one, yielding the noise-like error sequence $e[n]$. The separation should be done such that $y_L[n]$ and $e[n]$ are orthogonal to each other. While the linear subsystem will be described by its frequency response $H(e^{j\omega})$, the other one is characterized by the power density spectrum $\Phi_{\mathbf{ee}}(e^{j\omega})$ of its output sequence. The so-called noise-loading method, to be explained now, yields samples of estimates of these two functions at $\omega_k = k \cdot 2\pi/N$ [5].

We perform a sequence of measurements, using for excitation members $\tilde{v}_\lambda[n]$ of an ensemble of signals, being periodic for $n \geq 0$. In the simplest case, used here, they are generated as

$$\tilde{v}_\lambda[n] = \text{IFFT}\{V_\lambda[k]\} = \frac{1}{N}\sum_{k=0}^{N-1} V_\lambda[k] w_N^{-kn} \in \mathbb{R} \tag{A-1}$$

where the spectral values

$$V_\lambda[k] = |V| e^{j\varphi_\lambda[k]} \tag{A-2}$$

always have the same magnitude for all k and all λ, while $\varphi_\lambda[k]$ is a random variable uniformly distributed in $[-\pi, \pi)$ and statistically independent with respect to k and λ. The condition $\varphi_\lambda[k] = -\varphi_\lambda[N-k]$ has to be observed to get a real signal $\tilde{v}_\lambda[n]$. It turns out that the sequences $\tilde{v}_\lambda[n]$ are approximately normally distributed.

The system under test will be excited by these $\tilde{v}_\lambda[n]$, $\lambda = 1:L$, where L is the number of trials to be chosen such that the desired accuracy of the result is achieved. After the transient time the output sequence of the system will be periodic, described by

$$\tilde{y}_\lambda[n] = \tilde{y}_{L\lambda}[n] + \tilde{e}_\lambda[n] \tag{A-3}$$

Note that in the digital system, to be tested here with a periodic excitation, the error sequence $\tilde{e}_\lambda[n]$ will be periodic as well. The term $\tilde{y}_{L\lambda}[n]$ on the right-hand side can be expressed as

$$\tilde{y}_{L\lambda}[n] = \text{IFFT}\left\{H(e^{j\omega_k})V_\lambda[k]\right\} \tag{A-4}$$

where the $H(e^{j\omega_k})$ are samples of the frequency response of the linear subsystem to be determined. Minimizing

$$\sigma_{\mathbf{e}}^2[n] = \mathcal{E}\left\{|\tilde{e}_\lambda[n]|^2\right\} = \mathcal{E}\left\{\left|\tilde{y}_\lambda[n] - \frac{1}{N}\sum_{k=0}^{N-1} H(e^{j\omega_k})V_\lambda[k]w_N^{-nk}\right|^2\right\} \tag{A-5}$$

with respect to the $H(e^{j\omega_k})$ yields after some calculations for the special case described by (A-2)

$$H(e^{j\omega_k}) = \mathcal{E}\left\{\frac{Y_\lambda[k]}{V_\lambda[k]}\right\} \tag{A-6}$$

where $Y_\lambda[k] = \text{FFT}\{\tilde{y}_\lambda[n]\}$. Due to the orthogonality principle the sequence $\tilde{y}_{L\lambda}[n]$ is indeed orthogonal with regard to $\tilde{e}_\lambda[n]$ if the linear subsystem is described by these $H(e^{j\omega_k})$.

The expected value can be calculated approximately by averaging over L trials:

$$H(e^{j\omega_k}) \approx \hat{H}(e^{j\omega_k}) = \frac{1}{L}\sum_{\lambda=1}^{L}\frac{Y_\lambda[k]}{V_\lambda[k]} \tag{A-7}$$

Now the power density spectrum of the noise

$$\Phi_{\mathbf{ee}}(e^{j\omega_k}) = \frac{1}{N}\mathcal{E}\left\{|\text{FFT}\{\tilde{e}_\lambda[n]\}|^2\right\} \tag{A-8}$$

can be computed as

$$\Phi_{\mathbf{ee}}(e^{j\omega_k}) = \frac{1}{N}\mathcal{E}\left\{|Y_\lambda[k] - H(e^{j\omega_k})V_\lambda[k]|^2\right\} \tag{A-9}$$

$$\approx \frac{1}{N}\left[\mathcal{E}\{|Y_\lambda[k]|^2\} - |\hat{H}(e^{j\omega_k})|^2|V|^2\right] \tag{A-10}$$

An estimation will be obtained as

$$\hat{\Phi}_{\mathbf{ee}}(e^{j\omega_k}) \approx \frac{1}{N}\left[\frac{1}{L}\sum_{\lambda=1}^{L}|Y_\lambda[k]|^2 - |\hat{H}(e^{j\omega_k})|^2 \cdot |V|^2\right] \tag{A-11}$$

It can be shown that the results found with (A-6) and (A-11) are at least asymptotically unbiased and consistent. This means that the accuracy can be increased by increasing the number L of measurements.

The flow diagram in Fig. 7.11 describes the required steps. Furthermore, the MATLAB program `nlm(b,a,cs,w,N,L)` is given for the investigation of a system described by the coefficient vectors **b** and **a** and scaling coefficient(s) c_s, and is implemented with the wordlength w. On line 25 it calls the program `fxtfilter(.)`, an implementation of the transposed version of direct form II. Changes are required here especially if other systems are to be investigated.

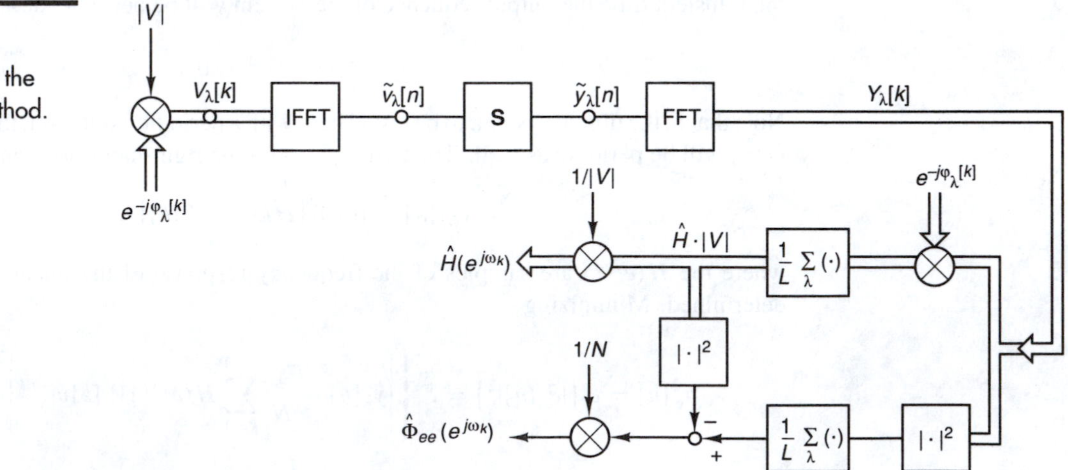

Figure 7.11
Flow diagram for the noise-loading method.

```
    function [H,PDS,Ne] = nlm(b,a,cs,w,N,L);
%  NLM   Noise loading method.
%
%  [H,PDS,Ne] = nlm(b,a,cs,w,N,L)
%  The program determines estimates of the complex frequency response
%  H(exp(j*om)) and the power density spectrum PDS(exp(j*om)) of the
%  noise at N equally spaced points omk = k*2*pi/N at the output of a
%  filter, implemented with a wordlength w. PDS as well as the noise
%  figure Ne are given in dB.
%  The method uses an averaging over L independent measuring results.
%
%  b,a = coefficients of the system under test.
%  cs  = scaling coefficient at the input to avoid overflow;

   sumH = zeros(1,N);                      % H(k) accumulation
   sumY2 = sumH;                           % |Y(k)|^2 accumulation

 for i = 1:L;
   fprintf('current iteration: %g\n', i);
   phi = 2*pi*rand(1,N/2-1);               % Generation of the periodic
   phi = [0 phi 0 -phi(N/2-1:-1:1)];       % input signal
   Vp = exp(j*phi);
   vp = real(ifft(Vp));
   v = [vp vp];
   y = fxtfilter(b,a,v,cs,w);              % applying v to the system under
   y = y(N+1:1:2*N);                       % test; selecting and transfor-
   Yp = fft(y);                            % ming the last period;
   sumH = sumH + Yp./Vp;                   % Accumulation
   sumY2 = sumY2 + real(Yp.*conj(Yp));
 end;
   H = sumH/L;                             % postprocessing.
   PDS = ((sumY2/L) - real(H.*conj(H)))/N;
   Q = 2^(1-w);
   Ne = 10*log10(mean(PDS)/(Q^2/12));
   PDS = 10*log10(PDS);
```

chapter 8

DISCRETE-TIME FILTER DESIGN

OVERVIEW

One of the most powerful operations of discrete-time signal processing (DSP) is that of filtering. As the name implies, a filter tries to separate parts of a signal according to some criterion (e.g. separating a desired signal from noise, or separating two radio stations). One might want to separate the weekly stock market price variations from the yearly variations.

There are two types of discrete-time filters: FIR (finite impulse response) filters and IIR (infinite impulse response) filters. For both types there are two distinct parts in the design problem: the *approximation problem*, where one tries to approximate a desired filter characteristic by an allowed one, and the *realization problem*, where one implements a transfer function in hardware or software. In this chapter we deal with the approximation problem; Chapters 5 and 7 cover realization. We assume that the reader is familiar with the basic notions of difference equations, frequency response, rational z-transforms, and so on. If not, please consult one of the appropriate earlier chapters.

For many practical problems, the specifications for a filter are given in terms of its frequency response. The approximation part of the filter design process can be broken down into four related stages.

- Choose a desired ideal response, usually in the frequency domain.
- Choose an allowed class of filters (e.g., a length-L FIR filter).
- Choose a measure or criterion of how good the approximation is.
- Develop a method to find the "best" member of the allowed class of filters according to the criterion of approximation.

These four steps are often repeated several times to get an acceptable filter. After the "best" filter is designed and evaluated, the desired response or the allowed class or the measure of quality might be changed and the filter redesigned.

There are three commonly used approximation error measures: least squared error, Chebyshev, and maximally flat. The average squared error is important because it uses the power or energy as a measure of the size of the error. The Chebyshev error has important physical meaning because it is the maximum of the difference between what you want and what you have. While the squared error and the Chebyshev error measures are global measures, the Taylor's series approximation is a local method that maximizes the smoothness of the approximation by matching as many derivatives at a point as possible.

Most of the projects and exercises in the chapter will use the basic low-pass filter as an example, but the ideas extend to other desired frequency responses. Two forms of ideal frequency responses are examined: the first having a passband with unity transmission and a stopband with zero transmission, and the second with a transition band between the passband and stopband which will allow a much better approximation for a given length or order. It is very important in any approximation or optimization problem to choose the error criterion and the desired ideal response carefully *and explicitly*.

In this chapter we consider the properties of both FIR and IIR filters to develop insight and intuition into their characteristics. It is this insight that one should use to choose the allowed class of filters or to choose the appropriate criterion of approximation. We design and analyze a set of filters with the goals of understanding the basic properties of discrete-time filters, gaining insight into the design process, and learning the characteristics of several standard design methods.

BACKGROUND READING

The basic methods of filter design are covered in most general DSP textbooks. Notation and a general introduction to filters are provided in Chapter 1 of this book. Two older books with good chapters on the topic are by Rabiner and Gold [1] and Gold and Rader [2]. Two books dealing specifically with the filter design problem are those of Parks and Burrus [3] and Taylor [4]. Excellent coverage of analog or continuous-time filter designs often used as prototypes for discrete-time IIR filters is presented by Van Valkenburg [5]. A good book on discrete least-squared error approximation is that by Lawson and Hanson [6].

[1] L. R. Rabiner and B. Gold. *Theory and Application of Digital Signal Processing*. Prentice Hall, Englewood Cliffs, NJ, 1975.

[2] B. Gold and C. M. Rader. *Digital Processing of Signals*. McGraw-Hill, New York, 1969.

[3] T. W. Parks and C. S. Burrus. *Digital Filter Design*. John Wiley & Sons, New York, 1987.

[4] F. J. Taylor. *Digital Filter Design Handbook*. Marcel Dekker, New York, 1983.

[5] M. E. Van Valkenburg. *Analog Filter Design*. Holt, Rinehart and Winston, New York, 1982.

[6] C. L. Lawson and R. J. Hanson. *Solving Least Squares Problems*. Prentice Hall, Englewood Cliffs, NJ, 1974.

[7] C. S. Burrus, A. W. Soewito, and R. A. Gopinath. Least Squared Error FIR Filter Design with Transition Bands. *IEEE Transactions on Signal Processing*, SP-40(6):1327–1340, June 1992.

[8] A. V. Oppenheim and R. W. Schafer. *Discrete-Time Signal Processing*. Prentice Hall, Englewood Cliffs, NJ, 1989.

[9] R. B. Darst. *Introduction to Linear Programming*. Marcel Dekker, New York, 1991.

[10] P. P. Vaidyanathan. Design and Implementation of Digital FIR Filters. In D. F. Elliott, editor, *Handbook of Digital Signal Processing: Engineering Applications*, chapter 2, pages 55–170. Academic Press, San Diego, CA, 1987.

[11] E. W. Cheney. *Introduction to Approximation Theory*. McGraw-Hill, New York, 1966.

[12] L. R. Rabiner. Linear Program Design of Finite Impulse Response (FIR) Digital Filters. *IEEE Transactions on Audio and Electroacoustics*, AU-20(4):280–288, October 1972.

[13] G. Strang. *Linear Algebra and Its Applications*. Academic Press, New York, 1976.

[14] A. Grace. *Matlab Optimization Toolbox*. The MathWorks, Inc., Natick, MA, 1990.

[15] C. S. Burrus and T. W. Parks. Time Domain Design of Recursive Digital Filters. *IEEE Transactions on Audio and Electroacoustics*, AU-18(2):137–141, June 1970.

DISCRETE DESIGN OF FIR FILTERS

OVERVIEW

The discrete Fourier transform (DFT) of the impulse response of an FIR filter gives uniformly spaced samples of its frequency response. This suggests a method for designing a filter. Choose an ideal frequency response, sample this ideal with L equally spaced samples, and take the inverse DFT of these samples to give the impulse response of the filter. This method is called *frequency sampling design* and the frequency response of the designed filter will exactly interpolate the desired samples. The main shortcoming of this approach is the lack of control of the frequency response between the samples.

A second design method is formulated to use a larger number of frequency samples than the length of the filter. Under these conditions the actual response will generally not pass through all the specified samples, but one can easily design a filter whose response has the least average squared error over these sample frequencies. Since both of these design methods operate on discrete samples of the frequency response, they are both developed and investigated in this section.

BACKGROUND READING

Details of frequency sampling design can be found in [1], and [3] and discussions of discrete least squared error approximation can be found in [6].

PROJECT 1: FIR FILTER DESIGN BY FREQUENCY SAMPLING

This method designs a filter whose frequency response passes exactly through specified samples of the desired frequency response and therefore is an interpolation technique. Since the DFT of the impulse response of an FIR filter is a set of equally spaced samples of its frequency response, the inverse DFT of the samples should be the impulse response of the filter. That is indeed the case, and it is the basis of this widely used filter design method.

Since this is the first project on filter design, the first exercise includes considerable help and detail. It is expected that the experience gained in going through that exercise will enable the student to carry out the remaining ones with much less help or detail provided.

Project description

The frequency response of a length-L FIR filter is given by the discrete-time Fourier transform (DTFT) of the impulse response

$$H(e^{j\omega}) = \sum_{n=0}^{L-1} h[n]\, e^{-j\omega n} \tag{1-1}$$

The length-L DFT of $h[n]$ is the set of L evenly spaced samples of $H(e^{j\omega})$ over ω from zero to 2π given by

$$H_k = H(e^{j2\pi k/L}) = \sum_{n=0}^{L-1} h[n]\, e^{-j2\pi nk/L} \tag{1-2}$$

with $k = 0, 1, 2, \ldots, L-1$. Since the length of the filter is equal to the number of frequency samples, the IDFT of the samples of the desired frequency response $H_d(e^{j\omega})$ is the impulse response

$$h[n] = \mathcal{IDFT}\{H_d(e^{j2\pi k/L})\} = \frac{1}{L}\sum_{k=0}^{L-1} H_k\, e^{j2\pi kn/L} \tag{1-3}$$

The frequency response of this filter will exactly interpolate the samples of the desired frequency response.

For the general case, both $h[n]$ and $H(e^{j\omega})$ are complex valued, which means that there are $2L$ degrees of freedom and $2L$ equations needed to determine the $2L$ unknowns. The samples of the frequency response are evenly spaced over the frequency range of ω from $-\pi$ to π or from 0 to 2π.

Most practical filter design problems have constraints. The impulse response $h[n]$ is usually real, which means that the real part of $H(e^{j\omega})$ must be an even function and the imaginary part must be odd. Thus there are only L degrees of freedom. If the frequency response has linear phase, the impulse response is symmetric and therefore has about half as many degrees of freedom as its length. There is an inherent difference in the frequency response of even- and odd-length filters that can be important and must be understood. The frequency response of an even-length linear-phase filter must be zero at $\omega = \pi$. The group delay (or phase slope) is an integer for an odd-length filter and an odd multiple of one half for an even length.

Frequency Response

The frequency response of a filter can be expressed in several ways. $H(e^{j\omega})$ can be decomposed into its real part and its imaginary part, both real-valued functions of ω, but this is usually not what is desired. In most cases, certainly for linear-phase filters, the magnitude with its associated phase is the preferred form. This is

$$H(e^{j\omega}) = |H(e^{j\omega})|\, e^{j\phi(\omega)} \tag{1-4}$$

where

$$|H(e^{j\omega})| = \sqrt{\mathrm{Re}\{H(e^{j\omega})\}^2 + \mathrm{Im}\{H(e^{j\omega})\}^2} \tag{1-5}$$

and

$$\phi(\omega) = \tan^{-1}\left(\frac{\mathrm{Im}\{H(e^{j\omega})\}}{\mathrm{Re}\{H(e^{j\omega})\}}\right) \tag{1-6}$$

This magnitude–phase description of the complex-valued $H(e^{j\omega})$ has problems when $H(z)$ has zeros exactly on the unit circle. $|H(e^{j\omega})|$ will not be analytic; it will have cusps at its zeros, and the phase will have discontinuities equal to an odd multiple of π. These problems can be eliminated by using the amplitude $A(\omega)$ rather than the magnitude $|H(e^{j\omega})|$, where

$$H(e^{j\omega}) = A(\omega)\, e^{j\theta(\omega)} \tag{1-7}$$

and

$$A(\omega) = \pm|H(e^{j\omega})| \tag{1-8}$$

is real but may take positive or negative values determined by what is necessary to make $A(\omega)$ smooth and to remove the discontinuities of π in the phase. $\theta(\omega)$ is the phase response consistent with $A(\omega)$ and is equal to $\phi(\omega)$ with the discontinuities removed. If $h[n]$ is real, $A(\omega)$ and $|H(e^{j\omega})|$ are even real functions of ω and $\phi(\omega)$ and $\theta(\omega)$ are odd functions. An important distinction between the two descriptions for linear-phase filters is that $H(e^{j\omega})$ and $|H(e^{j\omega})|$ are periodic in ω with period 2π. But $A(\omega)$ is periodic with period 2π if L is odd and with period 4π if L is even.

The usual definition of a linear phase filter is one whose amplitude–phase description has the phase given by

$$\theta(\omega) = K\,\omega \tag{1-9}$$

A strictly linear phase filter defined in terms of the magnitude with $\phi(\omega) = K\omega$ is usually too restrictive. The more useful definition in terms of the amplitude is

$$H(\omega) \;=\; A(\omega)\, e^{-jK\omega} \tag{1-10}$$

which allows working with the real function $A(\omega)$ and places all of the phase information in the one number K, the group delay.

Frequency-Sampling Design Formulas

Rather than use the general IDFT, special design formulas can be derived that include the restrictions of $h[n]$ being real and the phase being linear. These are in terms of L samples of the amplitude $A(\omega)$ over ω from zero to 2π and are given by

$$A_k = A\left(\frac{2\pi k}{L}\right) \tag{1-11}$$

The impulse response for L odd can be derived from (1-3) and (1-7) to be

$$h[n] = \frac{1}{L}\left[A_0 + \sum_{k=1}^{M} 2A_k \cos(2\pi(n-M)k/L)\right] \tag{1-12}$$

where the phase constant of linearity in (1-9) is given by

$$K = -M = -\tfrac{1}{2}(L-1) \tag{1-13}$$

giving

$$H_k = A(2\pi k/L)\, e^{-j2\pi k(L-1)/L} \tag{1-14}$$

as the samples of the frequency response in terms of samples of the amplitude and samples of the linear phase.

If the length is even, the impulse response is

$$h[n] = \frac{1}{L}\left[A_0 + \sum_{k=1}^{L/2-1} 2A_k \cos(2\pi(n-M)k/L)\right] \tag{1-15}$$

with the same phase constant of linearity as (1-13).

The schemes discussed above assume that the frequency samples are $\omega_k = 2\pi k/L$. The other possible evenly spaced sampling scheme is $\omega_k = (2k+1)\pi/L$, which gives the design formula for an odd length as

$$h[n] = \frac{1}{L}\left[(-1)^n A_M + \sum_{k=0}^{M-1} 2A_k \cos\left(2\pi(n-M)\left(k+\frac{1}{2}\right)/L\right)\right] \tag{1-16}$$

again with $M = (L-1)/2$. The impulse response for an even-length filter is

$$h[n] = \frac{1}{L}\left[\sum_{k=0}^{L/2-1} 2A_k \cos\left(2\pi(n-M)\left(k+\frac{1}{2}\right)/L\right)\right] \tag{1-17}$$

Design by Solving Simultaneous Equations

If the amplitude has unequally spaced samples, neither the IDFT nor the formulas can be used; instead, the set of simultaneous complex equations formed from

$$H(e^{j\omega_k}) = A(\omega_k)\, e^{-jM\omega_k} = \sum_{n=0}^{L-1} h[n]\, e^{-j\omega_k n} \tag{1-18}$$

must be solved. If the filter has linear phase and the length is odd, the equations are real and given by

$$A(\omega_k) = \sum_{n=0}^{M-1} 2h[n]\cos(\omega_k(M-n)) + h(M) \tag{1-19}$$

with $M = (L-1)/2$. Since there are $M+1$ unknown $h[n]$ values, there must be $M+1$ equations which require $M+1$ samples of $A(\omega)$ given.

If the filter has linear phase and the length is even, the equations are real and given by

$$A(\omega_k) = \sum_{n=0}^{L/2-1} 2h[n]\cos(\omega_k(M-n)) \tag{1-20}$$

This case requires $L/2$ samples and equations.

In this project we consider three methods of frequency sampling design of linear phase FIR filters: the use of the IDFT with (1-3), the use of explicit formulas with (1-12) through (1-20), and the solution of simultaneous equations with (1-19), (1-20). Each has its advantages and disadvantages.

Most textbooks call the odd- and even-length linear-phase FIR filters types I and II filters. If the phase response is linear plus a constant $\pi/2$, the odd- and even-length FIR filters are called types III and IV. These are used for differentiators and Hilbert transformers where the constant plus linear phase response is desired. They have design formulas similar to those of types I and II but with sine expansions [3].

It is somewhat surprising to note that the analysis and design formulas (1-19) and (1-12) are the discrete cosine transform and its inverse [3]. Indeed, the whole family of discrete cosine and discrete sine transforms are the same as the analysis and design formulas for linear-phase FIR filters with even or odd lengths using different sampling schemes.

Hints

In using the IDFT for the frequency-sampling filter design method with MATLAB, the `fft` command is used. It is fairly fast if the length L is composite (very fast if $L = 2^M$), but slow if the length is prime. The inverse FFT is implemented by an M-file that calls `fft`. If unequally spaced samples are used, the equations given by (1-19) and (1-20) are written as $A = Fh$, where A is the column vector of samples of the desired frequency response amplitude, F is the square matrix of cosines from (1-19) or (1-20), and h is the unknown vector of half the impulse response. These are solved in MATLAB via `h = F \ A`.

When plotting any function of an integer, such as the impulse response of a filter $h[n]$, use `stem`. The frequency response of the filters designed in this chapter can be calculated via `freqz`. In most cases a fairly large number of frequency response values should be calculated to give a smooth graph when plotted. In most cases both the magnitude and phase should be calculated and plotted.

The location of the zeros of the transfer function of the filter can be calculated and plotted easily in MATLAB. The command `z = roots(h)` will factor the polynomial with $h[n]$ as coefficients, and `plot(z,'o')` will plot the imaginary part versus the real part and place small circles at the location of the zeros of the polynomial.[1]

Most of the exercises will consider an ideal linear phase low-pass filter with an amplitude response of

$$A(\omega) = \begin{cases} 1 & \text{if } 0 \le \omega \le \omega_0 \\ 0 & \text{if } \omega_0 < \omega \le \pi \end{cases} \tag{1-21}$$

and some may include a transition band between the passband and stopbands such as

$$A(\omega) = \begin{cases} 1 & \text{if } 0 \le \omega \le \omega_p \\ \dfrac{\omega_s - \omega}{\omega_s - \omega_p} & \text{if } \omega_p < \omega < \omega_s \\ 0 & \text{if } \omega_s < \omega \le \pi \end{cases} \tag{1-22}$$

where ω_p is the edge of the passband, ω_s the edge of the stopband, and ω_0 the average band edge $\omega_0 = (\omega_p + \omega_s)/2$. These and perhaps others must be sampled in order to design the filter.

The first exercise illustrates the explicit form of the necessary MATLAB commands. Some of these should be put in the form of a function to be used in the other exercises or in other projects. The later exercises expect you to create the appropriate command or command sequence. Recall that MATLAB starts its addressing of elements in a vector at 1, yet many mathematical formulas start their variables at zero. Be careful to take this difference into account.

The length L of an FIR filter is the total number of coefficients. This is sometimes confused with the order of the filter, which is $L-1$. The number of zeros of the z-transform transfer function is equal to the order of the filter, which is one less than its length.

There is often confusion of cycle-per-second frequency in hertz, usually denoted f, with radian-per-second frequency, usually denoted ω. Recall that they are related by $\omega = 2\pi f$. If you find an error of around 6, you probably used f for ω, or vice versa.

EXERCISE 1.1

Design a Low-pass Filter

Design a length-23 linear-phase FIR low-pass filter to approximate an ideal response that has a passband edge of $\omega_0 = 0.3\pi$. Assume a unity sampling rate, which gives a Nyquist frequency of $\omega = \pi$.

a. Form a vector of the samples of the ideal amplitude over frequencies from zero to π. This amplitude vector will have ones for frequencies from zero to the band edge and zero from there up to π. Required symmetries determine the amplitude for frequencies from π to 2π. Show that the following MATLAB commands do this.

```
pass = fix(w0*L/(2*pi)) + 1;
if rem(L,2)==0, s = -1; else s = 1; end;
Ad = [ones(1,pass), zeros(1,L-2*pass+1), s*ones(1,pass-1)];
```

Plot `Ad` to see the ideal amplitude frequency response. What happens if one of the samples falls on ω_0? Explain why the line with the `if` statement is necessary.

b. Create a phase vector that when multiplied point by point by the amplitude vector gives the complex linear-phase frequency response. Recall that $j = \sqrt{-1}$. Explain how this is done by

---

[1] See `help zplane.m`.

```
M = (L-1)/2;
k = [0:L-1];
p = exp(2*pi*j*(-M)*k/L);
```

The sampled frequency response vector is simply a term-by-term product of this phase vector and the amplitude vector and is done by `H = Ad.*p;`.

c. Design the filter by using the IDFT with the MATLAB command

```
h = ifft(H)
```

This should be the real, symmetric impulse response $h[n]$. Remove any small or zero imaginary part by `h = real(h);`. If the imaginary part is not small, you have made a mistake, probably in the symmetries of H. Plot $h[n]$ using the `stem` command.

d. Test the filter by calculating its magnitude frequency response. This can be done over a large number of frequencies by

```
Mag = abs(fft(h,512));
w = [0:255]*pi/256;
plot(w, Mag(1:256));
```

Does it look like a good low-pass filter with the correct cutoff frequency? Why do we plot only half of `Mag`? Plot the ideal amplitude response from `Ad` in part (a) superimposed with this plot of `Mag` to see if the actual frequency response interpolates the samples of the ideal response as claimed. Try using the `freqz` to calculate the frequency response. Which method is easier and/or more versatile?

e. Calculate and plot the amplitude response by removing the linear phase from $H(e^{j\omega})$. This is done from (1-7) as $A(\omega) = H(e^{j\omega})\, e^{-j\theta(\omega)}$ with $\theta(\omega) = -M\,\omega = -((L-1)/2)\,\omega$ from (1-13). Show that this can be done by

```
M = (L-1)/2;
k = [0:511];
p = exp(2*pi*j*M*k/512);
Amp = real(p.*fft(h,512));
plot(Amp(1:256));
```

The magnitude and amplitude should be the same except where the amplitude is negative. Is that true for your plots?

f. An alternative method to calculate the frequency response can be used to test if the frequency response passes through the specified points. This is done by appending a multiple of 23 zeros to $h[n]$ and taking the DFT. Some of the calculated values will be at the frequencies of the original desired samples. Compare the frequency response of your designed filter at these points to see if the magnitude is the appropriate one or zero. Plot the magnitude or amplitude and the ideal magnitude or amplitude on the same graph to show the interpolation.

g. Plot the phase versus frequency to see if it is linear with the appropriate constant of linearity. Because phase is ambiguous modulo 2π and because a change of sign for the amplitude is equivalent to a phase shift of π, the phase plot may have surprising jumps. Notice and explain the size and location of the jumps or discontinuities in the phase plot. Also notice the unpredictable phase where the magnitude is zero. Investigate the use of the `unwrap` command to remove some of these jumps. Explain.

h. Plot the location of the zeros on the complex z-plane with `zplane.m` or via

```
plot(roots(h),'o');
```

Relate the zero locations to the shape of the frequency response plots. Relate the number of zeros to the length of the filter.

EXERCISE 1.2

Phase Response

If one calculates the phase response using the `phase` command, it is the $\theta(\omega)$ phase that goes with the magnitude that is produced. As was discussed in the project description above, this phase has jumps of π each time the amplitude changes sign even though a linear phase filter is being analyzed. This was observed in Exercise 1.1(g). For this exercise write an M-file function that removes the jumps of π from $\theta(\omega)$ to give $\phi(\omega)$. Remember that jumps of 2π can occur because of the mathematical ambiguity of angle modulo 2π. The `unwrap` command tries to remove these jumps. You may want to look at the unwrap M-file to see how that is done. Test your function on the phase curve of a linear phase filter and show that it indeed produces a linear phase.

EXERCISE 1.3

Use the Wrong Phase

Experiment with designing filters with the same desired amplitude as the first exercise, but with a phase that is not consistent with the filter length. Note especially the frequency response between samples.

a. Use a zero phase shift. In other words, use $H(e^{j\omega}) = A(\omega)$. How does the filter $h[n]$ compare with the one designed with the proper phase in Exercise 1.1? How does the magnitude response compare? Does the magnitude interpolate the desired samples the way frequency sampling design should? Plot the locations of the zeros of the transfer function and compare them with the locations when the proper phase is used in Exercise 1.1.

b. Use a phase shift as if the length were 11, but design a length-23 filter as in Exercise 1.1. How does the filter $h[n]$ compare with the one designed with the proper phase in Exercise 1.1? How does the magnitude response compare? Does it interpolate the desired samples? Plot the zero locations.

c. What phase shifts always give the same magnitude responses? Why?

d. Try other phase responses and discuss the effects on the final design and its actual amplitude response.

EXERCISE 1.4

Design of Even-Length Filters

To see the difference in odd- and even-length FIR filters, repeat Exercise 1.1 for $L = 22$. In this case the constant of linearity for the phase $M = (L - 1)/2$ is not an integer. It is an odd multiple of 0.5. Calculate and plot the magnitude and phase frequency response. Check values at the sample points. Plot the location of the zeros. Why does a symmetric even-length linear-phase FIR filter always have a zero at $\omega = \pi$? Compare the amount of overshoot near the band edge with the design for $L = 23$.

EXERCISE 1.5

Derive Design Formulas

Derive (1-12) from the linear phase condition and (1-3). Derive (1-16).

EXERCISE 1.6

Design by Formula

Since formulas can also be used for frequency sampling design of FIR filters, use (1-12) to design an FIR filter with the same specifications as given in Exercise 1.1. Do you get exactly the same filter as using the DFT? How would one choose which method to use?

EXERCISE 1.7

Design by Solving Simultaneous Equations

Design a length-23 FIR filter with the same specifications as used in Exercise 1.1 but solve for the filter coefficients by directly solving the simultaneous equations in (1-19) or (1-20). How many equations and samples are necessary? Remember that the number of equations and unknowns should be equal; that is, the matrix of cosines should be square. Are any redundant and can be removed? Are there any differences in designing even- and odd-length filters other than using the correct set of equations? How does this design method compare with the use of formulas or the use of the DFT?

EXERCISE 1.8

Alternative Sampling

Use the alternative sampling scheme implemented in formula (1-16) to design a low-pass filter with the same specifications as given above. What is the difference in the frequency response of the filters designed with the usual sampling scheme and with the alternative scheme? Which gives a closer approximation to the desired band edge ω_0. Explain why for some ω_0 the usual scheme is better, and for the others, the alternative scheme is better. For the two sampling methods and for even and odd lengths, when is there a sample at $\omega = 0$? When at $\omega = \pi$? Can you devise a method to use the IDFT and achieve the alternative sampling scheme design? Can you devise a method to use the solution of simultaneous equations with this sampling scheme?

EXERCISE 1.9

Gibbs Phenomenon

A Gibbs-type phenomenon occurs in the frequency sampling design of filters, much the same as it does with a direct Fourier expansion of a desired frequency response with a discontinuity. In the case of the Fourier transform, the peak of the overshoot is approximately 9% of the size of the discontinuity. What is the size of the corresponding overshoot in the frequency response of a filter designed by the frequency sampling method?

■ ■ PROJECT 2: USE OF A TRANSITION BAND IN THE FILTER SPECIFICATIONS

The exercises in the first project all use ideal amplitudes that are one or zero at each frequency sample. This means that the transition from passpass to stopband takes place in one frequency sampling interval, which is as fast a transition as can be specified using frequency sampling. The sharp transition causes all the filters to have a rather large amount of oscillation in the frequency response near the band edge.

In this project we consider the effects of introducing a transition band between the passband and stopband to give more flexibility in stating specifications and to reduce the size of the oscillation (Gibbs-type effect) in the amplitude response.

Project description

There are two ways to introduce a transition band when using the frequency-sampling design method. One is to specify a transition *function* for the desired ideal amplitude response. This ideal response is uniformly sampled and the filter is designed as was done in the preceding project by inverse DFT, formulas, or solving simultaneous equations.

The second approach uses nonuniform sampling by placing no samples in the transition band. This sets no constraints on the actual response in the transition band, and therefore it could do something unexpected. This second method causes the transition band to be a "don't care" region. Because of the nonuniform spacing of the samples, the only method to carry out this design is to solve simultaneous equations. The DFT or formulas cannot be used. Since the "don't care" transition band causes nonuniform spacing of frequency samples anyway, the samples in the passband and stopband do not have to be spaced uniformly as they do in the first approach.

Hints

The design of filters with a desired transition band using uniformly spaced samples uses the same methods as were used for Project 1. Design with nonuniformly spaced samples is trickier and care must be taken to place the samples where you want them and to have a consistent set of equations. In all the exercises we use the same specifications as in Exercise 1.1: $L = 23$ and $\omega_0 = 0.3\pi$.

EXERCISE 2.1

Transition Band

Modify the transition band of the ideal amplitude response (1-21) used earlier, which changes from passband to stopband in one frequency sampling interval to an ideal similar to (1-22). In this exercise use two sampling intervals by changing the sample nearest the band edge to one-half rather than 1 or zero and design an FIR filter with the same specifications as those used in Exercise 1.1. This should be consistent with a passband edge being approximately $\omega_p = 0.174\pi$ and the stopband edge being approximately $\omega_s = 0.347$. Compare the frequency response of this filter with that in Exercise 1.1 in terms of the rate of dropoff between the passband and stopband and in terms of the overshoot. Use both the inverse DFT and the formulas as was done in Project 1 and make sure that they give the same results. Notice the trade-off of these two characteristics.

EXERCISE 2.2

Relation of Transition Bandwidth to Overshoot

To further investigate the relationship of transition bandwidth and overshoot or Gibbs effect, make the transition three sample intervals wide, with the samples in the transition band of 0.667 and 0.333. This results from the passband edge being approximately $\omega_p = 0.174\pi$ and the stopband edge being approximately $\omega_s = 0.435$ and a linear transition band function. This gives a wider or slower transition and less overshoot. Try a transition bandwidth of four and five sample intervals width and discuss the effects on the overshoot.

EXERCISE 2.3

Optimize in the Transition Band

Create a transition band between the passband and stopband using the same specifications as in Exercises 2.1 and 2.2. Rather than use the linear transition function shown in (1-22), experiment with values at the frequency samples in the transition band to reduce the overshoot or maximum oscillations in the passband and stopband. What changes from the use of a straight line do you find? How much reduction in overshoot can you obtain? Do this for the case with one sample in the transition band and for two samples. In the first case you have one sample value to adjust, and in the second you have two. (This process could be automated by using linear programming to minimize the overshoot [1].)

EXERCISE 2.4

Don't-Care Transition Band

In this exercise we do not adjust the values of the frequency samples; we adjust where they are located. Create a transition band between the passband and stopband in the ideal frequency response that is the equivalent of first two frequency sample intervals, then three, using the same specifications as in the previous exercises. Within the passband and stopband, set the samples evenly spaced with as close to the same spacing in the passband and stopband. Within the transition band, set no samples.

This will cause the frequency samples to be spaced unevenly over the total frequency range and therefore prevent the use of the IDFT or the formulas in (1-12) through (1-17). Use formula (1-19) to obtain equations that can be solved by MATLAB to design the filter. How do the frequency response and zero locations compare with the filters designed by the IDFT or formulas in Exercise 2.3?

EXERCISE 2.5

Use of Unequal Sampling Densities

This exercise is a continuation of Exercise 2.4. Try using twice the sampling density in the stopband as in the passband but with the same total number of 23. What is the result? Try using a closer spacing of the samples near the transition band. Describe (only in a qualitative way) the effect.

PROJECT 3: FIR FILTER DESIGN BY DISCRETE LEAST-SQUARED-ERROR APPROXIMATION

The square of a signal or the square of an error is a measure of the power in the signal or error. This is clear if the signal is a voltage, current, force, or velocity. The time integral of the power of a signal is its energy and is often an important measure of the signal. In many practical problems, the integration cannot be carried out mathematically and is, therefore, approximated by a finite summation. It is this finite sum of the square of the difference between the desired frequency response and the actual frequency response that we will use as our approximation measure in this project. For equally spaced frequency samples, Parseval's theorem states that an optimal frequency-domain approximation implies an optimal time-domain approximation.

Project description

The discrete squared error measure is defined by

$$\epsilon = \frac{1}{N} \sum_{k=0}^{N-1} \left| H_d(e^{j\omega_k}) - H(e^{j\omega_k}) \right|^2 \tag{3-1}$$

where $H_d(e^{j\omega})$ is the desired ideal frequency response, $H(e^{j\omega})$ is the actual response of the length-L filter given by (1-2) and (1-19), and N is the number of frequency points over which the error is calculated. If the number of independent filter coefficients $h[n]$ is equal to the number of requirements or equations set in (1-2), it is possible to choose the $h[n]$ such that there is no error. This is what was done in the frequency sampling design method in Project 1. By choosing $N \gg L$, the summed squared error in (3-1) (appropriately normalized) approaches the integral squared error, which is often what is actually wanted in approximating $H_d(e^{j\omega})$.

Using Parseval's theorem, one can show that symmetrically truncating a length-N filter designed by frequency sampling will give a length-L filter whose frequency response is an optimal approximation to $H_d(e^{j\omega})$ in the sense that ϵ in (3-1) is minimized. This is true only for equally spaced samples because that is the requirement of Parseval's theorem. This result is similar to the fact that a truncated Fourier series is an optimal least squared error approximation to the function expanded. This exercise considers only linear-phase FIR filters; therefore, the long filter that is to be truncated may be designed by using the IDFT or the formulas in (1-12) and (1-15).

If the frequencies are not equally spaced, truncation will not result in an optimal approximation. If $N > L$, the equations given by (1-19) and (1-20) are overdetermined and may be written in matrix form as

$$A = Fh \tag{3-2}$$

where A is the length-N vector of samples of $A(\omega)$, F is the $N \times L$ matrix of cosines, and h is the length-L vector of the filter coefficients. Because of $A(\omega)$ being an even function, only $L/2$ terms are needed in A. Although these equations are overdetermined, they may be solved approximately in MATLAB with `h = F \ A`. MATLAB implements this operation with an algorithm that minimizes the error in (3-1).

The goal of this project is to learn how to design filters that minimize the discrete squared error, to understand the properties of this design method, and to examine the properties of the filters so designed.

Hints

The DFT implemented in the MATLAB function `fft` or an M-file implementation of formulas (1-12) through (1-17) will be used to design the long filter to be truncated for most of the exercises in this project. In some cases we will solve sets of overdetermined equations to obtain our approximations. Read the manual and use `help` on \ and / to learn about the approximate solution of overdetermined equations. It might be helpful to read about least squared error methods in references such as [6].

Many of the design and analysis methods used in this project are an extension of those in Project 1. Review the description and discussion in that project. To analyze, evaluate, and compare the filters designed by the various methods in this project, magnitude or amplitude frequency response plots, plots of zero locations, and plots of the filter itself should be made. It would be efficient to create special M-file functions that efficiently make these plots.

EXERCISE 3.1

Design an Odd-Length Low-Pass Filter

Design a length-23 linear-phase FIR low-pass filter to approximate an ideal response that has a passband edge of $\omega_0 = 0.3\pi$. Assume a unity sampling rate that gives a Nyquist frequency of $\omega = \pi$. Use the frequency sampling method described in Project 1 to design three filters of lengths 45, 101, and 501. Truncate them to symmetric filters of length-23 and compare them with each other. Compare the frequency responses and zero locations.

EXERCISE 3.2

How Many Samples Should Be Used?

If one really wants to minimize the integral squared error but must use the discrete least squared error method, what is the ratio of N, the length of the filter to be truncated, to L, the length of the filter, to obtain close results? This can be determined by running an example for a carefully chosen set of lengths.

EXERCISE 3.3

Residual Error and Parseval's Theorem

Verify Parseval's relation of the time-domain and frequency-domain sums of squares by calculating the approximation error of the length-23 filter designed in Exercise 3.1 using a frequency sampling length of 101. Do this in the frequency domain using (3-1) and the time domain from the sum of the squares of the truncated terms of $h[n]$.

EXERCISE 3.4

Use of Overdetermined Simultaneous Equations

Design a length-23 filter with the specification from Exercise 3.1, but use the solution of overdetermined simultaneous equations of (1-19) rather than truncation of a longer filter. Do this for the three values of N of 45, 101, and 501. You should get the same results as in Exercise 3.1. How does the design time compare? Are there numerical problems?

EXERCISE 3.5

Use a Transition Band

Design a length-23 filter with specifications similar to those in Exercise 3.1 but with a transition band. Let the passband be the range of frequencies $\{0 \leq \omega \leq 0.25\pi\}$, the stopband be the range $\{0.35\pi \leq \omega \leq \pi\}$, and the transition be a straight line connecting the two as described in (1-22). Use the same least discrete squared error criterion used in Exercise 3.1 for the same number of frequency samples. Compare the results with each other and with those in Exercise 3.1. Plot the frequency response of the three filters on the same graph to compare. Comment on the reduction of passband and stopband ripple versus the increase of transition band width.

EXERCISE 3.6

Don't-Care Transition Band

Design a length-23 filter with the same specifications as Exercise 3.5, but with no frequency samples in the transition band. This means that the frequency samples are not equally spaced and simultaneous equations from (1-19) will have to be solved approximately as was done in Exercise 3.4 or 2.4. Do this for the same three numbers of frequency samples. Compare with the results of Exercise 3.5 by plotting the frequency responses on the same graph. What difference would an even length cause?

EXERCISE 3.7

Weighting Functions

It is possible to use a weighting function in the definition of error.

$$\epsilon = \sum_{k=0}^{L-1} W_k \left| H_d(e^{j\omega_k}) - H(e^{j\omega_k}) \right|^2 \tag{3-3}$$

Derive a matrix formulation for the set of overdetermined simultaneous equations describing this problem. Design a length-23 filter with the same specifications as in Exercise 3.6 but with 10 times the weight on the stopband squared error as on the passband squared error. Discuss the result and the design process.

EXERCISE 3.8

Numerical Problems

If long filters with wide transition bands are designed, the simultaneous equations to be solved will be nearly singular. This ill-conditioning seems to be a function of the product of the filter length and the transition bandwidth. For what length–bandwidth product does this start to occur? Plot the frequency response of some filters designed by solving ill-conditioned equations. What are their characteristics?

LEAST-SQUARES DESIGN OF FIR FILTERS

OVERVIEW

The use of an approximation measure that is the integral of the square of the error is often used in filter design since it is in some ways a measure of the energy of the error. It is also attractive because Parseval's theorem states that an optimal approximation in the frequency domain using the integral squared error criterion will also be an optimal approximation in the time domain. In general, one cannot analytically solve this optimization problem, but in several important cases, analytical solutions can be obtained. In this section we investigate the design of low-pass, high-pass, bandpass, and band reject filters, with and without transition bands. We then look at the use of window functions to reduce the Gibbs effect that is sometimes undesirable.

While the main purpose of the first project was to learn how to design an FIR filter by a particular method and to understand the characteristics of that method, the goal of this project is one of comparisons. We want to see what characteristics are important and what trade-offs result from the different design methods. To that end we consider the integral squared error, the Chebyshev (maximum difference) error, the transition bandwidth, and the filter length.

BACKGROUND READING

The basic least integral squared error of the no-transition-band low-pass filter is discussed in most DSP textbooks as the Fourier series expansion method. It is discussed in [3] as a least-squared error method. The use of spline transition bands is developed in [3] and [7].

PROJECT 1: FIR FILTER DESIGN BY LEAST INTEGRAL SQUARED ERROR APPROXIMATION

In many filter design problems, it is the least integral squared error approximation that is desired. Although in most of these cases, the problem cannot be solved, there are a few important cases that do have analytical solutions. In this project we examine the ideal linear-phase low-pass filter with no transition band which gives a $\sin(x)/x$-form impulse response. We will also consider ideal frequency response with transition bands and will compare results with those obtained numerically using discrete least-squared error methods.

Project description

For the complex frequency response we use the discrete-time Fourier transform as was developed in the project description for the section *Discrete Design of FIR Filters*. This is

$$H(e^{j\omega}) = \sum_{n=-\infty}^{\infty} h_d[n]\, e^{-j\omega n} \tag{1-1}$$

For FIR length-L filters this is

$$H(e^{j\omega}) = \sum_{n=0}^{L-1} h_d[n]\, e^{-j\omega n} \tag{1-2}$$

For linear-phase FIR filters we define the amplitude $A(\omega)$ as a real-valued function and the phase $\psi(\omega) = M\omega$ as a continuous linear function such that

$$H(e^{j\omega}) = A(\omega)\, e^{jM\omega} \tag{1-3}$$

where $M = (L-1)/2$ gives the least phase shift for a causal filter. Read the earlier project description for details.

The integral square error measure is defined by

$$\epsilon = \frac{1}{2\pi}\int_{-\pi}^{\pi} \left|H_d(e^{j\omega}) - H(e^{j\omega})\right|^2 d\omega \tag{1-4}$$

where $H_d(e^{j\omega})$ is the desired ideal frequency response and $H(e^{j\omega})$ is the actual frequency response of the length-L filter. Because of the orthogonality of the basis functions of the discrete-time Fourier transform, Parseval's theorem states that this same error can be given in the time domain by

$$\epsilon = \sum_{n=-\infty}^{\infty} |h_d[n] - h[n]|^2 \tag{1-5}$$

where $h_d[n]$ is the inverse DTFT of $H_d(e^{j\omega})$ and $h[n]$ is the length-L impulse response of the filter being designed.

If $h_d[n]$ is infinitely long, but symmetric, and $h[n]$ is of length L, for an odd length (1-5) can be written in two parts as

$$\epsilon = \sum_{n=-M}^{M} |h_d[n] - h[n]|^2 + \sum_{n=M+1}^{\infty} 2h_d^2[n] \tag{1-6}$$

It is clear that choosing the unknown $h[n]$ to be equal to the given $h_d[n]$ minimizes ϵ in (1-6) and, therefore, in (1-4). In other words, symmetric truncation of $h_d[n]$ gives the optimal approximation of the frequency response. The only problem in using this result is the fact that the inverse DTFT (IDTFT) of a desired frequency response can often not be calculated analytically.

The ideal, no-transition-band low-pass filter has a frequency response with amplitude described by

$$A_d(\omega) = \begin{cases} 1 & \text{if } 0 \le \omega \le \omega_0 \\ 0 & \text{if } \omega_0 < \omega \le \pi \end{cases} \tag{1-7}$$

where ω_0 is the band edge between the passband and stopband. The inverse DTFT of $A_d(\omega)$ is

$$h_d[n] = \frac{1}{2\pi}\int_{-\pi}^{\pi} A_d(\omega)\, e^{j\omega n}\, d\omega = \frac{\sin(\omega_0 n)}{\pi n} \tag{1-8}$$

Since we inverted $A_d(\omega)$ rather than $H_d(e^{j\omega})$, this impulse response is noncausal and infinite in both positive and negative time. If we invert $H_d(e^{j\omega})$ and truncate, we have the optimal result of

$$h[n] = \begin{cases} \dfrac{\sin(\omega_0(n-M))}{\pi(n-M)} & \text{for } 0 \le n \le L-1 \\ 0 & \text{otherwise} \end{cases} \tag{1-9}$$

If a transition band is included in the ideal frequency response, a transition function must be specified. If that transition function is a straight line (first-order spline), the ideal amplitude is

$$A(\omega) = \begin{cases} 1 & \text{if } 0 \le \omega \le \omega_p \\ \dfrac{\omega_s - \omega}{\omega_s - \omega_p} & \text{if } \omega_p < \omega < \omega_s \\ 0 & \text{if } \omega_s \le \omega \le \pi \end{cases} \tag{1-10}$$

where ω_p is the edge of the passband, ω_s is the edge of the stopband, and ω_0 is the average band edge $\omega_0 = (\omega_p + \omega_s)/2$. The IDTFT of $A(\omega)$ is

$$h_d[n] = \frac{\sin(\omega_0 n)}{\pi n} \left[\frac{\sin(\Delta n/2)}{\Delta n/2} \right] \tag{1-11}$$

where $\Delta = \omega_s - \omega_p$ is the transition bandwidth. This has the form of the ideal no-transition-band impulse response of (1-8) multiplied by a wider envelope controlled by the transition bandwidth. The truncated and shifted version of this has a frequency response which is an optimal approximation to (1-10).

Although these formulas are described here for odd-length FIR filters, they hold for even lengths as well. Other transition functions also can be used but must be chosen such that the IDTFT can be taken. For cases where the integral in the IDTFT cannot be carried out, one must use the numerical methods described in the projects on discrete methods.

Hints

This project will require designing filters whose coefficients are given by formulas. These should be programmed in M-files so they may easily be used on different specifications. The resulting filters will be examined by plotting their magnitude or amplitude response, transfer-function zero locations, and impulse response. This should be reviewed from the projects on discrete methods.

An important analysis of filter length can be made by plotting the approximation error in (1-4) versus the filter length L. This can be calculated by approximating the integral by a summation over a dense grid of frequency samples or by calculating the major part of the second term in (1-6). An approximation-error-evaluating function should be written to calculate this error efficiently. It can be put in a loop to calculate the needed error versus length data.

When we say "examine" or "compare," it should be in terms of the integral squared approximation error, the maximum difference between the desired frequency response and the actual one, the width of the transition band, and the length of the filter. Look for and describe the trade-off of these characteristics for the various design methods.

EXERCISE 1.1

Design an Odd-Length Low-Pass Filter

Design a length-23 linear-phase FIR low-pass filter to approximate an ideal response that has a passband edge of $\omega_0 = 0.3\pi$ using the formula derived in (1-9). Assume a unity sampling rate that gives a Nyquist frequency of $\omega = \pi$. Plot the impulse response with the `stem` command. Plot the magnitude or amplitude frequency response of the filter. Plot the transfer-function zero locations. How do these compare with the designs of filters with the same specifications in the projects using discrete methods? The results should be fairly close to those of the discrete squared error designs with large numbers of frequency samples but should be noticeably different for designs with fewer samples or with the frequency sampling method. The pure Gibbs phenomenon predicts a maximum overshoot of approximately 9% of the discontinuity. Is that observed in the frequency response of this design?

EXERCISE 1.2

Design a Longer FIR Filter

Design a length-51 FIR filter to the same specifications used in Exercise 1.1. Plot the impulse response, magnitude and phase frequency response, and zero location of the transfer function. You may want to plot the log of the magnitude response for longer filters to better see details in the stopband. How do they compare with similar plots from Exercise 1.1? How does the maximum overshoot (Chebyshev error) compare? How does the integral squared error of the two filters compare?

EXERCISE 1.3

Approximation Error

Plot the approximation error of an optimal length-L filter designed to the specifications of Exercise 1.1 versus the length of the filter. Derive an empirical formula relating the error to the filter length.

EXERCISE 1.4

Use a Transition Band

Design a length-23 filter with specifications similar to those in Exercise 1.1, but with a transition band. Let the passband be the range of frequencies $\{0 \leq \omega \leq 0.25\pi\}$, the stopband be the range $\{0.35\pi \leq \omega \leq \pi\}$, and the transition be a straight line connecting the two. Use formula (1-11). This filter should be compared to the no-transition-band design in Exercise 1.1 by plotting the frequency response. In particular, compare the overshoot, the rate of change from passband to stopband, and the approximation error.

EXERCISE 1.5

Approximation Error

Plot the approximation error of an optimal length-L filter designed to the specifications of Exercise 1.4 versus the length of the filter. Will this curve have regions that are relatively flat? Explain this curve by looking at formula (1-11) and considering the effects of truncation. Derive a formula for the locations of these flat regions from (1-11).

EXERCISE 1.6

Spline Transition Function

The simple straight-line transition function used in (1-10) can be generalized to a pth-order spline [3], which gives an ideal impulse response of

$$h_d[n] = \frac{\sin(\omega_0 n)}{\pi n}\left(\frac{\sin(\Delta n/2p)}{\Delta n/2p}\right)^p \tag{1-12}$$

Design three length-23 filter to the specifications of Exercise 1.4 using values of $p = 1$, 2, and 10 and one filter with no transition band. Plot their amplitude response on the same graph. How does the value of p affect the frequency response? ($p = \infty$ is equivalent to no transition band.)

EXERCISE 1.7

Optimal Order Spline Transition Band

Use the spline transition function method of (1-12) to design length-L FIR filters with specifications of Exercise 1.4. Make a graph that contains plots of approximation error versus length for values of $p = 1, 2, 3, 4$, and 5.

EXERCISE 1.8

Error Formula

Plot the approximation error of an optimal FIR filter with a transition band versus the transition bandwidth Δ for a length of 23 and an average band edge of $\omega_0 = 0.3\pi$. Derive an empirical formula relating the error to the transition bandwidth. Does this result depend significantly on ω_0?

EXERCISE 1.9

Optimal Filter

Analyze the spline transition function method for various lengths, transition bandwidths, and values of p. From equation (1-12) and from the empirical evaluation of error versus length and other parameter curves, derive an empirical formula for an optimal value of p as a function of L and Δ. Write an M-file program that will design an optimal filter from L, ω_p, ω_s by choosing it own value of p and evaluating (1-12). Evaluate the designs.

EXERCISE 1.10

Comparison and Evaluation

Because the inverse discrete-time Fourier transform of many ideal responses cannot be evaluated analytically, the numerical methods of the project on discrete methods must be used. Compare truncated long discrete least-squared error designs to the designs by formula of this project. How many error samples should one take to give results close to the integral? How long can the filters be designed by the two methods?

PROJECT 2: DESIGN OF HIGH-PASS, BANDPASS, AND BAND-REJECT LEAST-SQUARED-ERROR FIR FILTERS

Earlier projects on discrete methods developed FIR filter design methods but illustrated them only on low-pass specifications. In this project we consider the approximation problems of high-pass, bandpass, and band-reject filters using techniques that convert low-pass designs. These techniques are interesting mostly for the analytic methods developed in Project 1, since the numerical methods in the projects on discrete methods can be applied directly to the new specifications. These methods will work only with unweighted least-squared-error designs.

Project description

The bandpass filter with no transition bands has an ideal amplitude response of

$$A_d(\omega) = \begin{cases} 0 & \text{for } 0 \le \omega \le \omega_1 \\ 1 & \text{for } \omega_1 < \omega < \omega_2 \\ 0 & \text{for } \omega_2 \le \omega \le \pi \end{cases} \tag{2-1}$$

where the lower passband edge is ω_1 and the upper is ω_2.

This can be obtained by subtracting the responses of two low-pass filters. It is the response of a low-pass filter with band edge at ω_1 minus the response of a low-pass filter with band edge at ω_2 where $\omega_1 > \omega_2$. Since the ideal responses are

$$A_{bp}(\omega) = A_{lp1}(\omega) - A_{lp2}(\omega) \tag{2-2}$$

and since the IDTFT is linear,

$$h_{bp}[n] = h_{lp1}[n] - h_{lp2}[n]. \tag{2-3}$$

This also holds if the ideal bandpass response has transition bands simply by using transition bands on the low-pass filters. Indeed, it allows different-width transition bands.

The bandpass filter can also be generated by modulating a low-pass filter. Multiplying the impulse response of a low-pass filter by sampled sinusoid will shift its frequency response. This property of the DTFT allows designing a prototype low-pass filter and then multiplying its impulse response by a sinusoid of the appropriate frequency to obtain the impulse of the desired bandpass filter. If used with transition bands, this method will not allow independent control of the two transition bands.

The ideal high-pass filter with no transition band has an amplitude response of

$$A_d(\omega) = \begin{cases} 0 & \text{for } 0 \leq \omega < \omega_0 \\ 1 & \text{for } \omega_0 \leq \omega \leq \pi \end{cases} \tag{2-4}$$

It can be generated by subtracting the response of a low-pass filter from unity or by multiplying the impulse response of a prototype low-pass filter by $(-1)^n$, which shifts the stopband of the low-pass filter to center around $\omega = \pi$.

The ideal band-reject filter with no transition bands has an amplitude response of

$$A_d(\omega) = \begin{cases} 1 & \text{for } 0 \leq \omega \leq \omega_1 \\ 0 & \text{for } \omega_1 < \omega < \omega_2 \\ 1 & \text{for } \omega_2 \leq \omega \leq \pi \end{cases} \tag{2-5}$$

where the lower reject band edge is ω_1 and the upper is ω_2.

This can be obtained by adding the response of a high-pass filter to that of a low-pass filter or by subtracting the response of a bandpass filter from unity. It can also be obtained through modulation by shifting the stopband region of a low-pass prototype filter to the reject band of the new filter.

Hints

The tools for working this project are the same as those for the earlier projects of this chapter. You will need to be able to design low-pass filters easily and reliably with and without transition bands. The "Hints" section of Project 1 in the section *Discrete Design of FIR Filters* and its Exercise 1.1 are very helpful for all the projects in this chapter.

EXERCISE 2.1

Design a Bandpass Filter

Design a least-squared-error linear-phase bandpass FIR filter with the lower passband edge at $\omega_1 = 0.2\pi$ and the upper passband edge at $\omega_2 = 0.3\pi$ using no transition bands and a length of 31. Use the design method that subtracts the designs of two low-pass filters. Plot the impulse response, the amplitude response, and the zero locations. Does it seem like a good approximation to the ideal? Show this by plotting the ideal and actual amplitude responses on the same graph. Do the location of the zeros make sense and agree with the frequency response?

EXERCISE 2.2

Design a Bandpass Filter Using Modulation

Design a bandpass filter using the same specifications as in Exercise 2.1, but use the modulation design method. That will require some care in choosing the band edge of the prototype low-pass filter and the frequency of the sinusoid to achieve the desired ω_1 and ω_2. Check by comparing with the design of Exercise 2.1.

EXERCISE 2.3

Design a Band-Reject Filter

Design a band-reject filter using the same band edges as the bandpass filter in Exercise 2.1. Analyze its impulse response, frequency response, and zero locations. Compare with the ideal.

EXERCISE 2.4

Design a High-Pass Filter

Design a length-23 high-pass FIR filter with a band edge at $\omega_0 = 0.3\pi$. Design it by both the subtraction and the shifting methods. Analyze the filter by plotting the impulse response, amplitude response, and zero locations. Show that you cannot design a high-pass even-length FIR filter. Why is this?

EXERCISE 2.5

Use a Transition Band

Design a length-31 bandpass filter with transition bands. Set the lower stopband as $\{0 \le \omega \le 0.08\pi\}$, the passband as $\{0.1\pi \le \omega \le 0.3\pi\}$, and the upper stopband as $\{0.4\pi \le \omega \le \pi\}$. Analyze its frequency response by plotting the amplitude response on the same graph as the ideal. Look at its other characteristics.

EXERCISE 2.6

Design a Multipassband Filter

Design a multipassband filter of length 51 with no transition bands. Set one passband as $\{0 \le \omega \le 0.2\pi\}$ and a second as $\{0.3\pi \le \omega \le 0.4\pi\}$. The first passband should have a gain of 1 and the second should have a gain of $\frac{1}{2}$. Plot the amplitude response and the ideal on the same graph.

PROJECT 3: FIR FILTER DESIGN USING WINDOW FUNCTIONS

Although the least-squared-error approximation design methods have many attractive characteristics, the Gibbs phenomenon, which is a relatively large overshoot near a discontinuity in the ideal response, is sometimes objectionable. It is the abrupt truncation of the infinitely long ideal impulse response that causes this overshoot, so the use of window functions to truncate the sequence more gently has been developed. The result is a hybrid method that starts out with a least-squared-error approximation but modifies it to reduce the Chebyshev error. The window function method is an alternative to defining $H_d(e^{j\omega})$ with transition bands as in Project 1. This project develops and analyzes several standard window-based FIR filter design methods.

Project description

The window method of FIR filter design starts with the design of a least-squared-error approximation. If the desired filter has a basic low-pass response, the impulse response of the optimal filter given in (1-8) is

$$\hat{h}_d[n] = \frac{\sin(\omega_0 n)}{\pi n} \qquad (3\text{-}1)$$

The shifted and truncated version is

$$h[n] = \begin{cases} \dfrac{\sin(\omega_0(n-M))}{\pi(n-M)} & \text{for } 0 \le n \le L-1 \\ 0 & \text{otherwise} \end{cases} \tag{3-2}$$

for $M = (L-1)/2$. The truncation was obtained by multiplying (3-1) by a rectangle function. Multiplication in the time domain by a rectangle is convolution in the frequency domain by a sinc function. Since that is what causes the Gibbs effect, we will multiply by a window function that has a smoother Fourier transform with lower sidelobes.

One method of smoothing the ripples caused by the sinc function is to square it. This results in the window being a triangle function, also called the Bartlett window; see `triang` and `bartlett` in MATLAB.[2]

The four generalized cosine windows are given by[3]

$$W[n] = \begin{cases} a - b\cos\left(\dfrac{2\pi n}{L-1}\right) + c\cos\left(\dfrac{4\pi n}{L-1}\right) & \text{for } 0 \le n \le L-1 \\ 0 & \text{otherwise} \end{cases} \tag{3-3}$$

The names of the windows and their parameters are:

Window	*MATLAB name*	*a*	*b*	*c*
Rectangular	`boxcar`	1	0	0
Hann	`hanning`	0.5	−0.5	0
Hamming	`hamming`	0.54	−0.46	0
Blackman	`blackman`	0.42	−0.5	0.08

A more flexible and general window is the Kaiser window given by

$$W[n] = \begin{cases} \dfrac{I_0(\beta\sqrt{1-[2(n-M)/(L-1)]^2})}{I_0(\beta)} & \text{for } 0 \le n \le L-1 \\ 0 & \text{otherwise} \end{cases} \tag{3-4}$$

where $M = (L-1)/2$, $I_0(x)$ is the zeroth-order modified Bessel function of the first kind and β is a parameter to adjust the width and shape of the window.

The generalized cosine windows have no ability to adjust the trade-off between transition bandwidth and overshoot and therefore are not very flexible filter design tools. The Kaiser window, however, has a parameter β which does allow a trade-off and is known to be an approximation to an optimal window. An empirical formula for β that minimizes the Gibbs overshoot is

$$\beta = \begin{cases} 0.1102(A-8.7) & \text{for } 50 < A \\ 0.5842(A-21)^{0.4} + 0.07886(A-21) & \text{for } 21 < A < 50 \\ 0 & \text{for } A < 21 \end{cases} \tag{3-5}$$

where

$$A = -20\log_{10}\delta \tag{3-6}$$

$$\Delta = \omega_s - \omega_p \tag{3-7}$$

$$L - 1 = \frac{A-8}{2.285\Delta} \tag{3-8}$$

with δ being the maximum ripple in the passband and stopband. Details for these formulas can be found in your textbook or one of the references.

---

[2]In MATLAB the `triang` and `bartlett` functions give *different*-length windows.

[3]As defined in MATLAB, the denominator is $L-1$; sometimes, it is taken as L.

Because the Bartlett, Hanning, and Blackman windows are zero at their endpoints, multiplication by them reduces the length of the filter by 2. To prevent this shortening, these windows are often made $L + 2$ in length. This is not necessary for the Hamming or Kaiser windows. These windows not only can be used on the classical ideal low-pass filter given in (3-1) or (3-2) but can be used on any ideal response to smooth out a discontinuity.

Hints

MATLAB has window functions programmed, but you will learn more and understand the process better by writing your own window M-files. However, you will find it instructive to examine the MATLAB M-files using the `type` command. The standard filter analysis tools described in Project 1 are useful.

Take care in the choice of the length used in the window function. For some it should be L and others $L + 2$. This difference is sometimes incorporated in the formulas for the windows. Check to see if that is the case.

EXERCISE 3.1

Design a Low-Pass Filter Using Windows

Design a length-23 linear-phase FIR low-pass filter with a band edge of $\omega_0 = 0.3\pi$ using the following windows:

a. Rectangular
b. Triangular or Bartlett
c. Hanning
d. Hamming
e. Blackman

Plot the impulse response, amplitude response, and zero locations of the four filters. Compare the characteristics of the amplitude response of the five filters. Do this in terms of the squared error, the Chebyshev error, and the transition bandwidth. Compare them to an optimal Chebyshev filter designed with a transition band and the least-squared-error filter designed with a spline transition function. How do you choose a transition bandwidth for a meaningful comparison?

EXERCISE 3.2

Design a Bandpass Filter Using Windows

Take the bandpass filter designed in Exercise 2.5 and apply the five windows. Analyze the amplitude response.

EXERCISE 3.3

Use the Kaiser Window

a. Plot the relationship in (3-5) to see the usual range for β. Why is $\beta = 0$ for $A < 21$?
b. Design a length-23 filter using the same specifications as in Exercise 3.1, but using a Kaiser window with $\beta = 4$, 6, and 9. Plot the impulse response, amplitude response, and zero locations of the three filters. Compare them with each other and with the results of Exercise 3.1. How does the trade-off of transition bandwidth and overshoot vary with β?

EXERCISE 3.4

Design of Bandpass Filters with a Kaiser Window

Apply the Kaiser window with the three values of β given in Exercise 3.3 to the bandpass filter as was done in Exercise 3.2. Analyze the amplitude response and compare with the results of Exercise 3.2.

EXERCISE 3.5

Chebyshev Error of the Kaiser Window

a. Set specifications of a length-23 low-pass FIR filter with passband in the range $\{0 \leq \omega \leq 0.3\pi\}$ and stopband in the range $\{0.35\pi \leq \omega \leq \pi\}$. Design a set of filters using the Kaiser window with a variety of values for β. Calculate the Chebyshev error over the passband and stopband using the `max` command in MATLAB. Plot this Chebyshev error versus β and find the minimum. Compare with the value given by the empirical formula in (3-5).

b. *Another Chebyshev error computation*: Repeat part (a) but use the squared error calculated only over the passband and stopband rather than the total $0 \leq \omega \leq \pi$. Run an experiment over various lengths and transition bandwidths to determine an empirical formula for β that minimizes the squared error.

CHEBYSHEV DESIGN OF FIR FILTERS

OVERVIEW

One of the most important error measures in the design of optimal FIR filters is the maximum difference between the desired frequency response and the actual frequency response over the range of frequencies of interest. This is called the Chebyshev error and is what is most obvious in a visual evaluation of a frequency response. When this error is minimized, the error takes on shape that oscillates with equal-size ripples. The minimization of this error in a filter design problem is usually done using the Parks–McClellan algorithm or linear programming. The characteristics of the solution are described by the alternation theorem. In this section we investigate the Remez exchange algorithm used by Parks and McClellan and the use of linear programming.

BACKGROUND READING

Details of the Parks–McClellan algorithm and the Remez exchange algorithm can be found in [1], [3], and [8]. The basic ideas of linear programming can be found in a number of books, such as [9], and the application to filter design can be found in [10]. Programs implementing the Parks–McClellan algorithm are included in the MATLAB command `remez`, and linear programming algorithms are implemented through simplex and quadratic programming, and Karmarkar's methods are available in the MATLAB optimization toolbox or other sources.

PROJECT 1: FIR FILTER DESIGN BY THE PARKS–MCCLELLAN METHOD

In this project we design filters using the Parks–McClellan method to observe the speed of the method, the characteristics of Chebyshev approximations, the extra ripple phenomenon, and the relationship between L, δ_p, δ_s, and $\Delta = \omega_s - \omega_p$. Here L is the length of the FIR filter, δ_p is the maximum of the magnitude of the difference between ideal and actual frequency response over the frequencies in the passband, δ_p is the same maximum value over the frequencies in the stopband, and Δ is the transition bandwidth.

Project description

Consider the characteristics of an optimal length-21 linear-phase low-pass FIR filter as the passband edge is changed. The optimal filter will have an amplitude response with ripples that oscillate around the desired response so that most of the error ripples are of equal magnitude and alternating sign in the passband and stopband. The frequencies where the error takes on these equal maximum sizes are called the *extremal frequencies*. In this project we consider the alternations, ripple, extremal frequencies, and root locations of this filter designed by the Parks–McClellan algorithm in MATLAB. The important band edge frequencies are given below. We have $L = 21$; therefore, $M = (L - 1)/2 = 10$.

The *alternation theorem* states that the optimal Chebyshev approximation must have at least $M + 2 = 12$ extremal frequencies. In the unusual case of the "extra ripple" filter, there will be $M + 3 = 13$ extremal frequencies. For the simple two-band low-pass filter, there can be no more than $M + 3$ extremal frequencies. From this theorem we can show that there will always be an extremal frequency at both band edges, ω_p and ω_s. There is always a ripple at $\omega = 0$ and $\omega = \pi$, but one of them may be a "small ripple" which is not as large as the other ripples, and therefore, 0 or π may or may not be an extremal frequency.

Hints

Use the MATLAB command `remez` to design the filter. You might also try the program developed in Project 2 and/or the linear programming approach of Project 3 to check answers and compare design speeds. Use the `help remez` statement to learn more about the remez command.

All filter design functions in MATLAB use a frequency scaling that is somewhat non-standard. When entering the cutoff frequencies ω_p and ω_s (units of radians), the values must be divided by π. Thus a cutoff frequency specified as $\omega_p = 0.22\pi$ would be entered as `0.22` in MATLAB. This scaling is not the usual normalized frequency where the sampling frequency is 1. Therefore, if the specifications call for a cutoff frequency of 1000 Hz when the sampling frequency is 5000 Hz, you must enter `0.4` in MATLAB, because the cutoff frequency in radians is $\omega_c = 2\pi(1000/5000) = 2\pi(0.2) = 0.4\pi$. In this example, the normalized frequency would be $1000/5000 = 0.2$, so the MATLAB frequency is twice the normalized frequency.

MATLAB uses the order of an FIR filter as a specification rather than the length. You should use $L - 1$ rather than L. When plotting the amplitude frequency response, use a scale that allows seeing the relative size of the ripples to determine if the ripples are of maximum size or not.

EXERCISE 1.1

Design a Length-21 FIR Filter

Use the MATLAB command `h = remez(20, [0,0.4,0.5,1], [1,1,0,0])` to design a length-21 filter with a passband from 0 to $\omega_p = 0.4\pi$ and a stopband from $\omega_s = 0.5\pi$ to π with a desired response of 1 in the passband and zero in the stopband. Plot the impulse response, the zero locations, and the amplitude response. How many "ripples" are there? How many extremal frequencies are there (places where the ripples are the same maximum size)? How many "small ripples" are there that do not give extremal frequencies, and if there are any, are they in the passband or stopband? Are there zeros that do not contribute directly to a ripple? Most zero pairs off the unit circle in the z-plane cause a maximum-size ripple in the passband or stopband. Some cause only a "small ripple," and some cause no ripple.

EXERCISE 1.2

Characteristics versus Passband Edge Location

Design and answer the questions posed in Exercise 1.1 for a set of different passband edges ($f_p = \omega_p/\pi$). Use `h = remez(20, [0,fp,0.5,1], [1,1,0,0])`. Do this for passband edges of

```
fp = 0.2000, 0.207, 0.2100, 0.222, 0.2230, 0.224, 0.225, 0.230
     0.2310, 0.240, 0.2900, 0.300, 0.3491, 0.385, 0.386, 0.401
     0.4015, 0.402, 0.4049, 0.412, 0.4130, 0.4999
```

Each of these cases is chosen to illustrate some characteristic of Chebyshev FIR filters. These frequencies include points where the characteristics change from one type to another. Discuss these in light of the alternation theorem. Note the number of ripples, number of extremal frequencies, zero locations (which are extra ripple filters), and so on. Which satisfy the alternation with $M+2$ and which with $M+3$ extremal frequencies? Which have all the zeros contributing to ripples and which do not? Present your results in a table.

EXERCISE 1.3

Characteristics of Narrowband Filters

If the bandwidth of a filter is on the order of the distance between two ripples, the passband may contain no actual oscillations (yet it does have extremal frequencies). It may have no zeros off the unit circle. Design a family of narrowband filters with different passband widths and discuss how their characteristics compare with the filters above and how they change with a changing passband edge.

PROJECT 2: ALTERNATION THEOREM AND REMEZ EXCHANGE ALGORITHM

The *alternation theorem* [11] states that an optimal Chebyshev approximation will have an error function which oscillates with a given number of equal-magnitude ripples that alternate in sign. The *Remez exchange algorithm* is a clever method of constructing that equal-ripple Chebyshev approximation solution. Rather than directly minimizing the Chebyshev error, this algorithm successively exchanges better approximations for the locations of error ripples. It is guaranteed to converge to the optimal equiripple solution under rather general conditions and it is the basis of the Parks–McClellan algorithm for designing linear-phase FIR filters. This project examines the mechanics and characteristics of this important and powerful algorithm.

Project description

The frequency response of a length-L FIR filter with impulse response $h[n]$ is given by the DTFT as

$$H(e^{j\omega}) = \sum_{n=0}^{L-1} h[n]\, e^{-j\omega n} \tag{2-1}$$

For an odd-length, linear-phase FIR filter, the impulse response has even symmetry and (2-1) becomes

$$H(e^{j\omega}) = e^{-jM\omega} \sum_{n=0}^{M} a[n] \cos(\omega n) \tag{2-2}$$

where $M = (L-1)/2$ is the group delay of the filter and the constant of linearity for the phase. This can be written

$$H(e^{j\omega}) = e^{-jM\omega} A(\omega) \tag{2-3}$$

where

$$A(\omega) = \sum_{n=0}^{M} a[n] \cos(\omega n) \tag{2-4}$$

is an even real-valued function called the amplitude. The $a[n]$ coefficients of the cosine terms are related to the impulse response by

$$a[n] = \begin{cases} h(M) & \text{for } n = 0 \\ 2h(n - M) & \text{for } 0 < n \le M \\ 0 & \text{otherwise} \end{cases} \tag{2-5}$$

This can be written in matrix form as

$$\mathbf{A} = \mathbf{C}\ \mathbf{a} \tag{2-6}$$

with $\mathbf{A}$ being a vector of samples of $A(\omega)$, $\mathbf{C}$ being a matrix of cosines terms defined in (2-4), and $\mathbf{a}$ being the vector of filter coefficients defined in (2-5).

The Chebyshev approximation filter design problem is to find the $a[n]$ (and from (2-5) the $h[n]$) which minimize the error measure

$$\epsilon = \max_{\omega \in \Omega} |A_d(\omega) - A(\omega)| \tag{2-7}$$

where Ω is a compact subset of the closed frequency band $\omega \in [0, \pi]$. It is the union of the bands of frequencies the approximation is over. These bands are the passband and stopband of our filter and may be isolated points.

The *alternation theorem* from Chebyshev theory states that if $A(\omega)$ is a linear combination of r ($r = M + 1$ for M odd) cosine functions [e.g., (2-4)], a necessary and sufficient condition that $A(\omega)$ be the unique optimal Chebyshev approximation to $A_d(\omega)$ over the frequencies $\omega \in \Omega$ is that the error function $E(\omega) = A(\omega) - A_d(\omega)$ have *at least* $r + 1$ (or $M + 2$) extremal frequencies in Ω. These extremal frequencies are points such that

$$E(\omega_k) = -E(\omega_{k+1}) \qquad \text{for } k = 1, 2, \ldots, r \tag{2-8}$$

where

$$\omega_1 < \omega_2 < \cdots < \omega_r < \omega_{r+1} \tag{2-9}$$

and

$$|E(\omega_k)| = \delta = \max_{\omega \in \Omega} |E(\omega)| \qquad \text{for } 1 \le k \le r + 1 \tag{2-10}$$

The alternation theorem states that the optimal Chebyshev approximation necessarily has an equiripple error, has enough ripples, and is unique.

The *Remez exchange algorithm* is a method that constructs an equiripple error approximation which satisfies the alternation theorem conditions for optimality. It does this by exchanging old approximations to the extremal frequencies with better ones. This method has two distinct steps. The first calculates the optimal Chebyshev approximation over $r + 1$ $(M + 2)$ distinct frequency points by solving

$$A_d(\omega_k) = \sum_{n=0}^{r-1} a[n] \cos(\omega_k n) + (-1)^k \delta \qquad \text{for } k = 1, 2, \ldots, r + 1 \tag{2-11}$$

for the r values of $a[n]$ and for δ. The second step finds the extremal frequencies of $A(\omega)$ over a dense grid of frequencies covering Ω. This is done by locating the local maxima and minima of the error over Ω. The algorithm states that if one starts with an initial guess of $r + 1$ $(M + 2)$ extremal frequencies, calculates the $a[n]$ over those frequencies using (2-11), finds a new set of extremal frequencies from the $A(\omega)$ over Ω using (2-4), and iterates these calculations, the solutions converge to the optimal approximation.

Hints

This project has three groups of exercises. In Exercises 2.1 and 2.2 we step manually through the Remez exchange algorithm to observe the details of each step. Exercises 2.1 and 2.3 through 2.5 will develop a sequence of functions that implement the complete

Remez exchange algorithm, pausing between each iteration to plot the frequency response and observe how the algorithm converges. Exercises 2.6 through 2.11 are generalizations.

The Remez algorithm has two basic steps which are iterated. In Exercise 2.1 we develop the first step as a function `cheby`, where (2-11) is solved to give an optimal Chebyshev approximation over the $M + 2$ frequencies in the vector `f`. In Exercise 2.3 we develop a function `update` which finds a new set of extremal frequencies by searching for the local maxima of the error over a dense grid of frequencies in the passband and stopband.

In a practical implementation, the number of grid points would be chosen approximately 10 times the length of the filter. In this project, to simplify calculations, frequency will be normalized to π rather than given in radians per second; the number of grid points will be set at 1000 so that a frequency value of $f_p = \omega_p/\pi = 0.25$ will correspond to an address in the frequency vector of 250. Remember that MATLAB uses 1 as the address of the first element of a vector and we often want zero.

The results of this project will be checked by using the built-in MATLAB `remez` function.

EXERCISE 2.1

Basic Chebyshev Alternation Equations

Create a MATLAB M-file function that computes the best Chebyshev approximation over $r + 1$ $(M + 2)$ frequencies by completing the following code. Put the proper statements in place of the ?????????.

```
function [a,d] = cheby0(f,Ad)
% [a,d] = cheby0(f,Ad)  calculates the a(n) for a
%  Chebyshev approx to Ad over frequencies in f.
%  For an odd length L linear phase filter with
%  impulse response h(n) and delay M = (L-1)/2, the
%  M+1  values of cosine coefficients a(n) are
%  calculated from the  M+2  samples of the desired
%  amplitude in the vector  Ad  with the samples
%  being at frequencies in the vector  f.  These
%  extremal frequencies are in bands between 0 and
%  0.5 (NORMALIZED FREQUENCY).  The max error, delta, is d.
%
M = length(f) - 2;                 %Filter delay
C = cos(2*pi*f'*[0:M]);            %Cosine matrix
s = ???????????;                   %Alternating signs
C = [C,s'];                        %Square matrix
a = C\Ad';                         %Solve for a(n) and delta
d = ??????;                        %Delta
a = ?????????;                     %a(n)
```

Test this function for $L = 13$ by applying it to

```
f = [0 .1 .2 .25 .3 .35 .4 .5];
Ad = [1 1 1 1 0 0 0 0];
a = cheby0(f,Ad);
```

Plot the amplitude frequency response of this filter together with the ideal and the interpolated initial `Ad` with the following program:

```
A = real(fft(a,1000)); A = A(1:501);                %calculate A
plot(A); hold;                                      %plot A
```

```
plot([0 fp*1000 fs*1000 501],[1 1 0 0]);          %plot Ad
plot(f*1000,A(f*1000+1),'o');                      %plot A at EP
plot([0,1000*fp],[1+d,1+d]); plot([0,1000*fp],[1-d,1-d]);
plot([1000*fs,500],[d,d]); plot([1000*fs,500],[-d,-d]);
pause;    hold off;
```

Note that the plot of $A(\omega)$ over the eight values of `f` is optimal in the sense that it satisfies the alternation theorem with equal-value alternating sign error. The plot of $A(\omega)$ over the dense grid of 1000 frequency samples is not optimal, and it is from this plot that new values for the extremal frequencies are obtained.

EXERCISE 2.2

Design an Odd-Length FIR Filter

Design a length-11 linear-phase FIR filter using the Remez exchange algorithm. Use an ideal lowpass $A_d(\omega)$ with a passband edge of $f_p = \omega_p/\pi = 0.2$ and a stopband edge of $f_s = \omega_s/\pi = 0.25$.

a. Form a vector `f` of $M + 2 = 7$ initial guesses for the extremal frequencies. Form a vector `Ad` of 7 ones and zeros at samples of the ideal amplitude response at the frequencies in `f`. Plot `Ad` versus `f`. Note that the frequency normalization is for a sampling rate of one per second rather the usual MATLAB convention of two per second.

b. Solve the Chebyshev approximation over the seven frequencies with the function created in Exercise 2.1. Plot the total amplitude response of the resulting filter using `plot(A)`. Plot the samples of `A` at the initial guesses of extremal frequencies.

c. Visually locate the seven new extremal frequencies where $|A(\omega) - A_d(\omega)|$ has local maxima over the passband and stopband, and exchange these new estimates of the extremal frequencies for the old ones in f. Always include the edges of the passband and stopband: `fs = 0.2` and `fp = 0.25`. Include $\omega = 0$ and/or $\omega = \pi$, whichever has the larger error. Solve for a new set of $a[n]$ over these new estimates of extremal frequencies using the `Cheby` function from Exercise 2.1. Plot the amplitude response.

d. Repeat this update of `f` and recalculation of $a[n]$ until there is little change.

How many iteration steps were needed for the algorithm to converge? How close are the results to those obtained from the built-in Remez command?

EXERCISE 2.3

Remez Exchange

In this and following exercises of this project, a full Remez exchange algorithm will be developed and investigated. Create a MATLAB function that will automatically find the new extremal frequencies from the old extremal frequencies by completing the following M-file code.

```
function [f,Ad] = update(fp,fs,a)
% [f,Ad] = update(fp,fs,a)  Finds the (L-1)/2 + 2 new extremal
% frequencies f and samples of Ad(f) consistent with f by
% searching for the extremal points over the pass and stopbands
% of A(f) calculated from the (L-1)/2+1 = D values of a(n).
%  For odd length-L and even symmetry h(n).
%
A = real(fft(a,1000));   A = A(1:501); %Amplitude response
kx = [];   Ad = [];
kp = fp*1000+1;                        %Address of passband edge
E = abs(A(1:kp) - ones(1:kp));         %Passband error
```

```
for k = 2:kp-1                          %Search passband for max.
  if (E(k-1)<E(k))&(E(k)>E(k+1))        %Find local max. in PB
    kx = [kx,k];  Ad = [Ad,1];          %Save location of max.
  end
end
ks = 1000*fs+1;                         %Address of stopband edge
kx = [kx,kp,ks];  Ad = [Ad,1,0];        %Add transition bandedges
for k = ks:500                          %Search stopband
  if (abs(A(k-1))<abs(A(k)))&(abs(A(k))>abs(A(k+1)))
    kx = [kx,k]; Ad = [Ad,0];           %Save location of max. in SB
  end
end
???????????????                         %Several lines of code for
???????????????                         % extremal freq. at
???????????????                         % f = 0 and/or f = .5
f = (kx-1)/1000;                        %Normalizes extremal freqs.
```

Test this function by applying it to data similar to that used in Exercise 2.2 to see if it does what you want in a variety of examples.

EXERCISE 2.4

Initial Extremal Frequencies

Write a MATLAB function called `init.m` that will take the specifications of filter length L, passband edge $f_p = \omega_p/\pi$, and stopband edge $f_s = \omega_s/\pi$, and generate the initial guesses of extremal frequencies as a vector `f` and the associated vector of samples `Ad` [i.e., $A_d(\omega)$]. Test it on several example specifications. Plot the results.

EXERCISE 2.5

Remez Filter Design Program

Write a MATLAB program that executes the `init.m` function to calculate the initial extremal frequencies and `Ad` followed by a loop which executes the Chebyshev approximation in `cheby0` and the updating of the extremal frequencies in `update`. Put a plot of the amplitude response in the loop to see what each step of the algorithm does. If you are running this on a very fast computer, you may want to put a pause after the plot. Design a length-13 linear-phase FIR low-pass filter with passband edge $f_p = \omega_p/\pi = 0.25$ and stopband edge $f_s = \omega_s/\pi = 0.3$ using this design program. Comment on how the extremal frequencies and δ change with each iteration. How do you decide when to stop the iterations?

Apply the design program to the same specifications but with $f_p = 0.27, 0.28, 0.29$. This should test the code written in Exercise 2.3. Discuss the differences in these cases.

EXERCISE 2.6

How Robust Is the Remez Exchange Algorithm?

Rewrite the `init.m` function to start with all of the extremal point guesses in the stopband (excluding $\omega = 0$ and ω_p). This will illustrate how robust the algorithm is and how it moves the excess extremal points from the stopband to the passband. Run this for several different passband and stopband edges, and discuss the results.

EXERCISE 2.7

Weighting Functions

Generalize the `cheby.m` function to include a weighting function. Design the filter in Exercise 2.5 but with a weight of 10 on the stopband error.

EXERCISE 2.8

Design Even-Order Filters

Write a new set of functions to design even-length filters. Recall that an even-length linear-phase FIR filter must have a zero at $\omega = \pi$, therefore, cannot have an extremal point there. Design a length-12 low-pass filter with the band edges given in Exercise 2.5. Comment on the convergence performance of the algorithm.

EXERCISE 2.9

Design Type III and IV Filters

Write a set of functions that will design type III and IV filters which have an odd-symmetric $h[n]$ and an expansion in terms of sines rather than the cosines in (2-4) and (2-11). Design a length-11 and a length-10 differentiator with these programs. Discuss the results. Note the problems with the odd-length differentiator.

EXERCISE 2.10

Design Bandpass Filters

Write new `init.m` and `update.m` functions that will design a bandpass filter. For the simple two-band low-pass filter, there are usually $M+2$ extremal frequencies but possibly $M+3$ for the "extra ripple" case. For more than two bands as in the bandpass filter, there may be still more extremal frequencies.

EXERCISE 2.11

How Good Are the Programs?

What are the maximum lengths these programs can design? How does the maximum length depend on ω_p and ω_s? How does the execution time compare with the built-in Remez function?

PROJECT 3: FIR FILTER DESIGN USING LINEAR PROGRAMMING

Linear programming has proven to be a powerful and effective optimization tool in a wide variety of applications. The problem was first posed in terms of economic models having linear equations with linear inequality constraints. In this project we investigate how this tool can be used to design optimal Chebyshev linear-phase FIR filters.

Project description

The Chebyshev error is defined as

$$\varepsilon = \max_{\omega \in \Omega} |A(\omega) - A_d(\omega)| \tag{3-1}$$

where Ω is the union of the bands of frequencies that the approximation is over. The approximation problem in filter design is to choose the filter coefficients to minimize ε.

It is possible to pose this in a form that linear programming can be used to solve it [10, 12]. The error definition in (3-1) can be written as an inequality by

$$A_d(\omega) - \delta \le A(\omega) \le A_d(\omega <) + \delta \tag{3-2}$$

where the scalar δ is minimized.

The inequalities in (3-2) can be written as

$$A \le A_d + \delta \tag{3-3}$$

$$- A \le -A_d + \delta \quad (3\text{-}4)$$

or

$$A - \delta \le A_d \quad (3\text{-}5)$$

$$- A - \delta \le -A_d \quad (3\text{-}6)$$

which can be combined into one matrix inequality using (2-6) by

$$\begin{bmatrix} C & -1 \\ -C & -1 \end{bmatrix} \begin{bmatrix} a \\ \delta \end{bmatrix} \le \begin{bmatrix} A_d \\ -A_d \end{bmatrix} \quad (3\text{-}7)$$

If δ is minimized, the optimal Chebyshev approximation is achieved. This is done by minimizing

$$\varepsilon = \begin{bmatrix} 0 & 0 & \cdots & 1 \end{bmatrix} \begin{bmatrix} a \\ \delta \end{bmatrix} \quad (3\text{-}8)$$

which, together with the inequality of (3-7), is in the form of the dual problem in linear programming [9] and [13].

Hints

This can be solved using the `lp()` command from the MATLAB Optimization Toolbox [14], which is implemented in an M-file using a form of quadratic programming algorithm. Unfortunately, it is not well suited to our filter design problem for lengths longer than approximately 11.

EXERCISE 3.1

Formulate the FIR Filter Design Problem as a Linear Program

A MATLAB program that applies its linear programming function `lp.m` to (3-7) and (3-8) for linear-phase FIR filter design is given below. Complete the program by writing the proper code for the lines having `??????????` in them. Test the program by designing a filter having the same specifications as given in Exercise 2.1 in the project on the Remez exchange algorithm. Compare the this design with the one done using Remez or by using the Parks–McClellan algorithm in the MATLAB command `remez`.

```
%  lpdesign.m  Design an FIR filter from L, f1, f2, and LF using LP.
%  L is filter length, f1 and f2 are pass and stopband edges, LF is
%  the number of freq samples.  L is odd.  Uses lp.m
%         csb 5/22/91
L1 = fix(LF*f1/(.5-f2+f1));  L2 = LF - L1;     %No. freq samples in PB, SB
Ad = [ones(L1,1); zeros(L2,1)];                %Samples of ideal response
f  = [[0:L1-1]*f1/(L1-1), ([0:L2-1]*(.5-f2)/(L2-1) + f2)]'; %Freq samples
M  = (L-1)/2;
C  = cos(2*pi*(f*[0:M]));                      %Freq response matrix
CC = ??????????????????????                    %LP matrix
AD = [Ad; -Ad];
c  = [zeros(M+1,1);1];                         %Cost function
x  = lp(c,CC,AD);                              %Call the LP
d  = x(M+2);                                   %delta or deviation
a  = ????????????                              %Half impulse resp.
h  = ????????????                              %Impulse response
```

EXERCISE 3.2

Design a Bandpass Filter

Design a bandpass FIR filter with the specifications given in Exercise 2.3 in the Remez project above but using linear programming. Compare the solutions.

EXERCISE 3.3

Analysis of the Speed of Linear Programming

For the specification of a low-pass filter used in Exercise 3.1, design filters of odd length from 5 up to the point the algorithm has convergence problems or takes too much time. Time these designs with the `clock` and `etime` commands. From a plot of time versus length, determine a formula that predicts the required time. This will differ depending on whether the simplex, Karmarkar, or some other algorithm is used. Try this on as many different algorithms as you have access to and discuss the results.

DESIGN OF IIR FILTERS

OVERVIEW

The infinite-duration impulse response (IIR) discrete-time filter is the most general linear signal processing structure possible. It calculates each output point via a recursive difference equation:

$$y[n] = -\sum_{k=1}^{N} a_k\, y[n-k] + \sum_{m=0}^{M} b_m\, x[n-m] \tag{0-1}$$

which assumes that a_k and b_m are not functions of the discrete-time variable n. Notice that the FIR filter is a special case when $N = 0$ and only past inputs are used. The design problem is to take a desired performance in the form of specifications, usually in the frequency domain, and to find the set of filter coefficients a_k and b_m that best approximates or satisfies them. The IIR discrete-time filter is analogous to the RLC circuit continuous-time filter or the active RC filter. Indeed, one of the standard methods of IIR filter design starts with the design of a continuous-time prototype which is converted to an equivalent discrete-time filter.

The advantage of the IIR filter over the FIR filter is greater efficiency. One can often satisfy a set of specifications with a significantly lower-order IIR than FIR filter. The disadvantages of the IIR filter are problems with stability and with quantization effects, the impossibility of exactly linear-phase frequency response, and the more complicated design algorithms. Most of the good and bad characteristics come from the feedback inherent in the IIR filter. It is the feedback of past output values that can cause the filter to be unstable or amplify the effects of numerical quantization, and it is feedback that gives the infinite duration response.

Several design methods for IIR filters use a continuous-time (analog) filter as a prototype which is converted to a discrete-time (digital) IIR filter. Some of the exercises concern characteristics of analog filters, but the goal is digital filter design. The purpose of this section is to become familiar with the characteristics of IIR filters and some of the standard design methods. You should not only evaluate the performance of these filters, you should compare them with the alternative, the FIR filter. It is customary to use order rather than length when setting specifications for IIR filters.

BACKGROUND READING

All DSP textbooks have some coverage of IIR filters. Particularly good discussions can be found in [1], [3], and [4], and the details of analog filter design useful as prototypes for discrete-time filters can be found in [5].

PROJECT 1: CHARACTERISTICS OF IIR FILTERS

This project will use MATLAB to design quickly several types of IIR discrete-time filters and analyze their characteristics. There are three descriptions of a filter that must be understood and related. First there is the impulse response, which is the most basic time-domain input–output description of a linear system. Second, there is the magnitude and phase frequency response, which is the most basic frequency-domain input–output description of a linear, time-invariant system. Third, there is the pole–zero map in the complex plane which is the most basic transfer function description. In this section we do not consider state-space descriptions since they are more related to implementation structures than approximation.

Project description

There are four classic IIR filters and their analog counterparts: (1) Butterworth, (2) Chebyshev, (3) Chebyshev II, and (4) elliptic function. They represent four different combinations of two error approximation measures. One error measure uses the Taylor's series. This method equates as many of the derivatives of the desired response as possible to those of the actual response. The other approximation method minimizes the maximum difference between the desired and actual response over a band of frequencies.

Since we will be working with analog (continuous-time) prototype filters together with digital (discrete-time) filters, we will denote continuous-time frequency by the uppercase omega (Ω) and discrete-time frequency by the lowercase omega (ω). The MATLAB analog filter design programs all normalize the band edge to $\Omega_0 = 1$. The concept of amplitude is not as useful as for the FIR filter and will not be used here.

1. The analog *Butterworth* filter is based on a Taylor's series approximation in the frequency domain with expansions at $\omega = 0$ and $\omega = \infty$. This filter is also called a maximally flat approximation since it is optimal in the sense that as many derivatives as possible equal zero at $\omega = 0$ and $\omega = \infty$. This approximation is local in that all the conditions are applied at only two points and it is the smoothness of the response that influences its behavior at all other frequencies. The formula for the magnitude squared of the normalized frequency response of an Nth order analog Butterworth lowpass filter is given by

$$|H(\Omega)|^2 = \frac{1}{1+\Omega^{2N}} \tag{1-1}$$

   This response is normalized so that the magnitude squared is always $\frac{1}{2}$ at $\Omega = 1$ for any N. Replacing Ω by Ω/Ω_0 would allow an arbitrary band edge at Ω_0.
2. The analog *Chebyshev* filter has a minimum maximum error over the passband and a Taylor's approximation at $\Omega = \infty$. The maximum error over a band is called the Chebyshev error. This terminology can be confusing since the Chebyshev filter minimizes the Chebyshev error only over one band. One of the interesting and easily observed characteristics of a Chebyshev approximation is the fact that the error oscillates with equal-size ripples. A Chebyshev approximation is often called an equal-ripple approximation, but that can be misleading since the error must not only be equal ripple, but there must be enough ripples.
3. The analog *Chebyshev II* filter (sometimes called the *inverse Chebyshev* filter) is a Taylor's series approximation at $\Omega = 0$ and has minimum Chebyshev error in

the stopband. This is often a more practical combination of characteristics than the usual Chebyshev filter.

4. The analog *elliptic* function filter (sometimes called the *Cauer* filter) uses a Chebyshev approximation in both the passband and the stopband. The Butterworth, Chebyshev, and Chebyshev II filters have formulas that can calculate the location of their poles and zeros using only trigonometric functions; however, the elliptic function filter requires evaluation of the Jacobian elliptic functions and the complete elliptic integrals, and therefore, the theory is considerably more complicated. Fortunately, MATLAB does all the work for you and all are equally easy to design.

These four optimal analog filters can be transformed into optimal digital filters with the bilinear transform that is investigated in the next project. The IIR filter design programs in MATLAB take care of the analog filter design and the bilinear transformation into the digital form automatically.

Hints

One can calculate samples of the frequency response of an IIR filter via `freqz`, which computes

$$H(e^{j\omega_k}) = H(e^{j2\pi k/L}) = \frac{DFT\{[\,b_0\; b_1\; \cdots b_M\,]\}}{DFT\{[\,a_0\; a_1\; \cdots a_N\,]\}} \tag{1-2}$$

Plot the pole locations on the complex z plane by using `plot(roots(a),'x')`, where `a` is a vector of the denominator coefficients; the zeros can be plotted with `plot(roots(b),'o')`.[4] The impulse response can be generated via `filter` and plotted with `stem`.

The `buttap`, `cheby1ap`, `cheby2ap`, and `ellipap` commands design analog prototypes which must be transformed into a digital IIR form using the `bilinear` command. The `butter`, `cheby1`, `cheby2`, and `ellip` commands design digital (discrete-time) IIR filters by automatically prewarping the specifications, calling the appropriate prototype design program and then applying the bilinear transformation. You should read the MATLAB manual about these commands as well as using the `help` command. You should also examine the M-file programs that implement the designs.

All discrete-time filter design functions in MATLAB use a frequency scaling that is somewhat nonstandard. When entering the cutoff frequencies ω_p and ω_s (units of radians), the values must be divided by π. Thus a cutoff frequency specified as $\omega_p = 0.22\pi$ would be entered as `0.22` in MATLAB. This scaling is not the usual normalized frequency where the sampling frequency is one, but assumes a sampling frequency of two samples per second. Therefore, if the specifications call for a cutoff frequency of 1000 Hz when the sampling frequency is 5000 Hz, you must enter `0.4` in MATLAB, because the cutoff frequency in radians is $\omega_c = 2\pi(1000/5000) = 2\pi(0.2) = 0.4\pi$.

Many of the exercises ask for a design followed by plots of impulse response, frequency response, and pole–zero location plots. It would be helpful to write an M-file to carry out this group of operations.

EXERCISE 1.1

How Is the Analog Butterworth Filter Optimal?

From (1-1), show how many derivatives of the magnitude squared are zero at $\Omega = 0$ for a fifth-order analog Butterworth filter.

---

[4] See `help` for `zplane.m`, which does all this and draws the unit circle.

EXERCISE 1.2

Analyze a Fifth-Order Digital Butterworth Low-Pass IIR Filter

Use the MATLAB command `butter` to design a fifth-order low-pass IIR filter with a sampling frequency of 2 Hz and a band edge of 0.6 Hz. Plot the magnitude and phase frequency responses using `freqz`. Plot the pole and zero location diagram. Plot the significant part of the impulse response using `filter` to give around 20 output values. Discuss how each pair of these three plots might be inferred from the third.[5]

EXERCISE 1.3

Analyze Fifth-Order Chebyshev and Elliptic Filters

Use the MATLAB command `cheby1` to design a fifth-order low-pass IIR filter with a sampling frequency of 2 Hz, a band edge of 0.6 Hz, and a passband ripple of 0.5 dB. Plot the magnitude and phase frequency responses. Plot the pole and zero location diagram. Plot the significant part of the impulse response. Discuss how each pair of these three plots might be inferred from the third.

Repeat for a fifth-order elliptic filter using `ellip` with a passband ripple of 0.5 dB, band edge of 0.6 Hz, and stopband ripple that is 30 dB less than the passband response.

EXERCISE 1.4

Compare the Order of the Four Designs

The filtering specifications for a particular job had a sampling rate of 2 Hz, passband ripple of 0.1, passband edge of 0.28 Hz, stopband edge of 0.32 Hz, and stopband ripple below 30 dB. What order Butterworth, Chebyshev, Chebyshev II, and elliptic filters will meet these specifications? Use the `buttord`, `cheb1ord`, `cheb2ord`, and `ellipord` commands. Discuss these results and experiment with other specifications. Why does the elliptic filter always have the lowest order?

EXERCISE 1.5

Compare IIR and FIR Filters

After working Exercise 1.4, design a Chebyshev FIR filter to meet the same specifications. What length filter designed by the Parks-McClellan method will meet them? What length filter designed using the Kaiser window will meet them? Discuss these comparisons.

PROJECT 2: USING THE BILINEAR TRANSFORMATION

The four classical IIR filter designs are usually developed and implemented by first designing an analog prototype filter, then converting it to a discrete-time filter using the bilinear transformation. This is done for two reasons. First, analog filter methods use the Laplace transform, where frequency is simply the imaginary part of the complex Laplace transform variable. The approximations are much more easily developed and described in the rectangular coordinates of the Laplace transform than in the polar coordinates of the z-transform. Second, one method can be used to design both discrete- and continuous-time filters. This project investigates this method.

Project description

The bilinear transform is a one-to-one map of the entire analog frequency domain $-\infty \leq \Omega \leq \infty$ onto the discrete-time frequency interval $-\pi \leq \omega \leq \pi$. The mapping is given by

[5] See also Chapter 5.

$$s = \frac{2}{T}\frac{z-1}{z+1} \tag{2-1}$$

Substituting this equation into a Laplace transfer function gives a discrete-time z-transfer function with the desired map of the frequency response. The analog and discrete-time frequencies are related by

$$\Omega = \frac{2}{T}\tan\left(\frac{\omega T}{2}\right) \tag{2-2}$$

where T is the sampling period in seconds. From this formula it is easily seen that each frequency in the half infinite analog domain is mapped to a frequency in the zero to π discrete-time domain. This map is nonlinear in the sense that near $\omega = 0$, the analog and discrete-time frequencies are very close to each other. However, as the discrete-time frequency nears π, the analog frequency goes to infinity. In other words, there is a nonlinear warping of the analog frequency range to make it fit into the finite discrete-time range. There is an ever-increasing compression of the analog frequency as the digital frequency nears π. The effects of this frequency warping of the bilinear transform must be taken into account when designing IIR filters by this method. Thus the bilinear method is suited for mapping frequency responses that are piecewise constant over frequency bands, such as low-pass and bandpass filters. It will not work well in designing the group delay or matching a magnitude characteristic such as a differentiator.

The design of IIR filters using the bilinear transformation has the following steps:

1. Modify the discrete-time filter design specifications so that after the bilinear transform is used, they will be proper. This is call *prewarping*.
2. Design the analog prototype filter using the prewarped specifications.
3. Convert the analog transfer function into a discrete-time z-transform transfer function using the bilinear transform.

This process is done automatically in the MATLAB design programs; however, to understand the procedure, we will go through the individual steps.

An alternative approach is to start with an analog prototype filter with its break frequency normalized to 1. The effects of prewarping are included in the bilinear transformation step. MATLAB does this.

Hints

Remember that frequency specifications are often given in f with units of hertz or cycles per second, but the theory is often developed in ω with units of radians per second. The relation is $\omega = 2\pi f$. Also, recall that the discrete-time filter consists of the coefficients a_k and b_m. The sampling frequency and other frequency normalizations and transformations affect only the band edge.

The MATLAB functions for analog (discrete-time) prototypes will produce filters whose analog cutoff frequency is normalized to $\Omega_c = 1$ (see `buttap`, `cheb1ap`, `cheb2ap`, and `ellipap`). The MATLAB function `bilinear` does both the prewarping and the transformation.

EXERCISE 2.1

Bilinear Transformation

Plot the relationship between the analog frequency Ω and the digital frequency ω specified by the bilinear transformation. Use several values for $(2/T)$ and plot the curves together. If an analog prototype has a cutoff frequency at $\Omega_c = 1$, how will the digital cutoff frequency change as T increases?

EXERCISE 2.2

Prewarp the Specifications

A fourth-order low-pass discrete-time Butterworth filter with a sampling frequency of 40 kHz is to be designed for a band edge of 8 kHz. What is the prewarped analog band edge?

EXERCISE 2.3

Design the Analog Prototype Butterworth Filter

Find the Laplace transform continuous-time transfer function for the fourth-order Butterworth filter in Exercise 2.2 using the prewarped band edge. Do this by hand or use the MATLAB `buttap`, giving a unity band edge.

EXERCISE 2.4

Apply the Bilinear Transformation

Find the z-transform discrete-time transfer function from the continuous-time transfer function in Exercise 2.3 using the bilinear transform. Do this by hand or use the MATLAB `bilinear` command. Compare this with the design done directly by MATLAB with `butter`.

PROJECT 3: DESIGN OF HIGH-PASS, BANDPASS, AND BAND-REJECT IIR FILTERS

It is possible to design a high-pass, bandpass, or band-reject IIR filter by a frequency transformation (change of variables) on a low-pass filter. This is possible for both analog and discrete-time IIR filters and is done by a different process than that used for an FIR filter.

Project Description

If one has a low-pass analog filter transfer function of the complex variable s, simply replacing s by $1/s$ will convert the low-pass filter into a high-pass filter. The replacement essentially replaces the "point" at infinity in the complex s plane by the origin and the origin by the point at infinite. It turns the s plane "inside out." This has the effect of converting a low-pass filter into a high-pass filter.

For a discrete-time z-transform transfer function, the point $+1$ in the complex z plane and the point -1 are interchanged to achieve the same conversion.

A low-pass analog filter can be converted into a bandpass filter by replacing s by

$$s \to \frac{s^2 + \omega_0^2}{s} \tag{3-1}$$

where $\omega_0 = \sqrt{\omega_1 \omega_2}$ is the geometric mean of the two band edges. This mapping doubles the order of the filter and is a one-to-two mapping of the frequency domain. The reciprocal of this transformation will produce a band reject filter from a low-pass filter.

Note. For the discrete-time case, the bandpass frequency transformations can be accomplished via all-pass functions that warp the ω axis. MATLAB does these frequency conversions automatically. In this project we analyze the characteristics of these filters.

EXERCISE 3.1

High-Pass Butterworth Filter

Use the MATLAB command `butter` to design a fifth-order high-pass Butterworth IIR filter with a sampling frequency of 2 Hz and a band edge of 0.7 Hz. Plot the magnitude and phase frequency responses. Plot the pole and zero location diagram. Plot the significant part of the impulse response. Compare with the low-pass design done above. Discuss how each pair of these three plots might be inferred from the third. Compare the impulse response to that of a Butterworth LPF.

EXERCISE 3.2

Design High-Pass Filters

Use the MATLAB command `cheby1` to design a fifth-order high-pass Chebyshev IIR filter with a sampling frequency of 2 Hz, a band edge of 0.7 Hz, and a passband ripple of 1 db. Plot the magnitude and phase frequency responses. Plot the pole and zero location diagram. Plot the significant part of the impulse response. Discuss how each pair of these three plots might be inferred from the third. Repeat for an elliptic filter with a stopband ripple of -40 dB, stopband edge of 0.25 Hz, and passband edge of 0.3 Hz.

EXERCISE 3.3

Bandpass Transformation

Plot the frequency mapping specified by (3-1). Explain how it transforms a low-pass prototype into a bandpass filter. Where are the band edges of the bandpass filter located if the low-pass filter has a cutoff at $\Omega_c = 1$? Also, explain what the result would be if the transformation were applied to a high-pass filter.

EXERCISE 3.4

Design Bandpass Filters

Use the MATLAB command `cheby1` to design a fifth-order bandpass Chebyshev IIR filter with a sampling frequency of 2 Hz, a lower band edge of 0.5 Hz, an upper band edge of 0.8 Hz, and a passband ripple of 1 dB. Plot the magnitude and phase frequency responses. Plot the pole and zero location diagram. Plot the significant part of the impulse response. Discuss how each pair of these three plots might be inferred from the third.

EXERCISE 3.5

Details

Design the bandpass filter in Exercise 3.4 by explicitly going through each step of the transformation using the `lp2bp`, `bilinear`, and `cheb1` commands.

PROJECT 4: IIR FILTER DESIGN IN THE TIME DOMAIN BY PRONY'S METHOD

Most IIR filters are designed from specifications given in the frequency domain, but there are situations where a time-domain design is desired. A particularly powerful method uses Prony's method (or Padé's method) to design an IIR filter by approximating or interpolating the first portion of a desired impulse response. Prony's method is a very useful tool for many applications and should be studied and understood.

Project description

The z-transform transfer function of an IIR filter defined by (0-1) is given by

$$H(z) = \frac{b_0 + b_1 z^{-1} + \cdots + b_M z^{-M}}{a_0 + a_1 z^{-1} + \cdots + a_N z^{-N}} = h[0] + h[1]\,z^{-1} + h[2]\,z^{-2} + \cdots \qquad \text{(4-1)}$$

where the impulse response, $h[n]$, has an infinite number of terms.

In the usual formulation of the approximation problem for filter design, one defines a solution error by

$$\epsilon_s(\omega) = \left[H(e^{j\omega}) - \frac{B(e^{j\omega})}{A(e^{j\omega})} \right] \tag{4-2}$$

and chooses the filter coefficients a_k and b_m to minimize some norm of ϵ_s. This is a linear problem for a polynomial, as we saw in the FIR filter design problem, but it is nonlinear here, where the coefficients enter into both the numerator and denominator of the rational function (4-1). This problem can be reformulated to become linear by defining an *equation error*:

$$\epsilon_e(\omega) = A(e^{j\omega})H(e^{j\omega}) - B(e^{j\omega}) \tag{4-3}$$

which can minimized by solving linear equations. The time-domain version of this is called Prony's method [3, 15] and the frequency-domain version is implemented by `invfreqz` in MATLAB and discussed in [3]. Clearly, in the time domain, since there are an infinite number of $h[n]$ terms, a finite number of them must be chosen. If the number of impulse samples used is the same as the unknown coefficients, $L = M + N + 1$, the problem can be solved exactly so that the impulse response achieved is exactly the same as what was desired for $0 \leq n \leq L$ and with nothing said about what happens for $n > L$. If $L > M + N + 1$, coefficients are found so that the sum of the squared equation error of (4-3) is minimized.

If coefficients can be found that make either ϵ_s or ϵ_e equal zero, the other is zero also. If the minimum error is not zero but is small, usually the coefficients that minimize $||\epsilon_e||$ are close to those that minimize $||\epsilon_s||$. If the minimum is not small, the solutions can be quite different. However, remember that the solution to both problems is optimal but according to different error measures. We usually want to minimize (4-2), but it is much easier to minimize (4-3), and that can be done by Prony's method. Notice that there is no control of stability.

Hints

Study the MATLAB M-file `prony` using `help` and `type`. This is discussed in [3] and [15].

EXERCISE 4.1

Exact Interpolation of Desired Response

It is desired to design an IIR filter that will exactly match the following 11 impulse response values:

```
h(n) = [1, 4, 6, 2, 5, 0, 1, 0, 0.1, 0, 0]
```

Find the IIR filter coefficients a_k and b_m using Prony's method for:

a. $M = 5$ and $N = 5$.

b. $M = 3$ and $N = 7$.

c. $M = 7$ and $N = 3$.

Using `filter`, calculate the first 20 values of the impulse response of these three filters designed by Prony's method. Do the first 11 match? Do the second 11 match? Try other combinations of M and N and discuss the results. How do the frequency responses compare? Analyze the pole and zero locations and discuss. Are the filters designed stable?

Repeat this exercise with the impulse response

```
h(n) = [1, 4, 6, 2, 5, 0, 1, 1, 0, 0.1, 0]
```

Compare the results using the two desired impulse responses.

EXERCISE 4.2

Minimum Equation Error Approximation

It is desired to design an IIR filter that will approximately match the following 20 impulse response values:

```
h(n) = [1, 4, 6, 2, 5, 0, 1, 1, 0, 0.1, 0, 0, 0, 0, 0, 0, 0, 0, 0, 0]
```

Find the IIR filter coefficients a_k and b_m using Prony's method for:

a. $M = 5$ and $N = 5$.

b. $M = 3$ and $N = 7$.

c. $M = 7$ and $N = 3$.

Using `filter`, calculate the first 20 values of the impulse response of these three filters designed by Prony's method and compare them with the desired values of $h[n]$. Repeat with the last 10 values equal to 1. Analyze and compare with the results of Exercise 4.1.

EXERCISE 4.3

Parameter Identification by Prony's Method

Use the MATLAB command `butter` to design a fifth-order low-pass IIR filter with a sampling frequency of 2 Hz and a band edge of 0.6 Hz, as was done in a Exercises 1.2 and 1.3. Calculate the first 11 terms of the impulse response using `filter`. Apply Prony's method for $M = 5$ and $N = 5$ to this impulse response and compare the designed filter coefficients with those that generated the impulse response. Try taking 15 terms of the impulse response and using $M = 7$ and $N = 7$. What are the results? Discuss your observations of this case.

EXERCISE 4.4

Effects of Noise or Error

Prony's method is somewhat sensitive to error or noise. Repeat Exercise 4.3 but add noise to the impulse response of the Butterworth filter before applying Prony's method. Do this by adding `K*rand(10,1)` to $h[n]$ for several values of K, ranging from values that make the noise very small compared to the impulse response up to values that make the noise comparable to the impulse response. Repeat this for the case where 15 terms are used. Repeat this for still larger numbers of terms. What conclusion can you draw?

PROJECT 5: SPECIAL TOPICS

This project is loosely structured to investigate two less well known methods of IIR filter design. The first is a frequency-domain version of Prony's method where the squared equation error is minimized. This is a very effective method but requires a complex desired frequency response (both magnitude and phase). The second is a method borrowed from the spectral estimation world and applied to filter design. The Yule-Walker method can be applied to a desired magnitude without requiring the phase to be specified, much as was done with the Butterworth and other classical methods. It gives an approximate least squared error approximation.

Project description

The least equation error method is similar to Prony's method but has the advantage that in the frequency domain, one can sample all of the region $0 \leq \omega \leq \pi$ and not have to truncate terms as was necessary in the time domain. It has the disadvantage of requiring complex values of the frequency-domain samples, where Prony's method used real values [3]. This method is implemented by `invfreqz` for discrete-time filters and `invfreqs` for continuous-time filters. The problems with using this method for many filter design cases is the requirement of providing the phase.

In this project we use the complex frequency response of an already designed filter upon which to first apply our method. This will give us a realistic phase with which to work. Notice from the name of the MATLAB command that the method is, in a sense, an inverse to the `freqz` command.

The advantage of the Yule–Walker method is that it requires only the magnitude of the desired response. The disadvantage is that it finds the optimal solution only approximately. It is implemented in MATLAB by `yulewalk`.

EXERCISE 5.1

Inverse Frequency Response

Use the MATLAB command `butter` to design a fifth-order low-pass IIR filter with a sampling frequency of 2 Hz and a band edge of 0.3 Hz as was done in Exercise 1.4. Calculate the complex frequency response using `freqz` at 100 equally spaced points. Apply the design method using `invfreqz` to this sampled frequency response and compare the designed filter coefficients with the Butterworth filter whose frequency response we used. Try a different number of frequency points and comment on the results.

EXERCISE 5.2

Least Equation Error Filter Design Method

As was done in Exercise 5.1, use the MATLAB command `butter` to design a fifth-order low-pass IIR filter with a sampling frequency of 2 Hz and a band edge of 0.6 Hz. Calculate the magnitude and phase frequency response of the filter from the output of `freqz`. Using that magnitude and a new desired phase curve that is linear in the passband and constant in the stopband (close to the designed phase), create a new complex frequency response and apply `invfreqz` to it. Look at the magnitude of the frequency response of the resulting filter and comment on it. Try several other piecewise linear-phase curves in an effort to get a "good" approximation to the magnitude and to the linear phase. Discuss the results.

EXERCISE 5.3

Second Least Equation Error Design

This exercise requires working Exercise 5.2. Take the best linear phase obtained and couple it with a desired magnitude of 1 in the passband and zero in the stopband. If you have numerical problems here, add a transition band and/or use a small but nonzero value in the stopband. Experiment with various desired ideal frequency responses in an effort to understand how this new method works. Take your best design and analyze the time-domain, pole–zero location, and frequency response characteristics as was done for the four classical methods. Discuss your observations.

EXERCISE 5.4

Yule–Walker Method

Design a fifth-order IIR filter to the same specifications as in the exercises above using the Yule–Walker method in the MATLAB command `yulewalk`. It may be necessary to have a small transition band to avoid too rapid a change in magnitude from 1 to zero. Experiment with this and compare the results with those designed by the least equation error and by the four classical methods. Design a thirteenth-order IIR filter to the same specifications and analyze its characteristics. Note, in particular, the zero locations.

chapter 9

DFT AND FFT ALGORITHMS

OVERVIEW

The fast Fourier transform (FFT) is the name of a family of algorithms for efficiently calculating the discrete Fourier transform (DFT) of a finite-length sequence of real or complex numbers. The various forms of Fourier transforms have proven to be extraordinarily important in almost all areas of mathematics, science, and engineering. While the Laplace transform, Fourier transform, Fourier series, z-transform, and discrete-time Fourier transform are important in analytical work, it is the DFT that we can actually calculate. Indeed, the DFT and its efficient implementation by the FFT are among the most important tools in digital signal processing.

The basic idea behind the FFT is the elimination of redundant calculations in direct evaluation of the DFT from its definition. It is the clever organization, grouping, and sequencing of operations that can give the minimum amount of arithmetic. In algorithm theory, this process is sometimes called "divide and conquer," but that name is misleading because simply dividing a problem into multiple small ones does not necessarily reduce the work. The process should be called "organize and share" to point out it is the sharing of common operations that eliminates redundant operations and saves arithmetic. In working the exercises in this chapter, one should look for these ideas and try to understand them in the most general way.

In this chapter we consider several approaches to calculating the DFT. The first step in deciding what algorithm to use is to pose a question carefully. The answer might be very different if you want all of the DFT values or if you want only a few. A direct calculation or the Goertzel algorithm might be the best method if only a few spectral values are needed. For a long prime length DFT, the chirp z-transform might be the best approach. In most general cases and for many lengths, the Cooley–Tukey FFT or the prime factor FFT will probably be the best choice. Understanding these ideas and learning how to answer questions with other considerations are the goals of this chapter.

There are three organizations used by most FFT algorithms; one requires that the length of the sequence to be transformed have relatively prime factors. This is the basis of the prime factor algorithm (PFA) and the Winograd Fourier transform algorithm (WFTA). The second can be applied to any factoring of the length into prime or composite factors but requires additional arithmetic. This is used in the Cooley–

Tukey fixed-radix FFT and mixed-radix FFT and also in the split-radix FFT. Still another approach converts the DFT into a filter. The chirp z-transform, Rader's method, and Goertzel's algorithm do that. These approaches are all covered in the following projects and exercises. It should be remembered that while most of the exercises in this chapter address the Fourier transform, the ideas also apply to many other transforms and signal processing operations.

BACKGROUND READING

Most general discrete-time signal processing textbooks will have one or more chapters on the DFT and FFT. They will certainly cover the basic Cooley–Tukey FFT. More specific information can be found in the books by Burrus and Parks [1], Brigham [2], Blahut [3], McClellan and Rader [4], or in Chapter 4 in Lim and Oppenheim [5]. An excellent overview article on modern FFT techniques has been published by Duhamel and Vetterli [6]. Programs can be found in Burrus and Parks [1] and in the IEEE DSP Program Book [7]. MATLAB itself has an interesting implementation of the FFT in its `fft` command (version 3.5 or later).

[1] C. S. Burrus and T. W. Parks. *DFT/FFT and Convolution Algorithms: Theory and Implementation*, John Wiley & Sons, New York, 1985.

[2] E. O. Brigham. *The Fast Fourier Transform and Its Applications*, Prentice Hall, Englewood Cliffs, NJ, 1988. Expansion of the 1974 book.

[3] R. E. Blahut. *Fast Algorithms for Digital Signal Processing*, Addison-Wesley, Reading, MA, 1985.

[4] J. H. McClellan and C. M. Rader. *Number Theory in Digital Signal Processing*, Prentice Hall, Englewood Cliffs, NJ, 1979.

[5] J. S. Lim and A. V. Oppenheim. *Advanced Topics in Signal Processing*, Prentice Hall, Englewood Cliffs, NJ, 1988.

[6] P. Duhamel and M. Vetterli. Fast Fourier transforms: A tutorial review and a state of the art. *Signal Processing*, 19(4):259–299, April 1990.

[7] DSP Committee, *Programs for Digital Signal Processing*, IEEE Press, New York, 1979.

[8] T. W. Parks and C. S. Burrus. *Digital Filter Design*. John Wiley & Sons, New York, 1987.

[9] A. V. Oppenheim and R. W. Schafer. *Discrete-Time Signal Processing*. Prentice Hall, Englewood Cliffs, NJ, 1989.

[10] H. V. Sorensen, M. T. Heideman, and C. S. Burrus. On calculating the split-radix FFT. *IEEE Transactions on Acoustics, Speech, and Signal Processing*, ASSP-34:152–156, February 1986.

[11] I. Niven and H. S. Zuckerman. *An Introduction to the Theory of Numbers*. John Wiley & Sons, New York, fourth edition, 1980.

[12] Oystein Ore. *Number Theory and Its History*. McGraw-Hill, New York, 1948.

[13] Donald E. Knuth. *The Art of Computer Programming*, Vol. 2, *Seminumerical Algorithms*. Addison-Wesley, Reading, MA, second edition, 1981.

DIRECT CALCULATION OF THE DFT

OVERVIEW

Calculating the discrete Fourier transform (DFT) of a number sequence from the definition (1-1) of the DFT is the most basic direct method. Since there are N DFT values, each calculated from N input values, there will be on the order of N^2 floating-point multiplications and additions required. In this section we investigate two approaches to this direct calculation.

BACKGROUND READING

Details of direct DFT calculations and Goertzel's algorithm can be found in [8,9].

PROJECT 1: CALCULATION OF THE DFT FROM THE DEFINITION

If only a few values of the DFT of a number sequence are needed, direct calculation of the DFT may be best. As you better understand the FFT, you will see that the efficiency gained by "sharing" operations is mainly lost when there are only a few final values to share the calculations. Also, if the length of the DFT is prime or has few factors, the FFT does not give any or much advantage, although the chirp z-transform covered in Project 2 of the section General Length FFTs will be considerably faster for large prime lengths.

Project description

MATLAB uses an interpreted language with very efficient individual commands, such as vector products or matrix multiplications but with fairly inefficient execution of `for` loops. Consequently, it can be used to simulate the hardware operation of a vector architecture computer which has very fast vector operations and slower scalar operations because of pipelining. In this project we measure the execution time and the number of floating-point operations (flops) of a direct calculation of the DFT using three different organizations of the algorithm.

The definition of the DFT is given by

$$X[k] = \sum_{n=0}^{N-1} x[n]\, W_N^{nk} \tag{1-1}$$

for

$$W_N = e^{-j2\pi/N} = \cos(2\pi/N) - j\sin(2\pi/N) \tag{1-2}$$

and $k = 0, 1, \ldots, N-1$.

This can be viewed as two nested loops of scalar operations consisting of a complex multiplication and a complex addition (or accumulation). It can also be viewed as a single loop which calculates each $X[k]$ by an inner product of the data vector made up of $x[n]$ and an N-point basis vector made up of W_N^{nk} for $n = 0, 1, \ldots, N-1$ and k fixed. Equation (1-1) can also be thought of as a single matrix multiplication [where the matrix is generated by MATLAB in `dftmtx` or as `fft(eye(N))`].

Hints

The evaluation of the algorithms in this project will be in terms of the execution time and in terms of the number of floating-point operations (flops) required. The timing can be done by preceding the operation being timed by the MATLAB statement: `t0 = clock;` and following it by `time = etime(clock,t0);`. The evaluation of the number of flops required is done in a similar manner using `f0 = flops;` before and `f1 = flops - f0;` after the operation. There may be inaccuracies with the timings on a time-shared computing system which will have to be dealt with by averaging several runs. Generate a test sequence of random complex numbers using: `x = (rand(1,N) + j*rand(1,N));`. The lengths to be used depend on the speed and memory size of the computer.

Remember that MATLAB starts the addressing of all vectors and arrays at 1, whereas the DFT formulas start at zero. Care must be taken in writing an M-file program or function from a mathematical formula to use the proper indexing. Because of the way that MATLAB counts flops, there will be some differences in the number of flops for evaluation of `exp()`, `cos()`, and `sin()`.

EXERCISE 1.1

Two-Loop Program

Write a program (script M-file) or function in MATLAB to evaluate the DFT in (1-1) using two nested `for` loops, with the inner loop summing over n and the outer loop indexing over k. Time the program for several lengths using the `clock` and `etime` commands mentioned above. Evaluate the number of flops required for several lengths and compare with what you would expect from the formula. Check the manual and use the `help` command to see how `flops` counts operations. Compare the times and flops of your DFT program with the built-in MATLAB command `fft` for the same lengths. Take into account that the `flops` command will count all arithmetic operations: exponential computation and index arithmetic as well as data arithmetic.

EXERCISE 1.2

One-Loop Program

Write a DFT program using one loop which steps through each value of k and executes an inner product. Time the program and evaluate the number of flops as was done for the two-loop program and compare the results for the same set of lengths. Explain the results obtained.

EXERCISE 1.3

No-Loop Program

Write a DFT program using a single matrix multiplication. Write your own DFT matrix rather than using the built-in `dftmtx`. Use the `exp` command with the exponent formed by an outer product of a vector of `n = 0:(N-1)` and a vector of `k = 0:(N-1)`. Time and evaluate flops for this program as was done for the two previous programs. Experiment with an evaluation that includes the formation of the DFT matrix and one that includes only the matrix multiplication. This can be done by precomputing the matrix for a given N. Comment on the differences and on the comparisons of the three implementations of the DFT formula. How many flops are used in generating the complex exponentials?

EXERCISE 1.4

Time and Flops versus Length

For each of the formulations above, write a program with a loop that will calculate the DFT with lengths from 1 up to a maximum value determined partially by the speed and memory of your computer. In each program, form a vector of times and a vector of flops. A plot of these vectors gives a picture of execution time versus length and of flops versus length. The theory predicts an N^2 dependence. Is that the case? Given that a single complex multiplication requires four real multiplications and two real additions, can you account for all the measured flops?

EXERCISE 1.5

Formula for Flops versus Length

Use the MATLAB command `polyfit` on the flop versus length data to determine the form and specific coefficients of the flop count.

PROJECT 2: GOERTZEL'S ALGORITHM

The Goertzel algorithm is a "direct" method to calculate the DFT, which also requires order N^2 operations. However, in some cases it uses approximately half the number of multiplications used by the direct methods investigated in Project 1 and illustrates a different organization.

Project description

The evaluation of the length-N DFT can be formulated as the evaluation of a polynomial. If a polynomial is formed from the data sequence by

$$X(z) = \sum_{n=0}^{N-1} x[n]\, z^n \tag{2-1}$$

it is easily seen from the definition of the DFT in (1-1) that the kth value of the DFT of $x[n]$ is found by evaluating $X(z)$ at $z = W_N^k$. This data polynomial is similar to the z-transform of $x[n]$, but the use of positive powers of z turns out to be more convenient.

An efficient method for evaluating a polynomial is Horner's method (also called nested evaluation). Horner's method uses a grouping of the operations:

$$X(z) = [[[x[4]z + x[3]]z + x[2]]z + x[1]]z + x[0] \tag{2-2}$$

This sequence of operations can be written recursively as a difference equation in the form

$$y[m] = z\, y[m-1] + x[N-m] \tag{2-3}$$

with the initial condition $y[0] = 0$ and the evaluated polynomial being the solution of the difference equation at $m = N$:

$$X(z) = y[N] \tag{2-4}$$

The DFT value $X[k]$ is then the value of $y[n]$ when $z = W_N^k$ and $n = N$.

This means that $X[k]$ can be evaluated by a first-order IIR filter with a complex pole at $z = W_N^k$ and an input of the data sequence in reverse order. The DFT value is the output of the filter *after N iterations.*

A reduction in the number of multiplications required by Goertzel's algorithm can be achieved by converting the first-order filter into a second-order filter in such a way as to eliminate the complex multiplication in (2-3). This can be done by multiplying the numerator and denominator of the first-order filter's transfer function by $z - W_N^{*k}$. If the number of multiplications are to be reduced, it is important *not* to implement the numerator until the last iteration.

Hints

In this project we use the MATLAB `polyval` command as well as the `filter` command and a personally written M-file to implement the Goertzel difference equation. Details on the use of these commands can be found in the manual and by use of the `help` command. "External" MATLAB functions are actually M-files that can be examined with the `type` command. "Internal" non-M-file commands or functions are generally faster than M-files. The calculations can be timed with the `clock` and `etime` commands as done in Project 1 and the number of floating-point operations used can be evaluated by the `flops` command.

Remember that MATLAB starts the addresses of all vectors and arrays at 1 while the DFT formulas start at zero. Care must be taken in writing an M-file program or function from a mathematical formula to use the proper indexing.

EXERCISE 2.1

Horner's Method

Verify that equations (2-3) and (2-4) are implementations of Horner's polynomial evaluation in (2-2). Write a MATLAB M-file program to calculate the DFT of a sequence by using the command `polyval` to evaluate (2-1) at $z = W_N^k$. After this is tested to be correct, put the statements in a loop to evaluate all N DFT values. Write a version of this program that does not use a loop, but rather, calls `polyval` with a vector argument to do all the evaluations at once. Measure the flops of both versions for several values of N and compare the results with those from Project 1 and discuss.

EXERCISE 2.2

Use the `filter` Command

Write a program to evaluate the DFT at one k value using the MATLAB `filter` command. After this is tested to be correct, put the statements in a loop to evaluate all N DFT values. Compare the times and flops for several values of N to the results of the direct evaluation from Project 1 and Exercise 2.1. Explain any differences.

EXERCISE 2.3

Use a Difference Equation

To better understand the implementation of Goertzel's algorithm, write an M-file implementation of the difference equation (2-3) rather than using the `filter` command. The DFT values should be the same as in the filter implementation of Exercise 2.1 or 2.2 or a direct calculation from (1-1). After this implementation is tested to be correct and put in a loop (or written to operate on a vector of inputs) to give all DFT outputs, compare the flops with the results of Exercises 2.1 and 2.2. Are they what you would expect? Compare execution times with the results from Exercise 2.1, 2.2, and, perhaps, Project 1 for several different lengths.

EXERCISE 2.4

Trig Function Evaluations

Compare the number of trigonometric function evaluations of Goertzel's method with the direct method.

EXERCISE 2.5

Second-Order Goertzel's Method

The first-order Goertzel algorithm can be modified into a second-order filter that uses only real multiplications and therefore reduces the number of required multiplications. The details can be found in [9] or [8]. Write a second-order Goertzel realization by an M-file that implements the second-order difference equation and evaluate its timings and flops. It should have approximately the same number of additions and one-half the multiplications. Do you find that?

EXERCISE 2.6

Real and Complex Data

It is inefficient to use a general DFT program that can take complex inputs on real input data. Evaluate the first- and second-order Goertzel algorithms in terms of number of operations required to calculate the DFT of real data in comparison to complex data.

THE COOLEY–TUKEY FFT

OVERVIEW

The Cooley–Tukey FFT is the name of a family of algorithms that use the decomposition of the DFT described in the original paper by Cooley and Tukey. The basic idea behind the FFT is elimination of the redundant calculations in the direct evaluation of the DFT from its definition in (1-1). This is done by factoring the length N and calculating multiple short DFTs with lengths equal to the factors. The Cooley–Tukey FFT allows any factoring of the length. The factors may be all the same, as is the case when the length is $N = R^M$. Here R is called the radix and the FFT is called a radix-R FFT. If the factors are different, such as $N = N_1 N_2 N_3 \cdots N_M$, the algorithm is called a mixed radix FFT and the factors may or may not be relatively prime. This is in contrast to the prime factor FFT, which requires all factors to be relatively prime (see the section *Prime Factor FFTs*).

Since the Cooley–Tukey FFT allows any factoring of the length, it is more versatile than the prime factor FFT. The disadvantage is that the Cooley–Tukey FFT requires multiplication by what are called *twiddle factors* that the prime factor FFT does not. The FFT usually achieves its greatest efficiency when the length can be factored in the largest number of factors. The most popular algorithms are the radix-2 and radix-4 FFTs.

This set of projects investigates three approaches to efficient calculation of the FFT. All use the Cooley–Tukey structure, which requires twiddle factors, but each develops a different aspect or point of view.

BACKGROUND READING

The Cooley–Tukey FFT is covered in all DSP books, but details can be found in [1] and [2].

■ ■ PROJECT 1: RECURSIVE DERIVATION OF THE FFT

A powerful and versatile algorithmic strategy is the use of recursion. The idea of recursion is both descriptive and enlightening when formulating the class of algorithms that implement the "divide and conquer" strategy. It can also be a very compact way of programming an algorithm but is sometimes inefficient in implementation. This project will use the decomposition possible with a composite-length DFT to derive the fundamentals of the FFT using a recursive realization. The original FFT was derived and programmed this way.

Project description

In this project we use basic properties of the DFT to write a recursive program that efficiently evaluates

$$X[k] = \sum_{n=0}^{N-1} x[n]\, W_N^{nk} \tag{1-1}$$

for

$$W_N = e^{-j2\pi/N}$$

Let $N = P \times K$, where N is the length of the original data sequence $x[n]$, K is the sampling interval, and P is the length of the sequence of samples $x[Kn]$. The *sampling property* gives the length-P DFT of the sequence of samples in terms of K shifted and summed DFTs of the original sequence as

$$\mathcal{DFT}\{x[Kn]\} = \frac{1}{K} \sum_{m=0}^{K-1} X[k + Pm] \tag{1-2}$$

The *shift property* relates the DFT of a shifted sequence to the DFT of the original sequence by

$$\mathcal{DFT}\{x[n+S]\} = W_N^{-Sk} X[k] \tag{1-3}$$

Now take the case where $N = 2 \times N/2$ (i.e., $K = 2$). One can show that a length-N $= 2^M$ DFT can be calculated from two length-$N/2$ DFTs. The *sampling property* states that the length-$N/2$ DFT of the even terms of $x[n]$ is

$$\mathcal{DFT}\{x[2n]\} = \tfrac{1}{2}\,(X[k] + X[k + N/2]) \tag{1-4}$$

Applying the shift, then the sampling properties and noting that $W_N^{N/2} = -1$ gives the length-$N/2$ DFT of the odd terms of $x[n]$ as

$$\mathcal{DFT}\{x[2n+1]\} = \frac{W_N^{-k}}{2}\,(X[k] - X[k + N/2]) \tag{1-5}$$

Solving these equations for $X[k]$ gives

$$X[k] = \mathcal{DFT}\{x[2n]\} + W_N^k\, \mathcal{DFT}\{x[2n+1]\} \tag{1-6}$$

$$X[k + N/2] = \mathcal{DFT}\{x[2n]\} - W_N^k\, \mathcal{DFT}\{x[2n+1]\} \tag{1-7}$$

for k and $n = 0, 1, \ldots, N/2-1$. This states that the length-N DFT of $x[n]$ can be calculated in two length-$N/2$ parts from the half-length DFT of the even terms of $x[n]$ plus the DFT of the odd terms multiplied by an exponential factor. This particular formulation of the evaluation is called a decimation-in-time (DIT) algorithm because the input is divided into two parts by taking the even terms for one and odd terms for the other. The exponential factor is called a *twiddle factor* because it is part of an extra non-DFT operation necessary to account for the shift of time index by one.

An alternative set of relationships that uses a decimation-in-frequency (DIF) organization is given using length-$N/2$ DFTs by

$$X[2k] = \mathcal{DFT}\{x[n] + x[n + N/2]\} \tag{1-8}$$

$$X[2k+1] = \mathcal{DFT}\{[x[n] - x[n+N/2]]W_N^n\} \quad (1\text{-}9)$$

for $k = 0, 1, \ldots, N/2-1$. Both the DIT and DIF formulas define a length-2^M DFT in terms of two length-2^{M-1} DFTs and those are evaluated in terms of four length-2^{M-2} DFTs and if this process repeated until the length is 1, the original DFT can be calculated with M steps and no direct evaluation of a DFT. This formulation is perfect for recursive programming, which uses a program that calls itself.

Hints

MATLAB supports recursive functions. For a recursive program or function to execute properly, it must have a stopping condition. In our case for the DFT, after M steps, when the length is reduced to 1, the single DFT value is the signal value; otherwise, the length is reduced further. This can be realized in a program using an `if` control which will keep the program calling itself and reducing the length by a factor of 2 until the length is 1 where the DFT is set equal to the signal value. Although certainly not necessary, some reading about recursive programming might enable you to get more from this project and to apply the ideas to other algorithms. Recursion is fundamental to certain general-purpose programming languages such as Lisp or Scheme.

EXERCISE 1.1

Example of a Recursive Program

The following MATLAB function will compute a sum of the elements in a vector recursively. It is given as an example, so that you can analyze how to write a recursive function in MATLAB. For a length-10 vector, determine how many times `recsum` will be called.

```
function  out = recsum(in)
%RECSUM
%   recursive summation
%
if( isempty(in) )
  out = 0;
else
  out = in(1) + recsum( in(2:length(in)) );
end
```

EXERCISE 1.2

Derive the Recursive Formulas

Derive the sampling and shift properties of (1-2) and (1-3) from the definition of the DFT in (1-1). Derive the DIT recursive formulas of (1-6) and (1-7) from the sampling and shift properties.

EXERCISE 1.3

Recursive Decimation-in-Time FFT

Write a MATLAB M-file function to evaluate a length-2^M DFT by recursively breaking the data vector into two half-length vectors with the DIT approach. Using (1-6) and (1-7), construct the DFT vector from the half-length DFTs of the even and odd terms of the data vector. Time and measure the flops for this program for several values of $N = 2^M$.

EXERCISE 1.4

Number of Floating-Point Operations Used

Derive a formula for the number of floating-point operations used in the recursive program and show that it is of the form $N \log_2(N)$. Compare with the values measured in Exercise 1.3.

EXERCISE 1.5

Radix-3 and Radix-4 Recursive FFTs

Derive the recursive formulas for a length-$N = 3^M$ DFT in terms of three length-$N/3$ DFTs. Write and evaluate a recursive program for a length-3^M DFT. Derive the appropriate formulas and repeat for a length-4^M DFT. Compare and evaluate the results, noting any difference in the number of required flops using the radix-2 and radix-4 algorithms on a data sequence of the same length.

EXERCISE 1.6

Recursive Decimation-in-Frequency FFT

Derive the DIF recursive formulas of (1-8) and (1-9). Write a recursive decimation-in-frequency DFT program. Compare its timings and flop count with the decimation-in-time program and with theoretical calculations.

PROJECT 2: TWO-FACTOR FFT WITH TWIDDLE FACTORS

The basic ideas and properties of the Cooley–Tukey FFT can be shown and evaluated with a simple two-factor example. All of the commonly used methods for developing FFT algorithms for long lengths involve factoring the length of the transform and then using the factorization to reduce the transform to a combination of shorter ones. The approach studied in this project can be applied to any factorization, whether or not the factors of N are relatively prime.[1] The only requirement is that the length itself not be prime. If the length is $N = R^M$, the resulting algorithm is called a radix-R FFT; if the length is $N = N_1 N_2 \cdots N_M$, the resulting algorithm is called a mixed-radix FFT. The goal of this project is to understand the principles behind this approach through a change of index variables rather than the recursion used in Project 1.

Project description

If the proper index map or change of variables is used, the basic DFT of (1-1) can be changed into a form of two-dimensional DFT. If the length is composite (not prime), it can be factored as

$$N = N_1 N_2 \tag{2-1}$$

and an index map defined by

$$n = N_2 n_1 + n_2 \tag{2-2}$$

$$k = k_1 + N_1 k_2 \tag{2-3}$$

For the case of three or more factors, N_1 would be replaced by N/N_2 in (2-3).

Substituting these definitions into the definition of the DFT in (1-1) gives

$$X[k_1 + N_1 k_2] = \sum_{n_2=0}^{N_2-1} \sum_{n_1=0}^{N_1-1} x[N_2 n_1 + n_2] W_{N_1}^{n_1 k_1} W_N^{n_2 k_1} W_{N_2}^{n_2 k_2} \tag{2-4}$$

[1] Relatively prime or co-prime means that the factors have no common divisors (e.g., 8 and 9 are relatively prime although neither is individually prime).

with n_1 and $k_1 = 0, 1, 2, \ldots, N_1 - 1$ and n_2 and $k_2 = 0, 1, 2, \ldots, N_2 - 1$. Equation (2-4) is a nested double sum that can be viewed as multiple short DFTs. The inner sum over n_1 is evaluated for each value of n_2. It is N_2 length-N_1 DFTs. The resulting function of k_1 and n_2 is multiplied by the set of $W_N^{k_1 n_2}$, which are called *twiddle factors*. The outer sum over n_2 is N_1 length-N_2 DFTs. If the length-N_1 and length-N_2 DFT are done by direct methods requiring N^2 operations, the number of complex multiplications is

$$\# \text{ MULT} = N_2 N_1^2 + N_1 N_2^2 + N = N(N_1 + N_2 + 1) \tag{2-5}$$

where the last term of N accounts for the twiddle-factor multiplications.

If the length N has M factors, the process can be continued down to the complete factoring of N into its smallest prime factors. This will result in a larger number of nested summations and DFTs and a multiplication count of

$$\# \text{ MULT} = N(N_1 + N_2 + \cdots + N_M + (M - 1)) \tag{2-6}$$

This is clearly smaller than the N^2 operations that direct evaluation or Goertzel's algorithm would require. Indeed, the greatest improvement will occur when N has a large number of small factors. That is exactly the case when $N = 2^M$.

The goal of the following exercises is to examine this decomposition of the DFT into multiple short ones interleaved with necessary twiddle factors. The details of the index map itself are examined in a later project.

Hints

In a practical implementation, the short DFTs are called butterflies (because of the shape of the length-2 DFTs flow graph) and are directly programmed. In the following exercises, we will use MATLAB's built-in `fft` command for the short DFTs and will write programs to combine them. If `fft` is applied to a vector, it returns the DFT vector. If it is applied to a matrix, it returns a matrix with columns that are DFTs of the columns of the original matrix. This is exactly what we need for our decomposition. We will use the element-by-element operator of `.*` to multiply the arrays by the twiddle-factor array. The transpose of a matrix A is denoted in MATLAB by $A.'$, which makes the rows of A the columns of $A.'$. Note that A' is the complex-conjugate transpose of A.

EXERCISE 2.1

Length-15 Cooley–Tukey FFT

Calculate a length-15 DFT using five length-3 DFT's, three length-5 DFTs, and one set of twiddle-factor multiplications. First form a 3×5 matrix using the index map in (2-2) for $N_1 = 3$ and $N_2 = 5$, which gives $n = 5n_1 + n_2$ and $k = k_1 + 3k_2$. This array has the first five elements of the input data as its first row, the second five elements as its second row, and the third five as the last row. The MATLAB command `fft(A)` gives a matrix whose columns are the DFTs of the columns of A. The twiddle-factor multiplications are done by forming a matrix of exponentials with `T = exp(-i*2*pi*[0:2].'*[0:4]/15)` and point-by-point multiplying it times the matrix of column DFTs. This is followed by a second application of the `fft` function, but after transposing the matrix to change the rows into columns. Finally, the matrix is converted back into a vector using (2-3). The program should be written for two general factors but applied to $N = 15$ for this exercise. Check the results against those calculated by a direct method. Note that the input data are sequential in the rows of the two-dimensional array, but after the calculations, the DFT values are sequential in the columns. This is because the input is indexed by (2-2), but the output is indexed by (2-3). For a general FFT program, the output is in a scrambled order and must be ordered properly before using.

Time and measure the flops for this program. Time and measure the flops of the length-3 and length-5 DFTs, and from these results compare your program with what you would expect.

EXERCISE 2.2

Length-16 Cooley–Tukey FFT

Calculate a length-16 DFT using two stages of four length-4 DFTs and the appropriate twiddle factors in a way similar to that used in Exercise 2.1. Time and measure flops and compare to the results in Exercise 2.1. Repeat for several other composite lengths. Try several long examples and explain the results.

EXERCISE 2.3

Equivalent Formulas

Show that the formulas for combining the two half-length DFTs in (1-6) and (1-7) and the formulas for combining three one-third-length DFTs in Project 1 are the same as short length-2 and length-3 DFTs and the appropriate twiddle factors in this project.

EXERCISE 2.4

In-Place Calculation of the Cooley–Tukey FFT

Because the decomposition of (2-4) uncouples the row and column calculations, it is possible to write the results of each short DFT over its data since those data will not be used again. Illustrate that property with your FFT program. Turn in your code and label where this in-place calculation is done.

PROJECT 3: SPLIT-RADIX FFT

In 1984 a new FFT algorithm was described that uses a clever modification of the Cooley–Tukey index map to give a minimum arithmetic implementation of an FFT for $N = 2^M$. It is known that the split-radix FFT (SRFFT) uses the theoretical minimum number of multiplications for allowed lengths up to 16, and although not optimal in terms of multiplications, it seems to give a minimum number of total floating-point arithmetic operations for lengths above 16. This project examines the structure of the SRFFT.

Project description

All of the algorithms and index maps for the Cooley–Tukey FFT, mixed-radix FFT, PFA, and WFTA are organized by stages, where each stage is a certain number of certain-length DFTs. For the fixed-radix FFT, each stage is the same DFT; for the mixed-radix FFT and PFA, each stage is a different-length DFT; and for the WFTA, the stages are partitioned to an even greater number and permuted to nest and combine all of the multiplications between the stages of additions. The split-radix FFT applies a different strategy by using two different DFTs and two different index maps in each stage. A radix-2 index map is used on the even terms and a radix-4 on the odd terms. This is shown in the following reformulation of the definition of the DFT into a decimation-in-frequency form. For the even spectral values

$$X[2k] = \sum_{n=0}^{N/2-1} (x[n] + x[n + N/2])W_N^{2nk} \tag{3-1}$$

and for the odd terms

$$X[4k+1] = \sum_{n=0}^{N/4-1} ((x[n] - x[n + N/2]) - j(x[n + N/4] - x[n + 3N/4]))W_N^n W_N^{4nk} \tag{3-2}$$

and

$$X[4k+3] = \sum_{n=0}^{N/4-1} ((x[n] - x[n+N/2]) + j(x[n+N/4] - x[n+3N/4]))W_N^{3n} W_N^{4nk} \tag{3-3}$$

This decomposition is repeated until the length is 2 when a single length-2 DFT is necessary.

Although it might appear that allowing other lengths, such as eight or 16 might improve efficiency further, they do not. This simple mixture of two and four is the best that can be done using this general organization and it seems to be the best for reducing multiplications and additions of any organization if $N = 2^M$. A careful analysis of the SRFFT shows it to be only slightly more efficient than a highly optimized radix-4 or radix-8 FFT.

Hints

The same ideas of recursive programming used in Project 1 are needed here. Two stopping conditions will be needed: a simple $X[k] = x[n]$ if the length is 1 and a length-2 DFT if the length is 2; otherwise, the program should call itself. The multidimensional approaches used in Project 2 will have to be modified to use two different maps. There will be some problems in obtaining a realistic count of the flops because MATLAB counts multiplication by j as a multiplication. Details of the SRFFT can be found in [5], [6], and [10].

EXERCISE 3.1

Derive Basic Equations

Derive equations (3-1)–(3-3) from the definition of the DFT in (1-1) and the appropriate radix-2 and radix-4 index maps.

EXERCISE 3.2

Recursive SR FFT

Write a recursive implementation of a decimation-in-frequency SRFFT as described in (3-1)–(3-3) using the approach described in Project 1. Test and debug until it gives correct DFT values for all k. The recursion will have to be stopped before the last stage since the last stage is different from the general case.

EXERCISE 3.3

Compare the Recursive SR FFT and Cooley–Tukey FFT

Compare the execution times and number of flops of the recursive SRFFT with those of the recursive radix-2 FFT from Project 1. Remember that MATLAB counts multiplication by j as a multiplication, although it really isn't one.

EXERCISE 3.4

Decimation-in-Time SR FFT

Derive the appropriate equations and write a recursive decimation-in-time SRFFT.

EXERCISE 3.5

Nonrecursive SR FFT

Write a multidimensional formulation (i.e., nonrecursive) of the SRFFT as was done in Project 2.

PRIME FACTOR FFTs

OVERVIEW

There are two organizations used by most FFT algorithms. The first requires that the length of the sequence to be transformed have relatively prime factors. This is the basis of the prime factor algorithm (PFA) and the Winograd Fourier transform algorithm (WFTA). The second can be applied to any factoring of the length into prime or composite factors but requires additional arithmetic. This is used in the Cooley–Tukey fixed-radix FFT and mixed-radix FFT and also in the split-radix FFT. Still another approach converts the DFT into a filter. The chirp z-transform, Rader's method, and Goertzel's algorithm do that.

PROJECT 1: TWO-FACTOR PRIME FACTOR ALGORITHM FFT

Although the index map used in the Cooley–Tukey FFT can be used for all cases of a composite length, a special case occurs when the factors are relatively prime.[2] When the two factors of N have no common factors themselves, it is possible to choose an index map that will give multiple short DFTs as before, but this time there are no twiddle factors. This approach is used with the prime factor algorithm (PFA) and the Winograd Fourier transform algorithm (WFTA). We again consider the evaluation of the DFT as defined by

$$X[k] = \sum_{n=0}^{N-1} x[n] W_N^{nk} \tag{1-1}$$

where

$$W_N = e^{-j2\pi/N}$$

Project description

This project is a companion to the project on the Cooley–Tukey FFT; therefore, the description of that material is applicable here. Suppose that the transform length N factors as $N = N_1 \times N_2$. We will now use an index map of the form

$$n = (N_2 n_1 + N_1 n_2) \mod N \tag{1-2}$$

This mapping will change $x[n]$ from a one-dimensional vector into a two-dimensional matrix $\tilde{x}[n_1, n_2]$ by redefining the independent variable $n \rightarrow (n_1, n_2)$. For the frequency domain, we define a similar index map $k \rightarrow (k_1, k_2)$:

$$k = (K_3 k_1 + K_4 k_2) \mod N \tag{1-3}$$

where K_3 is a multiple of N_2 and K_4 is a multiple of N_1. These values can be chosen in a way to remove the twiddle factors and cause the short summations to be short DFTs. The details of this map will be investigated in Project 2 but will not be needed here. Applying these maps to the DFT in (1-1) gives

$$X[K_3 k_1 + K_4 k_2] = \sum_{n_2=0}^{N_2-1} \sum_{n_1=0}^{N_1-1} x[N_2 n_1 + N_1 n_2]\, W_{N_1}^{n_1 k_1} W_{N_2}^{n_2 k_2} \tag{1-4}$$

---

[2] *Relatively prime* or *co-prime* means the factors have no common divisors (e.g., 8 and 9 are relatively prime although neither is individually prime).

with n_1 and $k_1 = 0, 1, 2, \ldots, N_1 - 1$ and n_2 and $k_2 = 0, 1, 2, \ldots, N_2 - 1$. Equation (1-4) is almost exactly the same form as found in the project on the Cooley–Tukey FFT, but now there are no twiddle factors. We have reduced the amount of arithmetic and found a cleaner decomposition of the DFT, but it requires a somewhat more complicated index map. The idea can be extended to more factors as long as they are all relatively prime. Then a one-dimensional DFT is converted into a multidimensional DFT. The goal of this project is to understand how these more general index maps (1-2) can eliminate the twiddle factors.

Hints

We will have to evaluate some of the index calculations modulo N. That can be done in two ways:

```
if n > N, n = n - N; end;
```

or

```
n = rem( rem(n,N) + N, N );
```

The first form will work only because stepping by N_1 and N_2 in (1-2) never causes n to exceed $2N$, and therefore a single subtraction will always suffice. The second applies `rem` twice to take care of the case where the remainder is negative.

EXERCISE 1.1

Index Map for 3 by 5

The following time index map will be used for the length-15 DFT:

$$n = (5n_1 + 3n_2) \mod 15 \tag{1-5}$$

The corresponding frequency index map must be

$$k = (10k_1 + 6k_2) \mod 15 \tag{1-6}$$

a. For the time indexing, create the 3×5 matrix of indices describing the one- to two-dimensional conversion. Each matrix entry should be the value of n corresponding to (n_1, n_2). Observe the difference between this index map and the simple "concatenate into rows" strategy used for the Cooley–Tukey FFT.

b. Do the same thing for the frequency-domain indexing. Note that the output will be in a scrambled order, because neither the rows nor columns contain sequential k indices.

EXERCISE 1.2

Length-15 PFA FFT

Calculate a length-15 DFT using the index maps in (1-5) and (1-6). Two computation steps are needed: (1) DFT all the columns, then (2) DFT all the rows. No twiddle-factor multiplications are performed between the row and column DFTs, unlike the mixed-radix FFT.

a. Implement this specific PFA FFT, and measure its flops. Compare to an implementation using twiddle factors. The next project shows why this approach cannot remove the twiddle factors in a length-16 DFT.

b. Use a test sequence that is `ifft([1:N])` to verify that the output is in scrambled order.

c. Show that for the PFA, the short transforms can be done in either order, five-point row DFTs first, then three-point column DFTs, or vice versa.

d. For the mixed-radix FFT, where the index map is concatenated into columns, show that the row DFTs must be computed first.

EXERCISE 1.3

Decimation-in-Time PFA FFT

Reverse the time and frequency maps in Exercise 1.2 to be

$$n = (10n_1 + 6n_2) \bmod 15 \tag{1-7}$$

$$k = (5k_1 + 3k_2) \bmod 15 \tag{1-8}$$

and show that it still works. This is similar to the DIF and DIT Cooley–Tukey FFTs. How do the number of flops compare with those of Exercise 1.2?

PROJECT 2: THE GENERAL LINEAR INDEX MAP

The index map or change of variable used to develop the Cooley–Tukey FFT is a rather straightforward application of decimation in time or frequency. The PFA and WFTA require a somewhat more complicated index map, but this clever reindexing removes all the twiddle factors. This project develops the general theory of the index map used for almost all types of FFTs and shows how it can be used to remove the necessity of an unscrambler or the equivalent of the bit-reversed counter. Although not necessary, some reading about basic number theory would give a deeper insight into some of the results [4,11,12]. Some practice with number-theoretic concepts is included in a later project.

Project description

The basic one-dimensional single-summation DFT of (1-1) for $N = N_1 \times N_2$ can be converted into a multidimensional nested summation form by a linear change of variables given by

$$n = \langle K_1 n_1 + K_2 n_2 \rangle_N \tag{2-1}$$

$$k = \langle K_3 k_1 + K_4 k_2 \rangle_N \tag{2-2}$$

where

$$n, k = 0, 1, 2, \cdots, N-1 \tag{2-3}$$

$$n_1, k_1 = 0, 1, 2, \cdots, N_1 - 1 \tag{2-4}$$

$$n_2, k_2 = 0, 1, 2, \cdots, N_2 - 1 \tag{2-5}$$

The notation $\langle n \rangle_N$ means the residue of n modulo N. In the following description, all the indices are evaluated modulo N. After a substitution of index variables, (1-1) becomes

$$X[K_3k_1 + K_4k_2] = \sum_{n_2=0}^{N_2-1} \sum_{n_1=0}^{N_1-1} x[K_1n_1 + K_2n_2] W_N^{(K_1n_1+K_2n_2)(K_3k_1+K_4k_2)} \tag{2-6}$$

$$= \sum_{n_2=0}^{N_2-1} \sum_{n_1=0}^{N_1-1} x[K_1n_1 + K_2n_2] W_N^{K_1K_3n_1k_1} W_N^{K_2K_3n_2k_1} W_N^{K_1K_4n_1k_2} W_N^{K_2K_4n_2k_2} \tag{2-7}$$

The question is, now: What values of K_i give interesting and useful results? There are three requirements to be considered.

1. The map $n \to (n_1, n_2)$ should be unique or one-to-one. That is to say, each pair (n_1, n_2) should correspond to a unique n, and vice versa. The same must be true for $k \to (k_1, k_2)$. This is necessary if (2-7) is to calculate all of the DFT values.
2. The map should result in a reduction of the required arithmetic compared to (1-1). This will occur if one or both of the middle two W_N terms in (2-7) becomes unity, which, in turn, happens if one or both of the exponents of W_N is zero modulo N.

3. The map should cause the uncoupled calculations to be short DFTs themselves. This is not necessary (or, in some cases, not possible), but gives a cleaner, more flexible formulation.

Hints

The notation $n = 0 \bmod L$ is equivalent to saying that n is a multiple of L.

EXERCISE 2.1

Index Mapping Function

Write a MATLAB function that will create an index map (2-1) for the two-factor case, $N = N_1 \times N_2$. The function must have four input arguments: N_1, N_2, K_1, and K_2, and one output—the matrix of indices. For example, the mapping in (1-5) gives the array

```
 0    3    6    9   12
 5    8   11   14    2
10   13    1    4    7
```

EXERCISE 2.2

Unique Index Mapping

The mapping $n \rightarrow (n_1, n_2)$ will be unique if the integer constants K_1 and K_2 are chosen wisely. The necessary and sufficient conditions for the map of (2-1) to be unique are stated in two cases, depending on whether or not N_1 and N_2 are relatively prime. First, notice that the constraint $0 < K_i < N$ obviously applies because all the indexing is reduced modulo-N. Thus there are at most $(N-1)^2$ possible mappings.

a. Write a MATLAB function that will test whether or not an index map is one-to-one. (*Hint*: The matrix of mapped indices can be converted to a vector and then sorted via the `sort` function, prior to a vector comparison.)

b. When $N = N_1 \times N_2$ it is possible to prove that either $K_1 = 0 \bmod N_2$ or $K_2 = 0 \bmod N_1$. Write a program that will generate all possible index maps for $N = N_1 \times N_2$ and determine which ones are one-to-one. Use the function from part (a) and the index map generating function you wrote in Exercise 2.1.

c. Now take the case $N = 15 = 5 \times 3$, and try all possible values for K_1 and K_2 to see which ones yield a unique map. Of the 196 different possible maps, how many are one-to-one? Verify that each satisfies the property in part (b).

d. Repeat for the case $N = 16 = 4 \times 4$.

e. When N_1 and N_2 are relatively prime[3] [i.e., $(N_1, N_2) = 1$] we can pick <u>both</u> K_1 and K_2 to be multiples of N_2 and N_1, respectively. Among all the possible choices generated for $N = 15$, how many fall into this category? Show that these (K_1, K_2) pairs also satisfy

$$(K_1, N_1) = (K_2, N_2) = 1 \qquad \text{(2-8)}$$

f. For the relatively prime case, it is not necessary that both $K_1 = 0 \bmod N_2$ and $K_2 = 0 \bmod N_1$. However, it will still be the case that (2-8) is true. Verify for all the pairs generated in part (c).

g. When N_1 and N_2 are not relatively prime [i.e., $(N_1, N_2) > 1$], we *cannot* have both $K_1 = 0 \bmod N_2$ and $K_2 = 0 \bmod N_1$, simultaneously. Instead, there are two cases:

$$K_1 = aN_2 \text{ and } K_2 \neq 0 \bmod N_1 \quad \Rightarrow \quad (a, N_1) = (K_2, N_2) = 1 \qquad \text{(2-9)}$$

[3]We will use the notation of (N, M) for the greatest common divisor of N and M.

or

$$K_1 \neq 0 \bmod N_2 \text{ and } K_2 = bN_1 \quad \Rightarrow \quad (K_1, N_1) = (b, N_2) = 1 \tag{2-10}$$

For the case of $N = 16$, verify that all candidate (K_1, K_2) pairs satisfy one of these conditions.

EXERCISE 2.3

Removing Twiddle Factors

The calculation of the DFT by (2-7) rather than (1-1) does not necessarily reduce the arithmetic. A reduction occurs only if the calculations are "uncoupled" by one of the middle W_N terms being unity. This occurs if one or both of the exponents are zero modulo-N, which, in turn, requires that

$$K_1 K_4 = 0 \bmod N \quad \text{and/or} \quad K_2 K_3 = 0 \bmod N \tag{2-11}$$

We must consider both the time and frequency maps together, so we must work with the quadruples $\{(K_1, K_2), (K_3, K_4)\}$, where the (K_3, K_4) pair produces a one-to-one mapping of $k \to (k_1, k_2)$.

a. If both the time and frequency index maps are unique, prove that one of the conditions in (2-11) can always be made true.

b. For the case of $N = 15$, both conditions can be satisfied. How many different quadruples $\{(K_1, K_2), (K_3, K_4)\}$ satisfy both? The resulting index map is called the *prime factor map*.

c. For the case $N = 16$, only one condition can be true. Verify this fact over the entire set of possible quadruples $\{(K_1, K_2), (K_3, K_4)\}$ that produce unique mappings. In this case, the map is known as the *common factor map*. How many satisfy one of the conditions?

EXERCISE 2.4

Reduction to Short DFTs

In order for the summations in (2-7) to be short DFTs, the following must also hold:

$$\langle K_1 K_3 \rangle_N = N_2 \quad \text{and} \quad \langle K_2 K_4 \rangle_N = N_1 \tag{2-12}$$

Under these conditions, (2-7) becomes either

$$X[K_3 k_1 + K_4 k_2] = \sum_{n_2=0}^{N_2-1} \sum_{n_1=0}^{N_1-1} x[K_1 n_1 + K_2 n_2] W_{N_1}^{n_1 k_1} W_N^{K_2 K_3 n_2 k_1} W_{N_2}^{n_2 k_2} \tag{2-13}$$

which is the mixed-radix FFT with twiddle factors; or

$$X[K_3 k_1 + K_4 k_2] = \sum_{n_2=0}^{N_2-1} \sum_{n_1=0}^{N_1-1} x[K_1 n_1 + K_2 n_2] W_{N_1}^{n_1 k_1} W_{N_2}^{n_2 k_2} \tag{2-14}$$

which is a two-dimensional DFT with no twiddle factors.

a. For the $N = 15$ case, determine how many mappings satisfy all three criteria above, including the prime factor map condition, which leads to (2-14).

b. Do the same for $N = 16$, which is the common factor map case (2-13).

In the remaining exercises we examine some different FFTs based on these index maps—one case is the traditional Cooley–Tukey FFT, the other the PFA form.

EXERCISE 2.5

Factor an Integer

Write a program to factor an integer n into its prime factors. Use a loop that indexes possible divisors from 2 to $n - 1$ and test each with `rem(n,d) == 0`. This might look like

```
factor = [ ];
for d = 2:n-1
   for rem(n,d) == 0
      n = n/d;
      factor = [factor, d];
   end
end
if factor == [ ], factor = n; end
```

a. Explain why it is only necessary to test `d` up to the square root of n. Explain why it is only necessary to test 2 and the odd integers greater than 2 up to the square root of n. Explain why it is only necessary to test prime values of `d` up to the square root of n (although this would require a table of primes up to the square root of n to be precomputed and available).

b. Modify the original program to reflect these conditions and test it.

c. Demonstrate that the program above fails (sometimes) when n has repeated roots. Modify it to handle repeated factors correctly.

EXERCISE 2.6

Index Maps for the FFT

While the form of the index maps allows a wide (infinite) variety of coefficients, there are practical advantages to choosing the smallest positive values that satisfy the required conditions. For $N_1 = 8$ and $N_2 = 7$ (relatively prime), the time index map satisfying (2-8) with the smallest positive coefficients giving no twiddle factors uses $a = b = 1$, giving

$$n = \langle N_2 n_1 + N_1 n_2 \rangle_N \tag{2-15}$$

Find the frequency index map (2-2) with the smallest positive coefficients satisfying both equalities in (2-8) and both in (2-11).

EXERCISE 2.7

Cooley–Tukey Index Map

For $N = 56 = 8 \times 7$, if we use the same common factoring in Exercise 2.6, then for the smallest positive coefficients allowing twiddle factors we have $K_1 = N_2$ and $K_2 = 1$, so

$$n = \langle N_2 n_1 + n_2 \rangle_N \tag{2-16}$$

Find the smallest positive coefficients (K_3, K_4) for the frequency index map (1-3) satisfying both (2-8) and (2-11).

EXERCISE 2.8

DIF

Repeat Exercise 2.6, but for $k = \langle 7k_1 + 8k_2 \rangle_N$ and finding the time map (1-2).

EXERCISE 2.9

DIT

Repeat Exercise 2.7 for $n = \langle n_1 + 8n_2 \rangle_N$.

EXERCISE 2.10

In-Order, In-Place PFA FFT Index Maps

If both the time and frequency index maps are forced to be the same, there will be no scrambling of order caused by in-place calculations. Set

$$n = \langle 7n_1 + 8n_2 \rangle_N \tag{2-17}$$

$$k = \langle 7k_1 + 8k_2 \rangle_N \tag{2-18}$$

Show that the uncoupling conditions are both met but that the short DFT conditions cannot be met. On the other hand, show that the short transformations are simply DFTs with a permuted order.

EXERCISE 2.11

Twiddle-Factor Array

For $N = 4^M$, examine the twiddle-factor array for its general structure. Create a 16 by 4 array with entries W_{64}^{nk}. The 16 rows multiply each of the 16 length-4 DFTs after the first stage of a length-64 radix-4 FFT. The first row and first column will always be unity. There will also always be one entry of $j = \sqrt{-1}$, and four with the real part equal to the imaginary part. What are the locations of these special twiddle factors? Repeat for $N = 128$ and perhaps others. Give general locations for these special values.

EXERCISE 2.12

Remove Trivial Operations

For a $N = 4^M$ radix-4 FFT, how many multiplications by 1 or $\pm j$ exist? How many twiddle factors have equal real and imaginary parts? How can this be used to save multiplications and additions in a general FFT?

PROJECT 3: PRIME-LENGTH DFT METHOD AND SOME BASIC IDEAS FROM NUMBER THEORY

For a time after the FFT was discovered, it was thought that no major improvements could be made for a prime-length FFT over the Goertzel algorithm. In 1968, Charles Rader published a short paper showing how to convert a prime-length DFT into a length-(P - 1) cyclic convolution [4]. Later, Winograd used this same idea to design DFT algorithms which use the absolute minimum number of multiplications. These optimal algorithms turned out to be practical for only a few short lengths, but those are very important when used with an index map to achieve longer DFTs. This project shows how to convert a prime-length DFT into cyclic convolution using Rader's permutation of the sequence orders. It also develops ideas in number theory useful to signal processing.

Project description

The arithmetic system used for the indices in this theory is over a finite ring or field of integers. All indexing arithmetic operations are performed modulo some finite integer modulus. If the modulus is a prime number, all nonzero elements will have a unique multiplicative inverse (i.e., division is defined) and the system is called a *field*. If the modulus is composite, some elements will not have an inverse and the system is called a *ring*. The process of calculating the remainder of a number modulo another is called residue reduction and the relationship of numbers having the same residues is called a congruence. These ideas and definitions are discussed in any introductory book on number theory [4, 11, 12].

Several definitions are needed to develop the ideas of this chapter. Euler's totient function, $\phi(N)$, is defined as the numbers of integers in $\mathcal{Z}_N = \{n | 1 \leq n \leq N-1\}$ that are relatively prime to N. For example, $\phi(3) = 2, \phi(4) = 2, \phi(5) = 4$.

Fermat's theorem states that for any prime number P and for all nonzero numbers $\alpha \in \mathcal{Z}_P$,

$$\langle \alpha^{P-1} \rangle_P = 1 \tag{3-1}$$

and a more general form called Euler's theorem states that for any N and for all nonzero $\alpha \in \mathcal{Z}_N$ that are relatively prime to N,

$$\langle \alpha^{\phi(N)} \rangle_N = 1 \tag{3-2}$$

When it exists, the inverse of an integer $n \in \mathcal{Z}_N$ is the integer $m \in \mathcal{Z}_N$, denoted n^{-1}, where $\langle mn \rangle_N = 1$. Using Euler's theorem, the inverse can be calculated from

$$m = \langle n^{-1} \rangle_N = \langle n^{\phi(N)-1} \rangle_N \tag{3-3}$$

The integer α is called an Nth root of unity modulo M if

$$\langle \alpha^N \rangle_M = 1 \tag{3-4}$$

and

$$\langle \alpha^L \rangle_M \neq 1 \qquad \text{for } L < N \tag{3-5}$$

Other terminology for the Nth root of unity is that α is of order N or that α belongs to the exponent N. Notice from Euler's theorem that N exactly divides $\phi(M)$ [i.e., $\phi(M)$ is an integer multiple of N]. If $N = \phi(M)$, α is called a primitive root. Primitive roots are important in several applications. It can be shown that they exist if and only if $M = 2, 4, P^r, or 2P^r$ and there are $\phi(\phi(M))$ of them.

If an integer N is prime, a primitive root r exists such that

$$m = \langle r^n \rangle_N \tag{3-6}$$

generates all of the nonzero integers between $m = 1$ and $m = N-1$ for $n = 0, 1, 2, \ldots, N-2$. There may be several primitive roots belonging to a modulus, each generating the same set of integers, but in a different order. In the finite field of integers $\mathcal{Z}_N = \{0 \leq n \leq N-1\}$, n is similar to a logarithm. Because this process generates a string of nonrepeating integers, a modification of it is sometimes used as a pseudo-random-number generator [13].

We now use this integer logarithm to convert a DFT into a cyclic convolution. The form of the DFT is

$$X[k] = \sum_{n=0}^{N-1} x[n] W^{kn} \tag{3-7}$$

and the form of cyclic convolution is

$$y[k] = \sum_{n=0}^{N-1} x[n] h[k-m] \tag{3-8}$$

with all indices evaluated modulo N.

The integer logarithm changes the product of k and n in the DFT into the difference in the cyclic convolution. Let

$$n = \langle r^{-m} \rangle_N \quad \text{and} \quad k = \langle r^s \rangle_N \tag{3-9}$$

with $\langle n \rangle_N$ denoting the residue of n modulo N. The DFT becomes

$$X[r^s] = \sum_{m=0}^{N-2} x[r^{-m}] W^{r^s r^{-m}} + x[0] \tag{3-10}$$

for $s = 0, 1, 2, \ldots, N-2$. and

$$X[0] = \sum_{n=0}^{N-1} x[n] \tag{3-11}$$

New functions are defined which are simply permutations in order of the old functions.

$$\tilde{x}[m] = x[r^{-m}] \tag{3-12}$$

$$\tilde{C}[s] = X[r^s] \tag{3-13}$$

$$\tilde{W}[n] = W^{r^n} \tag{3-14}$$

This results in the DFT being

$$\tilde{C}[s] = \sum_{m=0}^{N-2} \tilde{x}[m] \tilde{W}[s-m] + x[0] \tag{3-15}$$

which is a cyclic convolution of length $N-1$ (plus $x[0]$) of $x[n]$ and W^n in a permuted order.

Hints

All of the residue reductions can be calculated with the `rem()` function. Try to avoid using loops in implementing the various operations. Details of the ideas in this project can be found in [1], [3], [4], and [5]. A bit of reading of basic number theory, especially congruency theory [11, 12], would be helpful.

EXERCISE 3.1

Residue Reduction

The array of the residues

$$m = \langle \alpha^n \rangle_M \tag{3-16}$$

for $M = 5$ with rows for $\alpha = 1, 2, 3, \ldots, 6$ and columns for $n = 0, 1, 2, \ldots, 5$ is

$$m = \langle \alpha^n \rangle_M = \begin{bmatrix} 1 & 1 & 1 & 1 & 1 & 1 \\ 1 & 2 & 4 & 3 & 1 & 2 \\ 1 & 3 & 4 & 2 & 1 & 3 \\ 1 & 4 & 1 & 4 & 1 & 4 \\ * & 0 & 0 & 0 & * & 0 \\ 1 & 1 & 1 & 1 & 1 & 1 \end{bmatrix} \tag{3-17}$$

where $*$ is undefined. The second row is 2 raised to successively higher powers evaluated modulo 5. The second column is α raised to the first power. The fifth column is for the power or exponent equal to $n = \phi(5) = 4$, which illustrates Euler's theorem. From this array, determine the integer inverses of all nonzero elements of $\mathcal{Z}_5$. There are $\phi(\phi(5)) = 2$ primitive roots. One is $\alpha = 2$; what is the other? The elements of the rows and columns of this array are periodic. What is the period in α? What is the period in n? Study this array carefully to better understand the definitions and properties discussed in the project description or in your other reading.

EXERCISE 3.2

Roots of Unity

Form five arrays of the residues

$$m = \langle \alpha^n \rangle_M \tag{3-18}$$

for each $M = 6, 7, 8, 9,$ and 11. Let the arrays have rows for $\alpha = 1, 2, 3, \ldots, M+2$ and columns for $n = 0, 1, 2, \ldots, M+1$. Note that all α relatively prime to the modulus belong to some exponent. For each array indicate this exponent. All α relatively prime to the modulus have an inverse and it is the value just to the left of unity in each row of each array. Why? If β is the Nth root of unity modulo M, show in the arrays where β^q is the (N/p)th root of unity if q and M have a largest common factor of p.

EXERCISE 3.3

Primitive Roots

For the six arrays formed in Exercises 3.1 and 3.2, which moduli have primitive roots? Find the primitive roots in the arrays that have them and verify that there are $\phi(\phi(M))$ of them. When the modulus is prime, all nonzero elements of $\mathcal{Z}_M$ are generated by the primitive roots. When the modulus is composite and primitive roots exist, all nonzero elements are generated that are relatively prime to the modulus and these have inverses. This system is a ring. What is the pattern of the sequences generated by nonprimitive roots when the modulus is prime? when the modulus is composite and some primitive roots exist? when no primitive roots exist?

EXERCISE 3.4

Euler's Theorem

For each of the arrays formed in Exercises 3.1 and 3.2, indicate the cases where $\alpha^\phi = 1$. When does that not occur? Indicate the cases where $\alpha^N = 1$ for some $N < \phi(M)$. When does that not occur?

EXERCISE 3.5

Permutations

Using the arrays for $M = 5, 7, 11$, what are permutations of the data and permutations of the exponentials that will convert the DFT into cyclic convolution? For each M, how many choices are possible?

EXERCISE 3.6

Rader's Conversion

Convert the length-11 DFT matrix into a convolution matrix and calculate the DFT by convolution.

GENERAL LENGTH FFTs

OVERVIEW

There are some cases where one wants an algorithm for efficiently calculating the DFT of an arbitrary-length sequence. This is the case for the `fft()` command in MATLAB. In this packet we investigate the MATLAB FFT and the chirp Z-transform method for calculating the FFT.

PROJECT 1: EVALUATION OF THE MATLAB FFT

MATLAB has implemented a clever and general FFT algorithm in its `fft` function. The goal of this project is to analyze this function in such a way that using our knowledge of FFT algorithms will enable us to speculate on how it is implemented.

Project description

In earlier projects we have seen how composite-length DFTs can be implemented efficiently by breaking them down into multiple shorter ones. Here we will carefully time and measure the flops of the MATLAB `fft` function for a large number of lengths and then plot the time or flops versus the length. From this we can determine what kind of algorithm might be used.

Hints

Timing of the FFT is done by the following code: `x = rand(1,N); time = clock; fft(x); time = etime(clock,time);`.[4] This approach has two problems. Other users on a time-shared system can cause errors in the timing of your programs and the resolution of the clock may not be good enough to accurately time the shorter DFTs. These problems can be partially corrected by executing the FFT several times and averaging and/or by trying to do the timing when no one else is using the computer. A more consistent evaluation of an algorithm is the measurement of the number of floating-point operations (flops). For many algorithms, the floating-point operations take most of the time and therefore the timings and flops measurements are simply a multiple of each other. The flops are measured by the following code. `x = rand(1,N); f0 = flops; fft(x); f1 = flops - f0;`. The evaluation of the performance will involve generating a vector of time or flops for lengths varying from zero to several hundred or several thousand, depending on the speed and memory of your computer. The plots of time or flops versus length should not have the values connected by lines but should simply be "dots." If the vector of times (or flops) is the variable `t`, the plot should be made with the command `plot(t,'.')`.

To analyze the FFT it will be necessary to know the factors of the lengths. Although this can be done by hand for a small number of lengths, it would be helpful to write a program to factor an integer into its smallest prime factorization. A bit of reading or review of elementary number theory might be helpful [11,12].

[4] In the MATLAB `fft` there is a difference in the flop count depending on whether the input vector is real or complex; for example, try

```
x = rand(1,N) + j*rand(1,N);
```

Also, the flop counter may only be a 16-bit counter in some versions of MATLAB, so that it would overflow for large values of `N`.

EXERCISE 1.1

Execution Times and Flops for the MATLAB FFT

Create a vector of execution times for the MATLAB function `fft` by putting the timing of a single FFT in a `for` loop which steps through the lengths from zero to several hundred or several thousand, depending on the computing system being used. Also create a vector of flops in the same manner. These measurements could be made at the same time in the same loop. Plot the execution times and the number of flops in separate graphs versus the lengths of the FFT. Make the plots with simple points representing the times and flops rather than connected lines.

EXERCISE 1.2

Structure of Flops versus Length Plot

The plots made in Exercise 1.1 will have a very distinct structure that you should evaluate. The slowest times and largest numbers of flops will correspond to prime data lengths. What are the characteristics of the lengths of the other distinct time and flops groups? How do the lengths that are powers of 2 compare?

EXERCISE 1.3

What Is the Algorithm?

From the shape of the slowest times or greatest flops versus `N` plot, you should be able to conjecture what kind of algorithm is being used for a single prime-length FFT. From the next few distinct groupings, you should be able to determine if a decomposition is being used and if so, what kind. You should be able to tell if twiddle factors are being used by checking a length which is 3 to some high power, which, of course, must use twiddle factors. From all of this, write a program that calculates the times or flops from a formula based on your idea as to how the command is implemented. Compare the results of this formula with the measured data and correct until you have fairly good agreement. What is the algorithm being used?

EXERCISE 1.4

Formula for Flops

Based on the foregoing analysis of `fft`, develop a formula that will simulate the measured flops. Compare this with the theory.

PROJECT 2: CHIRP Z-TRANSFORM

The chirp z-transform is a method of evaluating the DFT using an FIR filter. The Goertzel method in Project 3 also used a filter, but the approach was very different. Rader's permutation used a cyclic convolution, but it could only be applied to prime-length DFTs (or with less efficiency to lengths that are a prime to a power). The chirp z-transform can be used on any data length and, while not as efficient as a Cooley–Tukey FFT or the PFA, it can be implemented using of the order of $N\log(N)$ arithmetic operations.

Project description

The Cooley–Tukey FFT, PFA, and Rader's method all used linear index maps to reorganize the basic DFT so that it could be evaluated more efficiently. Here we use a nonlinear index map to change the DFT into a noncyclic convolution. Applying the identity

$$(k-n)^2 = k^2 - 2kn + n^2 \tag{2-1}$$

$$nk = \left(n^2 - (k-n)^2 + k^2\right)/2 \tag{2-2}$$

to the definition of the DFT gives

$$X[k] = \left[\sum_{n=0}^{N-1} (x[n]\, W^{n^2/2}) W^{-(k-n)^2/2}\right] W^{k^2/2} \tag{2-3}$$

This has the form of first multiplying the data by a chirp sequence, then convolving that with the inverse of the chirp, and finally, multiplying the output of the convolution by the chirp. If posed as a filter, the impulse response of the filter is

$$h[n] = W^{n^2/2} \tag{2-4}$$

and the chirp transform of (2-3) becomes

$$X[k] = \left((x[k]h[k]) * h^{-1}[k]\right) h[k] \tag{2-5}$$

Care must be taken to implement the finite-length noncyclic convolution properly to obtain the correct length-N output sequence indicated in (2-3). It can be calculated directly or by use of the DFT. Indeed, the FFT is the way it is possible to improve efficiency over the Goertzel method.

Although discussed here as a means of calculating the DFT, the chirp z-transform is very flexible and can evaluate the z-transform on contours in the z-plane other than the unit circle and can efficiently evaluate a small number of values.

Hints

The implementation of the chirp z-transform can be done with the `conv` or the `filter` commands or with the `fft` and `ifft` commands. When using fast convolution with the FFT, be careful to take the proper part of the output. The evaluation can use the `flops` and/or the `clock` commands. Details of the methods can be found in [1] and [9].

EXERCISE 2.1

Write a Chirp DFT Program

Write a MATLAB function that will calculate the DFT using the chirp z-transform algorithm. Create a chirp vector with `n = 0:N-1; W = exp(-j*pi*n.*n/N);`. Multiply this times the data vector and convolve the result with a properly extended version of W using the `conv` command. The appropriate length-N segment of the output should be multiplied by W and that will be the DFT. Check the function against the `fft` function to make sure that it is correct. Measure the flops and execution times for a variety of lengths and compare with the FFT as was done in Project 1. Plot the number of required flops versus the length of the DFT and explain the numbers and shape.

EXERCISE 2.2

Use the FFT in the Chirp DFT

Write a MATLAB function to calculate the DFT using the chirp z-transform as was done in Exercise 2.1, but implement the convolution by using the MATLAB `fft` and `ifft` functions. Plot the flops and execution times versus length and compare with the FFT. Explain both the numbers of flops required and the dependency on N.

EXERCISE 2.3

Compare Chirp FFT with Other Methods

Make a comparison of the Goertzel algorithm, the MATLAB `fft` command, and the chirp z-transform implemented with the FFT from Exercise 2.2. Do this by making a plot of "flops" versus length for lengths up to 130 or more on a faster computer. Make the plot with "dots" for the `fft`, "x"'s for the chirp, and "o"'s for the Goertzel rather than connected lines. From these results, how would you implement a general-purpose FFT?

EXERCISE 2.4

Improvements for a Single Length

If one wants to execute a single DFT repeatedly on different data, a special-purpose program can be written that will precalculate the chirp in (2-4) and its FFT needed for the convolution. Write a MATLAB program that will count only the flops needed for execution assuming that all possible precalculations have been done. Compare these results with the `fft`. From this analysis over different lengths, how would you implement a special-purpose FFT?

EXERCISE 2.5

Calculation of Subset of the DFT values

The usual DFT is the set of N equally spaced samples of the z-transform of a length-N data sequence. Modify the chirp z-transform MATLAB function to calculate N samples of the z-transform on the unit circle between $\omega = 0$ and $\omega = \pi/2$. Does it require any more or less arithmetic that the version in Exercise 2.1 or 2.2?

EXERCISE 2.6

Use of the Chirp Z-Transform off the Unit Circle

Modify the chirp z-transform program of Exercises 2.2 and 2.3 to evaluate the z-transform of a length-N data sequence over L samples on a contour other than the unit circle. What contours can the method be used for? Derive a formula for the number of arithmetic operations needed as a function of N and L.

chapter 10

APPLICATIONS

OVERVIEW

In this chapter we present speech and radar applications. For the radar case, MATLAB is very effective in simulating radar echoes in noise. In the speech case, actual recordings can be processed within MATLAB. On machines with D-to-A capability, the processed speech can be played out for listening tests.

The first set of projects will introduce the basic measurements made in a radar system: range and velocity. Of particular importance is the linear-FM (LFM) chirp signal, its Fourier transform properties, and its processing via the "pulse compression" matched filter. The last project in the radar packet leads to implementation of a complete radar signal processor. One data file with an unknown target distribution is included as a "mystery" signal for testing. The objective of this project is to devise a processor that will detect all of the unknown targets and then estimate their range and velocity. Other data sets and test signals can be generated via the radar simulation written for MATLAB.

The processing of speech signals is one of the most fruitful areas of application of digital signal processing techniques. Applications such as speech enhancement, speech synthesis, digital speech coding, and speech recognition all present interesting opportunities for using digital signal processing algorithms. The second set of projects introduces speech processing with the goal of illustrating the basic properties of the speech waveform and the application of short-time analysis techniques. The third set of projects examines some of the attributes of a discrete-time system model for the production of the speech waveform. This model is composed of filters whose frequency response can be related to the physical features of the vocal tract and glottis. In addition, it serves as the basis for speech synthesis, speech coding, and speech recognition algorithms.

In the last section, A-to-D conversion of speech waveforms will be investigated. Sampling and quantization of speech waveforms is important because it is the first step in any digital speech processing system, and because one of the basic problems of speech processing is coding for digital transmission and/or storage. MATLAB provides a convenient environment for simulating and measuring the characteristics of quantization noise. It can also implement nonlinear operations, so that quantizers such as μ-law companding can be studied.

BACKGROUND READING

A description of the basic processing blocks needed in the radar system can be found in Chapter 5 of [1]. In addition, the book by Levanon is devoted to radar signals [2]. Appropriate background reading for the speech projects can be found in [3] to [6]. The quantization of waveforms is the topic of the book by Jayant and Noll [7].

[1] J. H. McClellan and R. J. Purdy. Application of digital signal processing to radar. In A. V. Oppenheim, editor, *Applications of Digital Signal Processing*, chapter 5, pages 239–330. Prentice Hall, Englewood Cliffs, NJ, 1978.

[2] N. Levanon. *Radar Principles*. John Wiley & Sons, New York, 1988.

[3] J. L. Flanagan. *Speech Analysis, Synthesis, and Perception*. Springer-Verlag, New York, second edition, 1972.

[4] L. R. Rabiner and R. W. Schafer. *Digital Processing of Speech Signals*. Prentice Hall, Englewood Cliffs, NJ, 1978.

[5] T. Parsons. *Voice and Speech Processing*. McGraw-Hill, New York, third edition, 1986.

[6] J. R. Deller, Jr., J. G. Proakis, and J. H. L. Hansen. *Discrete-Time Processing of Speech Signals*. Macmillan, New York, 1993.

[7] N. S. Jayant and P. Noll. *Digital Coding of Waveforms*. Prentice Hall, Englewood Cliffs, 1984.

[8] L. R. Rabiner and B. Gold. *Theory and Application of Digital Signal Processing*. Prentice Hall, Englewood Cliffs, NJ, 1975.

[9] A. E. Rosenberg. Effect of glottal pulse shape on the quality of natural vowels. *Journal of Acoustical Society of America*, 49(2):583–590, February 1971.

[10] G. Fant. *Acoustic Theory of Speech Production*. Mouton, The Hague, 1970.

RADAR SIMULATION

OVERVIEW

This set of projects will introduce methods for range and velocity measurements in a radar system. One waveform of particular importance is the linear-FM (LFM) chirp signal. Its processing via a matched filter will maximize SNR and enhance detectability. The matched filter also serves as a "pulse compression" operator whose output is extremely narrow and therefore is quite useful in resolving closely spaced targets. An interesting implementation of the digital matched filter can be done with FFT convolution to produce a processor that can run at radar sampling rates. Velocity is measured via the Doppler effect, which requires that the radar signal processor perform spectrum estimation. The combination of matched filtering and spectrum estimation forms the basis of a radar signal processor.

The last project in this section leads to implementation of a complete processor for extracting range and velocity information from radar echoes. One data file containing returns from an unknown target distribution is included as a "mystery" signal. A radar simulation (`radar.m` in Appendix A) has been written in MATLAB, so that additional signals can be generated for testing or as new mystery signals. The objective of this project is to write the M-files for processing the signals that are buried in receiver noise and clutter. The final implementation should be able to detect automatically all the unknown targets and then estimate their range and velocity.

BACKGROUND READING

It is impossible to provide enough background about radar in this introduction, so some outside reading will be needed. Not many books are devoted to the topic of radar signal processing, but one with a concise treatment is the text by Levanon [2]. Specific topics related to digital implementations can be found in Chapter 5 of [1] and Chapter 13 in [8].

PROJECT 1: PROPERTIES OF THE LFM CHIRP SIGNAL

In this project, the characteristics of the LFM chirp signal are investigated: its time-domain appearance and its Fourier transform. In Project 2 we investigate its autocorrelation function, which is the output of a "pulse compression" matched filter.

The *chirp* radar signal is defined by the formula

$$s(t) = e^{j\pi W t^2/T} \qquad -\tfrac{1}{2}T \leq t \leq \tfrac{1}{2}T \tag{1-1}$$

Since the phase of $s(t)$ varies quadratically versus t, and the derivative of phase determines the instantaneous frequency of the signal, the frequency changes linearly versus t. The signal is complex valued because it is the baseband form of the linear frequency modulation. The LFM signal is a pulse whose time duration equals T seconds. Over the life of the pulse, the changing frequency sweeps from $-\frac{1}{2}W$ to $+\frac{1}{2}W$ hertz.

An intuitive guess about the frequency spectrum $S(f)$ leads one to suspect that most of the energy in the frequency domain will be concentrated in the range $|f| < \frac{1}{2}W$. This is, in fact, true if the frequency sweeps slowly enough; equivalently, if T is large enough. In the examples that follow, the dependence of the Fourier transform on the time–bandwidth product (TW) is studied.

Hints

Since the chirp signal is complex-valued and is processed by a complex-valued matched filter, all plots must be made of either the real part of (1-1) or the magnitude of the Fourier transform to show the correct behavior.

Two types of chirps are considered in the exercises that follow: a discrete-time chirp and a continuous-time chirp. MATLAB can deal only with a sampled version of the LFM signal, so the analog chirp is simulated by oversampling $s(t)$. For the discrete-time case, the sampling frequency is taken nearly equal to the swept bandwidth W, but for the continuous-time case, oversampling by a factor of 5 or more is needed for an accurate simulation.

EXERCISE 1.1

Sampled Chirp Signal

In MATLAB the chirp signal must be represented as a discrete-time signal. Therefore, the formula for $s(t)$ in (1-1) must be sampled at the rate $f_s = 1/T_s$. The sampling rate can be tied to W, the swept bandwidth of the chirp. In many cases the chirp is more or less bandlimited to a frequency extent of W. Therefore, it is convenient to let $f_s = pW$, where $p \geq 1$ represents the oversampling factor.

a. Convert the continuous-time chirp formula (1-1) into a discrete-time signal by sampling at a rate $f_s = pW$. Give the equation for the discrete-time signal in the form

$$s[n] = \exp\left[j2\pi\alpha\left(n - \tfrac{1}{2}N\right)^2\right] \qquad 0 \leq n < N \tag{1-2}$$

Determine the correct formulas for α and N, and show that these parameters depend only on p and TW, the time–bandwidth product. (*Note*: It may not be possible to make the discrete-time chirp symmetric, depending on how the sampling times are defined. Starting at $t_n = -\frac{1}{2}T$ may not be the best strategy if $t = 0$ is not included in the sampling grid.)

b. Write an M-file to synthesize a discrete-time chirp. The function should have only two inputs, p and TW, and should return the complex-valued signal $s[n]$, as specified by the following comments.

```
function   s = dchirp( TW, p )
%DCHIRP      generate a sampled chirp signal
%  usage   s = dchirp( TW, p )
%      s :  samples of a digital "chirp" signal
%             exp(j(W/T)pi*t^2)   -T/2 <= t < +T/2
%     TW :  time-bandwidth product
%      p :  sample at p times the Nyquist rate (W)
```

c. Generate a sampled chirp signal whose time–bandwidth product is $TW = 50$. Plot the real and imaginary parts of the chirp signal, and observe how the apparent frequency changes versus time. The "chirped" pulse should have the characteristics given in Table 10.1.

TABLE 10.1

Desired Parameters for a Chirp Signal

Parameter	*Value*	*Units*
Pulse length	25	μs
Swept bandwidth	2	MHz
Sampling frequency	20	MHz
TW product	50	dimensionless

d. Redo the previous plot, but perform the sampling at $1.2W$. In this case the behavior of the chirp near its ends should exhibit the characteristics of a sampled sinusoid.

e. To show that the assumption of a "large" TW product is necessary, generate a chirp with $TW = 9$ and plot its Fourier transform. Use significant oversampling to simulate an analog chirp. Measure the fraction of the energy that lies outside the region $|f| < \frac{1}{2}W$.

f. *Optional*: Construct a plot of out-of-band energy versus TW over the range from 3 to 90. Although the out-of-band energy is quite small, it is possible to identify the value of TW where there is a "knee" in the curve.

EXERCISE 1.2

Fourier Transform of a Chirp

The Fourier transform magnitude of a chirp is *approximately* a rectangle in frequency *if the time–bandwidth product is large* (i.e., $TW > 50$). If we assume that $s(t)$ is the chirp and $S(f)$ is its Fourier transform, we can approximate $|S(f)|$ with a rectangle that extends from $f = -\frac{1}{2}W$ to $f = +\frac{1}{2}W$. Figure 10.1 shows that this rectangular approximation is not bad.

a. Find a formula for the height of the approximate spectrum $\tilde{S}(f)$ by using Parseval's theorem to equate the energies in the time and frequency domains.

b. Compute and plot the Fourier transform of the chirp for an oversampled case. Use the parameters in Table 10.1 from Exercise 1.1. This case should approximate the behavior of a continuous-time chirp, so use a long FFT (with zero padding) to get a smooth plot in the frequency domain. Scale the spectrum so that its magnitude near dc is correct.

c. Now consider the discrete-time case where the sampled chirp $s[n]$ is obtained by sampling just above the minimum rate—use $f_s = 1.2W$ and then $f_s = 2W$. Compute the DTFT of $s[n]$ and plot versus ω. Determine the appropriate cutoff frequency relative to $\omega = \pi$ in both cases. Make sure to use sufficient zero padding with the FFT to get a smooth plot of the DTFT.

d. Give a general formula for the height of the DTFT magnitude in terms of TW, N, and p. Again, Parseval's theorem should provide the link between the time and frequency domains.

e. The DFT of a chirp sometimes has remarkable properties, when $p = 1$. Since the number of nonzero samples in the discrete-time chirp signal is N, we can compute its N-point DFT. Show by example that whenever N is a multiple of 4 and $p = f_s/W = 1$, the DFT $S[k]$ is also an exact chirp. Determine a formula for the quadratic phase of $S[k]$, and verify that the magnitude of $S[k]$ is constant for all k.

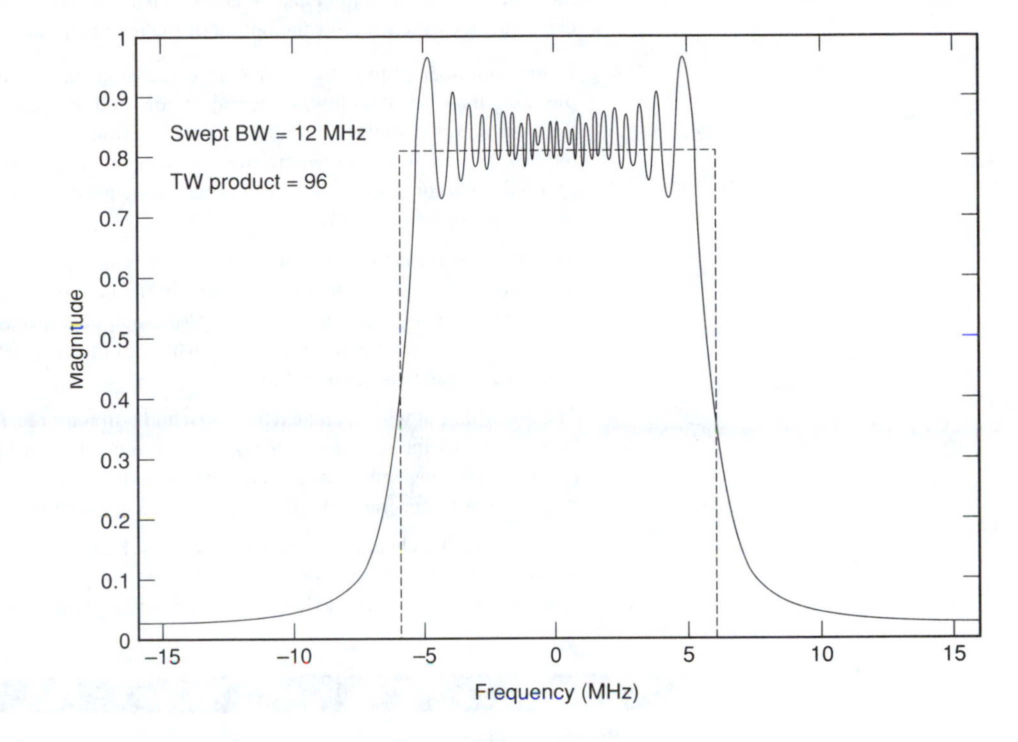

Figure 10.1
Fourier transform of a continuous-time chirp. Notice that most of the energy is concentrated in the frequency region between $-\frac{1}{2}W$ and $+\frac{1}{2}W$. The dashed line is the magnitude of an approximate transform $|\tilde{S}(f)|$, which is perfectly bandlimited.

PROJECT 2: RANGE PROCESSING

The transmitted signal in a radar must be designed differently for range estimation and for velocity estimation. In the case of range processing, the primary concern is to maximize the output SNR and range resolution from a matched filter, so large TW-product chirps are used. In this project, the LFM pulse and its matched filter are examined. The resulting output is a very narrow "compressed" peak that has large amplitude, which makes the detection of echoes easier. For a theoretical development of some equations describing the received radar signal, see the section *Theory of Radar Returns* in Project 4.

Hints

The matched filter involves a convolution that can be computed in either the time domain, via direct convolution (see `conv`), or in the frequency domain, via fast convolution with FFTs (see `fftfilt`).

EXERCISE 2.1

Pulse-Compression Matched Filtering

The matched filter is defined by either its frequency response, $H(f) = S^*(f)$, or by its impulse response, which, in turn, is determined by the transmitted waveform:

$$h(t) = s^*(-t) = e^{-j\pi Wt^2/T} \qquad -\tfrac{1}{2}T \le t \le +\tfrac{1}{2}T \tag{2-1}$$

The output of the matched filter is

$$y(t) = h(t) * Gs(t - T_d) \tag{2-2}$$

where T_d is a time delay due to a target at range $R = \frac{1}{2}cT_d$. Since the matched filter is time-invariant, it is sufficient to make plots for the case where $T_d = 0$.

A discrete-time matched filter would involve a sampled version of $h(t)$. Since the discrete-time matched filter is FIR, its output can be computed via direct convolution. The purpose of this exercise is to study the form for the output of the matched filter.

a. In the continuous-time case, derive an expression for the matched filter output (2-2) by plugging into the convolution integral. Show that $y(t)$ can be written with an envelope function that is a slightly modified "sinc" function. This result has to be true because if the Fourier transform were approximated with a rectangle $\tilde{S}(f)$ as in Fig. 10.1, the output of a pulse-compression matched filter would be a "sinc" function whose width would be inversely proportional to W.

b. To verify the form for the matched filter's output, generate an oversampled chirp signal with $TW = 50$ and $p = 8$ or 10. Use this signal in a matched filter, creating the output by direct convolution (e.g., via `conv`). Plot the output, especially near the peak, and verify that it has the correct mainlobe width (distance between first zero crossings). Make sure to label the time axis in correct units.

c. The entire output will be created by `conv` and will require a significant amount of computation, but only the region near the peak is of interest. Prove that the matched filter output (2-2) can also be written as the autocorrelation of $s(t)$. Then use the M-file `acf` from Appendix A to generate only the region near the mainlobe.

d. Generate the matched filter output for the $f_s = 1.2W$ case. Verify that the mainlobe width is correct, but observe that there are just a few samples on the mainlobe. For the general case ($f_s = pW$), determine how many samples will be on the mainlobe.

EXERCISE 2.2

Processing Gain

The matched filter enhances SNR and, as a result, it is able to detect chirp signals even when they are "buried" beneath the noise level.

a. Generate a signal vector that is 700 points long but which contains in the middle a chirp with a TW product of 64 and $f_s = 3W$. Add (complex) white Gaussian noise to the signal so that the SNR is -10 dB (i.e., standard deviation is $\sqrt{10}$ times the signal amplitude). Plot the real part of the raw signal.

b. Process the signal through a digital matched filter and plot the output versus n. Explain how the location of the peak in n is related to the beginning of the echo and the starting index of the digital matched filter. In other words, how would the peak location index be converted to a range measurement?

c. Calculate the processing gain (in dB) by subtracting the input SNR from the output SNR. Measure the peak output versus the noise floor to compute an output SNR. Relate this output signal-to-noise ratio to the TW product and the oversampling factor p. It might be necessary to experiment with different values of TW and p to uncover a simple formula for processing gain.

EXERCISE 2.3

Range Estimation

The compressed-pulse output of the matched filter is significant because its increased height makes it easier for the radar processor to detect the echo and estimate its location when calculating range, according to $R = \frac{1}{2}cT_d$. There is, however, some uncertainty in a range estimate due to the inherent uncertainty in locating the peak of the echo. In the discrete-time case, this uncertainty could be fairly large.

The time delay (T_d) due to the target range is irrelevant for the continuous-time case because the matched filter is time-invariant. However, in the discrete-time case where there is little or no oversampling, the relation between T_d and the sampling grid is crucial. The issue arises because there are so few samples on the mainlobe of the compressed pulse. If T_d is an integer multiple of $1/f_s$, one sample lies at the peak value and corresponds to the target location. Otherwise, the sampling of the compressed pulse straddles the peak location.

a. Generate a received signal and the corresponding discrete-time matched filter output for the case where T_d lies halfway between two sample times. Show that the output has two equal-height peaks in this case.

b. Generate a received signal containing three echoes at different time shifts plus (complex) noise with a standard deviation equal to twice the amplitude of the chirp signals. Use a TW product equal to 100, and sample at the rate $f_s = W$. Calculate the matched filter output $y[n]$ using `fftfilt`, or `conv`, and plot the magnitude of the output signal. Specify the time delay of the different signals so that the true peaks in $y[n]$ lie between sampling times (e.g., use $T_d \in \{5.5/f_s,\ 15.2/f_s,\ 22.7/f_s\}$).

c. For the output from part (b), make a visual identification of the peak locations. Then use the M-file `pkpicker` from Appendix A to extract the peaks. State a strategy for setting the threshold needed in the peak picking function.

d. Devise a peak location algorithm that interpolates to find a "best" estimate of the delay time from the three largest values picked from the peak. A polynomial interpolator should be sufficient. Test your algorithm by generating a noise-free input signal whose actual T_d lies somewhere between two sampling grid points. Determine the worst-case error in the time-delay estimate.

e. Apply your peak location algorithm to the signal from part (b) after it has been processed through `pkpicker`. Verify that the three peaks are identified and assigned the proper values of T_d.

EXERCISE 2.4

Mismatched Filtering

The matched filter has a frequency response that is the conjugate of $S(f)$. The rectangular shape in the frequency domain leads to sidelobes in the time domain. These sidelobes can be reduced by the usual technique of windowing, except that the window must now be applied by multiplying in the frequency domain; the result is called a mismatched filter, $\hat{H}(f) = W(f)S^*(f)$.

a. Repeat Exercise 2.1(b), but do the matched filter computation with FFT convolution. Make sure that you pad with zeros to avoid circular convolution. Plot the mainlobe region and include a few of the near-in sidelobes.

b. For the same case as in part (a), implement a "mismatched" filter by multiplying in the frequency domain with a Hamming window. This Hamming weighting can be implemented during the frequency-domain multiplication phase of the FFT convolution. The Hamming window should extend over the bandwidth of $S(f)$ only. Compare the "sidelobes" of the compressed pulse output in this case to the sidelobes for the compressed pulse from the *true* matched filter. Also compare the mainlobe widths.

c. Reconsider the answer to Exercise 2.1(d). Explain how the Hamming window affects the number of samples on the mainlobe of the compressed pulse.

PROJECT 3: VELOCITY PROCESSING

For velocity processing, a radar relies on the Doppler frequency shift caused by a moving target. The processor must perform a spectrum analysis, but the magnitude of the Doppler shift is so small that a different type of transmitted waveform is needed—a single pulse will not work. Usually, a burst waveform consisting of a coherent group of short pulses is employed in a Doppler radar. The spectral properties of the burst waveform will be investigated here. For a theoretical development of some equations describing the Doppler shift, see the section *Theory of Radar Returns* in Project 4.

Hints

When computing the Doppler spectrum, only a small number of signal values will be used. To get a smooth spectrum estimate, the FFT should be zero-padded.

EXERCISE 3.1

Fourier Transform of a Burst Waveform

The Doppler frequency shift due to a moving target, with velocity v, is

$$f_d = \frac{2v}{c} f_c$$

where f_c is the center frequency of the radar. Unless the target has an extremely large velocity v with respect to the velocity of propagation c, the magnitude of the Doppler shift is rather small, and impossible to detect from one radar pulse. A burst waveform containing repeated pulses $p[n]$ is needed to measure Doppler frequency shifts. When considered as one long transmitted waveform, the burst is defined by the following formula:

$$s[n] = \sum_{\ell=0}^{L-1} p[n - \ell M]$$

where M is the interpulse period, N the pulse length, and L the number of pulses.

a. The effect of a Doppler shift (at the receiver) is to multiply $s[n]$ by a complex exponential $e^{j\omega_d n}$, where ω_d corresponds to the Doppler shift. Determine the relationship between ω_d and the velocity v. Assume that the sampling frequency f_s is known.

b. Use MATLAB to plot an example spectrum with $\omega_d = 0$ (i.e., a stationary target). Take $L = 8$, $N = 7$, and $M = 19$. Take the pulse $p[n]$ to be a simple boxcar. The magnitude spectrum should have noticeable spectral lines, which are narrow peaks with finite bandwidth. Determine the mainlobe width of these spectral lines, and their spacing. Which parameters (M, N, L) determine these features?

c. Now take a moving target whose velocity gives a Doppler shift of ω_d. Use MATLAB to plot an example spectrum with $\omega_d = 2\pi/31$. Take $L = 8$, $N = 7$, and $M = 19$. Observe the movement of the spectral lines.

d. Determine the minimum value of ω_d that would give a detectable movement of the spectral lines.

e. If the value of ω_d gets too large, the spectral lines will move to a position that is ambiguous (or aliased). Determine this maximum value of ω_d at which aliasing occurs and the corresponding maximum velocity.

f. Use the windowing and convolution properties of the DTFT to make a sketch of the DTFT of $s[n]e^{j\omega_d n}$. To avoid messy algebraic forms, consider that $s[n]$ is produced by convolving $p[n]$ with a finite impulse train spaced by M. This analysis would explain the MATLAB results from the previous parts.

g. *Optional*: Make a plot of the transform for the case where $p[n]$ is an LFM pulse.[1] Make the TW product greater than 40, and then take the spacing between pulses to be 2 or 3 times larger than TW. Use a burst containing about 10 pulses. Then let the fre-

[1]The LFM burst processing may not be possible if your version of MATLAB has restricted vector size.

quency shift be small and plot two different cases to determine the minimum detectable Doppler shift.

EXERCISE 3.2

Velocity Processing with Burst Waveforms

In reality, the Doppler processing need not involve a Fourier transform of the entire burst waveform. The computation of a spectrum estimate to measure velocity can be done by taking just one sample from each pulse. Therefore, a rather small amount of data can be analyzed to compute the Doppler spectrum.

a. Generate a burst waveform with $\omega_d = 2\pi/31$. Take $L = 8$, $N = 7$, and $M = 19$. Process the waveform for its Doppler frequency content by taking one sample from each pulse, at a regular spacing of M. Then compute the FFT of these L samples and search for the peak location. Give the relationship between this peak frequency axis and ω_d.

b. Explain how the processing of part (a) can be viewed as lowering the sampling rate, consistent with the aliasing that was discovered in part (e) of Exercise 3.1.

c. The Doppler spectrum analysis can be done in conjunction with the pulse compression matched filter.[1] Generate a 10-pulse burst of LFM pulses, each with a TW product equal to 64. Oversample by a factor of $p = 2$, and separate the pulses by 128 zeros, making the interpulse period 256 samples. Multiply by a complex exponential at $\omega_d = 2\pi(0.0007)$. Process the entire burst through a filter matched to one of the individual pulses. Then pick off one sample from each compressed pulse at a regular spacing of 256, and compute the Doppler spectrum. Use points near the peak of the compressed pulse. Explain the relationship between the measured peak frequency and the original Doppler shift frequency ω_d.

PROJECT 4: RADAR SYSTEM SIMULATION

This project requires that you develop a fairly complete radar processing system that will extract the range and velocity of several "point" targets. The objective of the project is to estimate the parameters of the various targets by processing the returned waveform versus time delay and frequency (Doppler). The targets have different relative amplitudes, so you should extract amplitude information as well. The parameters of the transmitted linear-FM radar signal are given in Table 10.2.

TABLE 10.2

Radar Parameters of the Transmitted Waveform

Parameter	*Value*	*Units*
Radar frequency	7	GHz
Pulse length	7	μs
Swept bandwidth	7	MHz
Sampling frequency	8	MHz
Interpulse period	60	μs
Number of pulses	11	none
Time delay to start receive window	25	μs
Time delay to		8
end receive window	50	μs

Scenario

The parameters of the radar simulation and the targets were chosen to match those that might be expected in an ATC (air traffic control) application. One possible application might be short- to medium-range tracking for controlling the final approach patterns of commercial aircraft near an airport. This scenario would require a burst waveform to measure velocity accurately, but the maximum range is not large, so a relatively high PRF (pulse repetition frequency) can be used in the burst. This "scenario" can be used to check velocities and ranges for reasonableness.

Theory of Radar Returns

The received signal in a radar can be expressed completely in terms of its complex envelope signal. If the transmitted signal has a complex envelope given by $s(t)$, which represents the phase (and amplitude) modulation, the transmitted RF signal would be

$$\text{Re}\left\{ e^{j2\pi f_c t} \sum_{\ell=0}^{N_p-1} s(t - \ell\Delta) \right\}$$

where f_c is the center (RF) frequency of the radar.

For a moving target the range to the target varies as a function of time (t). If we assume that the velocity is a constant (v), the expression for the range is

$$R(t) = R_\circ - vt$$

where $R_\circ$ is the range at a "reference" time ($t = 0$). The minus sign is a convention that means a target traveling toward the receiver will have a positive velocity and therefore a positive Doppler shift.

The time delay to the target is $2R(t)/c$, so the received signal is

$$\text{Re}\left\{ G_{\text{target}} \sum_{\ell=0}^{N_p-1} e^{j2\pi f_c(t-2R(t)/c)} s(t - 2R(t)/c - \ell\Delta) \right\}$$

The reflectivity of the target gives rise to a gain term G_{target}. Since the carrier term can be extracted by a quadrature demodulator, the complex envelope of the received signal will be

$$r(t) = \sum_{\ell=0}^{N_p-1} e^{j2\pi f_c(-2R_\circ+2vt)/c} s(t - 2R_\circ/c + 2vt/c - \ell\Delta)$$

Within the summation, time can be referenced to the start of each pulse by introducing a new time variable $t^\ell = t - \ell\Delta$. Then the expression becomes

$$r(t) = \sum_{\ell=0}^{N_p-1} e^{j2\pi f_c(-2R_\circ^\ell+2vt^\ell)/c} s(t^\ell - 2R_\circ^\ell/c + 2vt^\ell/c)$$

where $R_\circ^\ell = R_\circ - v(\ell\Delta)$ is the range at the start of the ℓth pulse.

For a linear-FM signal, the complex envelope of the transmitted signal, $s(t)$, is actually made up of two parts: a pulse $p(t)$ that gates the signal on and off, and the LFM phase modulation. For $p(t)$, the delay term is just a shift, but in the LFM phase, the delay must be applied to the argument of the quadratic term. In the `radar.m` M-file, this is done by the `polyval()` function in the exponent of the last term that makes up the output signal `y`. The quadratic phase was previously extracted by using a call to `polyfit()`, thus allowing the user to enter samples of the complex envelope rather than parameters describing the LFM modulation (i.e., T and W).

Receive Window

The radar returns will be spread over a very wide time span when the range coverage needs to be large. To avoid range ambiguities, the receive window must be limited to the time interval between successive pulses. Thus the maximum range is $R_{\max} = c\Delta/2$; the minimum range is dictated by the length of the pulse. In the simulation function `radar()`, a receive window can be specified so as to limit the number of returned samples that must be processed. This also limits the data set to a manageable size.

Noise Sources

After the received signals due to targets are created with `radar.m`, noise needs to be added to the data. Two forms of noise have been added:

1. *Receiver noise*, which is modeled as white Gaussian noise. It is present in all the returns and is completely uncorrelated from one pulse to the next and from one time sample to the next.
2. *Clutter*, which is really a distributed target with (near) zero velocity. On a pulse-to-pulse basis, this sort of noise is *highly* correlated and is usually removed by prefiltering with a canceler.

For many of the returned pulses, the SNR and the signal-to-clutter ratio will be less than 1. Therefore, the uncompressed pulse is well below the noise and can be identified only after the pulse-compression stage.

Hints

The data file (`r100.mat`) contains the weighted sum of 4 to 8 different targets. A burst waveform was used as the transmitted signal. Each pulse in the burst gives 201 received data samples in range, so the data matrix is 201×11. Table 10.2 gives the detailed characteristics of the transmitted signal.

This synthetic data file was produced by the M-file called `radar.m`. You should consult the listing of this file for detailed questions about how the simulated radar returns were actually derived. The function `lchirp`, which is needed by `radar.m`, can be found in Appendix A. The exact parameter values (for velocity and range) are, of course, unknown, but `radar.m` could be used to generate other synthetic data sets to validate your processing scheme. Furthermore, `radar.m` would be useful to instructors interested in producing new data sets with unknown target configurations.

To do the processing automatically, it will be necessary to form estimates by finding peak locations. The M-file `pkpicker` in Appendix A is available for this purpose.

EXERCISE 4.1

Signal Processing System

To create the processing program, it will be necessary to analyze the radar waveform for its delay and its Doppler content. This will require the following steps (but not necessarily in this order):

1. Process the returns with a matched filter to compress the LFM pulse that was used by the transmitter. This requires that you resynthesize the transmitted *chirp* waveform according to the parameters given in Table 10.2. From this transmitted pulse, the impulse response of the matched filter can be defined.
2. The transmitted signal is a burst waveform consisting of 11 identical LFM pulses. If MATLAB imposes memory limitations, each pulse should be processed separately by the matched filter.
3. Velocity analysis requires a Fourier transform of the data across the 11 pulses of the burst to extract the Doppler frequency. This will require a DFT method of spectrum analysis,

but could be restricted to those ranges where there are likely to be targets. Identify the major velocity components from peaks in the frequency-domain plot. Make sure that you consider the possibility of positive and negative velocities.

4. The returned radar signal contains a very large clutter component, *so you must implement some sort of preprocessing to reduce the clutter return.*
5. The valid peaks in range and/or Doppler need to be identified, preferably by a peak picking algorithm. Visual identification of the peaks would be a first cut, but you should state where you set the threshold (visually). Automatic peak picking would require that you define a threshold that adapts its level depending on a local measure of the additive noise.

EXERCISE 4.2

Processed Results

Once the radar signal processing M-files are debugged, process the data file `r100.mat`. In addition, the simulation M-file `radar.m` can be used to produce other target returns to check out the M-files.

a. Determine how many moving targets are present, and for each make an estimate of the range (in km) and the velocity in m/s.

b. Be careful to convert all range plots and Doppler frequency plots to the correct units (i.e., hertz, meters/second, kilometers, etc.) This involves the use of the sampling frequency (in range or Doppler) and the length of the FFT.

c. If your machine has no memory restrictions, process the entire data set for velocity at all possible ranges. Collect this information into one large matrix and make a contour plot of the log magnitude versus R and v. Peaks in this "range-Doppler spectrum" should correspond to targets. Use the optional arguments to `contour` to label the axes correctly.

d. Which of the "targets" corresponds to a clutter return?

e. Due to the presence of noise, the estimate values are not correct. Determine the uncertainty in your measured values, and express this uncertainty as a percentage (i.e., 1%, 0.1%, ..., or one part in a million). State how the signal parameters affect this uncertainty.

RADAR.M (help section only)

```
function y = radar( x, fs, T_0, g, T_out, T_ref, fc, r, a, v )
%RADAR       simulate radar returns from a single pulse
%  usage:
%    R = radar( X, Fs, T_0, G, T_out, T_ref, Fc, R, A, V )
%      X:       input pulse (vector containing one pulse for burst)
%      Fs:      sampling frequency of input pulse(s)      [in MHz]
%      T_0:     start time(s) of input pulse(s)          [microsec]
%      G:       complex gains; # pulses = length(g)
%      T_out:   2-vector [T_min,T_max] defines output
%                 window delay times w.r.t. start of pulse
%      T_ref:   system "reference" time, needed to simulate
%                  burst returns. THIS IS THE "t=0" TIME !!!
%      Fc:      center freq. of the radar.                [in MHz]
%      R:       vector of ranges to target(s)          [kilometers]
%      A:       (complex) vector of target amplitudes
%      V:       vector of target velocities (optional)  [in m/sec]

%  note(1): VELOCITY in meters/sec !!!
%           distances in km, times in microsec, BW in MegaHz.

%  note(2): assumes each pulse is constant (complex) amplitude
%  note(3): will accommodate up to quadratic phase pulses
%  note(4): vector of ranges, R, allows DISTRIBUTED targets
```

TEST_RADAR.M

```
%
%   EXAMPLE of calling the function radar()
%     make one radar return for a burst of LFM pulses
%
clear,  format compact
T = 10     % microsec
W = 5      % MHz
fs = 10    % MHz, oversample by 2
s = lchirp(T, W, fs/W);

Np = 7;                 % 7 pulses
jkl = 0:(Np-1);
T_0 = 200*jkl;          % in usec
g = ones(1,Np);         % gains
T_out = [100 150];      % in usec
T_ref = 0;              % why use anything else?
fc = 10000;             % 10 GHz

Ntargets = 1;
ranges = 20;       % in km ???
amps = 1;
vels = 30;         % in m/sec

y = radar(s,fs,T_0,g,T_out,T_ref,fc,ranges,amps,vels);
```

INTRODUCTION TO SPEECH PROCESSING

The processing of speech signals is one of the most fruitful areas of application of digital signal processing techniques. Basic speech processing problems such as speech enhancement, speech synthesis, digital speech coding, and speech recognition all present interesting opportunities for application of digital signal processing algorithms. In such applications it is very important to understand the properties of the speech signal in order to make intelligent use of DSP techniques. This set of projects is intended as an introduction to speech processing with the goal of illustrating the properties of speech and the application of DSP techniques in a short-time analysis framework. Background for these projects can be found in [3] to [6].

PROJECT 1: SPEECH SEGMENTATION

Segmentation and phonetic labeling of speech waveforms is a fundamental problem in speech processing. In general, this is very difficult to do automatically, and even computer-aided segmentation by human beings requires a great deal of skill and knowledge on the part of the analyst. Nevertheless, it is very instructive to attempt to identify the parts of the speech waveform that correspond to the different phonemes of the utterance.

Hints

While MATLAB is far from an ideal tool for this purpose, plotting functions such as `plot( )`, `subplot( )`, and `stem( )` can be used for looking at short segments of speech waveforms. You will find that in some versions of MATLAB the basic plotting function `plot( )` has a vector-length limitation of 4094. For this reason the M-file `striplot( )` is available in Appendix A for plotting long vectors in a multiline format. An example of a speech waveform plotted using `striplot( )` is shown in Fig. 10.2. This plot shows the waveform of an utterance of the sentence *Oak is strong and also gives shade.* The waveform was sampled with a sampling rate of 8000 samples/s.

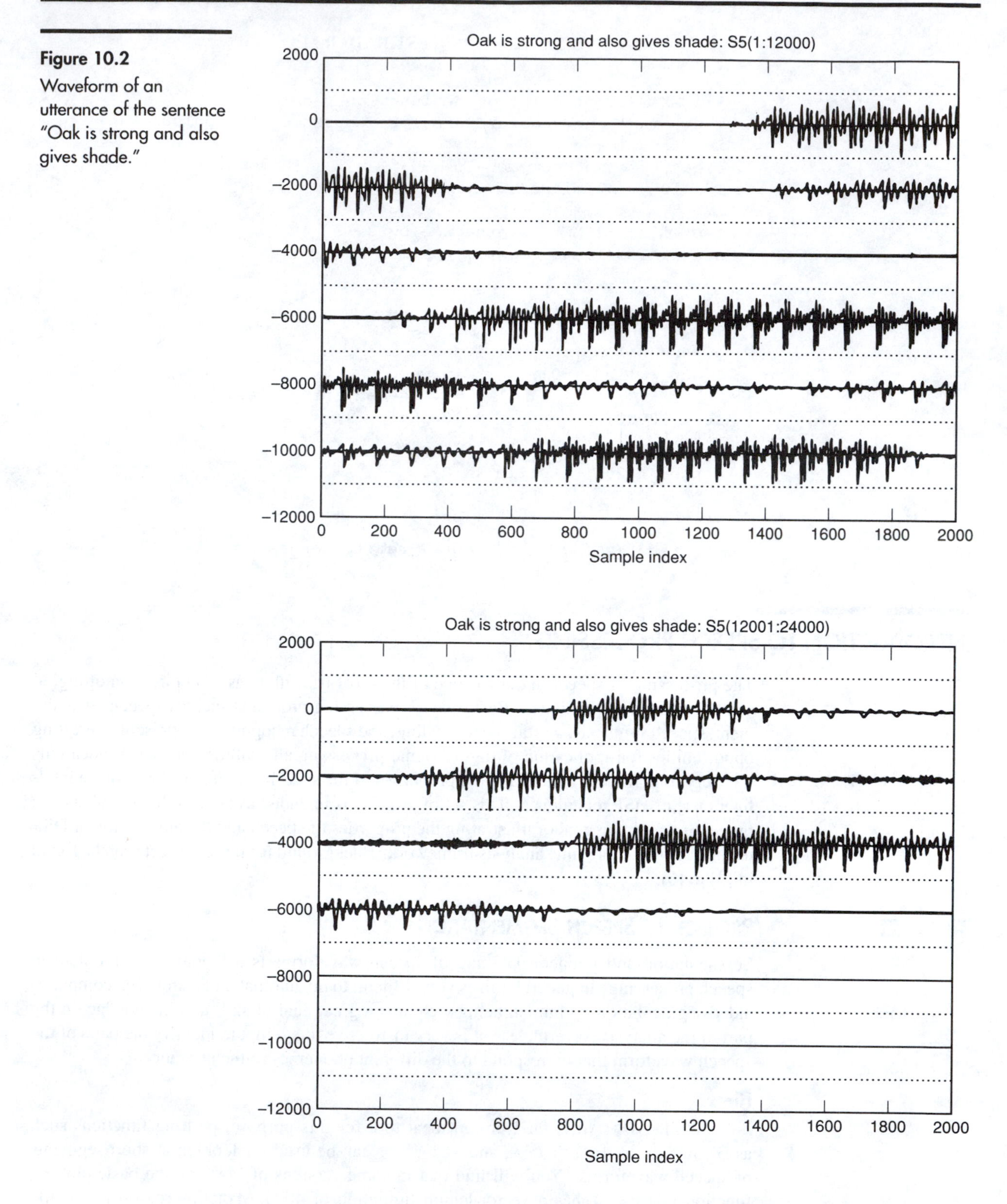

Figure 10.2 Waveform of an utterance of the sentence "Oak is strong and also gives shade."

Speech waveforms are represented in MATLAB by vectors of samples usually taken at a sampling rate of at least 8000 samples/s. This sometimes presents a problem due to the inherent variable-length limitation of some versions of MATLAB. With MATLAB 4.0 and 5.0, there is no problem in representing speech signals of several seconds' duration. For these cases several speech waveforms are provided in the distribution package. These are named `S1.MAT-S6.MAT`. Each of these files represents a complete utterance of length 24,576 samples at a sampling rate of 8000 samples/s. These files can be loaded into MATLAB with the command `load s5`, for example. For other versions of MATLAB, these sentences have been segmented into separate files of length 1000 samples using the M-file `chopfile` from Appendix A. Thus the waveform contained in file `S5.MAT` is also contained in the files `S5_1.MAT–s5_25.MAT`. Using these files, smaller sections of the waveform can be held in MATLAB's variable space for processing. Although this is obviously a bit awkward, it is necessary when only a limited version of MATLAB is available.

EXERCISE 1.1

Phonetic Representation of Text

First write out a phonetic representation of the sentence *Oak is strong and also gives shade* using the "ARPABET" system of phonetic symbols defined in Table 10.3.

TABLE 10.3
ARPABET Phonetic Symbol System

ARPABET	*Example*	*ARPABET*	*Example*	*ARPABET*	*Example*
IY	beat	AY	buy	F	fat
IH	bit	OY	boy	TH	thing
EY	bait	Y	you	S	sat
EH	bet	W	wit	SH	shut
AE	bat	R	rent	V	vat
AA	Bob	L	let	DH	that
AH	but	M	met	Z	zoo
AO	bought	N	net	ZH	azure
OW	boat	NX	sing	CH	church
UH	book	P	pet	JH	judge
UW	boot	T	ten	WH	which
AX	about	K	kit	EL	battle
IX	roses	B	bet	EM	bottom
ER	bird	D	debt	EN	button
AXR	butter	H	get	DX	batter
AW	down	HH	hat	Q	(glottal stop)

EXERCISE 1.2

Phonetic Labeling Using Waveform Plots

Use the plots of Fig. 10.2 together with the plotting features of MATLAB to examine the waveform in the file `S5.SP`, and make your best decisions on where each phoneme of the utterance begins and ends. Be alert for phonemes that are missing or barely realized in the waveform. There may be a period of "silence" or "noise" at the beginning and end of the file. Be sure to mark the beginning and end of these intervals, too, and label the interval with the corresponding ARPABET symbol. Make a table showing the phonemes and the starting and ending samples for each.

EXERCISE 1.3

Listening (*Optional*)

If you have D/A capability available on your machine, it is instructive to listen to the signal with some of the phonemes removed. Use the basic capabilities of MATLAB to construct two vectors corresponding to the following

Utterance 1: `Oa_ i_ ___o__ a__ a__o _i___ __a__`

Utterance 2: `__k _s str_ng _nd _ls_ g_ves sh_de`

That is, utterance 1 has only the sounds corresponding to the vowel letters in the sentence "Oak is strong and also gives shade," and utterance 2 has only the sounds corresponding to the consonant letters. (*Hint*: *If you have utterance 1 and the original utterance, how can you easily compute utterance 2*?) Write files in a format appropriate for available D to A facilities, and listen to them.

From the text representations of utterances 1 and 2, it appears that it would be easy to decode the sentence from the consonants but much more difficult to obtain it from only the vowels. After listening to the two utterances, which seems to be "most intelligible?"

PROJECT 2: PREEMPHASIS OF SPEECH

Speech signals have a spectrum that falls off at high frequencies. In some applications it is desirable that this high-frequency falloff be compensated by "preemphasis." A simple and widely used method of preemphasis is linear filtering by a "first difference" filter of the form

$$y[n] = x[n] - \alpha x[n-1] \tag{2-1}$$

where $x[n]$ is the input speech signal and $y[n]$ is the output "preemphasized speech" and α is an adjustable parameter.

EXERCISE 2.1

Preliminary Analysis

Determine analytical expressions for the impulse response, system function (z-transform of the impulse response), and the frequency response of the linear time-invariant system represented by (2-1). Use `freqz( )` to plot the frequency response of the preemphasis system for $\alpha =$ 0.5, 0.9, and 0.98. Plot all three functions together, and label the frequency axis appropriately for an 8-kHz sampling rate. How should α be chosen so that the high frequencies will be "boosted"?

EXERCISE 2.2

MATLAB Implementation

Use the MATLAB functions `filter( )` and `conv( )` to implement the preemphasis filter for $\alpha = 0.98$. What is the difference in the outputs for the two methods?

If you have not been able to read all of the waveform into MATLAB in one piece, what would you have to do at the edges of the subpieces to implement the preemphasis filter across the entire waveform?

EXERCISE 2.3

Plotting the Preemphasized Signal

Use `striplot( )` to plot the waveform of the preemphasized speech signal for $\alpha = 0.98$. If you are able to use long vectors, make plots comparable to Fig. 10.2. If your vector length is limited, plot as much as you can for comparison to Fig. 10.2. How does the preemphasized speech waveform differ from the original? What characteristics are unchanged by preemphasis?

Optional: Write an M-file to plot one segment of the input signal followed by the corresponding segment of the output signal, followed by the next segment of the input, and so on. You should be able to do this by creating a new vector from the input and output, and then calling the `striplot( )` M-file.

EXERCISE 2.4

Listening *(Optional)*

If you have D to A capability, create a file in appropriate format containing the original speech followed by a half-second of silence followed by the preemphasized speech. Listen to this waveform and describe the qualitative difference between the two versions of the same utterance.

PROJECT 3: SHORT-TIME FOURIER ANALYSIS

The short-time Fourier transform (STFT) is defined as

$$X_n(e^{j\lambda}) = \sum_{m=-\infty}^{\infty} w[n-m]x[m]e^{-j\lambda m} \tag{3-1}$$

$$= e^{-j\lambda n} \sum_{m=-\infty}^{\infty} w[-m]x[n+m]e^{-j\lambda m} = e^{-j\lambda n}\tilde{X}_n(e^{j\lambda}) \tag{3-2}$$

where $-\infty < n < \infty$ and $0 \le \lambda < 2\pi$ (or any other interval of length 2π). The concept of the time-varying spectrum underlies many of the most useful discrete-time processing algorithms for speech signals.

We can evaluate the STFT at a discrete set of frequencies $\lambda_k = 2\pi k/N$ and at a fixed time n through the use of the DFT (and FFT). If we assume that the window is such that $w[-m] = 0$ for $m < 0$ and $m > L-1$, a simple manipulation of (3-2) gives

$$X_n[k] = X_n(e^{j(2\pi/N)k}) = \sum_{m=n}^{n+L-1} w[n-m]x[m]e^{-j(2\pi/N)km} \tag{3-3}$$

$$= e^{-j(2\pi/N)kn} \sum_{m=0}^{L-1} w[-m]x[n+m]e^{-j(2\pi/N)km} = e^{-j(2\pi/N)kn}\tilde{X}_n[k] \tag{3-4}$$

where if $\tilde{w}[m] = w[-m]$,

$$\tilde{X}_n[k] = \sum_{m=0}^{L-1} \tilde{w}[m]x[n+m]e^{-j(2\pi/N)km} \qquad k = 0, 1, \ldots, N-1 \tag{3-5}$$

Note that $X_n[k]$ and $\tilde{X}_n[k]$ differ only by the exponential phase factor $e^{-j(2\pi/N)kn}$ and therefore $|X_n[k]| = |\tilde{X}_n[k]|$. Equation (3-5) simply states that $\tilde{X}_n[k]$ can be computed by the following steps:

a. Select L samples of the signal at time n; $\{x[n], x[n+1], \ldots, x[n+L-1]\}$. (For symmetric windows, it may be convenient to assume that n is at the center of the window interval.)

b. Multiply the samples of the speech segment by the window samples forming the sequence $\{\tilde{w}[m]x[n+m]\}$, $m = 0, 1, \ldots, L-1$.

c. Compute the N-point DFT of the "windowed speech segment" (padding with zeros if $N > L$).

d. Multiply by $e^{-j(2\pi/N)kn}$ (this can be omitted if only the magnitude of the STFT is to be computed).

e. Steps (a)–(d) are repeated for each value of n.

EXERCISE 3.1

Effect of Window Length

The length of the window is a key parameter of the the STFT. If the window is short compared to features in the time waveform, the STFT will track changes in these features. If the window is relatively long, changes with time will be blurred, but the STFT will have good resolution in the frequency (k) dimension. The following is an M-file from Appendix A for demonstrating the effect of window length on the DFT of a segment of speech.

```
function      speccomp(x,ncenter,win,nfft,pltinc)
%       speccomp(x,ncenter,win,nfft,pltinc)
%               x=input signal
%               ncenter=sample number that windows are centered on
%               win=vector of windows lengths to use;
%                   should use odd lengths e.g., [401,201,101]
%               nfft=fft size
%               pltinc=offset of plots (in dB)
%
%       Plots spectra with different window lengths all centered
%       at the same place.
%
if( (ncenter - fix(max(win)/2) < 1) | (ncenter + fix(max(win)/2) > length(x)) )
        disp('Window too long for position in input segment')
        return
end
nwins=length(win);
X=zeros(nfft,nwins);
con=1;
coninc=10^(pltinc/20);
for k=1:nwins
        n1=ncenter-fix(win(k)/2);
        n2=ncenter+fix(win(k)/2);
        X(:,k)=con*fft(x(n1:n2).*hamming(win(k)),nfft);
        con=con/coninc;
end
f=(0:nfft/2)*(8000/nfft);
X=sqrt(-1)*20*log10(abs(X(1:nfft/2+1,:)))+(ones(nwins,1)*f).';
plot(X)
xlabel('frequency in Hz'),ylabel('log magnitude in dB')
title( 'Short-Time Spectra with Different Window Lengths')
```

This M-file computes the DFT of windowed segments of the input signal. All of the windowed segments are centered on the same sample of the signal. All window lengths should be odd to maintain symmetry around this point of the waveform. Figure 10.3 shows an example output from this program.

Study the MATLAB M-file above to determine how it works and what it does. Note the use of the complex data feature of `plot( )` to make it convenient to plot multiple spectra on the same graph.

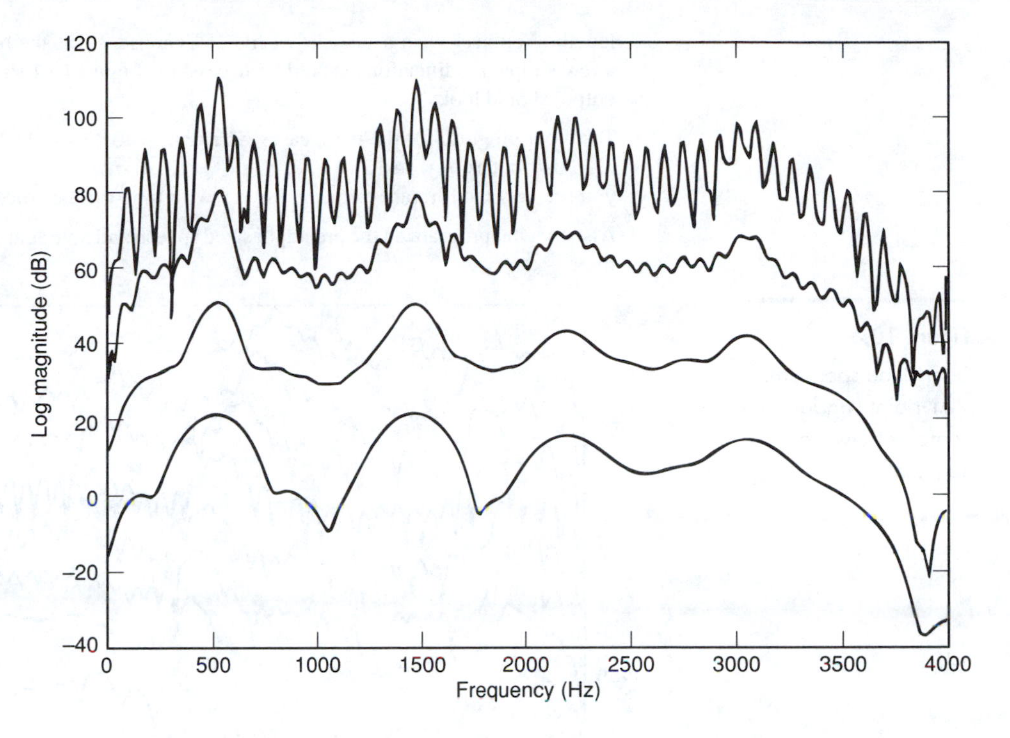

Figure 10.3
Comparison of spectral slices for windows of length 401, 201, 101, and 51 samples.

Run this program on the speech signal `S5.MAT` selecting as the center point the three cases `ncenter = 3750, 16100, 17200`. Use your results in Project 1 to determine the phonemes that occur at these three times. Use several different window lengths: for example, `[401,201,101,51]` and `nfft = 512`. What is the effect of shortening the window? Repeat for the preemphasized speech of Project 2 and compare to the results for the original speech waveform. Are the results as predicted in Exercise 2.1? Use the plots to estimate the formant frequencies of the two voiced segments of speech. Try other segments of the waveform if you have time.

EXERCISE 3.2

Effect of Window Position

Now write an M-file to compute and plot the STFT as a function of k for several equally spaced values of n. Your M-file should have the following calling sequence:

```
function        stspect(x,nstart,ninc,nwin,nfft,nsect,pltinc)
%       stspect(x,nstart,ninc,nwin,nfft,nsect,pltinc)
%               x=input signal
%               nstart=sample number that first window is centered on
%               ninc=offset between windowed segments
%               nwin=window length (should be odd)
%               nfft=fft size
%               nsect=number of sections to plot
%               pltinc=offset of spectra in plot (in dB)
%
%       Plots sequence of spectra spaced by ninc and starting with
%       window centered at nstart.
```

Your program should create a plot like that of Exercise 3.1 with frequency on the horizontal axis, but this time each spectrum corresponds to a different time rather than a different window length. You may wish to use the M-file of Exercise 3.1 as the basis for your program. Only a few simple modifications should be necessary. Figure 10.4 shows an example of how your output should look.

Test your program for the three cases `nstart = 3750, 16100, 17200` as in Exercise 3.1. Use values of `nsect = 10`, `ninc = 200`, `nwin = 401`, and `nfft = 512`. Can you see how the formant frequencies vary with time for the voiced segments?

Also try your program on the preemphasized speech and note again the effect of the preemphasis filter.

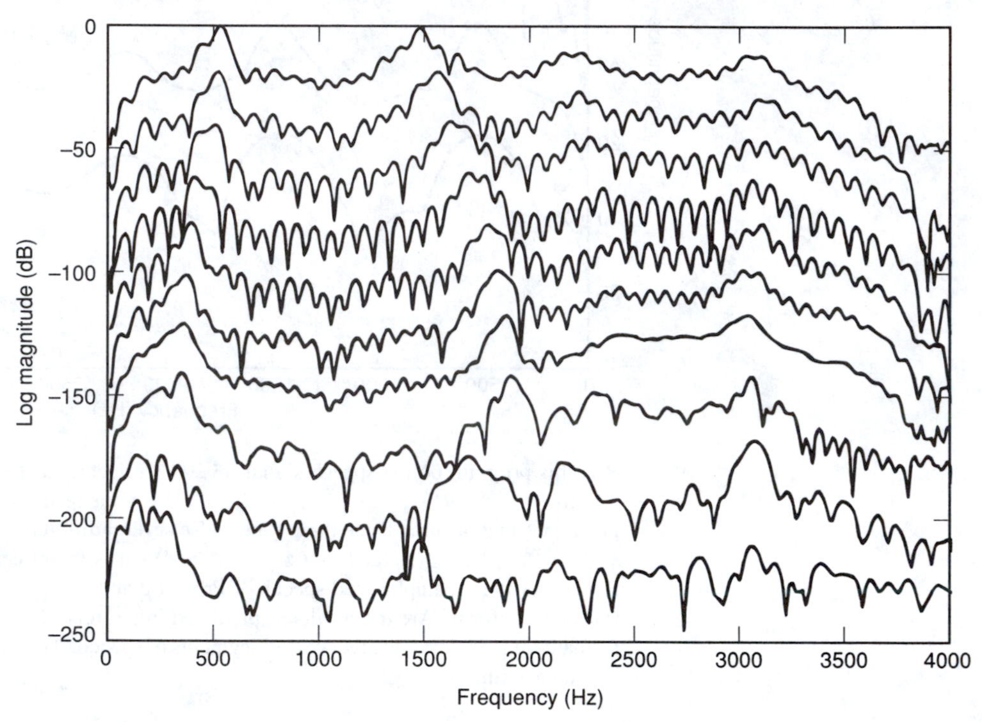

Figure 10.4
Short-time spectrum: 201-point window, 200 samples between segments.

SPEECH MODELING

The basis for most digital speech processing algorithms is a discrete-time system model for the production of samples of the speech waveform. Many useful models have been used as the basis for speech synthesis, speech coding, and speech recognition algorithms. The purpose of this set of projects is examine some of the details of the model depicted in Fig. 10.5.

PROJECT 1: GLOTTAL PULSE MODELS

The model of Fig. 10.5 is the basis for thinking about the speech waveform, and in some cases such a system is used explicitly as an speech synthesizer. In speech production, the excitation for voiced speech is a result of the quasi-periodic opening and closing of the opening between the vocal cords (the glottis). This is modeled in Fig. 10.5 by a combination of the impulse train generator and the glottal pulse model filter. The shape of the pulse affects the magnitude and phase of the spectrum of the synthetic speech output of the model. In this project we study the part labeled "Glottal Pulse Model $G(z)$" in Fig. 10.5.

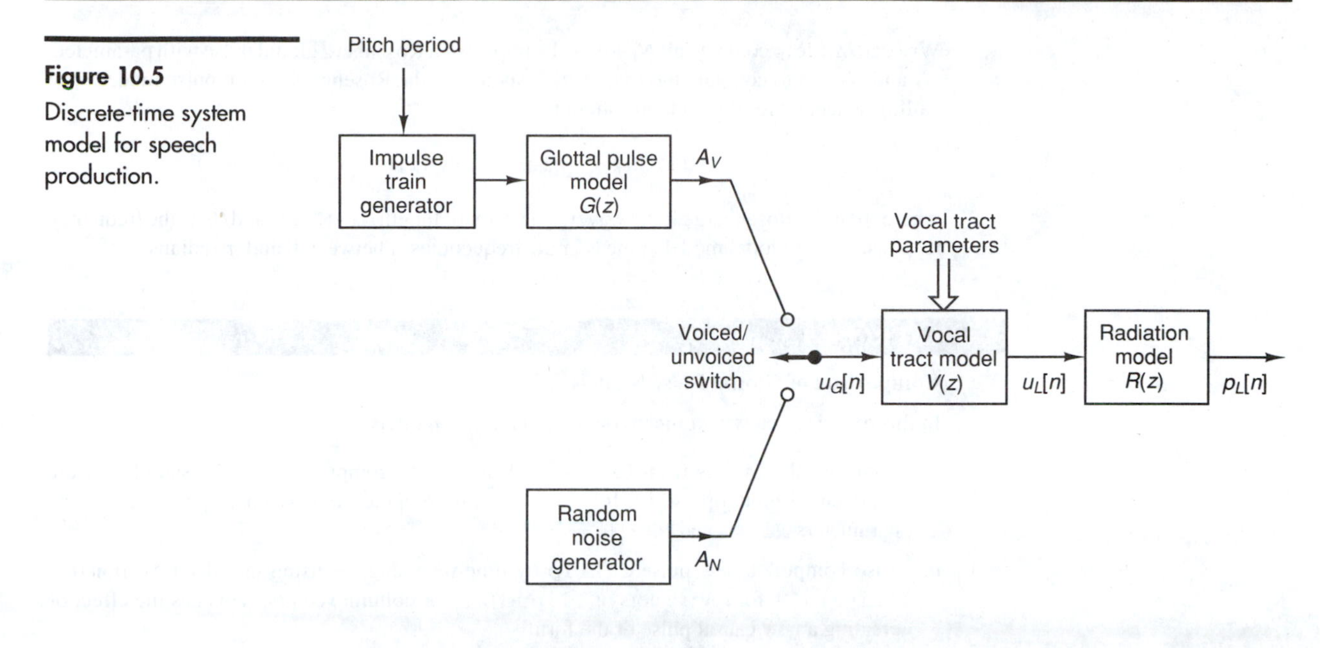

Figure 10.5 Discrete-time system model for speech production.

EXERCISE 1.1

Exponential Model

A simple model that we will call the *exponential model* is represented by

$$G(z) = \frac{-ae\ln(a)\,z^{-1}}{(1-az^{-1})^2} \tag{1-1}$$

where $e = 2.71828\ldots$ is the natural log base. Determine an analytical expression for $g[n]$, the inverse z-transform of $G(z)$. [The numerator of (1-1) is chosen so that $g[n]$ has maximum value of approximately 1.) Write an M-file to generate `Npts` samples of the corresponding glottal pulse waveform $g[n]$ and compute the frequency response of the glottal pulse model. The calling sequence for this function should be

```
[gE,GE,W]=glottalE(a,Npts,Nfreq)
```

where `gE` is the exponential glottal waveform vector of length `Npts`, and `GE` is the frequency response of the exponential glottal model at the `Nfreq` frequencies `W` between 0 and π radians. You will use this function later.

EXERCISE 1.2

Rosenberg Model

Rosenberg [9] used inverse filtering to extract the glottal waveform from speech. Based on his experimental results, he devised a model for use in speech synthesis, which is given by the equation

$$g_R[n] = \begin{cases} \frac{1}{2}[1-\cos(\pi n/N_1)] & 0 \le n \le N_1 \\ \cos[\pi(n-N_1)/(2N_2)] & N_1 \le n \le N_1+N_2 \\ 0 & \text{otherwise} \end{cases} \tag{1-2}$$

This model incorporates most of the important features of the time waveform of glottal waves estimated by inverse filtering and by high-speed motion pictures [3, 9].

Write an M-file to compute all N_1+N_2+1 samples of a Rosenberg glottal pulse with parameters N_1 and N_2 and to compute the frequency response of the Rosenberg glottal pulse model. The calling sequence for this function should be

```
[gR,GR,W]=glottalR(N1,N2,Nfreq)
```

where `gR` is the Rosenberg glottal waveform vector of length `N1+N2+1`, and `GR` is the frequency response of the glottal model at the `Nfreq` frequencies `W` between 0 and π radians.

EXERCISE 1.3

Comparison of Glottal Pulse Models

In this exercise you will compare three glottal pulse models.

a. First, use the M-files from Exercises 1.1 and 1.2 to compute `Npts=51` samples of the exponential glottal pulse `gE` for `a=0.91` and compute the Rosenberg pulse `gR` for the parameters `N1=40` and `N2=10`.

b. Also compute a new pulse `gRflip` by time-reversing `gR` using the MATLAB function `fliplr( )` for row vectors or `flipud( )` for column vectors. This has the effect of creating a new causal pulse of the form

$$g_{\mathrm{Rflip}}[n] = g_R[-(n - N_1 - N_2)] \tag{1-3}$$

Determine the analytical relationship between $G_{\mathrm{Rflip}}(e^{j\omega})$, the Fourier transform of $g_{\mathrm{Rflip}}[n]$, and $G_R(e^{j\omega})$, the Fourier transform of $g_R[n]$.

c. Now plot all three of these 51-point vectors on the same graph using `plot( )`. Also plot the frequency response magnitude in dB for all three pulses on the same graph. Experiment with the parameters of the models to see how the time-domain wave shapes affect the frequency response.

d. Write an M-file to plot Rosenberg pulses for the three cases $N_2 = 10, 15, 25$ with $N_1 + N_2 = 50$ all on the same graph. Similarly, plot the Fourier transforms of these pulses together on another graph. What effect does the parameter N_2 have on the Fourier transform?

e. The exponential model has a zero at $z = 0$ and a double pole at $z = a$. For the parameters `N1=40` and `N2=10`, use the MATLAB function `roots( )` to find the zeros of the z-transform of the Rosenberg model and also the zeros of the flipped Rosenberg model. Plot them using the M-file `zplane( )`. Note that the Rosenberg model has all its zeros outside the unit circle (except one at $z = 0$). Such a system is called a *maximum-phase* system. The flipped Rosenberg model, however, should be found to have all its zeros inside the unit circle, and thus it is a *minimum-phase* system. Show that, in general, if a signal is maximum-phase, then flipping it as in (1-3) produces a minimum-phase signal, and vice versa.

PROJECT 2: LOSSLESS TUBE VOCAL TRACT MODELS

One approach to modeling sound transmission in the vocal tract is through the use of concatenated lossless acoustic tubes as depicted in Fig. 10.6.

Using the acoustic theory of speech production [3, 4, 10], it can be shown that the lossless assumption and the regular structure lead to simple wave equations and simple boundary conditions at the tube junctions, so that a solution for the transmission properties of the model is relatively straightforward and can be interpreted as in Fig. 10.7a, where $\tau = \Delta x/c$ is the one-way propagation delay of the sections. For sampled signals with

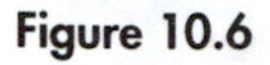

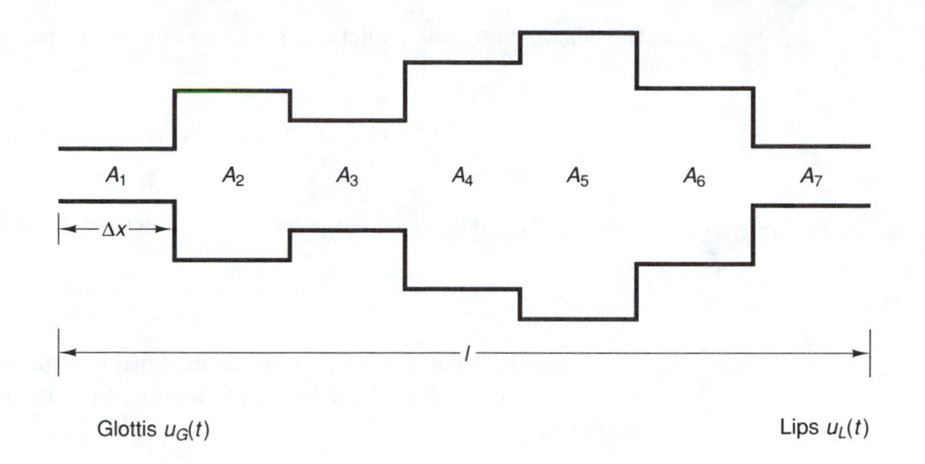

Figure 10.6
Concatenation of ($N = 7$) lossless acoustic tubes of equal length as a model of sound transmission in the vocal tract.

sampling period $T = 2\tau$, the structure of Fig. 10.7a (or equivalently Fig. 10.6) implies a corresponding discrete-time lattice filter [4] as shown in Fig. 10.7b or c.

Lossless tube models are useful for gaining insight into the acoustic theory of speech production, and they are also useful for implementing speech synthesis systems. It is shown in [4] that if $r_G = 1$, the discrete-time vocal tract model consisting of a concatenation of N lossless tubes of equal length has system function

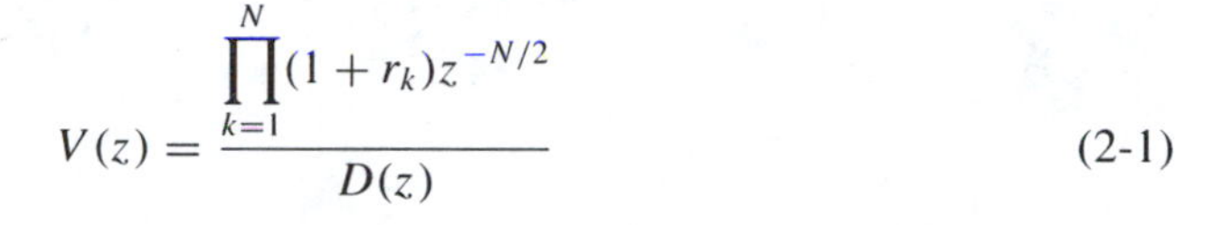

$$V(z) = \frac{\prod_{k=1}^{N}(1 + r_k)z^{-N/2}}{D(z)} \tag{2-1}$$

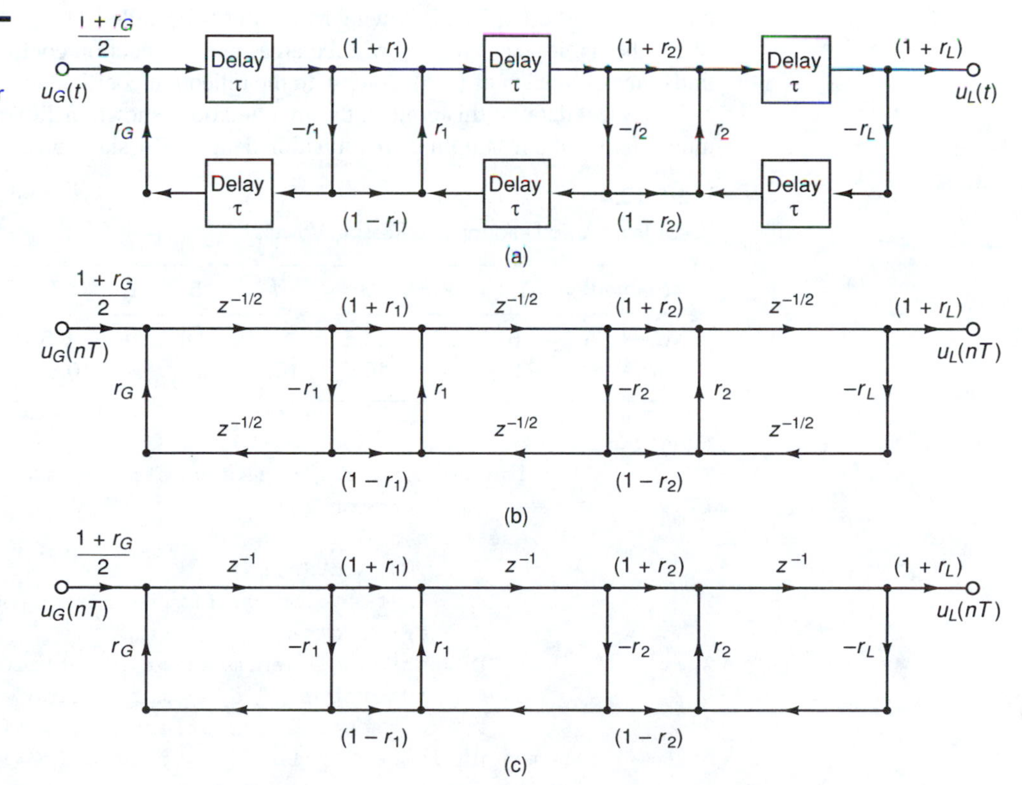

Figure 10.7
(a) Signal flow graph for lossless tube model ($N = 3$) of the vocal tract; (b) equivalent discrete-time system; (c) equivalent discrete-time system using only whole-sample delays in ladder part.

The denominator polynomial $D(z)$ in (2-1) satisfies the polynomial recursion [4]

$$\begin{aligned} D_0(z) &= 1 \\ D_k(z) &= D_{k-1}(z) + r_k z^{-k} D_{k-1}(z^{-1}) \qquad k = 1, 2, \ldots, N \\ D(z) &= D_N(z) \end{aligned} \tag{2-2}$$

where the r_k's in (2-2) are the reflection coefficients at the tube junctions,

$$r_k = \frac{A_{k+1} - A_k}{A_{k+1} + A_k} \tag{2-3}$$

In deriving the recursion in (2-2), it was assumed that there were no losses at the glottal end ($r_G = 1$) and that all the losses are introduced at the lip end through the reflection coefficient

$$r_N = r_L = \frac{A_{N+1} - A_N}{A_{N+1} + A_N} \tag{2-4}$$

where A_{N+1} is the area of an impedance-matched (no reflections at its end) tube that can be chosen to introduce a loss in the system [4].

Suppose that we have a set of areas for a lossless tube model, and we wish to obtain the system function for the system so that we can use the MATLAB `filter( )` function to implement the model; that is, we want to obtain the system function of (2-1) in the form

$$V(z) = \frac{G}{D(z)} = \frac{G}{1 - \sum_{k=1}^{N} \alpha_k z^{-k}} \tag{2-5}$$

[Note that in (2-5) we have dropped the delay of $N/2$ samples, which is inconsequential for use in synthesis.] The following MATLAB M-file called `AtoV.m` implements (2-2) and (2-3); that is, it takes an array of tube areas and a reflection coefficient at the lip end and finds the parameters of (2-5) along with the reflection coefficients.

As test data for this project, the area functions shown in Table 10.4 were obtained by interpolating and resampling area function data for Russian vowels as given by Fant [10].

TABLE 10.4
Vocal Tract Area Data for Two Russian Vowels.

Section	1	2	3	4	5	6	7	8	9	10
vowel AA	1.6	2.6	0.65	1.6	2.6	4	6.5	8	7	5
vowel IY	2.6	8	10.5	10.5	8	4	0.65	0.65	1.3	3.2

```
function          [r,D,G]=AtoV(A,rN)
%         function to find reflection coefficients
%         and system function for
%         lossless tube models.
%                   [r,D,G]=AtoV(A,rN)
%                   rN = reflection coefficient at lips (abs value < 1)
%                   A = array of areas
%                   D = array of denominator coefficients
%                   G = numerator of system function
%                   r = corresponding reflection coefficients
%         assumes no losses at the glottis end (rG=1).
[M,N] = size(A);
if(M~=1)  A = A';  end      %-- make row vector
```

```
N = length(A);
r = [];
for m=1:N-1
      r = [r (A(m+1)-A(m))/(A(m+1)+A(m))];
end
r = [r rN];
D = [1];
G = 1;
for m=1:N
      G = G*(1+r(m));
      D = [D 0] + r(m).*[0 fliplr(D)];
end
```

EXERCISE 2.1

Frequency Response and Pole–Zero Plot

a. Use the M-file `AtoV( )` to obtain the denominator $D(z)$ of the vocal tract system function, and make plots of the frequency response for each area function for `rN=0.71` and also for the totally lossless case $rN = 1$. Plot the two frequency responses for a given vowel on the same plot.

b. Factor the polynomials $D(z)$ and plot the poles in the z-plane using `zplane( )`. Plot the roots of the lossy case as `o`'s and the roots of the lossless case as `x`'s. (See `help zplane` from Appendix A.) Where do the roots lie for the lossless case? How do the roots of $D(z)$ shift as `rN` decreases away from unity? Convert the angles of the roots to analog frequencies corresponding to a sampling rate of $1/T = 10,000$ samples/s, and compare to the formant frequencies expected for these vowels [3, 4, 10]. For this sampling rate, what is the effective length of the vocal tract, in centimeters?

EXERCISE 2.2

Finding the Model from the System Function

The inverse problem arises when we want to obtain the areas and reflection coefficients for a lossless tube model given the system function in the form of (2-5). We know that the denominator of the system function, $D(z)$, satisfies (2-2). In this part we use (2-2) to develop an algorithm for finding the reflection coefficients and the areas of a lossless tube model having a given system function denominator.

a. Show that r_N is equal to the coefficient of z^{-N} in the denominator of $V(z)$ (i.e., $r_N = -\alpha_N$).

b. Use (2-2) to show that

$$D_{k-1}(z) = \frac{D_k(z) - r_k z^{-k} D_k(z^{-1})}{1 - r_k^2} \qquad k = N, N-1, \ldots, 2$$

c. How would you use the results of parts (a) and (b) to find r_{N-1} from $D_N(z) = D(z)$?

d. Using the results of parts (a), (b), and (c), state an algorithm for finding all of the reflection coefficients $r_k,\ k = 1, 2, \ldots, N$ and all of the tube areas $A_k,\ k = 1, 2, \ldots, N$. Are the A_k's unique? Write a MATLAB function to implement your algorithm for converting from $D(z)$ to reflection coefficients and areas. This M-file should adhere to the following definition:

```
function        [r,A]=VtoA(D,A1)
%       function to find reflection coefficients
%       and tube areas for lossless tube models.
%        [r,A]=VtoA(D,A1)
%               A1 = arbitrary area of first section
```

```
%                D = array of denominator coefficients
%                A = array of areas for lossless tube model
%                r = corresponding reflection coefficients
%        assumes no losses at the glottis end (rG=1).
```

[This new M-file can be similar in structure to `AtoV( )`.] For the vowel /a/, the denominator of the 10th-order model should be (to four-digit accuracy)

$$\begin{aligned} D(z) &= 1 - 0.0460z^{-1} - 0.6232z^{-2} + 0.3814z^{-3} + 0.2443z^{-4} + 0.1973z^{-5} \\ &\quad + 0.2873z^{-6} + 0.3655z^{-7} - 0.4806z^{-8} - 0.1153z^{-9} + 0.7100z^{-10} \end{aligned}$$

Use your MATLAB program to find the corresponding reflection coefficients and tube areas and compare to the data for the vowel /a/ in Table 10.4. If your program is working, there may still be small differences between its output and the data of Table 10.4. Why?

PROJECT 3: VOWEL SYNTHESIS

For voiced speech, the speech model of Fig. 10.5 can be simplified to the system of Fig. 10.8. The excitation signal $e[n]$ is a quasi-periodic impulse train and the glottal pulse model could be either the exponential or the Rosenberg pulse. The vocal tract model could be a lattice filter of the form of Fig. 10.7c, or it could be an equivalent direct-form difference equation as implemented by MATLAB.

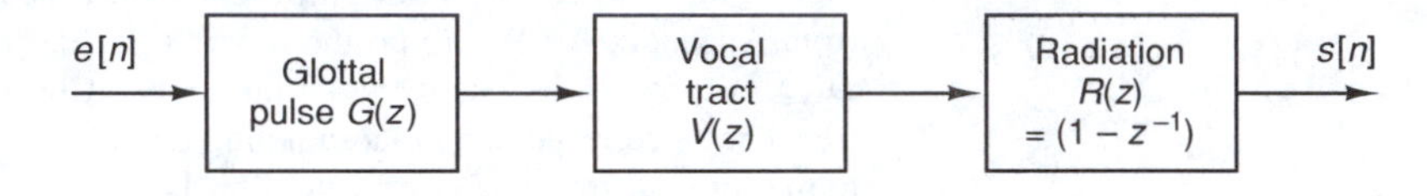

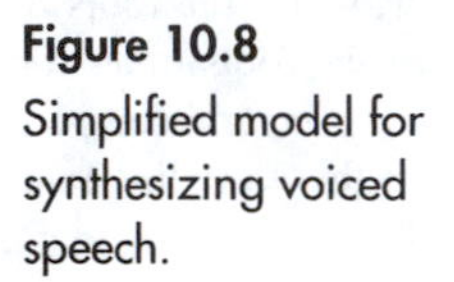

Figure 10.8 Simplified model for synthesizing voiced speech.

Hints

In this project we use the M-files written in Projects 1 and 2, together with the `filter( )` and `conv( )` functions to implement parts of the system of Fig. 10.8 and thereby synthesize periodic vowel sounds. A periodic pulse train can be synthesized by using the M-file `zerofill( )` from Appendix A, together with the MATLAB function `ones( )`.

EXERCISE 3.1

Periodic Vowel Synthesis

Assume a sampling rate of 10000 samples/s. Create a periodic impulse train vector `e` of length 1000 samples, with period corresponding to a fundamental frequency of 100 Hz. Then use combinations of `filter( )` and `conv( )` to implement the system of Fig. 10.8.

Use the excitation `e` and radiation system $R(z) = (1 - z^{-1})$ to synthesize speech for both area functions given above and for all three glottal pulses studied in Project 2. Use `subplot( )` and `plot( )` to make a plot comparing 1000 samples of the synthetic speech outputs for the exponential glottal pulse and the Rosenberg minimum-phase pulse. Make another plot comparing the outputs for the two Rosenberg pulses.

EXERCISE 3.2

Frequency Response of Vowel Synthesizer

Plot the frequency response (log magnitude in dB) of the overall system with system function $H(z) = G(z)V(z)R(z)$ for the case of the Rosenberg glottal pulse, $R(z) = (1 - z^{-1})$, and vocal tract response for the vowel /a/. Save your result for use in Exercise 3.3.

EXERCISE 3.3

Short-Time Fourier Transform of Synthetic Vowel

Compute the DFT of a Hamming-windowed segment (401 points) of the synthetic vowel and plot the log magnitude on the same graph as the frequency response of the synthesizer.

EXERCISE 3.4

Noise Excitation (Whispered Speech)

In producing whispered speech, the vocal tract is excited by turbulent airflow produced at the glottis. This can be modeled by exciting only the cascaded vocal tract and radiation filters with random noise. Using the function `randn( )`, excite the cascaded vocal tract/radiation filters for the vowel AA with a zero-mean Gaussian noise input. Plot the waveform and repeat Exercises 3.2 and 3.3 for the "whispered" vowel.

EXERCISE 3.5

Listening to the Output (Optional)

If D to A facilities are available on your computer, create files of synthetic voiced and whispered vowels of length corresponding to 0.5 s duration in the proper binary format, and play them out through the D to A system. For a 16-bit D to A converter you should scale the samples appropriately and use `round( )` to convert them to integers (of magnitude $\leq$ 32767) before writing the file. Does the synthetic speech sound like the desired vowels?

SPEECH QUANTIZATION

OVERVIEW

Sampling and quantization (or A-to-D conversion) of speech waveforms is important in digital speech processing because it is the first step in any digital speech processing system, and because one of the basic problems of speech processing is digital coding of the speech signal for digital transmission and/or storage. Sampling and quantization of signals is generally implemented by a system of the form of Fig. 10.9. In a hardware realization, the sample-and-hold circuit samples the input continuous-time signal and holds the value constant during the sampling period T. This gives a constant signal at the input of the A-to-D converter, whose purpose is to decide which of its quantization levels is closest to the input sample value. Every T seconds, the A-to-D converter emits a digital code corresponding to that level. Normally, the digital code is assigned according to a convenient binary number system such as two's-complement so that the binary numbers can be taken as numerical representations of the sample values.

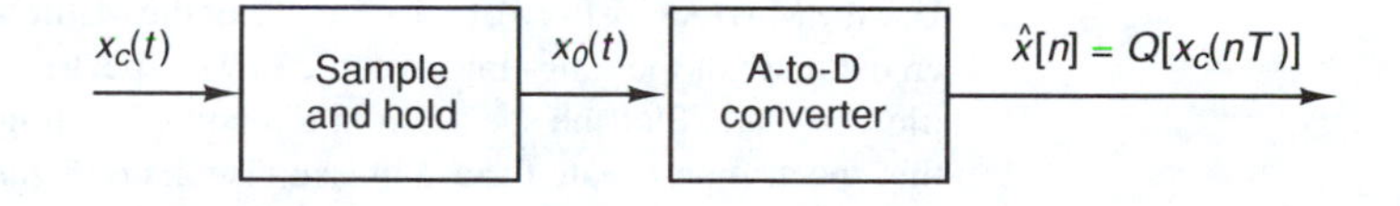

Figure 10.9 Representation of hardware for sampling and quantization of speech signals.

An equivalent representation of sampling and quantization is depicted in Fig. 10.10. This representation is convenient because it separates the sampling and quantization into two independent operations. The operation of the ideal sampler is well understood. The sampling theorem states that a bandlimited signal can be reconstructed precisely from samples taken at the rate of twice the highest frequency in the spectrum of the signal. In these projects it will be assumed that the speech signal has been low-pass filtered and

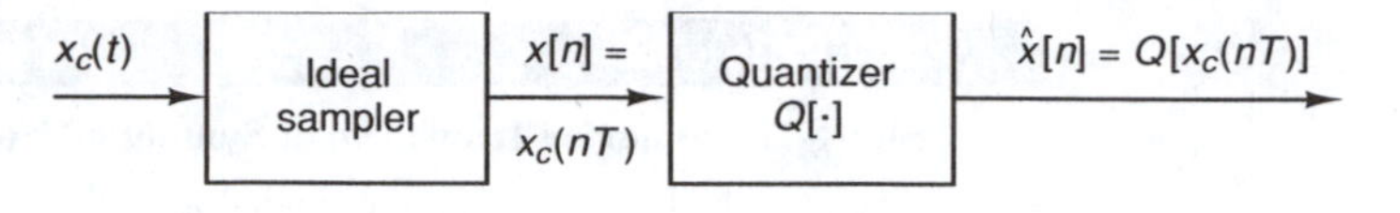

Figure 10.10 Representation of sampling and quantization that facilitates analysis and simulation.

sampled at a high enough sampling rate to avoid significant aliasing distortion. Therefore, it will be possible to focus solely on quantization of speech signal waveforms. Appropriate background reading for the projects can be found in [4] and [7].

PROJECT 1: SPEECH PROPERTIES

In this project you will use MATLAB tools to examine a particular speech waveform and verify some fundamental statistical properties of speech signals that are important for quantization.

Hints

The speech files `S1.MAT` – `S6.MAT` are available in Appendix A. The files were sampled with sampling rate 8000 samples per second and originally quantized to 12 bits. Subsequently, the samples were multiplied by 16 to raise the amplitude levels to just under 32767 (i.e., the maximum value for a 16-bit integer). Thus, these files are 12-bit samples pretending to be 16-bit samples. This will generally not be a problem in this project.

EXERCISE 1.1

Speech Waveform Plotting

First, load the file `S5.MAT` and create a vector of length 8000 samples, starting at sample 1200. *Divide the sample values by 32768 so that all samples have value less than 1.* Plot all 8000 samples with 2000 samples/line using the plotting function `striplot( )`.[2]

EXERCISE 1.2

Statistical Analysis

Compute the minimum, maximum, average, and mean-squared value of the 8000 samples from the file `S5.MAT`. Use the MATLAB function `hist( )` to plot a histogram of the 8000 samples. Experiment with the number and location of the histogram bins to obtain a useful plot. The histogram should show that the small samples are more probable than large samples in the speech waveform. Is this consistent with what you see in the waveform plot? See [4] and [7] for discussions of continuous probability density function models for the distribution of speech amplitudes.

EXERCISE 1.3

Spectral Analysis

Use the MATLAB M-file `spectrum( )` or the M-file `welch( )` from Appendix A to compute an estimate of the long-time average power spectrum of speech using the 8000 samples from file `S5.MAT`. Plot the spectrum in dB units, labeling the frequency axis appropriately. Save this spectrum estimate for use in Exercise 2.3 of Project 2.

PROJECT 2: UNIFORM QUANTIZATION

Figure 10.11 shows the input–output relation for a 3-bit uniform quantizer in which the input samples are *rounded* to the nearest quantization level and the output saturates for samples

[2]If you are using the Student Version 3.5 of MATLAB, you will be limited to variables of about 1000 samples. If such signal lengths are used, you should expect greater statistical variability in your results in some of the exercises.

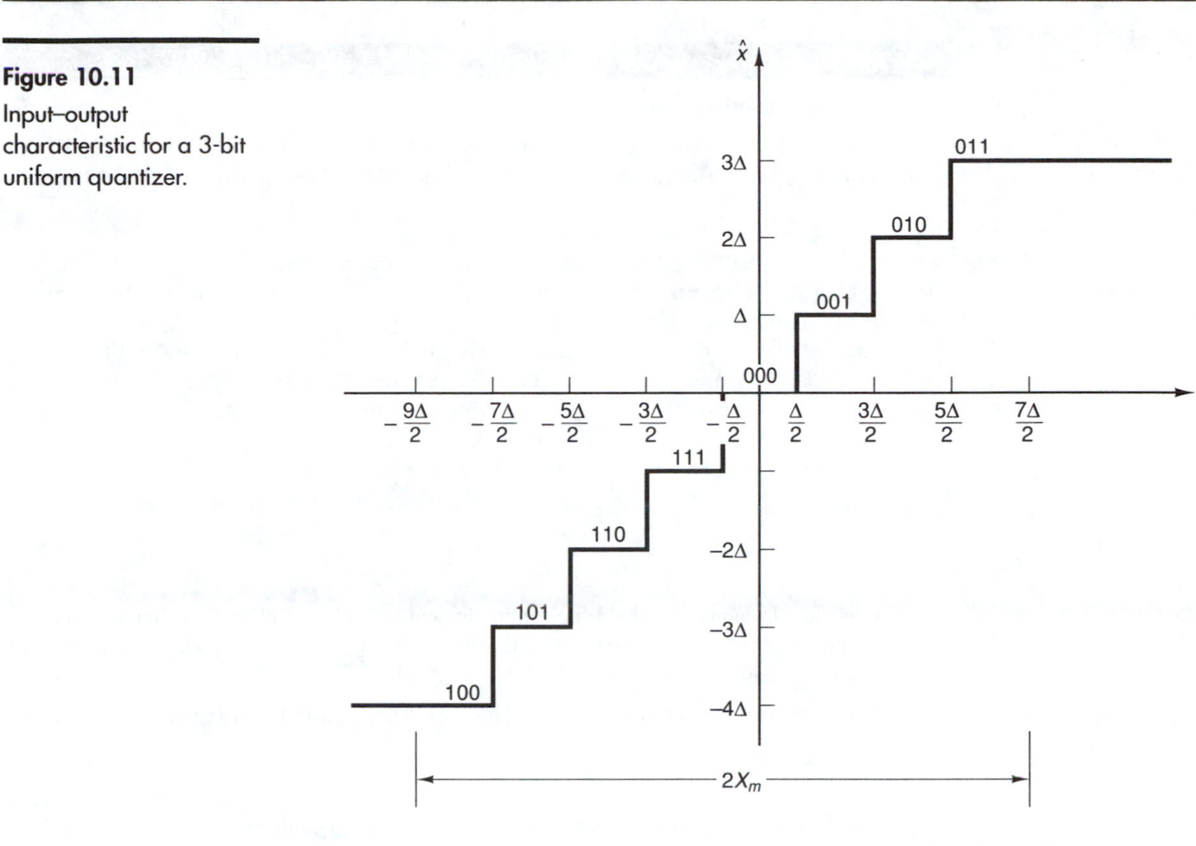

Figure 10.11
Input–output characteristic for a 3-bit uniform quantizer.

outside the range $-X_m - \Delta/2 \leq x < X_m - \Delta/2$. In discussing the effects of quantization it is useful to define the quantization error as

$$e[n] = \hat{x}[n] - x[n] \tag{2-1}$$

This definition leads to the additive noise model for quantization that is depicted in Fig. 10.12. If the signal sample $x[n]$ remains in the nonsaturating range of the quantizer, it is clear that the quantization error samples satisfy

$$-\Delta/2 \leq e[n] < \Delta/2 \tag{2-2}$$

Furthermore, speech is a complicated signal that fluctuates rapidly among the quantization levels, and if Δ is small enough, the amplitude of the signal is likely to traverse many quantization steps in one sample time. Under these conditions, it is found that the quantization error sequence is well described by the following model:

1. The error sequence $e[n]$ is uncorrelated with the unquantized sequence $x[n]$.
2. The error sequence has the properties of white noise: that is, it has a flat power spectrum, and the error samples are uncorrelated with one another.
3. The probability distribution of the error samples is uniform over the range of quantization error amplitudes.

These assumptions are tested in this project.

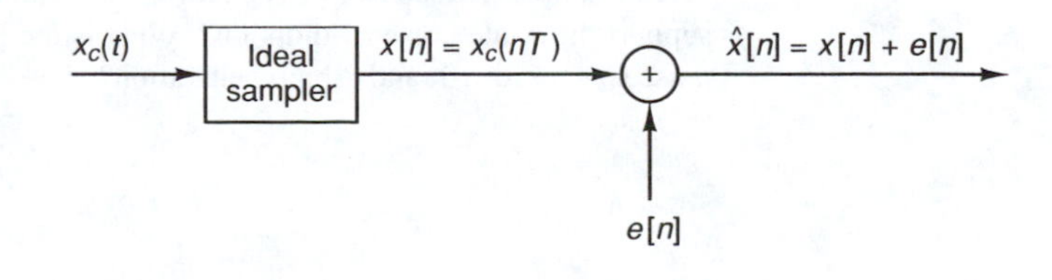

Figure 10.12
Additive noise model for sampling and quantization.

EXERCISE 2.1

Uniform Quantizer M-file

In this project you will use a uniform quantizer M-file `fxquant( )` from Appendix A to perform several quantization experiments. The comments from this M-file are as follows:

```
function X = fxquant( s, bit, rmode, lmode )
%X = fxquant( S, BIT, RMODE, LMODE )  simulated fixed-point arithmetic
%       fxquant  returns the input signal S reduced to a word-length
%       of BIT bits and limited to the range [-1,1). The type of
%       word-length reduction and limitation may be chosen with
%       RMODE:  'round'          rounding to nearest level
%               'trunc'          2's complement truncation
%               'magn'           magnitude truncation
%       LMODE:  'sat'            saturation limiter
%               'overfl'         2's complement overflow
%               'triangle'       triangle limiter
%               'none'           no limiter
```

As is clear from above, this M-file can implement a number of different quantizer functions. An important point is that the range of the quantizer is $[-1, 1)$. This is why the samples from the file `S5.MAT` were divided by 32,768.

To plot the input–output characteristics of this quantizer, type the following MATLAB statements:

```
x=-2:.001:2;
plot(x,fxquant(x,3,'round','sat'))
```

This displays the quantizer function for a 3-bit rounding quantizer with saturation. What is Δ for this quantizer, and over what range of x does the quantization error satisfy (2-2)? Now consider the statement

```
plot(x,fxquant(x,3,'round','sat')-x)
```

What is plotted in this case?

Change the parameters of the quantizer and repeat the plots to help understand the different ways that quantization can be implemented.

EXERCISE 2.2

Quantization Experiments

Use `fxquant( )` to quantize the 8000 input speech samples from the file `S5.MAT`. Using rounding and saturation, compute the quantization error sequences for 10-, 8-, and 4-bit quantization. Use the program `striplot( )` to plot these error sequences. What are the important differences among them? Do they look like they fit the white noise model? Make histograms of the quantization noise samples. Do they seem to fit the uniform amplitude distribution model?

EXERCISE 2.3

Spectral Analysis of Quantization Noise

Use `spectrum()` or `welch( )` to compute the power spectrum of the quantization noise sequences for 10, 8, and 4 bits. Plot these spectra on the same plot as the power spectrum of the speech samples. [*Remember*: The power spectrum in dB is $10\log_{10}(P)$.] Do the noise spectra support the white noise assumption? What is the approximate difference in dB between the noise spectra for 10- and 8-bit quantization? (See beginning of Project 4.)

EXERCISE 2.4

Quantization by Truncation

Set the parameter `RMODE` of `fxquant( )` to `'trunc'` and repeat Exercises 2.2 and 2.3. What is the main difference between the results for rounding and those for truncation?

PROJECT 3: μ-LAW COMPANDING

One of the problems with uniform quantization is that the maximum size of the quantization errors is the same no matter how big or small the samples are. For a coarse quantizer, low-level fricatives and other sounds may disappear completely because their amplitude is below the minimum stepsize. μ-Law compression/expansion is a way to obtain quantization errors that are effectively proportional to the size of the sample.

Hints

A convenient way of describing μ-law quantization is depicted in Fig. 10.13. In this representation, a μ-law compressor precedes a uniform quantizer. The combination (inside the dashed box) is a μ-law quantizer.

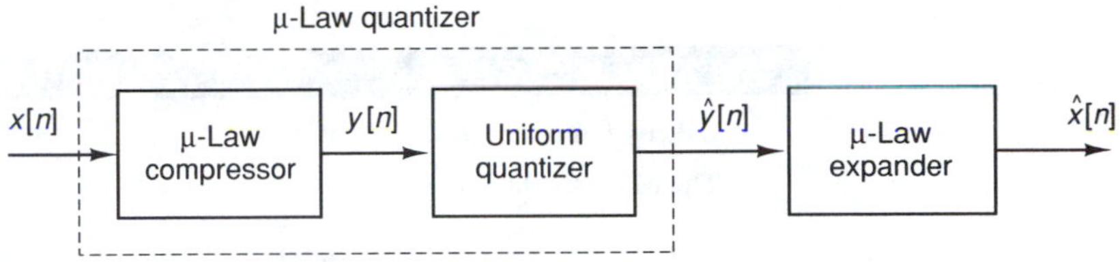

Figure 10.13 Representation of μ-law quantization.

The μ-law compressor is defined by the equation

$$y[n] = X_{\max} \frac{\log\left[1 + \mu \frac{|x[n]|}{X_{\max}}\right]}{\log(1 + \mu)} \cdot \operatorname{sign}(x[n])$$

The μ-law compressor is discussed in detail in [4, 7].

The following M-file implements the μ-law compressor on a signal vector whose maximum value is assumed to be $X_{\max} = 1$:

```
function        y=mulaw(x,mu)
%               function for mu-law compression
%         y=mulaw(x,mu)
%               x=input signal vector, column vector with max value 1
%               mu=compression parameter (mu=255 used for telephony)
sign=ones(length(x),1);
sign(find(x<0))=-sign(find(x<0));
y=(1/log(1+mu))*log(1*mu*abs(x)).*sign;
```

Note the use of the `find( )` function to locate the negative samples.

EXERCISE 3.1

μ-Law Compressor

This exercise is concerned with the μ-law compressor and its inverse.

a. Create a linearly increasing input vector `[0:0.0005:1]` and use it with the function `mulaw( )` to plot the μ-law characteristic for $\mu = 100$, 255, and 500 all on the same plot. $\mu = 255$ is a standard value used in telephony.

b. Using the segment of speech from file `S5.MAT` and a value of $\mu = 255$, plot the output waveform $y[n]$ of the μ-law compressor. Observe how the low-amplitude samples are increased in magnitude. Plot a histogram of the output samples and compare it to the histogram of the original samples.

c. To implement the system of Fig. 10.13, you must write an M-file for the inverse of the μ-law compressor. This M-file should have the following calling sequence and parameters:

```
function        x=mulawinv(y,mu)
%       function for inverse mulaw
%       x=mulawinv(y,mu)
%               y=input column vector Xmax=1
%               mu=mulaw compression parameter
%               x=expanded output vector
```

Use the technique used in `mulaw( )` to set the signs of the samples. Test the inverse system by applying it directly to the output of `mulaw()` without quantization.

EXERCISE 3.2

μ-Law Quantization

The MATLAB statement

```
yh=fxquant(mulaw(x,255),6,'round','sat');
```

implements a 6-bit μ-law quantizer. That is, it is the compressed samples that would be represented by 6 bits. When the samples are used in a signal processing computation or when a continuous-time signal is reconstructed, the samples must be expanded. Hence, the quantization errors will also be expanded, so that to determine the quantization error, it is necessary to compare the output of the inverse system to the original samples. That is, the quantization error would be `e=mulawinv(yh,255)-x;`. With this in mind, repeat all the exercises of Project 2 for the system of Fig. 10.13.

PROJECT 4: SIGNAL-TO-NOISE-RATIOS

A convenient way of comparing quantizers is to compute the ratio of signal power to quantization noise power. For experiments in MATLAB, a convenient definition of SNR is

$$\text{SNR} = 10 \log \left(\frac{\sum_{n=0}^{L-1} (x[n])^2}{\sum_{n=0}^{L-1} (\hat{x}[n] - x[n])^2} \right) \tag{4-1}$$

Note that the division by L required for averaging cancels in the numerator and denominator.

Hints

Under the assumptions of the noise model given in Project 2, it can be shown that the signal-to-noise ratio for a uniform quantizer with 2^{B+1} levels (B bits plus sign) has the form [4, 7]

$$\text{SNR} = 6B + 10.8 - 20 \log_{10} \left(\frac{X_m}{\sigma_x} \right) \tag{4-2}$$

where X_m is the clipping level of the quantizer (in our case $X_m = 1$) and σ_x is the rms value of the input signal amplitude. Thus (4-2) shows that the signal-to-noise ratio increases 6 dB per bit added to the quantizer word length. Furthermore, (4-2) shows that if the signal level is decreased by a factor of 2, the signal-to-noise ratio decreases by 6 dB.

EXERCISE 4.1

Signal-to-Noise Computation

Write an M-File to compute the signal-to-noise ratio as defined in (4-1). Its calling sequence and parameters should be

```
function        [s_n_r,e]=snr(xh,x);
%       function for computing signal-to-noise ratio
%       [s_n_r,e]=snr(xh,x)
%               xh=quantized signal
%               x=unquantized signal
%               e=quantization error signal (optional)
%               s_n_r=snr in dB
```

Use your SNR function to compute the SNRs for uniform quantization with 8 and 9 bits. Do the results differ by the expected amount?

EXERCISE 4.2

Comparison of Uniform and μ-Law Quantization

An important consideration in quantizing speech is that signal levels can vary with speakers and with transmission/recording conditions. This can result in significant variations of signal-to-noise ratio for a fixed quantizer. The following M-file from Appendix A compares uniform and μ-law quantization for a fixed quantizer with inputs of decreasing amplitude (by factors of 2). Using the M-files that were written in Projects 2 and 3 and the M-file `qplot( )`, make a plot for 10 bits with $\mu = 255$ over a range of 10 factors of 2. Explain the shape of the two curves. The program `qplot( )` plots the signal-to-noise ratios of a uniform and a μ-law quantizer for the same number of bits. Modify the program so that quantizers with several different numbers of bits can be compared on the same plot. Use the modified M-file to create a plot for 10, 8, 6, and 4 bits with $\mu = 255$ over a range of 10 factors of 2.

```
function        qplot(s,nbits,mu,ncases)
%       function for plotting dependence of signal-to-noise
%       ratio on decreasing signal level
%       qplot(s,nbits,mu,ncases)
%               s=input test signal
%               nbits=number of bits in quantizer
%               mu=mu-law compression parameter
%               ncases=number of cases to plot
%
P=zeros(ncases,2);
x=s;
for i=1:ncases
sh=fxquant(x,nbits,'round','sat');
P(i,1)=(i-1)+sqrt(-1)*snr(sh,x);
y=mulaw(x,mu);
yh=fxquant(y,nbits,'round','sat');
```

```
xh=mulawinv(yh,mu);P(i,2)=(i-1)+sqrt(-1)*snr(xh,x);
x=x/2;
end
plot(P)
title(['SNR for ',num2str(nbits),'-bit Uniform and ',num2str(mu),...
'-Law Quantizers'])
xlabel('power of 2 divisor');ylabel('SNR in dB')
```

Note how the complex plotting feature of `plot( )` *is used as a convenience in plotting multiple graphs on the same axes.*

Your plots should show that the μ-law quantizer maintains a constant signal-to-noise ratio over an input amplitude range of about 64:1. How many bits are required for a uniform quantizer to maintain at least the same signal-to-noise ratio as that of a 6-bit μ-law quantizer over the same range?

PROJECT 5: LISTENING TO QUANTIZED SPEECH (*optional*)

If your computer has D-to-A capability, it is instructive to listen to the quantized speech. Use MATLAB to create a binary file for your D-to-A system in the form

```
(quantized speech) (0.5 s silence)
(original speech)  (0.5 s silence)
(quantized speech)
```

Remember that the quantizer M-file `fxquant( )` *requires a maximum value of 1. You should multiply the samples by the appropriate constant (probably 32,768) and convert to integer before writing the file.* Listen to this file. Can you hear the quantization noise?

Another interesting experiment is to listen to the quantization noise. Form a file in the following format:

```
(quantized speech) (0.5 s silence)
(original speech) (0.5 s silence)
(quantization noise)
```

In this case the quantization noise should be scaled up more than the speech signal itself in order to hear the noise at the same level as the speech. Does the quantization noise sound like "white noise"? Does the noise have any of the characteristics of the speech signal?

chapter 11

SIGNAL MODELING

OVERVIEW

In this chapter we present a variety of special topics related to signal modeling and estimation. In the first section the widely used technique of linear prediction is studied. Several problems based on actual signals are posed, including one project on the prediction of stock market data. In the second section the application of linear prediction to speech modeling is presented. In this case, the linear predictor not only models the speech signal, but can also be used to resynthesize the signal from an all-pole model. In the third section the linear prediction methods are extended to the problem of exponential modeling. In this case a signal is represented by a weighted sum of complex exponentials with unknown exponents. The determination of the unknown exponents is done via Prony's method, which amounts to rooting the linear prediction polynomial. In addition, the problem of pole–zero modeling is studied using a variant of Prony's method called the Steiglitz–McBride iteration [1]. This algorithm also provides a superior time-domain match when fitting unknown exponentials to a signal. The fourth section examines the problem of interpolation from the viewpoint of least-squares signal estimation. The general theory presented there is applicable to the estimation of any linear functionals of a signal; we focus, however, on the problem of estimating samples of a subsampled signal (i.e., interpolation). The last section examines the problems of linear least-squares inversion and of the solution of inaccurate, insufficient, and inconsistent linear equations. The problem of noisy data is considered, and truncation of the singular value decomposition (SVD) expansion is proposed as a way to reduce the effect of noise (at the expense of resolution). The trade-off between noise and resolution is explored.

BACKGROUND READING

A number of advanced topics are presented in the edited collection [2] and in the text by Marple [3]. Material on the application of linear prediction to speech can be found in the text by Rabiner and Schafer [4]. Methods of linear inversion are presented in the paper by Jackson [5] and in Chapter 3 of the book by Lanczos [6].

[1] K. Steiglitz. On the simultaneous estimation of poles and zeros in speech analysis. *IEEE Transactions on Acoustics, Speech, and Signal Processing*, ASSP-25:229–234, June 1977.

[2] J. S. Lim and A. V. Oppenheim. *Advanced Topics in Signal Processing*. Prentice Hall, Englewood Cliffs, NJ, 1988.

[3] S. L. Marple. *Digital Spectral Analysis with Applications*. Prentice Hall, Englewood Cliffs, NJ, 1987.

[4] L. R. Rabiner and R. W. Schafer. *Digital Processing of Speech Signals*. Prentice Hall, Englewood Cliffs, NJ, 1978.

[5] D. D. Jackson. Interpretation of inaccurate, insufficient, and inconsistent data. *Geophysical Journal of the Royal Astronomical Society*, 28:97–109, 1972.

[6] C. Lanczos. *Linear Differential Operators*. Van Nostrand, New York, 1961.

[7] J. Makhoul. Linear Prediction: A Tutorial Review. *Proceedings of the IEEE*, 63(4):561–580, April 1975.

[8] C. L. Lawson and R. J. Hanson. *Solving Least Squares Problems*. Prentice Hall, Englewood Cliffs, NJ, 1974.

[9] J. R. Deller, Jr., J. G. Proakis, J. H. L. Hansen. *Discrete-Time Processing of Speech Signals*. Macmillan, New York, 1993.

[10] M. Golomb and H. F. Weinberger. Optimal approximation and error bounds. In R. E. Langer, editor, *On Numerical Approximation*, chapter 6, pages 117–190. The University of Wisconsin Press, Madison, WI, 1959.

[11] R. G. Shenoy and T. W. Parks. An optimal recovery approach to interpolation. *IEEE Transactions on Signal Processing*, ASSP-40(8):1987–1996, August 1992.

[12] D. G. Luenberger. *Optimization by Vector Space Methods*. John Wiley & Sons, New York, 1969.

[13] G. Oetken, T. W. Parks, and H. W. Schüßler. New results in the design of digital interpolators. *IEEE Transactions on Acoustics, Speech, and Signal Processing*, ASSP-23(3):301–309, June 1975.

[14] K. Aki and P. G. Richards. *Quantitative Seismology Theory and Methods*, Volume 2. W. H. Freeman and Co., San Francisco, 1980.

LINEAR PREDICTION

OVERVIEW

The idea of linear prediction is a powerful one in signal modeling. It is also directly connected to the use of all-pole models in spectrum estimation. The tutorial paper by Makhoul [7] provides an excellent overview of the subject. Many textbooks also treat the topic (e.g., Rabiner and Schafer [4] for speech processing). The next section deals with this important application.

In the prediction problem, we are given a signal $x[n]$ and we want to build a system that will predict future values. A *linear predictor* (Fig. 11.1) does this with an FIR filter.[1]

$$\hat{x}[n] = \sum_{k=1}^{P} (-a_k) x[n-k] \tag{0-1}$$

The best linear predictor will be one that minimizes an error such as least squares. If we want $\hat{x}[n]$ to be a "prediction" of the future value, $x[n+r]$, we minimize

$$E = \sum_{n} \left| x[n+r] - \hat{x}[n] \right|^2 \tag{0-2}$$

by choosing the predictor coefficients $\{a_k\}$. The range of the sum, to be specified later, leads to two different methods.

[1] The minus sign with the predictor coefficients $\{a_k\}$ is awkward but necessary to match the sign convention in MATLAB's `filter` function, and at the same time, express the prediction error $e[n]$ as a difference.

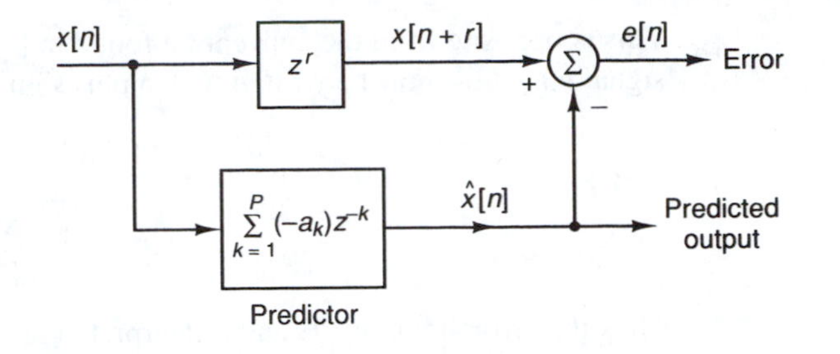

Figure 11.1
Block diagram for linear prediction. If $r = 0$, the predictor attempts to match the present value; if $r > 0$, it tries to predict a future value of $x[n]$.

After taking partials (or applying the orthogonality principle of least squares), the problem of minimizing E in (0-2) can be reduced to solving normal equations [8]. However, in MATLAB there is an easier way because the backslash operator (\) will solve a set of overdetermined linear equations in the least-squares sense. The predictor in (0-1) can be written out as a set of linear equations, with the minus sign moved to the left-hand side.

$$
\begin{array}{rcll}
-x[1+r] & \approx & a_1x[0]+a_2x[-1]+\cdots+a_Px[1-P] & (n=1)\\
\vdots & \vdots & \vdots & \\
-x[P+r] & \approx & a_1x[P-1]+a_1x[P-2]+\cdots+a_Px[0] & (n=P)\\
\vdots & \vdots & \vdots & \\
-x[L-1] & \approx & a_1x[L-2-r]+\cdots+a_Px[L-1-r-P] & (n=L-1-r)\\
\vdots & \vdots & \vdots & \\
-x[L-1+P+r] & \approx & 0+\cdots\cdots+0+a_Px[L-1] & (n=L-1+P)
\end{array}
\quad (0\text{-}3)
$$

This set of equations can be represented in matrix form as

$$-\mathbf{x} \approx \mathbf{X}\mathbf{a}$$

where the vector $\mathbf{x}$ and the matrix $\mathbf{X}$ contain known signal values. The squared error between the left- and right-hand sides will be minimized if the problem is solved in MATLAB via `a = -X \ x`. The resulting values for $\{a_k\}$ define the FIR linear predictor.

When $r = 0$ there are two methods of linear prediction, which are distinguished solely by which equations are included in the error sum (0-2).

1. *Autocorrelation method*: All possible equations from $n = 1$ to $n = L-1+P$ are included. Thus if the extent of the input data $x[n]$ is finite $0 \leq n < L$, the prediction distance is r, and the length of the predictor is P, there will be $L-1+P$ equations. In some cases the predictor will be trying to match 0, because $x[L] = 0,\ x[L+1] = 0,\ \ldots,\ x[L-1+P+r] = 0$.
2. *Covariance method*: Only those equations for which all values of $x[n]$ needed on both sides are present in the data [i.e., equations $(n = P)$ to $(n = L-1-r)$ in (0-3)]. This method uses fewer equations, only $L-P-r$, but does not predict past the end of the data.

For long input sequences, however, there should be essentially no difference in the solution, which can be obtained with the backslash operator in either case.

There is often confusion over the notation used for the predictor coefficients $\{a_k\}$, because there is no standard convention used in textbooks and papers. In this section the sign of the predictor coefficients $\{a_k\}$ will be taken consistent with MATLAB, so that the "prediction error filter" $A(z)$ will have plus signs for the a_k's.

$$A(z) = z^r + \sum_{k=1}^{P} a_k z^{-k} \quad (0\text{-}4)$$

This is opposite from the convention found in [7]. The notable difference is that the error signal $e[n]$ must now be written with a plus sign:

$$e[n] = x[n+r] - \hat{x}[n]$$
$$= x[n+r] + \sum_{k=1}^{P} a_k x[n-k]$$

Thus the error signal $e[n]$ can be interpreted as the output of an LTI system with transfer function $A(z)$ and input $x[n]$.

PROJECT 1: LINEAR PREDICTION OF DATA

In this project, linear prediction is applied to synthetic signals and to real data. The real data are from the Dow-Jones Industrial Average sampled weekly for about 94 years. With such a long sequence, the linear prediction method can be designed over one section of the data and then tested over other sections to evaluate its effectiveness as a predictor. Performance of the method on real data should illustrate some of the limits of the method imposed by the inherent assumption that the data fit an all-pole model. The book by Marple [3] also contains an interesting data set—sunspot numbers for the years 1845–1978.

Hints

You may find the MATLAB function `convmtx` useful, along with the backslash (`\`) operator, which can solve simultaneous linear equations in the least-squares sense. For plotting poles and zeros, use the M-file `zplane`.

EXERCISE 1.1

Function for Linear Prediction

Write two MATLAB functions to compute the prediction error filter coefficients $\{a_k : k = 1, 2, \ldots, P\}$, one for the "autocorrelation" method and the other for the "covariance" method. Each function should accept three input arguments: a vector of data (`x`), the order of the predictor (`P`), and the prediction distance (`r`). The output arguments should include the filter coefficients (`a`), the error sequence (`e`), and an index variable containing the sample numbers at which the error signal was computed (`I`). An example function shell is as follows:

```
function [a,e,I] = covpred(x, p, r)
% COVPRED          Covariance Method to predict x[n+r]
%   Usage:    [a, e] = covpred(x, p, r)
%
%       x : input signal
%       p : order of predictor ( = number of poles)
%       r : (OPTIONAL) prediction distance, i.e., predict x[n+r]
%       a : prediction error filter coefficients
%       e : prediction error signal over I
%           total error is E = sum(abs(e).^2)
%       I : range of error signal
%           (e.g., for covariance method I=p:(Lx-1+r))
%
% Example: x = filter(1, [1 0.2 0.3], [1 zeros(1,100)]);
%                [a,e] = covpred(x, 2, 0)
% The returned vector a should be [1  0.2  0.3]
```

Test your M-file on the example where $A(z) = 1 + 0.2z^{-1} + 0.3z^{-2}$ and $x[n]$ is the impulse response of $1/A(z)$.

EXERCISE 1.2

Process Synthetic Data

The MATLAB data file `ARdata.mat` contains three data sequences, `x1`, `x2`, and `x3`, and one AR prediction error filter, `a1`. The signals `x1` and `x2` were generated with the command `xi = filter(1, a1, xin)`, where `xin` is either an impulse or a white noise sequence. Sequence `x3` was generated as the impulse response of a rational filter $B(z)/A(z)$, so it can be used to uncover problems when the signal does not satisfy the all-pole model.

a. Use your functions from Exercise 1.1 to estimate the prediction error filter for `x1` using the both the autocorrelation method and the covariance method. The order of the predictor you design should be the same as the order of the filter `a1`. Compare the predictor coefficients you generated with those used to create the data. Are they the same?

b. Make pole–zero plots of the function $1/A(z)$ for the actual coefficients and for both sets of computed coefficients from part (a). Do the poles of your estimated system function exactly match those of the system function used to create the signal `x1`? Are some poles more closely matched than others? Explain any differences.

c. Use both functions to estimate the prediction error filter for `x2`. As in part (a), the order of the predictor you design should be the same as the order of the filter `a1`. Compare the predictor coefficients to `a1` and compare pole–zero plots.

d. Using the predictors you designed in parts (a) and (c), compute the mean-squared error for both methods (autocorrelation and covariance). Which one performs better in terms of the mean-squared error? Can you explain this performance difference?

e. The order of the filter used to generate `x3` is unknown; and it has zeros. Using the autocorrelation method, compute the mean-squared error E for orders $P = 1, 2, \ldots, 8$. Plot the mean-squared error versus order, and from your plot determine a good guess for the order of the filter that was used to generate `x3`. Repeat for the covariance method to see if you get the same value of P. Design the predictors for use in Exercise 1.3. Comment on any significant difference that you find between the two methods.

EXERCISE 1.3

Resynthesis of the Signal

Once the linear prediction coefficients are known, the polynomial $A(z)$ can be used to synthesize a signal with the same autocorrelation function as the original $x[n]$. Thus, when $r = 0$ in (0-4), the impulse response of the causal all-pole system

$$H(z) = \frac{G}{A(z)}$$

can be viewed as an approximate resynthesis of the signal $x[n]$. In this exercise we examine how well this synthesized signal $h[n]$ matches the original.

a. One issue is the scaling of the synthesized signal, or equivalently, the parameter G. A simple strategy for computing G is to make the total energy in $h[n]$ equal to the total energy in $x[n]$ [7, Eq. 35]. These energies are just the zero lag of the respective autocorrelation functions. Excite the system

$$H(z) = \frac{1}{A(z)}$$

with an impulse, using the `filter` command with the filter coefficients determined in Exercise 1.2(a). Compute the energy in $h[n]$ numerically and then calculate the value of G.

Scale the signal with G and compare the signal $Gh[n]$ to `x1`. Plot the impulse responses for both systems and the actual sequence `x1` on the same graph.

b. In the case of the signal `x2`, the same comparison is not possible, because this signal was produced by passing white noise through the filter defined by `a1`. Therefore, we must run the all-pole model with a white noise input to synthesize a signal to compare to the original. It does not make sense to compare signal values; instead, the autocorrelation functions must be compared. How many lags should match?

   It is possible to generate the impulse response $h[n]$ and then compute its autocorrelation. Compare this deterministic autocorrelation to that of `x1`. For a final cross-check use the function `acimp` from Appendix A, which computes the autocorrelation function directly from the transfer function.

c. For the signal `x3` compute the prediction coefficients via the autocorrelation and covariance methods. Then synthesize the impulse responses of the all-pole filters determined by the a_k's. Scale the responses correctly to match total energy. Plot the raw data and the impulse responses on the same axes. Explain differences in the match between the raw data and the synthesized signals. Consider the possibility that the raw data were created by a system that had both poles and zeros. Plot the roots of $A(z)$ for both methods. Compare the error E for both. On the basis of all these comparisons, which method gives the best results? (*Note*: See the section *Exponential Modeling* for ways to incorporate zeros into the modeling process to improve the time-domain match.)

EXERCISE 1.4

Stock Market Data

The MATLAB file `DJIAdata.mat` contains a sequence of Dow Jones Industrial Average weekly closing prices over a 94-year period (1897–1990). Suppose that you want to make your fortune on Wall Street and you are so confident in your knowledge of linear predictors that you decide to invest your money using the following strategy to choose between the stock market and a passbook savings account:

1. Use your predictor to estimate next week's closing price.
2. If the ratio of next week's *estimated* closing price to this week's *actual* closing price is greater than the weekly gain in the passbook savings account $(1 + 0.03/52)$, you should invest all your money (or remain invested).
3. If the ratio of next week's *estimated* closing price to this week's *actual* closing price is less than $(1 + 0.03/52)$, you should sell your investments, or remain in cash.

Assume that your investments perform as the DJIA (e.g., if the DJIA goes up by 2% in a week while you are invested, your investment appreciates by 2%). Assume that you keep cash in a passbook savings account earning 3% annual interest compounded weekly, so if you remain in cash your net worth increases by a factor of $(1 + 0.03/52)$ each week. Also assume that you use your connections to obtain free brokerage services, and avoid capital gains taxes.

a. Plot the DJIA data on both a linear and a semilogarithmic scale (see `semilogy`). For comparison, the consumer price index, which measures inflation, has risen by just under a factor of 20 over the same period.

b. Which method (autocorrelation or covariance) is most applicable to these data? It is easy to answer this question based on the nature of the data; no actual testing of a linear predictor is needed.

c. By plotting mean-squared error E versus predictor order P, estimate the order of a linear predictor that would perform well on the first decade of this data (i.e., on the first 521 weeks). Design the predictor, and plot the actual and predicted DJIA for the first decade on the same graph. Use the method you selected in part (a).

d. Given the predictor you designed for the first decade, test your investment strategy on the first decade. Begin with x =1000 dollars at the *end* of the first week, so you make 520 trading decisions per decade. Determine the maximum amount of money you could make in the first decade (i.e., if your strategy was always right). Calculate how much money you make if you follow a simple "buy and hold" strategy (where you always remain invested). How much money do you make using your predictor? Find the ratio of your gain to the maximum gain, and the ratio of your gain to the gain obtained by a "buy and hold" strategy.

e. Now, try the predictor you designed in part (b) on some of the other decades. This is a much more realistic test, as you cannot really design a predictor using knowledge of the future. For each decade under test, compute the maximum amount of money you could make if your strategy was always right, the amount you actually made, the ratio of your gain to the maximum gain, and the ratio of your gain to the "buy and hold" gain.

f. Design different predictors, one for each decade, and compare them. How different are the prediction coefficients? Plot the roots of each $A(z)$ to see how much variability there is among the different predictors. At first keep the model order the same as determined in part (c); but for some cases, recompute the optimal predictor order by plotting the mean-squared error E versus P.

g. For motivation, compute the maximum gain possible over the entire range of the data, assuming that your predictor always makes the "right" decision, and you start with an initial investment of, say, $x = 1000$ dollars. Now that you are motivated, experiment with new prediction strategies to see if you can do better than before. For example, use the capability to predict ahead one or two weeks to devise a more elaborate set of conditions for deciding when to "hold" and when to "fold." Alternatively, you might try updating the prediction coefficients more often that once per decade. These are just two ideas; more than likely, you can be more creative. Once you can consistently do better than the "buy and hold" strategy, see how close you can come to the maximum gain.

Project 2: Decorrelation by Linear Prediction

It is possible to remove unwanted correlation from a signal by doing prediction. In communication systems, this process is called *equalization* and is essential in reducing intersymbol interference due to channel characteristics. In this project, the autocorrelation function of a colored noise process will be measured, and then an FIR prediction filter will be designed, so that when it is excited by the colored noise, it produces an approximately white noise output. This is the problem of linear prediction for random signals, and it leads to the same normal equations as before. Hence, the "autocorrelation" and "covariance" methods of linear prediction still apply.

Hints

For computing the autocorrelation function, see the M-files `acf` and `acimp` in Appendix A.

EXERCISE 2.1

Decorrelation

The objective of this exercise is to demonstrate that the prediction process will decorrelate a correlated noise signal. Suppose that a signal has been filtered by a low-pass filter so that a majority of its high-frequency content has been attenuated (but not lost completely). Take the original input to be white Gaussian noise, and let the attenuating filter be a fourth-order Butterworth filter whose cutoff is at $\omega_p = \pi/18$. The output signal is correlated and is usually called colored noise.

a. Design the Butterworth filter using the function `butter` from the signal processing toolbox. Plot its frequency response magnitude and verify that the cutoff frequency is correct.

Plot, in addition, its impulse response and determine the length of the transient. Note that the Butterworth filter is not all-pole.

b. Process a white Gaussian noise input signal through the filter to generate the test signal for the prediction experiment. Generate a very long signal ($\approx$ 1000 points), but remove the transient at the beginning. Compute the autocorrelation sequence of the output and plot. Take enough lags in the autocorrelation to show where the correlation is significant.

c. Design a length-3 linear predictor that will predict ahead 2 samples (i.e., $x[n+2]$). Try the design with different portions of the test signal (e.g., segment lengths of $L = 1000$, 500, 200, 100, etc.). Do the design with either the autocorrelation method or the covariance method.

d. Plot the error signal out of the predictor: $e[n] = x[n+2] - \hat{x}[n]$.

e. Compute and plot the autocorrelation of the error signal. Compare to the theoretical autocorrelation function for white noise. Does the length-3 predictor completely decorrelate the input signal $x[n]$?

f. Repeat the predictor design for a longer predictor [e.g., a length-20 FIR predictor that still predicts two samples ahead ($r = 2$)].

EXERCISE 2.2

Equalization

If we view the "error" signal $e[n]$ as the desired output of the processor, the system $A(z)$ is called a *prediction error filter* (PEF). In Exercise 2.1, we demonstrated that linear prediction will decorrelate the input signal and produce an output that is essentially white noise. In this exercise we use the linear predictor to construct an equalizer or whitening filter. For the investigation of this exercise, we continue to use the synthetic test signal created via the fourth-order Butterworth filter in Exercise 2.1.

a. Design a length-5 linear predictor that will predict $x[n]$, (i.e., $r = 0$). Use one data segment that is 256 points long when computing the predictor coefficients $\{a_k\}$.

b. Since the signals $x[n]$ and $e[n]$ are random signals, we can compute their power spectra by an estimation technique such as the Welch–Bartlett method (see the section *FFT Spectrum Estimation* in Chapter 6). First, plot the power spectrum of the input $x[n]$.

c. Compute and display the power spectrum of the error signal, $e[n]$. Justify the statement that the prediction error filter can also be called a "whitening" filter.

d. Now derive the transfer function of the prediction error filter. This can be done by finding $E(z)/X(z)$ in terms of the $\{a_k\}$ in Fig. 11.1.

e. Plot the magnitude response of the PEF and compare with that of the Butterworth filter which first attenuated the data. State the mathematical relationship between the two frequency responses. Since the transfer function of the PEF tries to boost the high frequencies and undo the attenuation of the Butterworth filter, it is called an "equalizer."

f. Now compare the phase responses (or group delay). What is the implication of the fact that the PEF does not exactly cancel the phase of the Butterworth filter. Is it possible to obtain a better phase match with a longer FIR predictor?

g. It is also interesting to plot the frequency response of the linear predictor itself. Comment on the magnitude and group delay response of this FIR filter in light of the fact that prediction implies *negative* group delay.

LINEAR PREDICTION OF SPEECH

OVERVIEW

In this project you will study various aspects of the use of linear prediction in speech processing. This project follows closely the notation and point of view of [4], where speech is assumed to be the output of the linear system model shown in Fig. 11.2. In this figure, the input $e[n]$ is ideally either white noise or a quasi-periodic train of impulses. The linear system in Fig. 11.2 is assumed to be slowly time-varying such that over short time intervals it can be described by the all-pole system function

$$H(z) = \frac{G}{1 + \sum_{k=1}^{P} \alpha_k z^{-k}} \tag{0-1}$$

It is easily seen that for such a system, the input and output are related by a difference equation of the form

$$x[n] = -\sum_{k=1}^{P} \alpha_k x[n-k] + Ge[n] \tag{0-2}$$

(*Note*: The minus sign in front of the summation is consistent with the MATLAB `filter` function, but opposite from the notation in [4]).

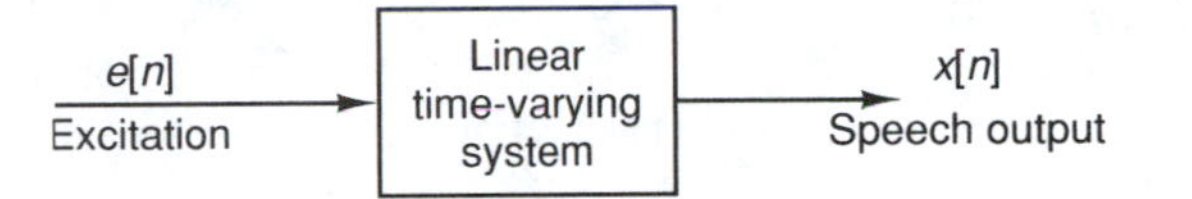

Figure 11.2 Speech model as a time-varying linear system.

Linear predictive (LP) analysis is a set of techniques aimed at finding the set of prediction coefficients $\{a_k\}$ that minimize the mean-squared prediction error between a signal $x[n]$ and a predicted signal based on a linear combination of past samples; that is,

$$\langle (f[n])^2 \rangle = \left\langle (x[n] + \sum_{k=1}^{P} a_k x[n-k])^2 \right\rangle \tag{0-3}$$

where $\langle \cdot \rangle$ represents averaging over a finite range of values of n. It can be shown that using one method of averaging, called the *autocorrelation method*, the optimum predictor coefficients $\{a_k\}$ satisfy a set of linear equations of the form

$$\mathbf{Ra} = -\mathbf{r} \tag{0-4}$$

where $\mathbf{R}$ is a $P \times P$ Toeplitz matrix made up of values of the autocorrelation sequence for $x[n]$, $\mathbf{a}$ is a $P \times 1$ vector of prediction coefficients, and $\mathbf{r}$ is a $P \times 1$ vector of autocorrelation values [4].

In using LP techniques for speech analysis, we make the assumption that the predictor coefficients $\{a_k\}$ are identical to the parameters $\{\alpha_k\}$ of the speech model. Then, by definition of the model, we see that the output of the prediction error filter with system function

$$A(z) = 1 + \sum_{k=1}^{P} a_k z^{-k} \tag{0-5}$$

is

$$f[n] = x[n] + \sum_{k=1}^{P} a_k x[n-k] \equiv Ge[n] \tag{0-6}$$

(i.e., the excitation of the model is defined to be the input that produces the given output $x[n]$ for the prediction coefficients estimated from $x[n]$). The gain constant G is therefore simply the constant that is required so that $e[n]$ has unit mean-squared value and is readily found from the autocorrelation values used in computation of the prediction coefficients [4].

It can be shown that because of the special properties of the LP equations (0-4), an efficient method called Levinson recursion[4] exists for solving the equations for the predictor parameters. However, for purposes of these exercises it is most convenient to use the general MATLAB matrix functions. Specifically, the following help lines are from an M-file `autolpc( )` from Appendix A that implements the autocorrelation method of LP analysis:

```
function   [A, G, r, a] = autolpc(x, p)
%AUTOLPC        Autocorrelation Method for LPC
%    Usage: [A, G, r, a] = autolpc(x, p)
%         x : vector of input samples
%         p : LPC model order
%         A : prediction error filter, (A = [1; -a])
%         G : rms prediction error
%         r : autocorrelation coefficients: lag = 0:p
%         a : predictor coefficients (without minus sign)
%--- see also ATOK, KTOA
```

PROJECT 1: BASIC LINEAR PREDICTION

The file `S5.MAT` contains the utterance *Oak is strong and also gives shade* sampled at 8 kHz. The phoneme SH in *shade* begins at about sample 15500 and ends at about 16750, while the phoneme AA in *shade* begins at about 16750 and ends at about 18800.

EXERCISE 1.1

12th-Order Predictor

Compute the predictor parameters of a 12th-order predictor for these two phonemes using a Hamming window of length 320 samples. For both phonemes, make a plot of the frequency response of the prediction error filter and the log magnitude response of the vocal tract model filter both on the same graph. Also use `zplane( )` to plot the zeros of the prediction error filter for both cases. *Hold onto the predictor information in both cases since you will need it for later exercises.*

What do you observe about the relationship between the zeros of the prediction error filter and the following: (1) the poles of the vocal tract model filter; (2) the peaks in the frequency response of the vocal tract model filter; and (3) the dips in the frequency response of the prediction error filter?

EXERCISE 1.2

Frequency Response of Model

In the two cases, compute the Fourier transform of the windowed segment of speech, and plot its magnitude in dB on the same plot as the vocal tract model filter. Use the parameter `G` (available from the prediction analysis) in the numerator of the model filter system function to get the plots to line up. What do you observe about the differences between the voiced AA and the unvoiced SH phoneme?

EXERCISE 1.3

Vary the Model Order

If you have time, it is instructive to look at other speech segments (frames) or to vary the window length and/or predictor order to observe the effects of these parameters. For example, compare the fit of the frequency response of the vocal tract model filter for $P = 8, 10, 12$, and 24 to the short-time Fourier transform of the speech segment.

EXERCISE 1.4

Include Preemphasis

Repeat Exercises 1.1 and 1.2 for the speech signal preemphasized with the two-point FIR filter:

```
y = filter([1, -0.98], 1, s5)
```

Compare the results with and without preemphasis.

EXERCISE 1.5

Prediction Error Filtering

Now use the prediction error filters to compute the prediction error sequence $f[n]$ for both phonemes. Use `subplot( )` to make a two-panel subplot of the (unwindowed) speech segment on the top and the prediction error on the bottom part of the plot.

What do you observe about the differences in the two phonemes? Where do the peaks of the prediction error occur in the two cases?

PROJECT 2: LINE SPECTRUM PAIR REPRESENTATIONS

A useful transformation of the LP coefficients is the *line spectrum pair* (LSP) representation [9]. The line spectrum pair polynomials are defined by the equations

$$P(z) = A(z) + z^{-(p+1)}A(z^{-1})$$
$$Q(z) = A(z) - z^{-(p+1)}A(z^{-1})$$

The LSP parameters are defined to be the angles of the roots of these two polynomials.

EXERCISE 2.1

M-File for Line Spectrum Pair

Write an M-file that converts the prediction error filter $A(z)$ to the LSP polynomials $P(z)$ and $Q(z)$. Its calling sequence should be as follows:

```
function [P, Q] = atolsp(A)
%ATOLSP    convert from prediction error filter to
%          line spectral pair (LSP) coefficients
%    Usage:
%          [P, Q] = atolsp(A)
%                A : column vector of prediction error filter
%          P and Q : column vectors of LSP polynomials
```

EXERCISE 2.2

Roots of LSP Polynomials

Use your M-file for both phonemes and use `zplane( )` to plot the roots of the two LSP polynomials with the roots of $P(z)$ plotted as x's and the roots of $Q(z)$ plotted as o's. Compare your plots to the plots of the zeros of the corresponding prediction error filters.

Observe the relationships among the roots of the two polynomials for each phoneme. In particular, note where all the roots lie radially, and note how the roots of the two polynomials interlace.

PROJECT 3: QUANTIZATION OF PARAMETERS

In using linear prediction in speech coding, it is necessary to quantize the predictor parameters for digital coding. One possibility is to quantize the predictor parameters [i.e., the coefficients of the predictor polynomial $A(z)$]. It is well known that these parameters are very sensitive to quantization. However, certain invertible nonlinear transformations of the predictor coefficients result in equivalent sets of parameters that are much more robust to quantization. One such set is the PARCOR parameters (or k-parameters), which are a by-product of the Levinson recursion method of solution of the LP equations [4]. Appendix A gives a pair of M-files called `atok( )` and `ktoa( )` which implement the transformation from predictor coefficients to PARCOR coefficients and the inverse, respectively. The help lines for these two M-files are given below.

```
function k = atok(a)
%ATOK      converts AR polynomial to reflection coefficients
%   Usage:  K = atok(A)
%        where each column of A contains polynomial coeffs
%          and   "     "     of K contains PARCOR coeffs
%
%      If A is matrix, each column is processed separately.

function a = ktoa(k)
%KTOA     converts reflection coefficients to AR polynomial
%   Usage:  A = ktoa(K)
%        where each column of A contains polynomial coefficients
%          and   "     "     of K contains PARCOR coefficients
```

In the following exercises you will compare the effects of quantization on the predictor parameters and the PARCOR parameters. If you look at the coefficients of the polynomial $A(z)$, you will find that they are probably less than or equal to 1. The coefficients can be quantized to a fixed number of bits using the quantizer M-file `fxquant( )` given in Appendix A. For coefficients that are less than one, `fxquant( )` can be used directly; that is, the statement

```
Ah = fxquant(A, 5, 'round', 'sat')
```

would quantize a coefficient `A` to 5 bits using rounding and saturation. (Of course, if the coefficient is less than 1, no additional error results from the saturation mode.) If the coefficient is greater than 1, but less than 2, the statement

```
Ah = 2*fxquant(A/2, 7, 'round', 'sat')
```

would be used for 7-bit quantization, where the location of the binary point would have to be specified as one bit to the right of the sign bit.

EXERCISE 3.1

Round Prediction Coefficients

Round the coefficients of $A(z)$ for the phoneme AA to 7 and 4 bits, respectively, using the method suggested above. Then make a plot of the the frequency responses of the original $1/A(z)$ and the two quantized vocal tract model filters, all on the same graph. Also check the stability of the quantized vocal tract filters by finding the roots of the polynomial $A(z)$.

EXERCISE 3.2

Round PARCOR Coefficients

Now take the polynomial $A(z)$ for the phoneme AA and convert it to PARCOR parameters using the function `atok( )`.

a. Round the PARCOR coefficients to 7 and 4 bits as in Exercise 3.1. Then convert the quantized PARCOR coefficients back to prediction coefficients and make a plot of the frequency responses of the original and the two PARCOR-quantized vocal tract filters as in Exercise 3.1. Also check the stability of the quantized vocal tract filters.

b. Compare the results of rounding $\{a_k\}$ versus rounding the PARCOR coefficients. Which quantized filters show the most deviation from the original frequency response? Were any of the resulting filters unstable? If not, try coarser quantization.

PROJECT 4: FORMANT TRACKING

In interpreting the prediction error filter, it is common to assume that the roots of $A(z)$ (i.e., the poles of the vocal tract filter) are representative of the formant frequencies for the segment of speech (frame) from which the predictor coefficients are computed. Thus, the angles of the roots expressed in terms of analog frequency are sometimes used as an estimate of the formant frequencies. For example, consider Fig. 11.3, which is a plot of all the pole angles as a function of speech frame index.

Figure 11.3 was obtained by the following algorithm:

1. Read as many samples of speech starting at sample number `nbeg` as you can comfortably work with into an array `x`. Set `n = 1`.
2. Compute the prediction coefficients for a Hamming windowed speech segment of length `nwin` samples, starting at sample `n`.
3. Find the magnitudes of the angles of the roots of the prediction error filter, and convert them to frequencies in hertz (assuming a sampling rate of speech is 8 kHz). Store the vector of frequencies as columns in a two-dimensional array (matrix) `F` where each column of `F` is the frequency vector of a "frame" of speech.
4. Set `n = n+ninc`, where `ninc` is the increment in samples between frames. While `n+nwin <= length(x)`, return to step 2 and repeat. Otherwise, quit.

After the matrix `F` is computed you can make a plot like Fig. 11.3 with `plot(F','*w')`.

EXERCISE 4.1

M-File for Formant Tracking

Write an M-file to implement the foregoing procedure. Use the following calling sequence:

```
function    F = formants(x, ninc, nwin, p)
%FORMANTS    Function to plot angles of roots of prediction
```

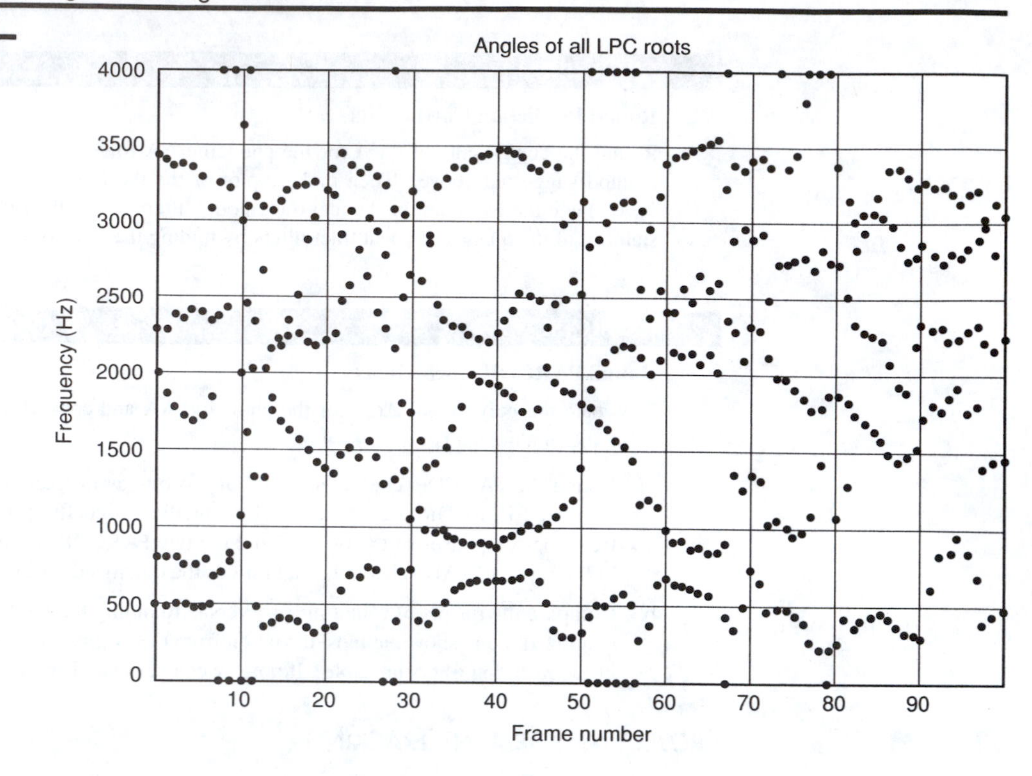

Figure 11.3 "Formant frequencies" estimated by linear prediction.

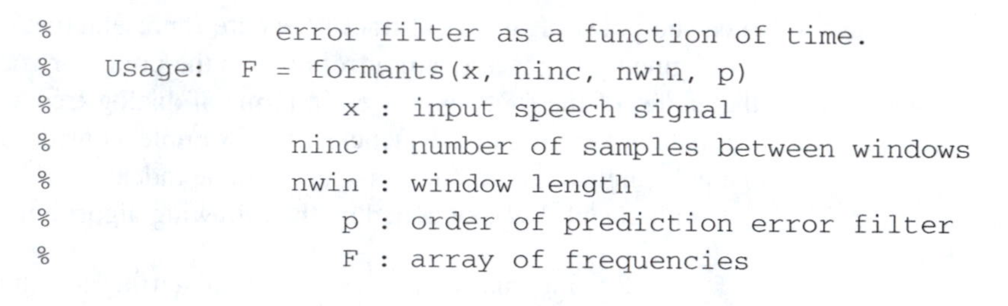

```
%               error filter as a function of time.
%   Usage:   F = formants(x, ninc, nwin, p)
%                     x : input speech signal
%                  ninc : number of samples between windows
%                  nwin : window length
%                     p : order of prediction error filter
%                     F : array of frequencies
```

Figure 11.3 shows an analysis of the samples `s5(1200:17200)`. In this case, 99 frames are separated by 160 samples with a window length of 320 samples. In this case, the speech was preemphasized with

```
y = filter([1, -0.98], 1, s5(1200:17200))
```

prior to the formant analysis. This preemphasis tends to remove the effect of the glottal wave. Since this involves 16000 speech samples, which is longer than can be read into a single array in PC-MATLAB, the processing might have to be done on a smaller section. A plot like Fig. 11.3 could be constructed by doing the analysis in pieces, but this is not essential for understanding the method. Simply test your program on speech segments of convenient length for your computing environment.

EXERCISE 4.2

Editing the Formant Tracks

You will note from Fig. 11.3 that the algorithm above plots the angles of all the roots, including the real roots which lie at $\omega = 0$ and $\omega = \pi$. Also, it might plot the angles of complex roots twice because these roots occur in complex-conjugate pairs. It is reasonable to eliminate these redundant roots, as well as any real roots, because they are obviously not formant frequencies.

It is also quite likely that roots whose magnitude is less than about 0.8 are not formants, so they should be eliminated also. Figure 11.4 shows that a much cleaner formant track is obtained when these extraneous roots are not included. Modify your M-file in Exercise 4.1 to perform this editing feature. In doing so, you should use the `find( )` function to locate the roots to be eliminated. Also, you will find that simply eliminating these roots from the frequency vectors would result in vectors of different lengths from frame to frame, and this would cause problems in making up the matrix `F`. A neat way to eliminate the desired roots from the plot is to replace them with MATLAB's object `NaN` (not a number). This would keep all the vectors the same length (`p`), but the `plot( )` function will automatically ignore the `NaN` values.

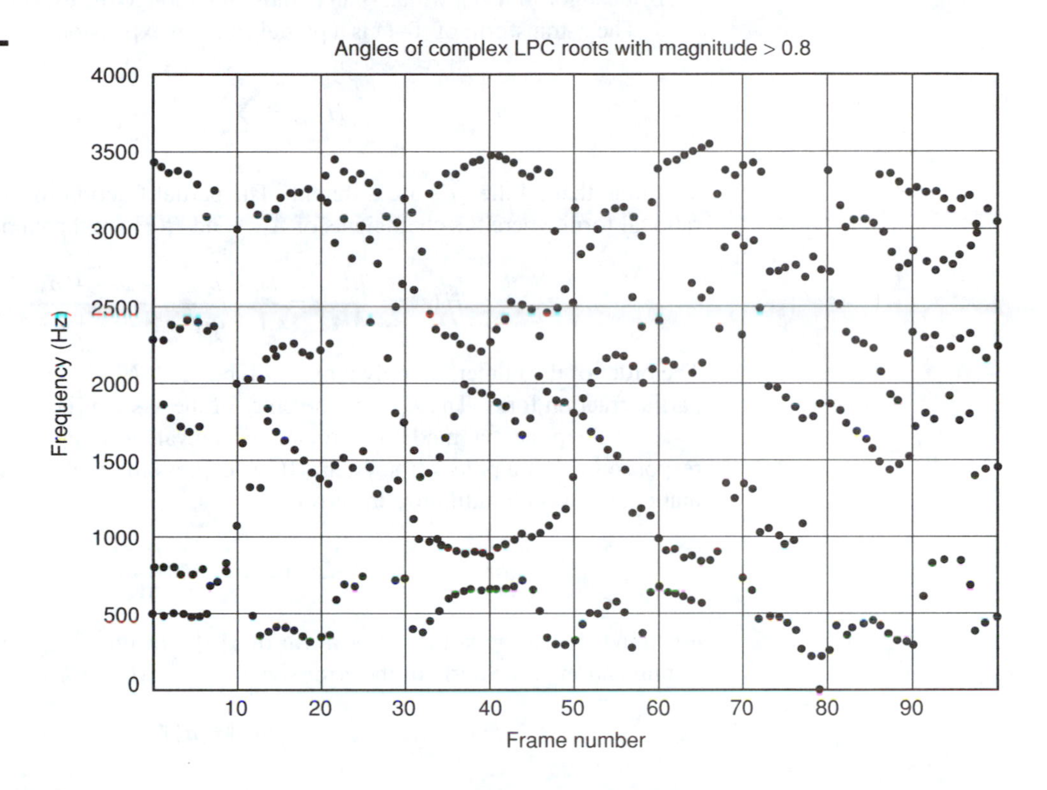

Figure 11.4 Cleaner "formant frequency" plot.

EXPONENTIAL MODELING

OVERVIEW

In this section we provide an introduction to exponential signal modeling. In this problem we seek to represent a signal $s[n]$ as a weighted sum of exponentials with unknown exponents:

$$s[n] \approx \sum_{k=1}^{N} c_k e^{\alpha_k n} = \sum_{k=1}^{N} c_k (\lambda_k)^n \tag{0-1}$$

where $\alpha_k = \log \lambda_k$ are the unknown exponents. If α_k is complex, (0-1) can represent a decaying sinusoid; if the real part of α_k is negative, the exponential decays.

Computing the parameters of the model $\{c_k, \ \lambda_k\}$ from the data $s[n]$ is a difficult task, because the representation (0-1) is nonlinear in the λ_k's. In fact, most algorithms for calculating the representation involve two steps: (1) determine the λ_k's, and (2) compute the c_k's, assuming that the λ_k's are already known. The important simplification of the exponential modeling problem lies in its connection to linear prediction. By using the covariance method of linear prediction, the problem of finding the λ_k's is reduced to a polynomial factoring operation. This sort of technique is usually called *Prony's method* when applied to exponential modeling.

In the first project, the basic idea behind Prony's method will be illustrated by showing that any exponential signal can be perfectly predicted by a linear predictor. Since the z-transform of the representation in (0-1) is rational but not all-pole, a complete solution to the exponential modeling problem requires a discussion of pole–zero modeling. Therefore, the second project takes up the full pole–zero modeling problem and develops the Steiglitz–McBride algorithm [1], which is an iterative solution to the exponential modeling problem.

The z-transform of (0-1) is a partial fraction expansion:

$$H(z) = \sum_{k=1}^{N} \frac{c_k}{1 - \lambda_k z^{-1}} \tag{0-2}$$

assuming that all the λ_k's are different. The partial fraction form can be combined into a rational form where the coefficients of $A(z)$ and $B(z)$ are the parameters of the model.

$$H(z) = \frac{B(z)}{A(z)} = \frac{b_0 + b_1 z^{-1} + \cdots + b_M z^{-M}}{1 + a_1 z^{-1} + \cdots + a_N z^{-N}} \tag{0-3}$$

The order of the numerator polynomial will be $M = N - 1$ because it is produced from the partial fraction form. The λ_k's are the poles of the system [i.e., the roots of $A(z)$].

The exponential modeling problem is equivalent to representing $s[n]$ as the impulse response $h[n]$ of a pole–zero system. If we express the problem as one of signal approximation, we want to minimize the error

$$E(z) = S(z) - \frac{B(z)}{A(z)}$$

assuming that $X(z)$ is the z-transform of $s[n]$. In the time domain this would require minimization of the norm of the error signal $e[n] = s[n] - h[n]$

$$\min_{\{b_\ell, a_k\}} \| e[n] \|$$

where $h[n]$, the inverse transform of $H(z)$, is the impulse response of the pole–zero system in (0-3).

In general, *direct* minimization of $\| e[n] \|$ requires solving a set of complicated nonlinear equations. The two-step procedure suggested by Prony's method simplifies the problem but also changes it somewhat. In effect, an *indirect* modeling problem is solved (i.e., linear prediction), and then this solution is used to approximate the direct solution.

Background Reading

More details on this approach to the pole–zero modeling problem can be found in Chapter 1 of [2] or in Chapter 11 of [3].

Project 1: Prony's Method

The basic idea underlying Prony's method is that an exponential signal can be canceled completely by a linear predictor. Thus the zeros of the canceling filter are the poles needed in (0-1).

EXERCISE 1.1

Cancel an Exponential

a. Generate 25 points of the signal $s[n] = a^n u[n]$, with $a = -0.88$. Make a `stem` plot of $s[n]$.

b. Process $s[n]$ through a 2-term FIR filter $G(z) = 1 + \gamma z^{-1}$. Compute and plot the output for the cases $\gamma = 0.9$ and -0.9.

c. Determine the value of γ so that the output signal will be exactly zero for $n \geq 1$.

d. Extend this idea to the second-order case. Let $s[n] = \sin(\pi n/4)\, u[n]$. Process $s[n]$ through a 3-term FIR filter $G(z) = 1 + \gamma_1 z^{-1} + \gamma_2 z^{-2}$, with $\gamma_1 = 0$ and $\gamma_2 = 1$.

e. Now select the coefficients γ_1 and γ_2 to make the output zero for $n \geq 2$.

f. Determine the zeros of $G(z)$ found in part (e). Explain the relationship between these zeros and the signal $s[n] = \sin(\pi n/4)\, u[n]$.

EXERCISE 1.2

Prony's Method

When the signal is composed of a large number of (complex) exponentials, a general approach is needed to design the FIR system that will cancel the signal. Refer to the pole–zero model given in (0-3). In the time domain, the relationship in (0-3) is just a linear difference equation with coefficients a_k and b_ℓ.

$$-\sum_{k=1}^{N} a_k y[n-k] + \sum_{\ell=0}^{M} b_\ell x[n-\ell] = y[n] \qquad n = 0, 1, 2, \ldots, L-1 \qquad \text{(1-1)}$$

where $x[n]$ is the input and $y[n]$ the output. If $s[n]$ is an impulse response that satisfies this difference equation, then for $n \geq N$ we get

$$-\sum_{k=1}^{N} a_k s[n-k] = s[n] \qquad n = N, N+1, \ldots, L-1 \qquad \text{(1-2)}$$

which is a set of simultaneous linear equations in the N unknowns $\{a_k\}$.

a. Rewrite equation (1-2) for $\{a_k\}$ in matrix form.

b. The M-file below will generate a signal $s[n]$ that is a sum of exponentials as in (0-1). Use the following parameters to generate $s[n]$ for $0 \leq n \leq 30$.

```
lambda  =  [ 0.9 0.7+0.7j 0.7-0.7j -0.8 ]
     c  =  [ 3.3 4.2 -4.2 2.7 ]
```

Make a `stem` plot of $s[n]$.

```
function  ss = pronysyn( lam, c, nn )
%PRONYSYN   synthesize a sum of exponentials
%   usage:     ss = pronysyn( lam, c, nn )
%     lam = vector of exponents
%       c = vector of weights
%      nn = vector of time indices
%      ss = output signal
%
N = length(lam);
```

```
ss = 0*nn;
for k = 1:N
   ss = ss + c(k)*exp(lam(k)*nn);
end
```

c. Form a polynomial $G(z)$ whose roots are given by `lambda` in Exercise 1.1 (see `help poly`). Demonstrate that processing $s[n]$ from part (b) through $G(z)$ will give a zero output for $n \geq n_0$. Determine the value of n_0.

d. The function `residuez` can be used to convert from a partial fraction representation to a rational $B(z)/A(z)$ form. Show that the same signal as in part (b) can be generated by using `filter` with the appropriate `b` and `a` vectors.

EXERCISE 1.3

Test Signal

The data file `EXPdata.mat` contains two signals that were generated via the exponential model in (0-1). The first, `sigclean`, is exactly in the form (0-1).

a. For this signal, determine an FIR system that will exactly cancel `sigclean` past a certain point. From this predictor calculate the exponents λ_k needed in the model for `sigclean`. Use the minimum number of equations from (1-2).

b. Once the correct values of the λ_k's are determined, write a set of simultaneous linear equations for the unknown gains c_k. This can be done by considering (0-1) to be a linear equation for each n. Since there are N c_k's, N equations in N unknowns should be sufficient. Write the equations in matrix form, and solve using the backslash operator \ in MATLAB. If you generate more equations than unknowns, do you still compute the same answer?

c. Write an M-file that will compute both the exponents and the gains, thus implementing what would be considered an extended form of Prony's method.

EXERCISE 1.4

Noisy Signals

The second signal in `EXPdata.mat` is `signoisy`, which is just the signal `sigclean` plus a small amount of additive noise.

a. Use the M-file from Exercise 1.3(c) to calculate λ_k and c_k for the noisy signal. Use the minimum number of equations needed from (1-2) and (0-1). Comment on the differences that you observe.

b. Redo part (a) with more equations than unknowns. In fact, use the maximum number of equations permitted by the length of the data set L. Comment on the difference between the λ_k's from this computation and those from part (a).

c. As in Exercise 1.2(d), determine values for the filter coefficients and resynthesize the signal from the pole–zero model. Create an impulse response of the model $h[n]$ that is the same length as `signoisy` and then plot both on the same graph. Include the noise-free signal `sigclean` for comparison.

d. What is the modeling error? Find the norm of the error ($\| e \|$) between the true signal $s[n]$ and your estimate $h[n]$.

PROJECT 2: POLE–ZERO MODELING

Complete modeling of an exponential signal requires a pole–zero model. Since Prony's method is unable to calculate the correct poles when the signal is noisy, the computation of the zeros will also be incorrect in Prony's method. However, in the technique known as *iterative prefiltering* [1], the denominator polynomial (i.e., poles) determined at one iteration is used to form a new problem in which a generalized form of the equations appearing in Prony's method is solved. The key feature of this method is that only linear equations have to be solved at any step. Furthermore, the method usually converges within 3 to 5 iterations if the signal is well matched by an exponential model.

The basic difficulty with Prony's method is that it does not minimize the true error between the given impulse response $s[n]$ and the model's impulse response $h[n]$. Instead, an "equation error" is minimized. The two errors can be described in the z-transform domain as

$$E_{\text{eq}}(z) = S(z)A(z) - B(z) \tag{2-1}$$

$$E_{\text{true}}(z) = S(z) - \frac{B(z)}{A(z)} = \frac{1}{A(z)}[A(z)S(z) - B(z)] \tag{2-2}$$

They are related via

$$E_{\text{true}}(z) = E_{\text{eq}}(z)/A(z) \tag{2-3}$$

So the basic idea is to develop a recursion in which the equation error is weighted so that it will be closer to the true error. This requires two distinct operations:

1. *Kalman's method*: A method for finding the pole–zero model of a system when the input signal and the output signal are both known.
2. *Iterative prefiltering*: Assuming that a computation of the poles has been done, a pseudo input–output problem is created; then Kalman's method is applied.

If the second step is carried out repeatedly and the answers for $B_i(z)$ and $A_i(z)$ converge, the error minimized is the true error.

EXERCISE 2.1

Kalman's Method

Assume that a rational system $H(z)$ has been tested such that both the input and output signals are known. The input does not have to be an impulse signal.

$$Y(z) = \frac{B(z)}{A(z)}X(z)$$

The rational system is to be approximated (maybe with zero error) by a pole–zero model (0-3). The number of poles and zeros must be fixed a priori. The objective is to determine the parameters $\{a_k\}$ and $\{b_\ell\}$, and to do so by solving only linear equations.

In (1-1) there are $M + N + 1$ unknowns, and the number of equations depends on the length of the data available for $x[n]$ and $y[n]$. Usually, there will be more equations than unknowns (if the modeling is to work). These overdetermined equations can then be solved in the least-squares sense (using the backslash operator in MATLAB). This is *Kalman's method.*

a. Write out *all* the equations in (1-1) for the specific case where $M = 1$ and $N = 2$ and the length of the data sequences is $L = 7$ (i.e., $x[n] \neq 0$ and $y[n] \neq 0$ only when $0 \leq n \leq 6$).

b. The equations (1-1) can be written in matrix form. Write a MATLAB function that will produce the coefficient matrix and the right-hand side for the equations in (1-1).

c. To test the function, generate 20 points of the output signal when the input to the following system is a random signal with unit variance:

$$\frac{2 + 4z^{-1} + 2z^{-2}}{1 + 0.8z^{-1} + 0.81z^{-2}}$$

Solve the resulting equations using backslash. Compare the estimated parameters with the true values of $\{a_k,\ b_\ell\}$.

d. To test the robustness of this solution, add some noise to the output $y[n]$. For the 20-point signal, add white Gaussian noise with a variance of 0.01; try it also with a variance of 0.1. Comment on the robustness of the answer.

e. Try a 100-point signal with additive noise. Is the answer more robust with the longer signal?

Note. Kalman's method, when used for the impulse response modeling problem, amounts to Prony's method to find $A(z)$ followed by cross-multiplication of $A(z) \times S(z)$ to get the numerator $B(z)$. This is usually not very robust.

EXERCISE 2.2

Prefiltering Equations

The iterative prefiltering equations will be developed, assuming that an estimate of the denominator polynomial $A_i(z)$ has already been done at the ith iteration. The all-pole filter $1/A_i(z)$ has an impulse response which we will call $h_i[n]$. We can also apply $1/A_i(z)$ to the given signal $s[n]$ and thus produce the following signal: $s_i[n] = s[n] * h_i[n]$. If we now apply Kalman's method with $h_i[n]$ playing the role of the input, and $s_i[n]$ the output, we can compute a new pole–zero model so that the following is approximately true:

$$A_{i+1}(z)S_i(z) \approx B_{i+1}(z)H_i(z) \tag{2-4}$$

In other words, we must solve the following set of overdetermined linear equations:

$$-\sum_{k=1}^{N} a_k s_i[n-k] + \sum_{\ell=0}^{M} b_\ell h_i[n-\ell] = s_i[n] \tag{2-5}$$

The improvement from step i to $i+1$ is based on the observation that the error being minimized in (2-4) can be written as

$$E_{\mathrm{K}}(z) = A_{i+1}(z)S_i(z) - B_{i+1}(z)H_i(z) = A_{i+1}(z)\frac{S(z)}{A_i(z)} - \frac{B_{i+1}(z)}{A_i(z)}$$

Therefore, if the Kalman error converges to zero, we get $A_i(z) \approx A_{i+1}(z) \rightarrow A(z)$ and $B_i(z)/A_i(z) \rightarrow S(z)$.

a. Write a function that will produce the matrix equations described in (2-5). Omit those that have zero entries due to going beyond the length of the signal.

b. Since the impulse response $h_i[n]$ can be produced for all $n \geq 0$, it is tempting to think that an arbitrary number of nonzero equations can be written. However, the convolution of $h_i[n]$ with $s[n]$ is not useful outside the finite range of the data given. Use this fact in limiting the number of equations for the least-squares solution.

EXERCISE 2.3

Steiglitz–McBride Iteration

Now comes the iterative part: Since a computation of Kalman's method yields a new denominator polynomial, we can redo the whole process with the new $A(z)$.

a. Write a MATLAB function `stmcbrid` that will implement the complete iteration, which is called the *Steiglitz–McBride* method.

b. Generate the same example as in Exercise 2.1(c). Use the Prony solution as the starting point and apply the Steiglitz–McBride method [1].

c. Use the function `pronysyn` to generate test signals that are weighted sums of exponentials. Verify that your function `stmcbrid` will compute the correct answer in the noise-free case. It might be convenient to have `stmcbrid` return the c_k and λ_k parameters instead of the a_k's and b_ℓ's.

d. Test the robustness of the technique by using `pronysyn` to generate a test signal and then add noise. Experiment with the SNR and the length of the signal L. Compare the pole positions λ_k with and without noise.

e. Apply your function to the unknown signal `signoisy` from the file `EXPdata.mat` in Exercise 1.4. Comment on the improvement obtained with the Steiglitz–McBride iteration versus Prony's method.

SIGNAL ESTIMATION

OVERVIEW

In this section we examine the problem of interpolation from the viewpoint of least-squares signal estimation. The general theory used here is applicable to the estimation of any linear functionals of a signal; we focus, however, on the problem of estimating samples of a subsampled signal (i.e., interpolation).

This section follows material in sections of the chapter by Golomb and Weinberger, "Optimal approximation and error bounds" [10]. See especially pp. 132–135, 140–143, and the Introduction, p. 117. The results we need are summarized below.

Given signal measurements (linear functionals)

$$F_i(\mathbf{u}) = f_i \qquad i = 1, \ldots, N$$

where the values of the linear functionals F_i are f_i for the signal $\mathbf{u}$, and given that $\mathbf{u}$ belongs to the signal class

$$C = \left\{\mathbf{u} \in \mathcal{H} : \langle \mathbf{u}, \mathbf{u} \rangle \leq r^2,\, F_i(\mathbf{u}) = f_i \; i = 1, \ldots, N\right\}$$

the best estimate of the signal, $\hat{\mathbf{u}}$, is a linear combination of the representers ϕ_i of the linear functionals, F_i.

$$\hat{\mathbf{u}} = \sum_{i=1}^{N} c_i \phi_i$$

where the coefficients c_i are chosen so that $\hat{\mathbf{u}}$ has the given values f_i of the linear functionals,

$$F_i(\hat{\mathbf{u}}) = f_i \qquad i = 1, \ldots, N$$

The signal estimate $\hat{\mathbf{u}}$ is best in the sense that

$$\max_{\mathbf{u} \in C} |F(\mathbf{u}) - F(\hat{\mathbf{u}})|$$

is minimized.

With this method we are not limited to interpolating an evenly decimated signal—we can interpolate nonuniformly subsampled signals, and we can extrapolate a signal in intervals in which we have no samples (such as in prediction of "future" samples). Additionally, we may obtain a bound on the error of our estimate. Throughout the section we assume that the signal to be interpolated is the output of a known finite-dimensional linear transformation with a known bound on the input norm. This is a special form of class C above. More details on the application of signal estimation to the interpolation problem can be found in the paper by Shenoy and Parks [11].

BACKGROUND READING

The fundamental material on linear functionals and Hilbert space operators can be found in the text by Luenberger [12]. This should serve as a solid foundation for reading the chapter by Golomb and Weinberger [10]. For the application of least squares to interpolation filter design, see the paper by Oetken, Parks, and Schüßler [13].

PROJECT 1: FINDING THE OPTIMAL ESTIMATE IN A FILTER CLASS

In this project we obtain an optimal estimate of a signal with missing samples. We assume that the subsampled signal is known to be in a certain class of signals, called a filter class. A signal $\mathbf{a}$ is input to a linear transformation $\mathbf{C}$, which outputs $\mathbf{u}$:

$$\mathbf{u}_{n\times 1} = \mathbf{C}_{n\times m}\mathbf{a}_{m\times 1}$$

If we let matrix multiplication by $\mathbf{C}$ represent a filtering operation of length-l impulse response $\mathbf{h}$ with a finite-length signal, we obtain

$$\begin{pmatrix} u_1 \\ u_2 \\ \vdots \\ u_n \end{pmatrix} = \begin{pmatrix} h_l & h_{l-1} & \cdots & h_1 & 0 & \cdots & 0 \\ 0 & h_l & \dots & h_2 & h_1 & \dots & 0 \\ \vdots & \vdots & \ddots & \vdots & \vdots & \ddots & \vdots \\ 0 & 0 & \dots & h_l & h_{l-1} & \dots & h_1 \end{pmatrix} \begin{pmatrix} a_1 \\ a_2 \\ \vdots \\ a_m \end{pmatrix}$$

Here we choose $\mathbf{h}$ to represent an averaging operation with an averaging length of l samples, given by

$$h(n) = \begin{cases} \dfrac{1}{l} & \text{if } 1 \leq n \leq l \\ 0 & \text{otherwise} \end{cases}$$

In this problem we suppose that a number of samples of $\mathbf{u}$ are unknown and we would like to estimate them as accurately as possible. To do this we need to use all of the information about $\mathbf{u}$ that we have. One thing we know about $\mathbf{u}$ is that it is the result of a signal that has been averaged (i.e., it is the left-hand side of the matrix equation above). This information, along with a suitable bound on the norm of the input signal $\mathbf{a}$, makes $\mathbf{u}$ a member of a *filter class*. Knowing that $\mathbf{u}$ is in a filter class, and given a certain set of known linear functionals of $\mathbf{u}$ (e.g., samples), we may estimate unknown linear functionals of $\mathbf{u}$ using the techniques of deterministic signal estimation described in Golomb and Weinberger [10]. In this section *the unknown linear functionals are the unknown samples of* $\mathbf{u}$. You will estimate these linear functionals with the knowledge that $\mathbf{u}$ has been created by an averaging operation.

Hints

You may find the following MATLAB functions useful: `convmtx`, `norm`, and backslash \. The backslash operator will solve a set of linear equations, it will compute the least-squares solution for an overdetermined set, and it will get a solution in the underdetermined case.

EXERCISE 1.1

Create the Averaging Matrix

In MATLAB, create an $n \times m$ averaging matrix $\mathbf{C}$ as depicted above. Use $n = 31$ and $l = 5$.

EXERCISE 1.2

Create Some Test Signals

Create a length-m input signal $\mathbf{a}$ of normally distributed noise. Let $\mathbf{u} = \mathbf{Ca}$. Create a vector of sample indices, xx, at which we know the signal $\mathbf{u}$. For now, use

```
xx = [ 3 10 14 18 21 29 ];
```

Let yy be the output signal sampled at xx, that is,

```
yy = u(xx);
```

(*Note*: In real life, we will not know $\mathbf{a}$ or $\mathbf{u}$, of course, but for the purpose of evaluating our method this will be very useful.)

EXERCISE 1.3

Create the Inner Product Matrix

Let

```
R = C*C';
Q = pinv(R);
```

R is the correlation matrix of the impulse response $\mathbf{h}$. Its kth column is the representor in the Q-inner-product space for the linear functional that extracts the kth sample of a signal. That is, if $\phi_k = \texttt{R(:,k)}$, then

$$\langle \phi_k, \mathbf{u} \rangle_Q = \mathbf{u}_k$$

where $\mathbf{u}_k$ is the kth sample of $\mathbf{u}$. We want to form the Q inner product matrix of the representors corresponding to known samples, that is, the matrix Φ with elements

$$\Phi_{ij} = \langle \phi_i, \phi_j \rangle_Q$$

where i and j are elements of xx. Call this matrix PHI. (*Hint*: Recall that $\langle \mathbf{x}, \mathbf{y} \rangle_Q = \texttt{x'*Q*y}$ for column vectors x and y. PHI can be created with one line of MATLAB code using only the variables R, Q, and xx).

EXERCISE 1.4

Calculating ubar, the Optimal Estimate

The best estimate ubar is a linear combination of the representors of the known linear functionals, that is,

```
ubar = R(:,xx)*c;
```

for some column c. We know that ubar goes through the known samples, that is,

$$\langle \phi_k, \texttt{ubar} \rangle_Q = \texttt{yy(k)}$$

for all k in xx. With this information we can use PHI and yy to solve for c.

Calculate ubar and plot it along with the actual signal $\mathbf{u}$. What is the norm of the error, $\|\texttt{ubar} - \mathbf{u}\|$?

EXERCISE 1.5

Q-Norm of ubar

For this exercise, add a constant to the filter input $\mathbf{a}$ on the order of twice the variance. For the resultant subsampled $\mathbf{u}$, calculate ubar and plot it along with $\mathbf{u}$. Is there an alarming property of the optimal estimate that you notice readily from this plot? Generate a number of inputs **a** (with nonzero mean) and for each, write down the `Q`-norm of the error, $\|$`ubar` $- \mathbf{u}\|_Q$, the `Q`-norm of `ubar`, $\|$`ubar`$\|_Q$, and the `Q`-norm of **u**, $\|\mathbf{u}\|_Q$. Compare $\|$`ubar`$\|_Q$ and $\|\mathbf{u}\|_Q$. Is one always larger than the other?

EXERCISE 1.6

Special Case: The Averaging Operation

Find `ubar` for the following case: $n = 30, l = 6$, and `xx = [1:3:n]`. What sort of interpolation does this very closely approximate? In light of this, plot the representors corresponding to samples that we know [with `plot(R(:,xx))`], and consider taking a linear combination of these vectors. Note that the optimal interpolation in this case (equally spaced outputs of an averaging operator) results from a very simple interpolation scheme. Are there other spacings in the vector `xx` that lead to the same simple interpolation scheme? Are there other n and l that lead to the same simple interpolation scheme? Try to answer these two questions by finding a relationship between n, l, and p (where `xx = [1:p:n]`) that, when satisfied, results in this sort of simple interpolation.

PROJECT 2: ERROR BOUNDS: HOW GOOD IS THE ESTIMATE?

Now that we have calculated the estimate `ubar`, we would like to understand how well the actual signal has been approximated. In this project we find the maximum error possible in each sample estimate, as well as worst-case signals in our filter class that achieve the maximum error at some samples.

EXERCISE 2.1

Calculating `ybar` ($\bar{\mathbf{y}}$)

For each unknown sample, $\bar{\mathbf{y}}$ is a unit vector (in the `Q`-norm) that can be scaled and added to `ubar` to yield a signal in our filter class that is "farthest away" from `ubar`. As there is a different $\bar{\mathbf{y}}$ for each unknown sample of **u**, we will create a matrix `ybar` whose kth column `ybar(:,k)` is $\bar{\mathbf{y}}$ for the kth sample of **u**. Thus in MATLAB, `ybar` will be an $n \times n$ matrix [note that $\bar{\mathbf{y}}$ is undefined for samples that we know, so we will just set `ybar(:,xx)` to zero)].

For an unknown sample k, `ybar(:,k)` is a linear combination of the representors of the known samples and of the representor of the kth sample, that is,

```
ybar(:,k) = R(:,[k xx])*cbar;
```

for some column `cbar`. We know that `ybar(:,k)` is orthogonal to the representors of the known samples, that is,

$$\langle \phi_i, \texttt{ybar(:,k)} \rangle_Q = 0$$

for all i in `xx`. With this information we can use `PHI`, `R`, `xx`, `k`, and `Q` to solve for `cbar` up to a constant factor. This constant factor is determined from the condition that `ybar(:,k)` is a unit vector in the `Q`-norm:

```
ybar(:,k) = ybar(:,k)/sqrt(ybar(:,k)'*Q*ybar(:,k));
```

Calculate `ybar(:,k)` for all k not in `xx`. For those k in `xx`, set `ybar(:,k)` to zero. [*Hint*: The most straightforward way to form the `ybar` matrix is columnwise inside a `for` loop, although it is possible to do so (without normalization) with a "one-liner" (one line of MATLAB code). To save time, consider inverting `PHI` one time only prior to entering the `for` loop.]

EXERCISE 2.2

Worst-Case Signals and Error Bounds

The vector `ybar(:,k)` has the property that when scaled by the factor

```
scale = sqrt(a'*a - ubar'*Q*ubar);
```

and added to or subtracted from `ubar`, the resultant signal is worst case in that it lies on the "edge" of our filter class. The factor `scale` is a measure of the distance from the center of our filter class to the boundary of our filter class. Here we use our knowledge of the energy of **a** to describe the boundary of our filter class; in an actual application this number is assumed known.

The signal

```
uworst(:,k) = ubar + scale*ybar(:,k);
```

has maximum possible error at sample k and it lies on the boundary of our filter class. Create a matrix `uworst` whose columns are the worst-case signals for different samples. What is the `Q`-norm of `uworst(:,k)`?

Create a vector of maximum errors by multiplying `scale` by the absolute value of the diagonal of the `ybar` matrix.

EXERCISE 2.3

Plotting the Estimate with Error Bounds

Plot the estimate `ubar` in the following manner:

a. Plot the upper and lower error bounds with a dashed line on the same plot. It is suggested that you use one plot statement so that MATLAB can scale the y-axis properly.

b. With the plot above held, plot `ubar` with a solid black line between the two bounds.

c. For comparison, plot the original signal **u** with plus or asterisk characters.

d. Finally, plot the worst-case signals on the same plot. Your plot will be very messy, so focus your attention on the worst-case signal for one particular sample. Does any worst-case signal achieve the maximum error at more than one sample index?

LEAST-SQUARES INVERSION

OVERVIEW

In this section we examine the problems of linear least-squares inversion and of the solution of inaccurate, insufficient, and inconsistent linear equations. The singular value decomposition (SVD) is used to produce a "solution" to a set of equations that may not have an exact solution. If an exact solution does not exist, a solution is found which minimizes the sum of the squared errors in the equations. If the equations do not have a unique solution, the minimum norm solution is used.

The problem of noisy data is considered and truncation of the SVD expansion is proposed as a way to reduce the effect of noise (at the expense of resolution). The trade-off between noise and resolution is explored.

BACKGROUND READING

This method of linear inversion is presented in the paper by Jackson [5] and Chapter 3 of the book by Lanczos [6]. Its application in geophysics is treated in Chapter 12 of the book by Aki and Richards [14]

PROJECT 1: LEAST-SQUARES INVERSION

In this project you study the least-squares solution of a set of linear equations. Throughout this project, explicit reference is made to the paper by Jackson [5]. An effort has been made to keep the notation identical to that in the paper, whenever possible, to avoid undue confusion.

The system shown in Fig. 11.5 is implemented in MATLAB. A signal, $\mathbf{x}$ (the model), is input to a linear transformation, $\mathbf{A}$, which outputs $\mathbf{y}$ (the measurement). Note that, in general, $n \neq m$:

$$\begin{pmatrix} y_1 \\ y_2 \\ \vdots \\ y_n \end{pmatrix} = \begin{pmatrix} a_{11} & a_{12} & \cdots & a_{1m} \\ a_{21} & a_{22} & \cdots & a_{2m} \\ \vdots & \vdots & \ddots & \vdots \\ a_{n1} & a_{n2} & \cdots & a_{nm} \end{pmatrix} \begin{pmatrix} x_1 \\ x_2 \\ \vdots \\ x_m \end{pmatrix}$$

or just

$$\mathbf{y}_{n\times 1} = \mathbf{A}_{n\times m}\mathbf{x}_{m\times 1}$$

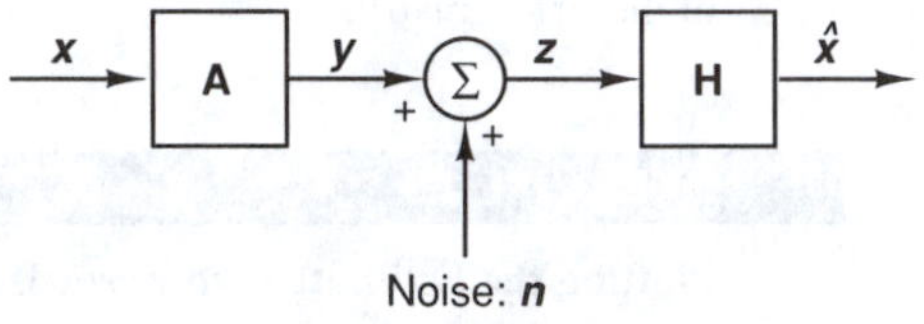

Figure 11.5
Linear system with additive measurement noise

The transformation, $\mathbf{A}$, is a sample-invariant averaging operator, where l is the number of samples averaged to produce each output sample:

$$y_k = \frac{1}{l}\sum_{j=k}^{k+l-1} x_j$$

equivalently, the entries of the matrix $\mathbf{A}$ are

$$a_{ij} = \begin{cases} \dfrac{1}{l} & \text{if } 0 \leq (j-i) < l \\ 0 & \text{otherwise} \end{cases}$$

In this project we examine the problem of inverting this operation and the resulting trade-offs in performance under different conditions.

The "true" measurement vector, $\mathbf{y}$, has a measurement error associated with it, modeled by the noise vector $\mathbf{n}$ in Fig. 11.5. The "observed" measurement, $\mathbf{z}$, is the signal available to analyze.

It is desirable to apply an inversion operation, $\mathbf{H}$, to $\mathbf{z}$ to produce an estimate $\hat{\mathbf{x}}$ of the model $\mathbf{x}$. The operator $\mathbf{H}$ may be designed by an SVD (singular value decomposition) procedure on $\mathbf{A}$. One of the inverses possible to construct in this manner, the Lanczos inverse, has several properties that may be desirable in an inverse:

- It always exists (this is not trivial).
- It is a least-squares solution (in the equation error $\mathbf{A}\hat{\mathbf{x}} - \mathbf{y} = \epsilon$).
- It is a minimum norm solution.
- Resolution of the model parameters is optimized (in some sense).

Refer to [5] for a thorough explanation of these properties. In addition to the Lanczos inverse, the SVD procedure may be modified (often referred to as a *truncated* SVD) to create additional inverses, with different performance trade-offs. In general, the optimality of the Lanczos inverse is traded off for lower variance in the model parameter estimates (see [5, p. 104]).

Hints

You will complete various MATLAB functions to implement the system shown in Fig. 11.5. Printouts of the partially completed functions are attached. Using test data and test noise provided for you, you will examine the performance of the inversion system under different conditions. Make sure that you look ahead to the last exercises in Project 4 (which contain questions), so that you understand what concepts you are expected to learn.

EXERCISE 1.1

Averaging Matrix

Complete the MATLAB function `MAKEAVR.M`, which creates the matrix **A**, so that it performs as advertised:

```
A = makeavr(n,m)
```

EXERCISE 1.2

Least-Squares Inverse

Complete the function `INVERT.M`, which performs the (full-rank or truncated) SVD-based inversion

```
B = invert(A,q)
```

where **B** is a generalized inverse of **A** and q is the rank of **B**. That is, for example, if $\text{rank}(\mathbf{A}) = p$ and $q = p$, then **B** is the Lanczos inverse of **A**.

EXERCISE 1.3

Completing the MODLEST Function

Complete the function `MODLEST.M`. This calls the functions `INVERT.M` and `MAKEAVR.M` to implement the system in Fig. 11.5:

```
[H,A,xhat] = modlest(x,n,q)
```

PROJECT 2: TESTING WITH NOISE-FREE SIGNALS

EXERCISE 2.1

Computing Test Signals

Load the test signals into the MATLAB workspace (these include two measurement noise vectors and four input signals):

```
load lslab
```

EXERCISE 2.2

Evaluation of the Least-Squares Inverse

Run `MODLEST.M` to perform Lanczos inversion (i.e., full rank: $q = p$). Use the following pairs of test input and test noise signals, respectively:

`x1` and `n1`

`x2` and `n1`

`x3` and `n1`

(Note that the "noise" signal `n1` is a zero vector used to simulate zero measurement noise.)

Analyze these signal combinations as thoroughly as you can with the functions you have written and any additional MATLAB (built-in) functions you think would be helpful. Be sure that you can answer the questions posed in the evaluation section of Project 4. (*Note*: The plots in `MODLEST.M` may not be scaled the same.)

PROJECT 3: INVERSION OF NOISY SIGNALS

EXERCISE 3.1

Noise Gain Calculations

Complete the function `VARIANC.M`, which calculates the noise gain of the matrix **H** (see [5, p. 98]). Notice that this will effectively compute the sensitivity of the inversion operation to measurement noise (why?):

```
[var] = varianc(H)
```

EXERCISE 3.2

Resolution Calculations

Complete the function `RESOLVE.M`, which calculates the resolving error of the model estimating system, defined as follows:

$$r_k = \sum_{j=1}^{m} \left[\left(\sum_{i=1}^{n} h_{ki}\, a_{ij} \right) - \delta_{kj} \right]^2 \qquad k \in [1, m]$$

where

$$\delta_{kj} = \begin{cases} 1 & \text{if } k = j \\ 0 & \text{otherwise} \end{cases}$$

(This equation is a corrected version of equation 21 in [5].) This is only one of many ways that model estimate resolution may be defined (a different definition is used in Exercise 3.4). Note that a higher error means lower resolution and that $\max\{r_k\} \leq 1$. (Why?) Your completed function should compute r_k for $k \in [1, m]$:

```
[r] = resolve(H,A)
```

EXERCISE 3.3

Evaluation with Noise

Using the following test input-test noise pair, examine the system performance for (at least) $q = 3, 5, 10, 15, 17, 20$:

`x4` and `n2`

Create $\mathbf{R} = \mathbf{H} * \mathbf{A}$, the "resolving" matrix, for each case. Plot it versus the identity matrix:

```
subplot(211);   mesh(R);   mesh(eye(R));
```

Using `RESOLVE.M` and `VARIANC.M`, calculate the resolution and variance for the index $k = 10$. Make a (hand) plot of these calculated values versus q [i.e., plot `var(10)` versus q and `r(10)` versus q together on one plot]. This plot should have the general characteristics of Fig. 1 in [5].

EXERCISE 3.4

Resolution Definitions

Using a different measurement of resolution, make a similar (hand) plot to the one mentioned above:

resolution = "width of estimated pulse at amplitude = 0.3"

That is, measure the (amplitude = 0.3) crossings of the *estimated* pulse, $\hat{\mathbf{x}}$, for the same range of q. [*Note*: Don't be too concerned about getting very accurate results here, as you'll be estimating the width from a graph. To aid in this, try

```
clg;  plot(xhat);  grid]
```

PROJECT 4: EVALUATION OF LEAST-SQUARES INVERSION

EXERCISE 4.1

Type of Equations

Using the *Jackson* terminology [5], what type of system does **A** represent [i.e., underdetermined, overconstrained, *strictly* overconstrained (overconstrained but not underdetermined), etc.]? How does this relate to the the quality of the estimated model parameters? That is, discuss uniqueness, exactness, and so on.

EXERCISE 4.2

Comparison of Performance

Relate the performance of the system in Exercises 2.2 and 3.3, to the relationship between the test input and test noise and the various vector spaces associated with **A** (i.e., $\mathcal{V}, \mathcal{V}_{\mathbf{o}}, \mathcal{U}, \mathcal{U}_{\mathbf{o}}$, etc., where $\mathbf{A} = \mathbf{U}\Lambda\mathbf{V}^t$ is the SVD of **A**). How could you generate these test signals and others like them? (It is not necessary to write any MATLAB code for this.)

EXERCISE 4.3

Relationship Between x and $\hat{\mathbf{x}}$

What is the relationship between **x** and $\hat{\mathbf{x}}$ in Exercise 2.2? Why are they equal in some cases and not equal in others? Relate this to the discussion in Exercise 4.2.

EXERCISE 4.4

Relationship Between x1 and x3

What is the relationship between **x1** and **x3**?

EXERCISE 4.5

Sample Invariance

Is **H** a sample-invariant operator? Is it *almost* sample-invariant? Explain.

EXERCISE 4.6

Effect of q

How does the difference between $\hat{\mathbf{x}}$ (with noise) and $\hat{\mathbf{x}}$ (without noise) in Exercise 3.3 change as q is varied? Explain.

EXERCISE 4.7

Choice of Test Signal

Note that **x4** is a pulse signal. Do you think that this is an appropriate test signal for Exercise 3.3? Why or why not?

MATLAB SCRIPT FILES AND INCOMPLETE FUNCTIONS

These MATLAB shells are available on the distribution disk under the names `invert.m`, `makeavr.m`, `modlest.m`, `resolve.m`, and `varianc.m`.

```
*********************************************  INVERT.M  ******
function B = invert(A, q)
%
% This routine finds the generalized inverse of the matrix A.
% The rank of the inverse is q, where q <= p and p = rank of A.
%
% The routine works the same as one computing the Lanczos
% inverse (which it will do if q = rank(A)), except that
% only those q largest (in absolute magnitude) singular values
% of A are used in creation of H.
%
[U,S,V] = svd(A);
%
% Check to make sure inversion of desired order is possible:
%  This code (from RANK.M) prevents having to do the SVD twice.
%
diag_S = diag(S);
tol = max(size(A))*diag_S(1)*eps;
rank_A = sum(diag_S > tol);
if (q > rank_A)
rank_of_A = rank(A)
q = q
error('The rank of A is insufficient to produce an inverse of rank q.');
end
%
%  Now resize the results so that the matrix S is square.
%  This is the standard notational convention for SVD
%  MATLAB is not standard in that it forces U and V to be square.
%       That is, MATLAB returns the vectors associated with zeros
%       singular values.
%       Simultaneously, change the size of U,S and V to accomodate
%       the reduced order inversion.
%

%============>              <===================
%============> ADD CODE HERE <===================
%============>              <===================
%
% Now create the inverse:
%
B = V*inv(S)*U';
*********************************************  MAKEAVR.M  ******
function A = makeavr(n, m)
%
%       This function creates a matrix that describes a
%       sample-invariant averaging operator.
%
```

```
%       Note that the averaging period = l = (m-n+1),
%       where A is (n x m).
%
%       The entries are normalized so that the sum across rows = 1
%       I.e.,
%               A(i,j)    =     1/(m-n+1)    0 <= (j-i) < (m-n+1)
%                               0            otherwise
%
% Note: an input with n > m returns an error.
%
if(n > m),  error('n > m input to MAKEAVER is not allowed');  end
%
%=============>                <====================
%=============> ADD CODE HERE <====================
%=============>                <====================

*********************************************  MODLEST.M  ******
function [H, A, xhat] = modlest(x, noise, q)
%
%  Inputs:
%        x : The input vector, length = m.
%    noise : The measurement noise vector, length = n.
%       q : The rank of the inverse.
%
% Outputs:
%       A : The input transformation matrix
%       H : A generalized inverse for A, rank = q
%    xhat : The model (estimate) of the input vector.
%
% Convert the inputs to column orientation:
%
[n,m] = size(x);
if (n == 1)   %--- I.e., if x is a row vector.
  x = x';
end
%
[n,m] = size(noise);
if (n == 1)     %--- I.e., if noise is a row vector.
  noise = noise';
end
%      Create the averaging matrix A:
%         The dimensions of A are variable, depending on the
%         dimensions of the signal and measurement noise
%
%=============>                <====================
%=============> ADD CODE HERE <====================
%=============>                <====================
%
%  Create the measurement vector, y, and the inversion matrix, H:
%
%=============>                <====================
```

```
%============> ADD CODE HERE <===================
%============>               <===================
%
%   Calculate the model estimate(s), xhat:
%      xhat_no_noise is a variable which is instructive to look at.
%      It is the xhat which would be produced if the measured
%      vector y had no noise on it. It is instuctive to plot,
%      but in practice, with a real system, you would not
%      have access to y (only to z) and thus could not look at
%      xhat_no_noise.
%
xhat_no_noise = H*y;
%
%============>               <===================
%============> ADD CODE HERE <===================
%============>               <===================
%
%  The following plots can be commented out:
%
clg
subplot(221), plot(x); title('model vector x');
plot(y);               title('measurement vector y');
plot(xhat_no_noise);   title('xhat (no noise)');
plot(xhat);            title('xhat');
pause
clg
**********************************************  RESOLVE.M  ******
function r = resolve(H, A)
%
%     Inputs:
%          A  : a matrix
%          H  : an inverse for A
%
% Outputs:
%       r  : a column vector, length = m, where m is the
%            dimension of the square matrix H*A = R
%            error(k) is the 2-norm of the error (squared)
%            in approximating an impulse with the kth row
%            of H*A = R.  If H is the Lanczos inverse for A,
%            this error is minimized for each k, over all
%            possible inverses for A.
%
%============>               <===================
%============> ADD CODE HERE <===================
%============>               <===================
%


**********************************************  VARIANC.M  ******
function var = varianc(H)
```

```
%
% Inputs:
%      H : a matrix
%
% Outputs:
%     var : a length m column vector where H is (m x n)
%           var(k) is the noise gain of the matrix
%        relative to the kth index inputs.
%      Refer to Jackson, "Interpretation of ...", pg. 98.
%
%============>               <===================
%============> ADD CODE HERE <===================
%============>               <===================
```

appendix A

SOFTWARE AND PROGRAMMING NOTES

OVERVIEW

In this appendix we present an overview of all the supporting M-files needed for the projects in this book. A brief listing of the help information for the M-files is provided to serve as a quick reference guide. These files have all been tested extensively under version 3.5 of MATLAB and on three different computer platforms: Macintosh, DOS, and UNIX. In addition, a list of all the data files is provided. Since the M-files are text, they are interchangeable among the various operating systems. Similarly, the binary data files have a universal format that allows any version of MATLAB to read them, regardless of where they were written. All of these files are available via anonymous FTP (`ftp.ece.gatech.edu or IP address 130.207.224.30`) in the directory `pub/MATLAB` and also as a disk distributed through The MathWorks, Inc., makers of the MATLAB software.

In addition to just listing all the utility functions, we present a few "programming tips" that, if followed, will enhance your use of MATLAB. It is not our objective to teach programming in this book, and certainly we cannot cover too much in this short appendix. However, all the exercises demand that programs be written for computation and plotting, so programming skill is needed to learn as much as possible from each project. Perhaps the most important point is that you must learn to exploit the vector nature of the MATLAB language.

BACKGROUND READING

Obviously, the most complete source of up-to-date information about the MATLAB software is the reference manual:

[1] MATLAB *Reference Guide: High-Performance Numeric Computation and Visualization Software,* The MathWorks, Inc., South Natick, MA, 1984–1992.

Furthermore, you should be aware that some information is available on FTP and through user groups that are coordinated on the INTERNET and also from MathWorks via e-mail: `info@mathworks.com`.

VERSION 5.0

At the time this book was first published, the MATLAB software was in transition from version 3.5 to version 4.0. Early in 1997 The MathWorks, Inc. released MATLAB version 5.0, which represented a major upgrade. Some of its new features were not compatible with the original M-files for this book. Furthermore, version 3.5 is rare and most users are now upgrading to version 5. We therefore decided to update the text and the distribution of M-files to the current form.

Throughout this book, the use of version 5.0 is assumed. Differences in the command syntax between MATLAB versions are pointed out in footnotes. Use the `help` command for their correct use. Some functions not available in version 3.5 are needed for the projects and exercises in this book. M-files providing the same functionality are included in the current software distribution. Since version 5.0 has different defaults for its graphical display, separate versions of the plotting routines are needed for version 4.x and older. These different versions are also included in subdirectories of the current distribution.

PROGRAMMING TIPS

In this section we present a few programming tips that should help improve your MATLAB programs. For more ideas and tips, study some of the functions provided in this appendix or some of the M-files in the toolboxes of MATLAB. Copying the style of other programmers is always an efficient way to improve your own knowledge of a computer language. In the hints below we discuss some of the most important points involved in writing good MATLAB code. These comments assume that you are both an experienced programmer and at least an intermediate user of MATLAB.

AVOID FOR LOOPS

There is temptation among experienced programmers to use MATLAB in the same fashion as a high-level language like FORTRAN or C. However, this leads to very inefficient programming whenever `for` loops are used to do operations over the elements of a vector (e.g., summing the elements in a vector). Instead, you must look for the MATLAB functions that will do the same operation with a function call—in the case of summing, there is a MATLAB function called `sum`.

An alternative strategy that also avoids `for` loops is to use vector operations. In the sum example, the trick is to recognize that the sum of the elements in a row vector can be obtained by multiplying by a column vector of all ones. In effect, an inner product operation computes the sum.

The primary reason for introducing these tricks is that a `for` loop is extremely inefficient in MATLAB, because it is an interpreted language. Macro operations such as matrix multiplies are about as fast as micro operations such as incrementing an index, because the overhead of interpreting the code is present in both cases. The bottom line is that `for` loops should be used only as a last resort, and then probably only for control operations, not for computational reasons. More than likely, 90% of the `for` loops used in ordinary MATLAB programs can be replaced with equivalent, and faster, vector code.

VECTORIZE

The process of converting a `for` loop into a matrix-vector operation could be referred to as *vectorizing*. Sometimes vectorizing appears to give a very inefficient answer in that more computation is done than in the `for` loop. Nonetheless, the resulting program will run much faster because one simple operation is applied repeatedly to the vector.

Repeating Rows or Columns

Often it is necessary to form a matrix by repeating one or more values throughout. If the matrix is to have all the same values, functions such `ones(M,N)` and `zeros(M,N)` can be used. But suppose that you have a row vector **x** and you want to create a matrix that has 10 rows, each of which is a copy of **x**. It might seem that this calls for a loop, but not so. Instead, the outer-product matrix multiply operation can be used. The following MATLAB code fragment will do the job:

```
X = ones(10,1) * x
```

If `x` is a length-L row vector, the matrix `X` formed by the outer product is $10 \times L$.

Vector Logicals

One area where slow programs are born lies in conditionals. Seemingly, conditional tests would never vectorize, but even that observation is not really true. Within MATLAB the comparison functions such as greater than, equal to, and so on, all have the ability to operate on vectors or matrices. Thus the MATLAB code

```
[1 2 3 4 5 6] < 4
```

will return the answer `[1 1 1 0 0 0]`, where `0` stands for `FALSE` and `1` represents `TRUE`.

Another simple example is given by the following trick for creating an impulse signal vector:

```
nn = [-20:80];    impulse = (nn==0);
```

This result could be plotted with `stem( nn, impulse )`. In some sense, this code fragment is perfect because it captures the essence of the mathematical formula, which defines the impulse as existing only when $n = 0$.

Vectorize a CLIP Function

To show an example of vectorizing at work, consider writing an M-file that will clip an input signal to given upper and lower limits. The code from a conventional language would look like the following in MATLAB:

```
function  y = clip( x, lo, hi )
% CLIP ---  threshold large and small elements in matrix x
%  ==========> SLOWEST POSSIBLE VERSION <================
%
[M,N] = size(x);
for m = 1:M
  for n = 1:N
    if x(m,n) > hi
       x(m,n) = hi;
    elseif x(m,n) < lo
       x(m,n) = lo;
end, end, end
y = x;
```

The problem with this first version is the doubly nested `for` loop which is used to traverse all the elements of the matrix. To make a faster version, we must drop the loop altogether and use the vector nature of logicals. Furthermore, we can exploit the fact that `TRUE` and `FALSE` have numerical values to use them as masks (via multiplication) to select parts of the matrix `x`. Note that `( [x<=hi] + [x>hi] )` is a matrix of all ones.

```
function y = clip( x, lo, hi )
% ============> FAST VERSION <=============
%  (uses matrix Logicals to replace Loops)
y = (x .* [x<=hi])  +  (hi .* [x>hi]);
y = (y .* [x>=lo])  +  (lo .* [x<lo]);
```

If you count the number of arithmetic operations done in the second version, you will find that it is much greater than the count for the first version. To see this, use a very large matrix for `x` and time the two functions with `etime` and `flops`. Even though you can generate cases where the second version requires 10 times as many operations, it will still run much faster—maybe 10 times faster!

COLON OPERATOR

One essential part of MATLAB that is needed to avoid `for` loops is the colon notation for selecting parts of matrices. The help for : is given below.

```
>>help :
: Colon. Used in subscripts, FOR iterations and possibly elsewhere.
J:K  is the same as [J, J+1, ... , K]
J:K  is empty if J > K.
J:I:K  is the same as [J, J+I, J+2I, ... , K]
J:I:K  is empty if I > 0 and J > K or if I < 0 and J < K.
The colon notation can be used to pick out selected rows,
columns and elements of vectors and matrices.
A(:) is all the elements of A, regarded as a single
column. On the left side of an assignment statement, A(:)
fills A, preserving its shape from before.
A(:,J) is the J-th column of A
A(J:K) is A(J),A(J+1), ... ,A(K)
A(:,J:K) is A(:,J),A(:,J+1), ... ,A(:,K) and so on.
For the use of the colon in the FOR statement, See FOR.
```

The colon notation works from the idea that an index range can be generated by giving a start, a skip, and then the end. Therefore, a regularly spaced vector of integers (or reals) is obtained via

```
iii = start:skip:end
```

Without the `skip` parameter, the increment is 1. Obviously, this sort of counting is similar to the notation used in FORTRAN DO loops. However, in MATLAB you can take it one step further by combining it with a matrix. If you start with the matrix `A`, then `A(2,3)` is the scalar element located at the second row and third column of `A`. But you can also pull out a 4 × 3 submatrix via `A(2:5,1:3)`. If you want an entire row, the colon serves as a wild card [i.e., `A(2,:)` is the second row]. You can even flip a vector by just indexing backward: `x(L:-1:1)`. Finally, it is sometimes necessary just to work with all the values in a matrix, so `A(:)` creates a column vector that is just the columns of `A` concatenated together. More general "reshaping" of the matrix `A` can be accomplished with the `reshape(A,M,N)` function.

MATRIX OPERATIONS

The default notation in MATLAB is matrix. Therefore, some confusion can arise when trying to do pointwise operations. Take the example of multiplying two matrices `A` and `B`. If the two matrices have compatible dimensions, `A*B` is well defined. But suppose that both are 5 × 8 matrices and that we want to multiply them together element by element. In fact, we cannot do matrix multiplication between two 5 × 8 matrices. To obtain pointwise multiplication we use the "point-star" operator `A .* B`. In general, when "point" is used with another arithmetic operator, it modifies that operator's usual matrix definition to a pointwise one. Thus we have `./` and `.^` for pointwise division and exponentiation. For example, `xx = (0.9) .^ [0:49]` generates an exponential of the form a^n, for $n = 0, 1, 2, \ldots, 49$.

SIGNAL MATRIX CONVENTION

Often it is necessary to operate on a group of signals all at once. For example, when computing the FFT on sections of a signal, it is convenient to put each section of the signal into one column of a matrix and then invoke the `fft` function to operate on the entire matrix. The result is that the 1-D FFT is computed down each column of the matrix. Another example along the same lines is the `sum` function, which when applied to a matrix returns a vector answer—each element of the vector result is a column sum from the matrix. What would `sum( sum(A) )` compute for the matrix `A`?

This convention is not universal within MATLAB. For example, the `filter` function, which is another workhorse DSP function, will process only one vector at a time.

POLYNOMIALS

Another convention that is used in MATLAB and is needed for DSP is the representation for polynomials. For the z-transform we often work with expressions of the form

$$H(z) = \frac{B(z)}{A(z)} = \frac{\sum_{\ell=0}^{M} b_\ell z^{-\ell}}{\sum_{k=0}^{N} a_k z^{-k}}$$

In MATLAB the polynomials $B(z)$ and $A(z)$ are represented by vectors `b` and `a` containing their coefficients. Thus `a = [1 -1.5 0.99]` represents the polynomial $A(z) = 1 - 1.5z^{-1} + 0.99z^{-2}$. From the vector form we can extract roots via the M-file `roots(a)`, and also perform a partial fraction expansion with `residuez`. In addition, the signal processing functions `filter` and `freqz` both operate on the rational system function $H(z)$ in terms of the numerator and denominator coefficients: $\{b_\ell\}$ and $\{a_k\}$.

```
yout = filter(b, a, xin)

[H, W] = freqz(b, a, Nfreqs)
```

SELF-DOCUMENTATION VIA HELP

MATLAB has a very convenient mechanism for incorporating help into the system, even for user-written M-files. The comment lines at the beginning of any function are used as the help for that function. Therefore, it behooves the programmer to pay attention to documentation and to provide a few introductory comments with each M-file. For example, if you type `help freqz`, the response is

```
>> help freqz

 FREQZ Z-transform digital filter frequency response.  When N is an integer,
  [H,W] = FREQZ(B,A,N) returns the N-point frequency vector W and the
  N-point complex frequency response vector H of the filter B/A:
```

```
                                  -1                   -nb
        jw       B(z)     b(1) + b(2)z   + .... + b(nb+1)z
     H(e   ) = ---- = --------------------------
                 A(z)             -1                   -na
                               1  +  a(2)z  + .... + a(na+1)z
   given numerator and denominator coefficients in vectors B and A. The
   frequency response is evaluated at N points equally spaced around the
   upper half of the unit circle. To plot magnitude and phase of a filter:
              [h,w] = freqz(b,a,n);
              mag = abs(h);  phase = angle(h);
              semilogy(w,mag), plot(w,phase)
   FREQZ(B,A,N,'whole') uses N points around the whole unit circle.
   FREQZ(B,A,W) returns the frequency response at frequencies designated
   in vector W, normally between 0 and pi. (See LOGSPACE to generate W).
   See also YULEWALK, FILTER, FFT, INVFREQZ, and FREQS.
```

You can also list the entire file `freqz.m` (by doing `type freqz`) to see that the help response consists of the initial comments in the file. If the M-file is a built-in, help is still available; for example, for the `filter` function:

```
>> help filter

FILTER Digital filter.
Y = FILTER(B, A, X) filters the data in vector X with the
filter described by vectors A and B to create the filtered
data Y.  The filter is a "Direct Form II Transposed"
implementation of the standard difference equation:

y(n) = b(1)*x(n) + b(2)*x(n-1) + ... + b(nb+1)*x(n-nb)
                 - a(2)*y(n-1) - ... - a(na+1)*y(n-na)

[Y,Zf] = FILTER(B,A,X,Zi) gives access to initial and final
conditions, Zi and Zf, of the delays.
See also FILTFILT.
```

PLOTTING

The graphical capabilities of MATLAB have grown considerably since version 3.5. When a user learns to combine plotting with the vector notation of the colon operator, many different types of displays can be created. Recent upgrades in MATLAB versions 4 and 5 have introduced graphical objects that can be manipulated via "handle graphics" to produce graphical user interfaces (GUIs) as well. For signal processing, the most important kind of plots are those of discrete-time signals (using `stem` in versions 4 and 5), and of frequency responses (generated via `freqz` or `dtft`). The plot function in MATLAB has considerable flexibility for plotting one or more functions. When making comparisons, different line types can be specified. In most cases, `plot` uses an autoscaling algorithm to set the axes of the plot, but this can be overridden with the `axis` command.

THREE-DIMENSIONAL PLOTS

For spectrograms and other moving window processes, a two-dimensional gray-scale display can be obtained using the `image` or `imagesc` command. Other possibilities for display of 3-D data are `contour` and `mesh`, and the variation on the `mesh` plot format called a "waterfall" plot (see `help waterf`). If each line of the waterfall plot is a spectrum, it is relatively easy to track variations of spectral peaks.

The syntax for the contour plot was changed after version 3.5. The order of the arguments is now contour(X,Y,Z), but it is still possible to use the optional argument V in contour(X,Y,Z,V) to force more contour lines at certain levels and thereby darken the plot for regions were there are peaks. The help on contour gives more details.

FIGURE WINDOWS

When making comparison between separate results, it is essential to have two or more plots on the screen simultaneously. One way to do this is to use the figure command, which opens up a new figure window for plotting. All plots go to the active figure window, which is the last one specified in a figure command or touched by the mouse. For example, two cosines can be plotted in separate windows via

```
figure(1)
plot(cos(2*pi*0.07*(0:20) - pi/2))
figure(2)
plot(sin(2*pi*0.07*(0:20)))
```

SUBPLOTS

Another way to make comparisons is via the subplot feature of MATLAB, which puts several plots on one figure. For example, to plot two Bessel functions on one page use

```
subplot(2,1,1); stem(bessel(1,[1:30]))

subplot(2,1,2); stem(bessel(3,[1:30]))
```

The argument of subplot(2,1,2) specifies that the subplot will be made of tiles arranged in a 2×1 array, and the integer n specifies that the next plot will be placed into the second (lower) tile.[1] Although subplot can produce $M \times N$ arrays with many tiles, there is a practical limit as to what can be seen. Finally, note the 'Position' feature of subplot, which permits exact placement of individual graphs.

The help comments are given as follows:

```
SUBPLOT Create axes in tiled positions
   SUBPLOT(m,n,p), or SUBPLOT(mnp), breaks the Figure window into
   an m-by-n matrix of small axes, selects the p-th axes for
   for the current plot, and returns the axis handle.  The axes
   are counted along the top row of the Figure window, then the
   second row, etc.  For example,

      SUBPLOT(2,1,1), PLOT(income)
      SUBPLOT(2,1,2), PLOT(outgo)

plots income on the top half of the window and outgo on the
bottom half.

   SUBPLOT('position',[left bottom width height]) creates an
   axis at the specified position in normalized coordinates (in
   in the range from 0.0 to 1.0).

   If a SUBPLOT specification causes a new axis to overlap an
   existing axis, the existing axis is deleted.  For example,
   the statement SUBPLOT(1,1,1) deletes all existing smaller
   axes in the Figure window and creates a new full-figure axis.
```

[1]The previous syntax subplot(212) is "grandfathered" and still supported in version 5.

SIGNAL PROCESSING PLOTS

Some of the M-files provided with this software are dedicated to special plotting formats needed in DSP. These include the function `dtft` for plotting the Fourier transform of a discrete-time signal, `striplot` for plotting extremely long signals such as speech signals, `fmagplot` for showing the Fourier transform of an "analog" signal, and the waterfall plot `waterf` mentioned earlier. The function `dtft` presents a frequency response plot over the range $-\pi < \omega < \pi$, with $\omega = 0$ centered. Another very widely used function is `zplane` for pole-zero plots. This particular function will set the aspect ratio to square and draw a unit circle prior to plotting the poles as $\times$'s and the zeros as o's.

DATA FILES

A number of the projects involve the processing of real or synthetic data. For example, the speech projects use recording of various sentences as found in the files `s1.mat` and `s5.mat`. These can be played out if your computer is equipped with a D-to-A system—the sentences were recorded at $f_s = 8000$ Hz and quantized to 12 bits. The two sentences are:

S1. *The pipe began to rust while new* (female speaker)
S5. *Oak is strong and also gives shade* (male)

For use with the student version of MATLAB, these long speech files have been chopped up into shorter `.mat` files. These shorter files are each 1000 points long and are named `S5_0.mat` through `S5_24.mat` to hold all 24,576 samples of the speech waveform. This was done with the M-file `chopfile.m` contained in this release. When processing with the student version, there are limits on the array size that have been increased in versions 4 and 5. The function `gluedata` is provided so that sections longer than 1000 samples can be extracted from several of the short `.mat` files.

The names of all the data files and the chapter(s) where they are used are listed below.

```
ARdata.mat      Chapter 11
bat.mat         Chapter 3  (speech)
BLIdata.mat     Chapter 4  (bandlimited interpolation)
b3pulses.mat    Chapter 3
DJIAdata.mat    Chapter 11 (Dow Jones data)
EXPdata.mat     Chapter 11
gdeldata.mat    Chapter 1  (group delay)
intfere.mat     Chapter 3
r100.mat        Chapter 10 (radar simulation)
s1.mat s5.mat   Chapter 10 and 11 (speech signals)
tonemyst.mat    Chapter 3
vowels.mat      Chapter 3  (more speech)
```

UTILITY FUNCTIONS

Most of the M-files needed for these computer projects will be available in the basic toolboxes of MATLAB. In the standard distribution of MATLAB, there is always a toolbox called `MATLAB` which contains many of the commonly used functions. This toolbox contains text files with the extension `.m`, and it extends the basic core functions of MATLAB.

Another toolbox of interest for these DSP projects is the `SIGNAL PROCESSING` toolbox. This toolbox is an optional extension to MATLAB, which is needed to do the projects in this book on topics such as filter design and spectrum estimation. Some, but not all of these signal processing M-files have been bundled with the student version of MATLAB. The notable exceptions are the windowing functions such as `kaiser`, although these are not too hard to program, as shown in Chapter 3 in the section *Spectral Windows.*

The M-files contained on the distribution disk under the `Functions` directory are, for the most part, utility functions that find widespread use in various projects throughout this book. They are also useful additions to the Signal Processing Toolbox; in fact, some of them may be incorporated into that toolbox in future releases of MATLAB.

```
==> Functions/AtoV.m <==
function  [r, D, G] = AtoV(A,rN)
%AtoV    find reflection coefficients & system function
%----    denominator for lossless tube models.
%   Usage:    [r, D, G] = AtoV(A, rN)
%      rN : reflection coefficient at lips (abs value < 1)

==> Functions/acf.m <==
function [ak, lags] = acf(x, m, w)
%ACF     compute autocorrelation function at m lags
%---     via Rader's method based on the FFT.
%        ==> works for complex-valued signals
%

==> Functions/acimp.m <==
function    p = acimp(b,a,N)
%ACIMP   Calculate autocorrrelation sequence of an impulse
%-----    response given the coefficients of H(z) = B(z)/A(z).
%   Usage:  p = acimp(b,a,N);
%         p : autocorrelation sequence of the impulse response

==> Functions/asinc.m <==
function y = asinc( x, L )
%ASINC   compute sin(Lx/2)/sin(x/2)    (for matrix x)
%   Usage:    y = asinc(x, L)
%      x : argument of asinc function
%      L : length of corresponding rectangular pulse

==> Functions/atok.m <==
function k = atok(a)
%ATOK        converts AR polynomial to reflection coefficients
%   Usage:   K = atok(A)
%         where each column of A contains polynomial coeffs
%           and    "    "    of K contains PARCOR coeffs

==> Functions/autolpc.m <==
function   [A, G, r, a] = autolpc(x, p)
%AUTOLPC       Autocorrelation Method for LPC
%   Usage: [A, G, r, a] = autolpc(x, p)
%        x : vector of input samples
%        p : LPC model order
```

```
==> Functions/ccf.m <==
function [kk, lags] = ccf(x ,y, m, w);
%CCF   compute cross-correlation function at a few lags
%---  in frequency domain via Ch.M. Rader's algorithm.
%     ccf =  SUM{ x[n+lag] y*[n] }  may be complex-valued
%

==> Functions/chopfile.m <==
function chopfile(fname, L)
%CHOPFILE    break a long speech file into several smaller files
%   Usage:  chopfile( 'fname', L )
%      'fname' : name of input file
%                (ASSUMES variable in 'fname' is also called fname)

==> Functions/convolm.m <==
function   H = convolm(x, num_zeros, pad)
%CONVOLM    Make convolution matrix, optionally padded with zeros
%   Usage:   H = convolm(X, P)
%        H :  convolution matrix with P columns
%             H = [ h(i,j) ], where h(i,j) = x(p+i-j)

==> Functions/db.m <==
function y = dB( x, dBrange, dBmax )
%DB      convert an array to decibels
%  Usage:   Y = dB( X, dbRANGE, dbMAX )
%          will compute 20 Log(X)
%          and then scale or clip the result so that

==> Functions/dtft.m <==
function [H, W] = dtft(h, N)
%DTFT   calculate DTFT at N equally spaced frequencies
%   Usage:   [H, W] = dtft(h, N)
%      h : finite-length input vector, whose length is L
%      N : number of frequencies for evaluation over [-pi,pi)

==> Functions/factorit.m <==
function factors = factorit(n)
%FACTORIT      factor an integer
%   Usage:  factors = factorit(n)
%         n : integer to factor
%   factors : vector containing all the factors

==> Functions/flipdtft.m <==
function     [G, Wflipped] = flipDTFT(H, W)
%FLIPDTFT   flip the DTFT:  G(w) = H(-w)
%    Usage: [G, Wflipped] = flipDTFT(H, W)
%         H : DTFT values (complex)
%         W : frequency samples

==> Functions/fmagplot.m <==
function   fmagplot( xa, dt )
%FMAGPLOT   Plot  Fourier Transform (Mag) of "ANALOG" signal
%   Usage:   fmagplot( xa, dt )
%       xa :   "ANALOG" signal
%       dt :    sampling interval for the simulation of xa(t)
```

```
==> Functions/fxquant.m <==
function X = fxquant( s, bit, rmode, lmode )
%FXQUANT       simulated fixed-point arithmetic
%   Usage:  X = fxquant( S, BIT, RMODE, LMODE )
%       returns the input signal S reduced to a word-length
%       of BIT bits and limited to the range [-1,1). The type of

==> Functions/gdel.m <==
function    [gd, w] = gdel(x, n, Lfft)
%GDEL    compute the group delay of x[n]
%   Usage:   [gd, w] = gdel( x, n, Lfft )
%       x :  Signal x[n] at the times (n)
%       n :  Vector of time indices

==> Functions/genint.m <==
function    yint = genint(N)
%GENINT   generate interference for TONE GENERATOR mystery signal
%   Usage:      Y = genint(N)
%       N : signal length of the interference
%       Y : output signal which has a continuously

==> Functions/gluedata.m <==
function    sigseg = gluedata(signal,n0,l)
%GLUEDATA   concatenate data segments of one long, segmented signal
%    Usage:  sigseg = gluedata('signal',n0,l)
%   signal:  the string 'signal' must be the name of data segments
%               available in MAT files on your MATLAB path

==> Functions/ktoa.m <==
function    a = ktoa(k)
%KTOA    converts reflection coefficients to AR polynomial
%   Usage:  A = ktoa(K)
%      where each column of A contains polynomial coefficients
%        and   "    "    of K contains PARCOR coefficients

==> Functions/lchirp.m <==
function  x = lchirp(T,W,p)
%LCHIRP  generate a sampled chirp signal with linear FM
%        exp(j(W/T)pi*t^2)  -T/2<=t<+T/2
%  Usage:  X=lchirp(T,W,<p>)
%

==> Functions/mod.m <==
function y = mod(x,N)
%MOD     Compute (x mod N) and x can be
%---       either positive or negative.
%   Usage:  y = mod(x,N);
%         y : remainder of (x/N)

==> Functions/mulaw.m <==
function y = mulaw(x, mu)
%MULAW     mu-law compression for signals with
%-----        maximum value of 32767
%   Usage:   y = mulaw(x, mu);
%      x : input signal, column vector with max value 32767
```

```
==> Functions/pkpicker.m <==
function  [peaks, locs] = pkpicker( x, thresh, number, sortem )
%PKPICKER        pick out the peaks in a vector
%   Usage:   [peaks,locs] = pkpicker( x, thresh, number, sortem )
%         peaks    :  peak values
%         locs     :  location of peaks (index within a column)

==> Functions/pronysyn.m <==
function  ss = pronysyn( lam, c, nn )
%PRONYSYN   synthesize a sum of exponentials
%   Usage:    ss = pronysyn( lam, c, nn )
%    lam = vector of EXPONENTS
%      c = vector of weights

==> Functions/pseudinv.m <==
function Ainv = pseudinv(A, r)
%PSEUDINV   Pseudo-inverse of rank r.
%   Usage:    Ainv = pseudinv(A, r)
%      produces the rank-r inverse of A, from the SVD of A.
%      Only r singular values are retained for the inverse

==> Functions/qplot.m <==
function   [snrunif, snrmu] = qplot(s, nbits, mu, ncases)
%QPLOT   for plotting signal-to-noise ratio of quantizers
%   Usage:  [snrunif, snrmu] = qplot(s, nbits, mu, ncases)
%               s : input test signal
%           nbits : number of bits in quantizer

==> Functions/radar.m <==
function y = radar( x, fs, T_0, g, T_out, T_ref, fc, r, a, v )
%RADAR       simulate radar returns from a single pulse
%  Usage:
%    R = radar( X, Fs, T_0, G, T_out, T_ref, Fc, R, A, V )
%      X:       input pulse (vector containing one pulse for burst)

==> Functions/speccomp.m <==
function   speccomp(x, ncenter, win, nfft, pltinc)
%SPECCOMP    Plots spectra with different window lengths
%--------          all centered at the same place.
%   Usage:  speccomp(x, ncenter, win, nfft, pltinc)
%           x : input signal

==> Functions/srexpand.m <==
function y = srexpand(x, L)
%SREXPAND    zero fills with L-1 zeros between each sample
%   Usage:  Y = srexpand(X, L)
%        insert L-1 zeros between each sample of the sequence X.
%        The output sequence Y has length equal to length(x)*L.

==> Functions/striplot.m <==
function xmax = striplot(x, fs, n, ntick, xmax)
%STRIPLOT    plot long signal in horizontal strips
%-------       ( good for multi-line speech wfms )
%  Usage:  striplot(X, FS, N)  plots waveform X with N pts/line
%         FS = sampling rate (Hertz); used only for labeling
```

```
==> Functions/test_rad.m <==
%
%   EXAMPLE of calling the function radar()
%     make one radar return for a burst of LFM pulses
%
clear,  format compact

==> Functions/tonegen.m <==
function  [y,code] = tonegen(digits, scale, yint)
%TONEGEN      generate "mystery" signal containing tones
%   Usage:      [Y, C] = tonegen(D, S, Xint)
%        D : vector of digits for a 5-element code
%            if length(D)<5, the function will pick random digits

==> Functions/waterf.m <==
function   peak_to_peak = waterfall(x, scale)
%WATERF            "Waterfall" Plot
%   Usage:    waterf(X)
%        plots the waveforms in X(time,rcvr) vs. "time"
%        Each trace is auto scaled so that the peak-to-peak value

==> Functions/welch.m <==
function  [P, c] = welch(x, N, M, wintype, nplot)
%WELCH   Power Spectrum Estimation by Welch's method
%   Usage:       [P,c] = welch(x,N,M,wintype,nplot)
%       P : power spectrum by Welch's method
%       c : correlation function = inverse of Welch power spectrum

==> Functions/zerofill.m <==
function Y_out = zerofill(X_in, L)
%ZEROFILL  is the "expander" operation used in multi-rate filters
%   Usage:   y = zerofill(x, L) creates an output vector y(n) as:
%                        /
%               y(n) = <   x(n/L),  for n = 0 modulo L
```

Index